Intrusive Thinking

From Molecules to Free Will

Strüngmann Forum Reports

Julia R. Lupp, series editor

The Ernst Strüngmann Forum is made possible through
the generous support of the Ernst Strüngmann Foundation,
inaugurated by Dr. Andreas and Dr. Thomas Strüngmann.

This Forum was supported by the
Deutsche Forschungsgemeinschaft

Intrusive Thinking

From Molecules to Free Will

Edited by

Peter W. Kalivas and Martin P. Paulus

Program Advisory Committee:

Aikaterini Fotopoulou, Rita Z. Goldstein,
Peter W. Kalivas, Julia R. Lupp, Martin P. Paulus,
Trevor W. Robbins, and Peter Tse

The MIT Press

Cambridge, Massachusetts
London, England

© 2020 Massachusetts Institute of Technology and
the Frankfurt Institute for Advanced Studies

Series Editor: J. R. Lupp
Editorial Assistance: A. Ducey-Gessner, C. Stephen, M. Turner
Photographs: N. Miguletz
Lektorat: BerlinScienceWorks

The book was set in TimesNewRoman and Arial.
Printed and bound in the United States of America.

Library of Congress Cataloging-in-Publication Data

Names: Kalivas, Peter W., 1952- editor. | Paulus, Martin P. (Martin Peter),
    1961- editor.
Title: Intrusive thinking : from molecules to free will / edited by Peter
    W. Kalivas and Martin P. Paulus.
Description: Cambridge, Massachusetts : The MIT Press, [2020] | Series:
    Strüngmann forum reports | Includes bibliographical references and
    index.
Identifiers: LCCN 2020025459 | ISBN 9780262542371 (paperback)
Subjects: LCSH: Intrusive thoughts. | Neuropsychology. | Cognitive
    neuroscience.
Classification: LCC RC531 .I587 2020 | DDC 616.8--dc23
LC record available at https://lccn.loc.gov/2020025459

10 9 8 7 6 5 4 3 2

Contents

Systems and Models

Interventions and Treatments

List of Contributors

Anderson, Michael C.   MRC Cognition and Brain Sciences Unit, University of Cambridge, Cambridge, CB2 7EF, U.K.

Aron, Adam   Dept. of Psychology, University of California San Diego, La Jolla, CA 92093, U.S.A.

Badre, David   Dept. of Cognitive, Linguistic, and Psychological Sciences, Brown University, Providence, RI 02912-1821, U.S.A.

Balleine, Bernard W.   School of Psychology, University of New South Wales, Sydney NSW 2052, Australia

Banich, Marie T.   Institute of Cognitive Science, Dept. of Psychology and Neuroscience, University of Colorado Boulder, Boulder, CO 80309, U.S.A.

Bershad, Anya K.   Semel Institute for Neuroscience and Human Behavior, Dept. of Psychiatry, University of California Los Angeles, Los Angeles, CA 90095, U.S.A.

Bonci, Antonello   Global Institutes on Addictions, Miami, FL 33132, U.S.A.

Bonsall, Michael B.   Dept. of Zoology, University of Oxford, Oxford OX1 3SZ, U.K.

Brady, Kathleen T.   Dept. of Psychiatry and Behavioral Sciences, Medical University of South Carolina, Charleston, SC 29425, U.S.A.

Brewer, Judson   Mindfulness Center and Dept. of Behavioral and Social Sciences, Brown University School of Public Health; Dept. of Psychiatry, The Warren Alpert Medical School of Brown University, Providence, RI 02903, U.S.A.

Bruchas, Michael R.   Center for Neurobiology of Addiction, Pain, and Emotion; Dept. of Anesthesiology and Pain Medicine; Dept. of Pharmacology; Dept. of Bioengineering, University of Washington, Seattle, WA 98195, U.S.A.

Buss, David M.   Dept. of Psychology, The University of Texas at Austin, Austin, TX 78712, U.S.A.

Cortese, Aurelio   Computational Neuroscience Laboratories, ATR Institute International, Kyoto 619-0288, Japan

Critchley, Hugo D.   Brighton and Sussex Medical School, University of Sussex, Brighton BN1 9XP, U.K.

Denys, Damiaan   Dept. of Psychiatry, University of Amsterdam, Amsterdam UMC, 1105 AZ Amsterdam, The Netherlands

de Wit, Harriet Dept. of Psychiatry and Behavioral Neuroscience, University of Chicago, Chicago, IL 60637, U.S.A.

Espinosa, Lisa Dept. of Clinical Neuroscience, Karolinska Institutet, SE-171 65 Solna, Sweden

Fedota, John R. Neuroimaging Research Branch, Cognitive Neuroscience and Psychopharmacology Section, National Institute on Drug Abuse, Intramural Research Program, Baltimore, MD 21224, U.S.A.

Flagel, Shelly B. Michigan Neuroscience Institute, Dept. of Psychiatry, University of Michigan, Ann Arbor, MI 48109, U.S.A.

Fotopoulou, Aikaterini Research Dept. of Clinical, Educational and Health Psychology, University College London, London WC1E 6BT, U.K

Frangou, Sophia Dept. of Psychiatry, Icahn School of Medicine at Mount Sinai, New York, NY 10029, U.S.A.

Friston, Karl J. Wellcome Centre for Human Neuroimaging, London WC1N 3AR, U.K.

Goldstein, Rita Z. Depts. of Psychiatry and Neuroscience, Icahn School of Medicine at Mount Sinai, New York, NY 10029, U.S.A.

Gourley, Shannon L. Yerkes National Primate Research Center, Emory University, Atlanta, GA 30329, U.S.A.

Haber, Suzanne N. Dept. of Pharmacology and Physiology, University of Rochester, Rochester, NY 14642, U.S.A.; Dept. of Neuroscience, McLean Hospital, Harvard Medical School, Belmont, MA 02478, U.S.A.

Hanlon, Colleen A. Clinical Neuromodulation Laboratory, Dept. of Cancer Biology, Wake Forest School of Medicine, Winston-Salem, NC 27157, U.S.A.

Heinz, Andreas Dept. of Psychiatry and Psychotherapy, Campus Mitte, Charité–Universitätsmedizin Berlin (corporate member of Freie Universität Berlin, Humboldt-Universität zu Berlin, and Berlin Institute of Health), 10117 Berlin, Germany

Holmes, Emily A. Dept. of Psychology, Uppsala University, SE-752 37 Uppsala, Sweden; Dept. of Clinical Neuroscience, Karolinska Institutet, SE-171 65 Solna, Sweden

Huys, Quentin J. M. Division of Psychiatry and UCL Max Planck Centre for Computational Psychiatry and Ageing Research, University College London, Russell Square House, London WC1B 5EH, U.K.

Kalivas, Peter W. Dept. of Neuroscience, Medical University of South Carolina, Charleston, SC 29464, U.S.A.

Lau, Hakwan   Dept. of Psychology and Brain Research Institute, University of California Los Angeles, Los Angeles, CA 90095, U.S.A; Dept. of Psychology and State Key Laboratory of Brain and Cognitive Sciences, University of Hong Kong, Hong Kong

Liu, Kayuet   Dept. of Sociology and California Center for Population Research, University of California Los Angeles, Los Angeles, CA 90095, U.S.A.

Maia, Tiago V.   Institute for Molecular Medicine, Faculty of Medicine, University of Lisbon, 1649-028 Lisbon, Portugal

McTeague, Lisa M.   Depts. of Psychiatry and Behavioral Sciences, Medical University of South Carolina, Charleston, SC 29425, U.S.A.

Milton, Amy L.   Dept. of Psychology, University of Cambridge, Cambridge CB2 3EB, U.K.

Monfils, Marie-Hélène   Dept. of Psychology, The University of Texas at Austin, Austin, TX 78712, U.S.A.

Paulus, Martin P.   Laureate Institute for Brain Research, Tulsa, OK 74136, U.S.A.

Phillips, Paul E. M.   Center for Neurobiology of Addiction, Pain and Emotion, Depts. of Psychiatry and Behavioral Sciences and Pharmacology, University of Washington, Seattle, WA 98195-7360, U.S.A.

Picciotto, Marina R.   Depts. of Psychiatry, Neuroscience and Pharmacology, Yale University, New Haven, CT 06508, U.S.A.

Robbins, Trevor W.   Behavioural and Clinical Neuroscience Institute, Dept. of Psychology, University of Cambridge, Cambridge CB2 3EB, U.K.

Roberts, Angela C.   Depts. of Physiology, Development and Neuroscience, University of Cambridge, Cambridge CB2 3DY, U.K.

Schiller, Daniela   Icahn School of Medicine at Mount Sinai, New York, NY 10029, U.S.A.

Schlagenhauf, Florian   Dept. of Psychiatry and Psychotherapy, Campus Mitte, Charité–Universitätsmedizin Berlin (corporate member of Freie Universität Berlin, Humboldt-Universität zu Berlin, and Berlin Institute of Health), 10117 Berlin, Germany

Schooler, Jonathan W.   Psychological and Brain Sciences, University of California Santa Barbara, Santa Barbara, CA 93106-9660, U.S.A.

Schwarzbach, Jens   Dept. of Psychiatry and Psychotherapy, University of Regensburg, 93053 Regensburg, Germany

Seamans, Jeremy K. Dept. of Psychiatry, The University of British Columbia, University Hospital–Koerner Pavilion, Vancouver V6T2B5, Canada

Singh, Laura Dept. of Psychology, Uppsala University, SE-752 37 Uppsala, Sweden

Stein, Elliot A. Neuroimaging Research Branch, Cognitive Neuroscience and Psychopharmacology Section, National Institute on Drug Abuse, Intramural Research Program, Baltimore, MD 21224, U.S.A.

Tse, Peter Dept. of Psychological and Brain Sciences, Dartmouth College, Hanover, NH 03755, U.S.A

Visser, Renée M. Dept. of Clinical Psychology, University of Amsterdam, 1018 WS Amsterdam, The Netherlands

Voss, Martin Dept. of Psychiatry and Psychotherapy, Campus Mitte, Charité–Universitätsmedizin Berlin (corporate member of Freie Universität Berlin, Humboldt-Universität zu Berlin, and Berlin Institute of Health); St. Hedwig Hospital, Berlin, 10115 Germany

Preface

Science is a highly specialized enterprise: one that enables areas of enquiry to be minutely pursued, establishes working paradigms and normative standards, and supports rigor in experimental research. All too often, however, "problems" are encountered that fall outside the scope of any one disciplinary area, and to progress, other perspectives are needed to expand conceptualization, increase understanding, and identify new trajectories for research to pursue.

The Ernst Strüngmann Forum was established in 2006 to address such problems. Founded on the tenets of scientific independence and the inquisitive nature of the human mind, we provide a platform for experts to scrutinize topics that require input from multiple areas of expertise. Our gatherings, or Forums, are best envisioned as intellectual retreats where disciplinary idiosyncrasies are put aside and existing perspectives questioned. Consensus is never the goal. Instead, participants work collectively to expose gaps in current understanding and formulate key questions that will propel research forward. The results of the entire process are disseminated through the Strüngmann Forum Report series.

This volume reports on the 30th Ernst Strüngmann Forum. It presents a synthesis of the ideas and perspectives that evolved over a two-year period and highlights questions that remain to be addressed through future work. For those seeking insight into the process, this brief overview is offered.

In 2017, Peter Kalivas and Martin Paulus approached me to discuss the possibility of proposing a theme on intrusive thinking. Having recently experienced another Ernst Strüngmann Forum on computational psychiatry (Redish and Gordon 2016), they were eager to subject the phenomenon of intrusive thinking to intense examination and sought our help to create the requisite dialogue. Their proposal provided the impetus, but as anyone who has been involved with our approach will tell you, one never knows where the discourse will ultimately lead as perspectives from other experts become available.

After the proposal was accepted, Katerina Fotopoulou, Rita Goldstein, Trevor Robbins, and Peter Tse joined us on the Program Advisory Committee to construct a framework that would support a dynamic, multidisciplinary discussion. The committee delineated discussion topics, identified potential participants, and formulated overarching goals:

- To explore the mechanisms of intrusive thinking across multiple levels of analysis: cells, circuits, and psychological processes
- To examine the role of intrusive thinking as a core symptom across psychiatric disorders
- To identify promising interventions for treating intrusive thinking

- To consider the societal and philosophical implications of intrusive thinking

In addition, four primary themes were established around which the discussion would unfold and key questions were proposed for each working group to consider. In advance, invited papers transmitted information on specific topics to a diverse group of experts (from behavioral neuroscience, cognitive neuroscience, neuroimaging, theoretical neurobiology, psychiatry, computational neuroscience/psychiatry, philosophy, and neuropsychopharmacology), who gathered in Frankfurt am Main, Germany, from June 14–19, 2019, for a most lively discussion.

A summary of this exchange is captured here in this volume. Organized around the primary themes, it contains the background papers in their finalized form (i.e., after peer review and revision) as well as the summary reports from each working group (Chapters 5, 9, 13, and 17).

As you approach this book, it is important to realize that a Forum is not a linear process. The framework put into place was designed to guide, not constrain, the discussion. Indeed, alternative perspectives were cultivated at each stage. Within this dynamic setting, replete with multiple and often divergent viewpoints, topics emerged that could not be resolved. These issues, highlighted in the individual chapters, are evidence of a fruitful exchange, and it is our hope that they will be used to stimulate future enquiry and action.

An endeavor of this kind creates unique group dynamics and puts demands on everyone who participates. Throughout, each contributor played an active role, and for their efforts and time, I express my gratitude. A special word of thanks goes to the Program Advisory Committee, to the authors and reviewers of the background papers, as well as to the moderators of the individual working groups (Antonello Bonci, Peter Tse, Trevor Robbins, and Rita Goldstein). Further, Shannon Gourley, Renée Visser, Angela Roberts, and Judson Brewer deserve special recognition: as rapporteurs for the working groups, they skillfully guided the preparation and finalization of their group's report. Importantly, I wish to extend my sincere appreciation to Peter Kalivas and Martin Paulus, whose vision and commitment were essential to the successful realization of this Forum.

The Ernst Strüngmann Forum is able to conduct its work due to the institutional stability provided by the Ernst Strüngmann Foundation, established by Dr. Andreas and Dr. Thomas Strüngmann in honor of their father, Dr. Ernst Strüngmann. As a tribute to his life's work, the Ernst Strüngmann Forum promotes scientific collaboration among international researchers, with the aim of expanding knowledge in basic science and identifying future research directions. Valuable partnerships accompany our work: The Scientific Advisory Board of the Ernst Strüngmann Forum ensures the scientific independence of the Forum. The Deutsche Forschungsgemeinschaft provided supplemental financial support for this Forum, and the Frankfurt Institute for Advanced Studies shares its vibrant intellectual setting with us.

Expanding the boundaries to knowledge is not an easy enterprise. Yet as limitations in current understanding start to come into focus, the act of formulating strategies to move forward becomes a most invigorating experience. Results from the brain-storming at this Forum have already started to influence ongoing research. It is our hope that this volume will spur further opportunities and lead ultimately to the creation of novel interventions that are capable of easing the suffering brought about by the phenomenon of intrusive thinking.

Julia R. Lupp, Director, Ernst Strüngmann Forum
Frankfurt Institute for Advanced Studies
Ruth-Moufang-Str. 1, 60438 Frankfurt am Main, Germany
https://esforum.de/

The Day Sue Almost
Got Mad at Me

Have you ever wondered why thoughts appear and can motivate you to do things you wouldn't otherwise do? One afternoon, Sue (my wife) called me at work to tell me she was going out with friends after work, and it was up to me to get dinner and do homework with the kids. Sue and her messages were (and remain to this day) very motivating for me, so I packed up and went down to my car. On the way to the car and during the drive home, I randomly thought about what I would cook for dinner, which kid was having a test and would need help, and various other plans for the evening. About halfway home, Kyle from the pub called. He was energized because a buddy from southern California had just flown in and thus Kyle was rounding people up for what sounded like a really fun night. I was going to tell him, "Sorry, I can't make it," but then the 20 beers on tap suddenly popped into my mind, as did the hilarious time we had the last time we all got together. Pretty soon I wasn't thinking about dinner plans or school tests any more. I was making plans to be with my friends and thinking about the latest release IPA from the local microbrewery. When we ended our call, I had told him that I wasn't sure what I would do. At this point, we might ask ourselves what is more motivating: going home to cooking and homework, or going to the pub for beer and friends. Each one of us faces these types of decisions, big and small, all the time, and what we do depends largely on how we organize our thoughts about each possible scenario and which ultimately seems most important to us. In this story, if you know Sue, you pretty much know already that I went home and had to hear about all of the pub hilarity the next morning. But, if I suffered from alcohol use disorder, thoughts about what would happen at the pub, the taste of the beer, and past really fun pub experiences would have inevitably intruded until they all but squeezed out thoughts of dinner and homework with my kids. My plans to go home would begin to fade until they were all but forgotten, or perhaps I would rationalize that stopping by for one of the special IPAs before going home would somehow work out.

The story above illustrates how thoughts of pub friends and beer can intrude in substance use disorder. However, the intrusion of traumatic events in post-traumatic stress disorder, rumination on negative outcomes in depression, or hearing voices in schizophrenia are all examples of thoughts generated by your brain that can contribute to debilitating psychiatric disorders. Of course, it is a natural and healthy adaptive process to produce thoughts either randomly or in association with the world we are experiencing, and then to use these thoughts to navigate successfully toward desired outcomes.

This volume explores and provides the best possible explanations for what this process is, how it gets usurped in psychiatric disorders, and what this knowledge of how the brain handles thoughts means for concepts of free will and one's responsibility for poor decisions, especially when a thought disorder exists. It addresses how the brain is organized to create thoughts that can be ignored or can build in motivational content, and how we then weigh thoughts to decide on behavior that best adapts us to the world. It also poses and attempts to answer a number of questions that are commonly asked: How do the mechanisms of thought intrusion and decision making get corrupted in psychiatric disorders to create intrusions that cannot be controlled? How are thought intrusions usurped by motivation to produce behavior that may be maladaptive, at least according to social norms? What is free will and what responsibility does free will (or lack of it) create for how we behave?

— Peter W. Kalivas

1

Intrusive Thinking

From Molecules to Free Will

Peter W. Kalivas and Martin P. Paulus

Intrusive thinking has been defined as "any distinct, identifiable cognitive event that is unwanted, unintended, and recurrent. It interrupts the flow of thought, interferes in task performance, is associated with negative affect, and is difficult to control" (Clark 2005:4). Intrusive thinking is ubiquitous: it occurs as random, sometimes annoying daily interruptions but can also be a profoundly disabling symptom in almost all psychiatric disorders. Clinical examples include, but are not limited to, craving in drug addiction, recall of life-threatening events in posttraumatic stress disorder (PTSD), rumination in depression, and hallucinations in schizophrenia. The transdiagnostic nature of intrusive thought in neuropsychiatry points to the importance of understanding the genesis of the intrusions and how they can be therapeutically monitored and regulated. The neuroplasticity and circuitry that underpin intrusive thinking need to be clearly characterized in order to identify sensitive biomarkers and targets useful for translation into novel pharmacological and psychosocial treatments.

This 30th Ernst Strüngmann Forum brought together experts from a wide range of scientific disciplines to evaluate the current state of research into intrusive thinking and to define the boundaries of what is known and unknown about this perplexing and ubiquitous phenomenon. Using a dynamic cross-fertilization strategy of discovery (for more information, see the Preface), teams of experts from neurobiology, neuropsychiatry, and cognitive neuroscience explored four thematic areas: (a) molecular and circuits, (b) psychological cognitive processes, (c) system approaches and models, and (d) interventions and treatments. Within each working group, discussion centered on advancing research and treatment strategies, on identifying challenges to be addressed by future research, on developing models of intrusive thinking, and on the social and philosophical implications beyond its relevance for mental health. This volume, organized around the thematic areas of the working groups, synthesizes the multifaceted deliberations that took place in Frankfurt, Germany, from June 14–19, 2019.

Molecular and Circuits

The first section focuses on identifying molecular and circuit approaches to intrusive thinking. In Chapter 2, Paul Phillips and Amy Milton emphasize that although complex, a biological understanding of intrusive thoughts is tractable based on our current knowledge of model-based and model-free systems and their operation. These systems reflect different approaches of how individuals learn about their environments (Daw et al. 2011). Phillips and Milton point out that it is important to appreciate that these circuits are not fixed and immutable, but rather it is likely that they undergo repeated rounds of plasticity and metaplasticity, which may be the source of intrusive events as a consequence of imbalance within the circuit.

Bernard Balleine (Chapter 3) reviews evidence of parallel circuits mediating the distinct forms of control associated with reflexive and volitional actions, and the interactions between these circuits in determining adaptive behavior. He conceptualizes intrusive thoughts and actions not as a failure of habitual or goal-directed control processes, but rather as a failure of the cooperative and competitive interactions between cortical and basal ganglia circuit processes that form the basis of habitual and goal-directed behaviors.

In Chapter 4, Michael Bruchas posits that advances in molecular technologies beyond standard optogenetic and chemogenetic strategies will provide unprecedented precision in understanding synaptic and circuit mechanisms of the thought generation and behavioral actions that characterize intrusive thinking. Using these approaches in animal models can provide an important bridge toward clinical translational of how brain circuits function and adapt. As examples, he highlights closed-loop sensing of neuronal activity (GCaMP or other) and optogenetic (channelrhodopsin-2 or halorhodopsin equivalents) or pharmacological manipulations in a wireless setting. These approaches can yield real-time sensing during behaviors defined to represent "intrusive thoughts" across species and, together with optogenetic and pharmacological control, can establish causality and mechanisms for intrusive thoughts.

Shannon Gourley et al. (Chapter 5) summarize the principal findings from their group's discussions. They highlight standard animal models, behavioral tests, and outcome measures that could be exploited to shed light on the neurobiological components of intrusive thought. They also propose a conceptual model which captures intrusive thoughts as an emergent property of multiple systems (emotional, cognitive, motor, and autonomic/somatic) that are represented in hubs throughout the brain. Finally, they suggest that when the choreography between these neural hubs and their corresponding nodes becomes disrupted, this creates a loss of homeostatic and/or cognitive control and leads to maladaptive thought intrusions and inappropriate behaviors.

Psychology and Cognition

In this section, the psychological and cognitive processes important for intrusive thinking are the focal point for discussion. Marie Banich (Chapter 6) begins with a review of different methods that can be used to examine and understand intrusive thought, such as self-report and diary measures, to capture the experience, duration, and intensity of intrusive thoughts. She also discusses self-report measures of the difficulty in controlling such thoughts. Moreover, she identifies a number of behavioral paradigms (e.g., the Think/No-Think paradigm) that specifically address mechanisms of memory retrieval and suppression, and pinpoints a significant challenge that remains; namely, the need to find and validate a method that can be used to corroborate the occurrence of intrusive thinking and its contribution to clinical diagnosis.

In Chapter 7, Marie Monfils and David Buss address the psychological aspects of intrusive thinking, highlighting instances in both clinical and nonclinical contexts. They emphasize the high prevalence of intrusive thinking across the population and the challenges this brings when attempts are made to identify unique meaning behind its occurrence. Further, they suggest that the possible adaptive nature of the phenomenon be accounted for in everyday and clinical contexts, when treatments strategies are considered.

The psychiatric literature is reviewed by Florian Schlagenhauf, Andreas Heinz, and Martin Voss in Chapter 8. They point out that although intrusive thought is part of the diagnostic criteria for PTSD and obsessive-compulsive disorder (OCD), it is also a prominent symptom in other psychiatric conditions, such as drug craving in addiction or rumination in depressive disorders. From a descriptive perspective, intrusive thought must be distinguished from thought insertion, observed in schizophrenia and related psychotic disorders. Schlagenhauf et al. raise issues that remain critically unsolved: Are there similar psychological and neurobiological mechanisms underlying intrusive thinking across diagnostic categories, despite striking phenomenological differences between, for instance, negative verbal thoughts and intrusive visual images and memories?

In Chapter 9, the summary of this working group, Renée Visser et al. begin by revisiting systematically definitions of intrusive thinking, addressing all circumstances in which intrusions might occur as well as their manifestations across health and disorders. The least constrained definition emphasized the interruptive, salient, and experienced nature of intrusive mental events as compared with a common definition that specifies unwanted and conscious as well as interruptive criteria. This process brought them to an alternative, more inclusive definition of intrusions as "*interruptive, salient, experienced mental events.*" They propose that clinical intrusive thinking differs from its nonclinical form with regard to frequency, intensity, and maladaptive reappraisal, and emphasize that recurrence is an important property, particularly for psychiatric manifestations. Visser et al. explore the neurocognitive processes which underpin intrusive thinking and its control and discuss its relationship to agency, meta-consciousness, and (mal)adaptiveness or desirability.

Systems and Models

To consider intrusive thinking within the framework of systems and models, John Fedota and Elliot Stein (Chapter 10) examine the neuroimaging literature, which shows that specific and related dysfunction in the calculation of salience involved multiple neuroanatomically and functionally linked regions. These regions involve both cerebral cortex and subcortical areas. These authors suggest that bias at each stage of salience attribution leads to both an overrepresentation of potentiated stimuli and an insensitivity to counterfactual evidence that normally signal the need to alter behavior. In addition, they raise questions to be taken up by future research such as: What are the limits of the neurobiological framework for intrusive thinking that is solely centered on cortical-subcortical-thalamic loops?

In comparison, David Badre (Chapter 11) takes a systems approach to address the following questions: Are there one, many, or any networks whose primary function is best described as cognitive control? Do the networks support cognitive control in the brain, and, if so, is their intrinsic and extrinsic organization "hub-like" or "hierarchical"? Are the networks for cognitive control modulatory or transmissive in the pathway from thought to action? Does controllability apply at the level of cognitive function or brain state? Badre proposes that answering these questions will constrain any conception of how intrusive thought can be controlled and lead to the development of novel mechanism-based interventions.

In Chapter 12, Kayuet Liu and Hakwan Lau advocate a framework for how conscious awareness of one's own intentions and emotions enables the formation of causal narratives about oneself and the world. They argue that these narratives determine one's sense of agency. Moreover, this sense of agency depends on the extent to which one is capable of forming culturally appropriate narratives. Liu and Lau evaluate different ways to characterize consciousness and focus on one that may prove most useful for the intersection between individual agency and intrusive thinking.

In the report of their discussions (Chapter 13), Angela Roberts et al. suggest a general scheme to identify essential elements of an intrusive experience, and where in this scheme dysregulation could occur to increase the likelihood of an intrusion. They find that computational models of intrusive thinking can be embedded in a Bayesian model of active inference, integrated with psychological and physiological models of interoception (i.e., the processing of the brain-body relationship), and informed by psychological and neurobiological models of working memory and associative learning. These models can be examined in terms of flexibility and stability of intrusive thinking. Finally, Roberts et al. conceptualize intrusive thinking as emerging from a deficit in the neuromodulatory mechanisms and dynamics that implement the top-down control of attention; namely, its sensory attenuation.

Interventions and Treatments

In this section, directions are explored for the development of interventions and treatments for intrusive thinking within the context of intrusive thinking as a transdiagnostic endophenotype of many psychiatric disorders. In Chapter 14, Emily Holmes and coauthors suggest focusing efforts on core clinical symptoms (i.e., intrusive imagery). They propose that different interventions may interfere with the reconsolidation of a memory on different levels. By looking specifically at intrusive imagery, rather than broad and fuzzy assemblies of symptoms clusters, they propose that one may be able to radically change current treatment approaches in mental health. Moreover, Holmes et al. highlight the need to develop psychological interventions that can prevent involuntary distressful images from intruding into a person's mind, while still enabling the person to recall voluntarily information about the event.

Harriet de Wit and Anya Bershad consider the challenges in studying intrusive thoughts as a unique entity in Chapter 15. They conclude that relatively little is known about the effects of psychiatric medications on intrusive thoughts, both within disorders separate from other symptoms as well as across disorders. In addition, they note that a wide range of medications are used to treat intrusive thoughts and that these target a range of different neurotransmitter systems. Thus, given the uncertainty about intrusive thinking as a singular entity, it is difficult to determine which medications are specifically effective for regulating intrusive thinking.

In Chapter 16, Colleen Hanlon and Lisa McTeague review transcranial magnetic stimulation as a tool to induce causal change in behavior, cortical excitability, and frontostriatal activity, thus providing an overview of the cortical and subcortical areas that are often implicated in intrusive thinking. They propose that interactions between clinical and preclinical neuroscience researchers, with both electrophysiology and pharmacology backgrounds, could further enhance the efficacy of neuromodulatory approaches.

Judson Brewer et al. consider the challenges in defining intrusive thoughts and the difficulty in distinguishing normal processes of cognition and emotion from indicators of dysfunction, from practical, neurobiological, and cultural points of view (Chapter 17). Throughout, they use the term *intrusive events* to encompass both thoughts and images that become intrusive. Brewer et al. review the behavioral, pharmacological, and emerging electromagnetic brain interventions to target intrusive events. In addition to a discussion of its neuroscientific basis and clinical relevance, they address intrusive events that impact society in multiple ways, and thus can be understood outside of a solely biological or medical perspective.

Summary

Several themes emerged throughout the Forum, which we feel important to highlight, as they could be developed into future programs of scientific and clinical discovery into the phenomenon of intrusive thought. First, all groups wrestled with the definition of intrusive thinking (e.g., whether it represents a singular process) and emphasized the need for bringing operational criteria (tightly related to the underlying neurobiology) to the definition of intrusive thinking. As a consequence, especially when the transdiagnostic nature of thought intrusions are considered, future investigation may aim to develop a psychometrically more refined delineation of intrusive thinking that links closely to the underlying neurobiology. Such an approach may be similar to what has been used across psychopathology (i.e., the use of questionnaires based on item banks) and might substantially reduce subject burden by using an adaptive measurement framework similar to the PROMIS system (Cella et al. 2010).

Second, there is some consensus that intrusive thinking is best understood in terms of a circuit dysfunction. Such circuits exist on a molecular, synaptic, cell, and systems level and may be best understood as dynamical systems, which when perturbed by disruptive events give rise to the intrusion. The precise delineation of these circuits will require sophisticated molecular and systems neuroscience approaches as well as complementary computational models to determine which processes contribute to intrusive thinking.

Third, although phenomenologically related symptoms (e.g., craving, intrusive images, ruminations, hallucinations) exist, it is far from clear whether these dysfunctions have a common neural substrate, involve similar neurochemical circuits, and can be modulated using similar interventions. Thus, future studies need to develop a transdiagnostic framework of intrusive thinking and sophisticated experimental approaches to determine the validity of intrusive thinking components and to assess where the construct converges and diverges.

Finally, intrusive thinking extends beyond psychopathology into everyday life. In a society where individuals are constantly exposed to distractions, it will be important to determine whether intrusive thinking emerges with increased frequency as a consequence. Moreover, the cultural framework of intrusive thinking frequently determines the valence and the urge to resist intrusions, and the expectation of free will is fundamentally challenged by the existence of unwilled intrusive thinking. Future investigations may need to bring more conceptual clarity to the definition of intrusive thinking in the context of agency, free will, and culture. Taken together, the study of intrusive thinking is fertile ground to delineate an important endophenotype that has thus far been neglected by mainstream basic and clinical neuroscience research but which offers enormous potential in finding interventions that can ease mental suffering.

Molecules and Circuits

2

What Are the Circuits That Mediate and Update Intrusive Thinking?

Paul E. M. Phillips and Amy L. Milton

Abstract

This chapter discusses psychological constructs considered to be central to the mediation of intrusive thinking and the neural circuits that underlie these processes. It assimilates intrusive thoughts with conditioned responses, discerns associate structures that can support these responses, and suggests how episodic information may be integrated with these associations. Mechanisms by which intrusive thoughts can be updated are explored, with a focus on extinction and memory reconsolidation. Intrusive thoughts ultimately engage many areas of the brain as they encompass sensory, cognitive, motor, and somatic processes. In this chapter, the focus is on specific circuits within the prefrontal-limbic network that are proposed to encode, update, and maintain the content of intrusions. These circuits include interconnecting pathways between the ventral tegmental area, nucleus accumbens, medial prefrontal and orbitofrontal cortices, hippocampus, and the amygdaloid complex.

Introduction

To begin to answer the question posed in the title, we must first consider what intrusive thinking is. Specifically, we need to address the nature of the content of an intrusive thought: Is it an episodic representation? Could it be a visceral urge, an emotive state that is devoid of episodic content? There is unlikely to be a single answer, as intrusive thinking is ultimately made up of different compositions of these extremes, specific to the underlying pathology. For example, delusions in schizophrenia are clearly episodic in nature, often composed of a detailed and elaborate narrative. By contrast, in obsessive-compulsive disorder (OCD), obsessions come more in the form of an urgency state that is often likened to anxiety. While there may be episodic aspects to this process, such as obsessing over specific contexts, the debilitating qualities of the obsession emerge from the anxiety-like state. Likewise, obsessive behavior in substance use disorders and other addictive disorders comes in the form of craving, an emotive state which, in its fundamental form, is devoid of episodic content. This state, however, is intimately attached to episodic representation of experiences related to the addictive behavior. Further, posttraumatic stress disorder (PTSD) is also associated with an anxiety-like state that is very clearly linked to fragmented episodic memories. This differentiation between the content of intrusive thoughts may, on the surface, seem subtle but is likely to be critical when its neural substrates are considered because the circuits that process episodic information are separable from those underlying emotive states. These differences fall into a common dichotomy in the control of behavior; namely, there are parallel systems that subserve cognitive functions. This dichotomy has a classic separation of processes that can be loosely summarized in the form of a speed–accuracy trade-off: fast, low-computational processes that amount to estimations are generated in parallel to more precise computations that have a higher cognitive demand. We and others have equated this dichotomy to parallel processes used in machine learning that are classified as model-free and model-based computations (Clark et al. 2012) as we believe this is an intuitive and tractable framework. The basis of this separation is the complexity of the information that is stored to support a learned association. A model-based computation establishes an environmental model that can be used to explore potential and inferred connections between stimuli and states. In contrast, a model-free computation is a single-dimensional value assigned to a stimulus based on the reliability of its association with a motivationally relevant outcome.

Central to this line of reasoning, intrusive thoughts are often triggered by environmental stimuli. For instance, in substance use disorders, drug craving can be elicited by drug cues and is posited to become stimulus bound during the transition to addiction (Tiffany and Carter 1998). Accordingly, drug seeking in rodent models, a proxy for craving, can be elicited by unconditioned stimuli (e.g., abused substances or stressors) or conditioned stimuli (e.g., drug cues or conditioned stressors). In PTSD, intrusive episodes are often linked to environmental stimuli in a manner consistent with overgeneralization (i.e., when otherwise neutral stimuli elicit a threat response). Hence, intrusive thoughts often take the form of Pavlovian-conditioned responses, which is our focus in this chapter. As an aside, it is worth noting that while intrusive thoughts can be triggered by unconditioned stimuli, they are unlikely to be *unconditioned responses* because they are not elicited in naïve individuals but rather develop with psychiatric pathology. Importantly, where tested, stimulus-driven intrusions exhibit similar neural signatures to those that are not triggered by an explicit external stimulus (M. C. Anderson, pers. comm.).

What neural circuits are necessary to support these associations? We will discuss circuits that support Pavlovian associations, both for emotive responses and those that incorporate representation of stimulus properties. We will also consider structures that mediate the storage and retrieval of episodic memories as well as the interactions between all of these circuits. We will make the case that circuitry, including midbrain dopamine neurons that project to the nucleus accumbens in the ventral striatum (mesolimbic pathway) or to the medial prefrontal cortex (mesocortical pathway), is necessary for at least a subset of emotive associations. Thereafter, discussion of the neural circuitry will be broadened to include substrates that support Pavlovian associations that can support inferential reasoning, with a specific focus on the central role of the orbitofrontal cortex (OFC) in these higher cognitive processes. We will also explore medial temporal and frontal structures implicated in the acquisition and consolidation of episodic memories and discuss circuits involved in the process of updating existing associations, both through extinction and memory reconsolidation.

Mesocorticolimbic Dopamine- and Emotive-Conditioned Responses

Many psychiatric disorders that are associated with intrusive thinking (e.g., schizophrenia, substance use, OCD, and PTSD) have been linked to perturbations in dopamine transmission in the striatum and/or the prefrontal cortex. These clues have driven extensive research on dopamine transmission with regard to psychiatric disorders, including those where intrusive thinking is a prominent feature.

Reward Prediction Errors

In the mid- to late 1990s, computational neuroscientists and computer scientists who shared an interest in learning algorithms came to the hypothesis that dopamine transmission provides a critical teaching signal for stimulus–reward associations in a model-free learning algorithm (Barto 1995; Montague et al. 1996). This notion was most famously linked to the empirical research of Wolfram Schultz et al. (1997). While this area has expanded enormously since then, the computational role of dopamine transmission in reward learning is most commonly ascribed to variants of the temporal difference algorithm (Sutton and Barto 1998). This algorithm is a time-derivative model that evolved from simple trial-by-trial learning models developed to account for animal behavior. The hypothesized role of dopamine neurons in the model is to signal discrepancies between reward expectation and rewards received to update the current expectation of reward: firing increases when rewards are larger than expected and decreases when rewards are smaller than expected. This model puts dopamine in a critical role in the acquisition and maintenance of stimulus–reward associations. In addition to this putative role in learning, mesolimbic dopamine transmission has consistently been shown to invigorate the enactment of conditioned responses (Flagel et al. 2011b; Ostlund and Maidment 2012).

Model-Free and Model-Based Processes

In a simple cue–reward learning task, where there is spatial separation of the cue and the reward, some animals will approach the cue during its presentation ("sign tracking") whereas others will approach the place where the reward will subsequently be delivered ("goal tracking"). This latter conditioned response requires cognitive representation of the spatial location of future reward, consistent with a model-based process. Interestingly, when animals emit these model-based conditioned responses, dopamine release in the nucleus accumbens does not follow the canonical prediction error signal (Flagel et al. 2011b). This suggests that the specific pattern of signaling is selective for model-free computations. Furthermore, while intact dopamine transmission is necessary to perform either the conditioned response or to acquire sign-tracking conditioned responses, it is not necessary to acquire goal-tracking responses. Similarly, disrupting dopamine transmission during the Pavlovian-to-instrumental transfer disrupts general transfer (invigoration) but it does not affect the cue's ability to bias action selection specifically toward the cue-associated reward, which requires a model-based representation. Recent experiments, however, have revealed a role for dopamine-encoded prediction error signals in some model-based processes (Sharpe et al. 2017). Specifically, it was demonstrated that phasic dopamine signals participate in the updating of the stimulus–stimulus association that takes place in the absence of motivationally valent stimuli and can be used to generate model-based inferences (sensory preconditioning). Thus, mesolimbic dopamine transmission has a somewhat nuanced role in Pavlovian processes, with some predilection for simple stimulus–reward associations. It appears to be important in the acquisition and updating of model-free stimulus–reward associations as well as some, but not all, model-based associations. In addition, mesolimbic dopamine has universal psychomotor-activating properties by which it invigorates the response to conditioned stimuli; however, through this, it is thought to convey only the emotive properties of the association rather than specific (sensory) properties of the conditioned or associated unconditioned stimulus.

Aversive Signaling

To date, the focus has been on associations with appetitive stimuli. Needless to say, intrusive thoughts are often evoked by aversive stimuli. The role of the mesolimbic and mesocortical dopamine systems in computations relating to aversive information has been much more controversial. In many cases, it is simply assumed that aversive stimuli will be computed in this learning model in a similar manner to stimuli that predict lower than previously expected rewards. However, the evidence for this type of encoding is mixed. Mirenowicz and Schultz (1996) observed minimal responses to mildly aversive stimuli (air puffs to the hand) in midbrain dopamine neurons of nonhuman primates. In contrast, Matsumoto and Hikosaka (2009) observed robust changes in the activity of dopamine cells on the presentation of aversive stimuli. In this latter work, the investigators reported some populations of dopamine neurons that encoded aversive information by changing their firing rates in the opposite direction to reward information. Specifically, the presentation of unexpected aversive stimuli or conditioned stimuli that increased the expectation of aversion resulted in reduced firing of these neurons. However, Matsumoto and Hikosaka also reported populations of putative dopamine neurons that increased their firing rate to predictions of aversion. These neurons tended to reside in dorsolateral aspects of the dopaminergic ventral midbrain (dorsolateral substantia nigra pars compacta). They responded similarly to stimuli that increased the expectation of aversion to those that increased the expectation of reward, an encoding pattern often referred to as an unsigned prediction error. However, this coding pattern is not without controversy. Specifically, Fiorillo (2013) has argued that the observed positive responses to aversive stimuli relate to their sensory properties rather than their motivational salience. Others have also questioned whether all of the recorded neurons in these studies are truly dopamine-containing neurons. To address this concern of neuronal-type specificity, Cohen et al. (2012) recorded the firing rates of genetically identified dopamine neurons in mice in response to presentations of appetitive- and aversive-related stimuli. They reported that modulation of the firing rates to aversive-related stimuli were consistently in the opposite direction to those for reward stimuli. These neurons, however, were exclusively recorded in the ventral tegmental area (i.e., the ventral medial aspect of the ventral midbrain) and so did not include the homologous anatomical region from which the unsigned prediction error signals were reported by the Hikosaka group. In neurochemical studies that measure dopamine levels in terminals, results with aversive stimuli have also been mixed. Studies measuring dopamine levels over minutes tend to report increases in dopamine to the presentation of aversive stimuli, especially in the prefrontal cortex (Young 2004; Butts et al. 2011). In contrast, measurements on the order of seconds reveal decreases in dopamine in the nucleus accumbens to aversive stimuli (Roitman et al. 2008). Some investigators addressing this issue have argued that increases in dopamine transmission following an aversive stimulus observed over minutes were responses to the relief from aversion at the offset of the stimulus rather than a response to the onset (Ungless 2004).

Integration of Appetitive and Aversive Information

With systematic interrogation of dopamine neurons in the ventral tegmental area based on identified afferent and efferent connectivity, Lammel et al. (2012)

proposed that appetitive and aversive processing by dopamine neurons is segregated into subcircuits. They demonstrated that activation of a circuit connecting the lateral habenula, ventral tegmental area, and the medial prefrontal
cortex produces an aversive state, while activation of another circuit connecting the laterodorsal tegmentum, ventral tegmental area, and lateral shell of
the nucleus accumbens produces an appetitive state. Complementary to these
data, Tye and colleagues recently reported that a set of prefrontal-projecting
dopamine neurons were selectively activated by aversive stimuli, whereas dopamine neurons that projected to the nucleus accumbens were activated by
rewards (Vander Weele et al. 2018). While this schema does not account for all
of the apparent discrepancies in the literature, it does move the field forward
toward a resolution. However, the issue of whether and how dopamine transmission integrates appetitive and aversive information in a manner that could
instantiate a single, unitary learning model is still an open question. Given
our current understanding, this integration could be through one of several
possibilities, including bidirectionality of appetitive and aversive information
by mesolimbic dopamine neurons (Roitman et al. 2008; Cohen et al. 2012),
gleaning appetitive and aversive information from different populations of dopamine neurons (Lammel et al. 2012; Vander Weele et al. 2018), or combining
appetitive information from dopamine neurons with aversive information from
other neural substrates (Daw et al. 2002).

Broader Circuity

When considering the entirety of the pathways which support these stimulus–stimulus associations, the complexity of the circuitry rapidly expands. In
addition to the laterodorsal tegmentum and the lateral habenula, the central
nucleus of the amygdala, in particular, has been implicated as important up-
stream circuitry to dopamine neurons in model-free processes (Clark et al.
2012). Nonetheless, it is important not to dismiss the rich convergent inputs
into the ventral midbrain from many areas of the brain (Geisler and Wise
2008). Indeed, an optimal temporal difference algorithm should have access to
all available sources of predictive information about rewards and punishers as
well as information on motivational states.

More Complex Stimulus–Stimulus Associations

Orbitofrontal Cortex

From a plethora of research, it is evident that the OFC encodes many different features of motivational stimuli (Thorpe et al. 1983; Padoa-Schioppa and
Assad 2006; Stalnaker et al. 2014). Indeed, it seems that just about any aspect of perception is encoded in about twenty percent of OFC neurons! This

multidimensional encoding seems to be especially important to support model-based associations that permit inferential reasoning. For example, OFC lesions do not affect the acquisition of simple Pavlovian associations or the ability for animals to avoid food that has been paired with illness or fed to satiety. However, unlike intact controls, OFC-lesioned animals continue to respond to stimuli that predict these devalued outcomes (Gallagher et al. 1999). These reinforcer devaluation studies demonstrate that the OFC contributes to the animal's ability to derive the updated expected value of a cue by linking the previously learned stimulus–reward association with the current incentive value of that outcome without having yet directly experienced the pairing of this cue with the devalued outcome. Consistent with the OFC playing a selective role for model-based computations, OFC lesions do not affect general Pavlovian-to-instrumental transfer, but they do disrupt the cue's ability to selectively enhance responding for the specific outcomes the cues predict (Ostlund and Balleine 2007a). While the computational role of the OFC in these processes is not yet precisely defined, a reasonable inference would be that the OFC is used to evaluate the content of model-based associations upon their retrieval rather than being directly involved in the acquisition or storage of this information per se. In this regard the OFC and dopamine may have parallel, complementary roles in response to conditioned stimuli with dopamine conveying emotive responses and OFC relaying stimulus-specific sensory information.

Episodic Information

While the above discussion differentiates associations that represent specific features of conditioned or unconditioned stimuli, these more complex associations do not necessarily incorporate episodic information. However, since intrusive thoughts are often episodic in nature, it is pertinent to expand our consideration to circuits involved in the processing of episodic information. The study of episodic memory is a particularly rich area of neuroscience and has primarily focused on the medial temporal lobe, including hippocampus formation as well as the prefrontal cortex and to some extent the frontal and parietal cortices. However, investigations of the interactions between Pavlovian associations and episodic memory is relatively sparse. Nonetheless, a particularly elegant series of experiments from Shohamy and colleagues have provided evidence that the hippocampus has central roles in associative processes. For example, they showed that hippocampal episodic information can provide a framework for model-based processes to permit inferential reasoning (Wimmer and Shohamy 2012). They also demonstrated that the hippocampus can drive dynamic corticostriatal network connectivity governing stimulus–reward associations (Gerraty et al. 2018). These innovative studies have started to build a foundation for understanding the use and integration of episodic information into learned associations. This platform could be particularly useful for studies of intrusive thinking.

Updating and Inhibition of Intrusive Thought

Extinction

To think about the updating of intrusive thoughts, we should consider the different plasticity processes that can be engaged following the consolidation of an intrusive thought or memory. Broadly speaking, the neural trace or "ensemble" that encodes a thought or memory can be triggered to induce retrieval, reconsolidation, or extinction of the trace under similar, but importantly different, conditions. Extinction has been the most extensively studied, having been originally described by Pavlov (1927). Extinction is operationally defined as the degradation of behavior that was previously supported by a learned association. It takes place when the reliability of the association is weakened, typically occurring after extensive exposure to the unreinforced cue. Extinction can be modeled as a reduction in the strength of an association between a condition and unconditioned stimulus in a simple bidirectional learning system, such as the model-free system putatively associated with dopamine transmission. However, for many associations, it has been argued that extinction learning is not simply the "unlearning" of an association but new, discriminative learning that associates different states of the world with new contingencies in a more complex model. While there are clear demonstrations that extinction *can* be new learning—most likely dependent on prefrontal cortical regions—it perhaps should not be assumed that this is a universal mechanism. At least at the level of the amygdala, molecules usually associated with the depotentiation of synapses increase in activity during Pavlovian extinction, and these molecules are also necessary for extinction to occur (Merlo et al. 2014). Thus, it remains a possibility that different associations have fundamentally different mechanisms of extinction. Regardless of mechanism, a defining feature of intrusive thinking is that the underlying associations are relatively resistant to extinction.

Reconsolidation

The resistance to extinction of intrusive memories may be especially maladaptive if, instead of engaging extinction, reactivation of the intrusive thought leads instead to strengthening of ensemble via reconsolidation mechanisms. Reconsolidation is a process, more recently described than extinction, that can also potentially act to update memories. Reconsolidation refers to the induction of memory lability (or, more mechanistically, memory "destabilization") under certain conditions of retrieval, and the subsequent "restabilization" of the trace in a strengthened or updated form. The restabilization phase is dependent upon protein synthesis in a similar manner to the original consolidation process when the memory was first stored. Therefore, preventing reconsolidation following memory retrieval (e.g., by administration of a protein synthesis inhibitor) can essentially erase the memory. Reconsolidation has been investigated in the context of fear memories (Nader et al. 2000), where it was shown to be dependent on protein synthesis in the basolateral amygdala. Likewise, the basolateral amygdala is necessary for the reconsolidation of drug-related associations that support place conditioning and conditioned reinforcement (Milton et al. 2008a, b; Théberge et al. 2010). Interestingly, protein synthesis in the nucleus accumbens core is also required for reconsolidation of associations supporting conditioned drug place preference but is not required to reconsolidate drug-cue associations that support conditioned reinforcement (Théberge et al. 2010). The extent to which the disruption of reconsolidation in one node of a motivational network can impinge on the function of other nodes in that network is a question that has been surprisingly understudied. Despite this, it has been hypothesized that reconsolidation could provide a mechanism by which memories can be strengthened (Lee 2008), generalized (Vanvossen et al. 2017), and integrated into wider memory networks (Hardt et al. 2010). Therefore, if reconsolidation mechanisms were engaged upon reactivation of an intrusive thought, it is possible that a form of mnemonic "positive feedback" could be established, by which a reactivated intrusive thought not only becomes strengthened when it restabilizes, but generalizes and integrates with other associative traces to lead to an increase in the number of cues and contexts that could trigger the intrusive thought. The consequent increase in the likelihood of triggering the thought would potentially increase the likelihood of subsequent reactivation, leading to further strengthening, generalization, integration, and so on.

Therapeutic Strategies Using Updating Mechanisms

In addition to providing a potential pathological mechanism underlying the persistent and recurrent nature of intrusive thought, reconsolidation could also provide a therapeutic target. While the administration of protein synthesis inhibitors to humans to disrupt reconsolidation is not straightforward, it is possible to capitalize on the updating function of memory reconsolidation in designing therapeutic approaches. For example, developing an approach first used in the preclinical literature (Monfils et al. 2009), Schiller et al. (2010) used a "retrieval-extinction" procedure in which they reactivated a fear memory and subsequently extinguished this memory while it was destabilized (in the "reconsolidation window"). Similarly, James et al. (2015) used visuospatial interference (playing the video game *Tetris*) to disrupt the reconsolidation of intrusive mental images produced by watching traumatic films clips. Although further research needs to be conducted to determine the mechanisms by which these procedures produce effects, including corroboration at the molecular level (Cahill and Milton 2019), these approaches hold potential for the development of new treatment for neuropsychiatric conditions characterized by intrusive thoughts. As a cautionary note, an important consideration for these types of approaches is that different types of stimulus–stimulus associations (i.e., emotive and cognitive) could exist in parallel, supported by independent neural circuits. Therefore, which memory or, more specifically, which aspect of the association drives the pathology may be critical. With some pathologies, for instance, extinguishing an intrusive episodic memory could be futile if the untreated emotive association reattaches to a new episodic memory.

Conclusions

The neural circuits that mediate and update intrusive thoughts are complex but potentially tractable based on our current understanding of model-based and model-free systems and their operation. It is important to appreciate, however, that these circuits are not fixed and immutable, but rather it is likely that they undergo repeated rounds of plasticity and metaplasticity, leading to imbalance within the circuit. In this way, we hypothesize that an adaptive physiological process, supported by functional neural circuitry, can become persistent, recurrent, and pathological.

Acknowledgments

This work was supported by NIH grants R01-DA039687, U01-AA024599, and P50-MH106428-5877 to Paul Phillips, and U.K. Medical Research Council Programme Grant MR/N02530X/1 to Amy Milton. Amy Milton is further supported by the Ferreras-Willetts Fellowship in Neuroscience at Downing College, Cambridge.

3

Corticostriatal Intrusions

Bernard W. Balleine

Abstract

The loop-like circuits which link the cortex and basal ganglia have been implicated in a range of functions; most recently in the precursors to movement, including planning and decision making. Damage to these circuits induced by various disease states have, therefore, been heavily implicated in a range of symptoms, including intrusive involuntary thoughts and actions associated with, for example, neurodegenerative and psychiatric conditions as well as addictions of various kinds. This chapter focuses on recent evidence of parallel circuits that mediate the distinct forms of control associated with reflexive and volitional actions, and the interactions between these circuits in determining adaptive behavior. It discusses two kinds of interaction important for understanding intrusive actions and thoughts: competitive interactions, whereby circuits controlling volitional actions regulate reflexive or habitual responses, and cooperative processes that allow the simulation of specific actions to become manifest in performance. It then explores the role of information derived from predictive learning in action selection and choice. The influence of such information is conveyed through a specific corticobasal ganglia circuit, damage to which has been implicated in compulsive action. The evidence considered generally suggests that intrusive thoughts and actions are the product of an imbalance between corticobasal ganglia circuits rather than dysfunction in any one circuit or its related control process.

Introduction

Intrusive thoughts and actions are highly debilitating symptoms associated with a range of psychiatric disorders and addictions. There are numerous theories regarding the psychological, behavioral, and neural determinants of such intrusions; however, in recent years, a growing theme has been to link these conditions to the processes associated with involuntary or reflexive actions, commonly known as habits (Everitt and Robbins 2016; Robbins et al. 2019). The distinction between volitional, goal-directed actions and reflexes is one with which students of adaptive behavior have grappled for millennia. If, today, this issue feels more tractable, it is because of advances in our understanding of the behavioral processes and, consequently, the neural networks that support these forms of action control. For the purposes of this chapter, I will focus on two aspects of these networks that appear to have important implications for future research into these issues. The first lies in the relationship between the neural networks that mediate goal-directed and habitual control: how they compete and cooperate to support our everyday activities, and the consequences of failures in these interactive processes. The second lies in the way an even more fundamental reflexive system mediating Pavlovian-conditioned reflexes interacts with goal-directed processes to provoke the initiation and performance of otherwise volitional actions in ways that can often appear maladaptive. Although both undoubtedly relate to the habitual control of intrusions, neither wholly depends on such factors. Both, however, highlight how actions and habits are integrated and, ultimately appear to suggest that failures, when they occur, are failures in that integrative process.

Actions and Habits

It is now well recognized that the performance of instrumental actions (i.e., those actions through which we manipulate the environment) is subject to two forms of control process, typically referred to as goal-directed and habitual control. Goal-directed actions are determined by their relationship to and the value of their consequences or outcome. They are rapidly acquired through a learning process sensitive to the causal relationship between action and outcome, and they uniquely engage a process of cognitive-emotional integration in linking these causal relations to goal values based on the motivational and emotional processes through which the incentive values of specific outcomes are established (Dickinson 1994; Dickinson and Balleine 1994; Balleine 2001). As such, goal-directed control is effortful and costly; it engages considerable cognitive and emotional resources but provides fast solutions that are highly flexible (i.e., they can be executed or withheld on demand). Habits, on the other hand, are determined by their antecedents rather than their consequences. They are acquired more slowly through a process of sensorimotor association with specific associations selected by reinforcement. They require few, if any, cognitive resources for their acquisition or performance and are emotionally subject to a reinforcement signal (during acquisition) and to the net level of arousal or drive (during performance) (Dickinson and Balleine 2002). Habits are released, rather than executed, by environmental events and are difficult to withhold once released. They are, however, highly organized actions that have a similar topography across repetitions; thus, they make an adaptive, low-cost solution to common or routine problems. Importantly, these two forms of action control have not only been found to be mediated by distinct learning rules, associative structures, and emotional feedback processes (cf. Dickinson 1994); they also engage distinct neural processes.

Common Tests of Goal-Directed Action

Much of the evidence for distinct forms of action control has been derived from the use of a battery of tests of goal-directed control through which the sensitivity of animals, whether rodent or human, to changes in the action–outcome relationship and in outcome value is evaluated (see Dickinson and Balleine 1994; Balleine and Dickinson 1998; Balleine 2005). One such test, the *contingency degradation test*, assesses sensitivity to the relationship between an action and its consequences by increasing the probability of access, p, to a specific outcome, O, in the absence of an action, A, or $p(O|noA)$. Given a specific probability that an action earns an outcome (i.e., $p(O|A)$), $p(O|noA)$ can be increased until the outcome is equally probable if the action is performed or not. At that point, the action is no longer causally related to the outcome and, if the actor is sensitive to this, then the actor should no longer perform the action (Hammond 1980). Increasing $p(O|noA)$ beyond that point means that the action prevents the outcome, which should reduce responding even faster, something demonstrated through the use of omission schedules (Davis and Bitterman 1971). Goal-directed action in humans and animals is exquisitely sensitive to these manipulations of the action–outcome relationship, largely because these manipulations alter the causal relationship between these events (Dickinson 2012). As a consequence, not only is performance affected, judgments as to how causal an action is, with respect to its specific outcome, are similarly modified by these contingency manipulations (Shanks and Dickinson 1991). Finally, these changes in contingency are highly selective; altering the relationship between one action and outcome does not affect performance based on the causal relationship between other actions and their outcomes (Balleine and Dickinson 1998).

The second commonly used test is called *outcome devaluation*, which assesses sensitivity of performance to changes in the value of the outcome of an action and is usually conducted without feedback (i.e., in extinction). Hence, having learned a number of specific action–outcome relationships (A1 → O1, A2 → O2, etc.), the value of one outcome can be altered, after which the subject is given a choice between the various actions. Goal-directed control reflects the ability to integrate the action–outcome relationship with the altered value of the outcome to modify performance of the action. Again, considerable evidence has demonstrated that goal-directed actions in both humans and rodents are sensitive to these kinds of treatment (e.g., Balleine and Dickinson 1998; Balleine and O'Doherty 2010).

Neural Bases

Studies of goal-directed action have found evidence that the acquisition and performance of such actions depend on the rich connections of the prefrontal cortex (PFC), including the human ventromedial PFC and medial orbitofrontal cortices, with the striatum, specifically the caudate nucleus (Balleine and O'Doherty 2010). The greater resolution allowed by rodent studies has significantly refined this general picture. Thus, it is now clear that the homologous region of the medial PFC in rats, the prelimbic cortex (area 32), is engaged during the early acquisition of goal-directed actions (Hart and Balleine 2016). Furthermore, whereas the input layers 2/3 contribute significantly to the consolidation of goal-directed learning, the output layers (particularly the intratelencephalic neurons in layer 5) are critical to this process. Activity in their bilateral projection is necessary for learning-related plasticity in the posterior dorsomedial striatum (one of their main targets) and thus for the acquisition of goal-directed actions (Hart et al. 2018a, b).

Using the above tests, we demonstrated, some time ago, that rodents with damage to the prelimbic cortex, the posterior dorsomedial, or the mediodorsal thalamus show deficits in their sensitivity to contingency degradation and outcome devaluation, and continued to respond as if neither the contingency nor outcome value had changed. Evidence that these structures share a high degree of interconnectivity led to the claim that they constitute the critical cortico-striatal-thalamo-cortical loop through which goal-directed actions are encoded (see Figure 3.1a; Balleine 2005). Importantly, this same insensitivity to shifts in the action–outcome relationship and in outcome value is found in habitual actions. Early in training, actions are sensitive to changes in contingency and value, whereas with continued practice, performance naturally becomes insensitive to these changes as action control shifts from the consequences of an action to antecedent environmental stimuli with which the action becomes associated (Dickinson et al. 1995, 1998). Chief among these stimuli is the context in which the action is trained (Thrailkill and Bouton 2015).

At a neural level, the acquisition and performance of habits depends on a corticostriatal circuit linking the sensorimotor regions of frontal cortex with the putamen or dorsolateral striatum (Figure 3.1a). Again, treatments causing degeneration to, or temporary inactivation of, structures in this circuit block the acquisition of habit learning and render habitual actions goal directed; that is, despite extensive training, they continue to show sensitivity both to shifts in the action–outcome relationship and outcome value (Yin et al. 2004, 2006). Generally, therefore, these findings have been interpreted as suggesting that the goal-directed and habitual control of instrumental actions is a competitive process: any reduction in goal-directed control increases the likelihood actions will be habitual, whereas any reduction in habitual control increases the likelihood actions will be goal directed. Thus, inactivation of the dorsomedial striatum immediately places actions under habitual control (Yin et al. 2005a), whereas inactivation of the dorsolateral striatum appears immediately to render actions goal directed (Yin et al. 2006), as if these two processes are always active and merely compete for control over performance (Balleine et al. 2009).

In fact, even under normal circumstances the goal-directed process can rapidly inhibit habitual control. This can be detected in our everyday activities. A

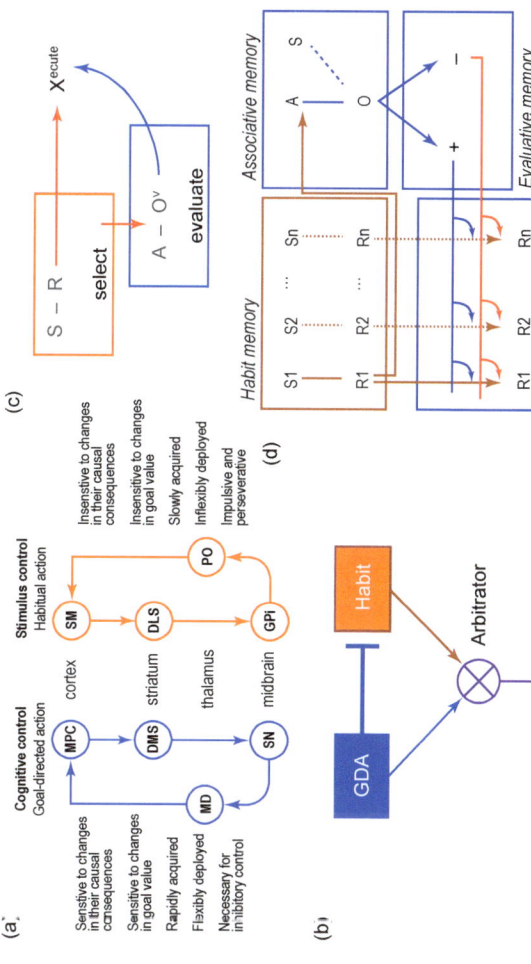

Figure 3.1 Competition and cooperation between actions and habits. (a) Schematic of neural circuits mediating goal-directed and habitual action control, as well as their primary characteristics: A loss of goal-directed control tends to yield dysregulated habitual actions, whereas a loss of habitual control yields dysregulated goal-directed actions. (b) One view of the competitive processes in goal-directed action (GDA) and habitual action control in which both compete for control of performance via some form of arbitration; habits emerge via arbitration but are heavily regulated. (c) Illustration of the cooperation between the stimulus-mediated selection and action-mediated evaluation processes necessary to generate action execution. Generally, stimuli (S) generate urges to respond (R) that initiate the retrieval of specific actions (A), their outcomes (O), and the value (v) of those consequences. If negative, this evaluation would check that urge; if positive, the urge would translate into an executed action. (d) This descriptive model can be further elaborated into an associative cybernetic model which views instrumental performance as the direct outcome of cooperation between S-R and A-O associative processes, whose joint influence sums at the motor system to drive motor output.

useful example is to consider the effect of seeing a police car in the rearview mirror while driving on the freeway: After a period of carefree, apparently cognitively disconnected driving, the sight of the highway patrol causes a very distinct change in our behavior. Do we carry on driving in such a carefree manner? Not likely! Even if we are within the speed limit and are generally obeying the rules of the road, our vigilance would dramatically increase and our driving would become far more deliberate. Thus, habitual control has been suppressed. Likewise, rats behaving habitually will rapidly transition to goal-directed control when the lever response is punished by the actual delivery of an aversive or noxious outcome (Dickinson et al. 1983, 1995). The rapidity of this adjustment is, however, severely curtailed by damage to or inactivation of the dorsomedial striatum (Yin et al. 2005b; Furlong et al. 2017), a finding that is consistent with the argument that the return of control to the goal-directed system is the source of the rapid suppression of habits in a punishment situation (see Figure 3.1b).

Loss of Control and Intrusive Action

Some forms of psychopathology that result in intrusive or compulsive actions find their source in a defective ability to suppress habitual actions; most notably in drug addiction. During the development of addiction, the pursuit of drugs of abuse rapidly becomes habitual, coming under the control of internal and external states and stimuli rather than the consequences of acting (Corbit et al. 2014; Furlong et al. 2016, 2017, 2018). It is important, however, to distinguish habitual drug seeking from other forms of habitual behavior. Under normal conditions, habit learning can be highly adaptive: habits allow us and other animals to relegate the control of routine behavioral responses to a system that uses few cognitive resources, freeing up a limited executive capacity. In contrast, habitual drug seeking is pathological: drug exposure increases the rate of acquisition of habitual actions and the influence of drug-associated contexts (Ostlund et al. 2010) and cues (Glasner et al. 2005) on their performance. Furthermore, a distinguishing feature of habitual drug seeking is the addict's loss of executive control over the habit. As is commonly noted, drug pursuit persists in the face of often severe, negative consequences. The compulsive pursuit of drugs can be viewed, therefore, as the product of (a) a drug-induced increment in habit and (b) a decrement in the addict's ability to exert control over these actions in the face of persistent, negative feedback (Ostlund and Balleine 2008). The need for these two processes to interact may help clarify why relatively few people who take drugs ultimately become addicted; although taking drugs may have some of the hallmarks of a habit, resilience in their goal-directed system to the effects of drug exposure may help to reduce the risk of those habits ultimately becoming dysregulated (Burt et al. 2016).

Consistent with this argument, we have found that exposure either to drugs themselves, such as cocaine or methamphetamine, or to contexts previously paired with drugs can render goal-directed action habitual in tests of outcome devaluation (Ostlund et al. 2010; Corbit et al. 2014; Furlong et al. 2017). Furthermore, actions in these contexts can persist when punished by the delivery of an aversive outcome (Furlong et al. 2018), something that we found was linked to reduced activity in D1-expressing spiny projection neurons (SPNs) in the posterior dorsomedial striatum (Furlong et al. 2017). There are two populations of SPNs in the striatum that account for roughly 95% of the neurons: direct path D1-expressing (dSPNs) and indirect path D2-expressing SPNs (iSPNs). Whereas dSPNs tend to increase functional output, and hence increase goal-directed control, iSPNs inhibit it (Gerfen and Surmeier 2011). As such, a loss of dSPN activity should be expected to have the effects we observed. In an attempt to rescue normal function, therefore, we attempted to redress the relative balance between the activity of dSPNs and iSPNs by inhibiting iSPNs. How we achieved this was complex. The binding of dopamine at D2 receptors on iSPNS inhibits the activity of these neurons and, as such, a D2 antagonist might be expected to be sufficient. However, D2 receptors are also expressed on multiple interneurons as well as iSPNs in the dorsal striatum, reducing the selectivity of this manipulation. Importantly, however, adenosine A2A receptors are only expressed on iSPNs in this region, and the inhibition of these receptors increases the effects of D2 binding only on iSPNs (Tozzi et al. 2007). We hypothesized, therefore, that local infusion of an A2A antagonist would (a) increase D2 receptor activity, (b) reduce iSPN activity, (c) restore the balance between dSPNs and iSPNs, and thus (d) allow the animals to exert behavioral control over their instrumental performance. This is indeed what we found: rats were now able to exert behavioral control over their habits and reduced performance in a punishment test to a similar degree whether they were tested in a drug-paired or unpaired context (Furlong et al. 2017).

Similar deficits in contingency degradation and outcome devaluation have been described in various psychiatric conditions linked to changes in the circuitry mediating goal-directed action control (Griffiths et al. 2014). We found, for example, that the causal judgements of a cohort of youths diagnosed with major depression were relatively insensitive to changes in the causal relationship between action and outcome (Griffiths et al. 2015, 2016a). Furthermore, reductions in causal awareness were correlated with size and shape changes of the globus pallidus (GP) and midline thalamic structures in the basal ganglia output circuit that feeds back to the cortex. Tractography confirmed the relationship between the dysfunctional area of GP and the striatum, on one hand, and the midline thalamus, on the other (Griffiths et al. 2015). This suggests, particularly given that subjects were at an early stage in illness progression, that such changes may reflect the precursors of later prefrontal cortical and corticostriatal deficits in depression, as has also been claimed by others with regard to schizophrenia (e.g., Simpson et al. 2010). Indeed, in a recent study, we found deficits in the flexibility of causal judgment in chronic schizophrenia similar to that observed in major depression (Morris et al. 2018). Furthermore,

B. W. Balleine

we found that schizophrenic subjects had a severe deficit in the sensitivity of their instrumental performance to outcome devaluation. We did not, however, find deficits in the ability of schizophrenic subjects to describe the immediate relationship between actions and their consequences, nor did we find that schizophrenic subjects differed from healthy controls in their ratings of outcome value after devaluation. Rather, it appeared that the schizophrenic subjects had difficulty integrating their knowledge of the action–outcome relationship with the changes in outcome value to permit them to make accurate choices. As might be expected, this effect was related to reduced activity in the dorsolateral prefrontal cortex and in the head of the caudate. Furthermore, whereas the deficits in neural activity were related to the severity of negative symptoms, particularly avolition and alogia, devaluation sensitivity correlated with functional measures, such as hours of paid employment over the last six weeks (Morris et al. 2015). Generally, therefore, with respect to its role in controlling intrusive actions, each of these effects suggests that dysfunction in the prefrontal-caudate-globus pallidus-thalamic feedback circuit compromises goal-directed control.

The Dysregulation of Goal-Directed Action

It would be interesting to know whether the same kind of dysregulation occurs when habitual control is inhibited. Does reduced habitual control result in dysregulated goal-directed action? If so, what would that dysregulation look like? There are, in fact, layers of complexity in attempting to understand the interaction of these apparently distinct action controllers. At one level it is clear that goal-directed and habit processes compete for control of performance (as reviewed above). There is also evidence that these processes cooperate in the integration of stimulus-mediated action selection with action evaluation, particularly during the performance of goal-directed actions. Perhaps the strongest evidence comes from studies of instrumental reinstatement: after a period of extinction, we have found that free delivery of the instrumental outcome reinstates performance of its associated action (Ostlund and Balleine 2007b). Importantly, the ability of the outcome to generate this effect does not depend on its value: devaluation of the freely paired outcome does not affect its ability to select its associated action in reinstatement tests. Nevertheless, the subsequent level of performance of that action is affected by the devaluation treatment: devaluing the outcome that serves as a goal for the retrieved action reduces the vigor of its performance but not the ability of the outcome to reinstate performance (Balleine and Ostlund 2007; de Wit et al. 2009).

Thus, in the ordinary course of events, this evidence suggests that the outcome controls actions in two ways: (a) through a form of stimulus–response association, according to which the stimulus properties of the outcome select its associated action, and (b) through the action–outcome association, through which the action retrieves the outcome as a goal. This behavioral evidence suggests that a selection–evaluation–execution sequence lies at the heart of goal-directed instrumental performance and that this control requires the cooperative integration of the goal-directed and habitual control processes (see Figure 3.1c; Balleine and Ostlund 2007; Balleine and O'Doherty 2010). From this perspective, two possibilities may arise. First, if maladaptive habitual control results in attenuated cooperation between the habitual and goal-directed control processes, then this would break the link between the urge to respond and the regulation of the effect of that urge on performance, potentially resulting in an urge to act but without the subsequent capacity to evaluate and, when necessary, to suppress that action. Such intrusive urges to act may be one source of obsessions and, if strong enough, of compulsions (as discussed above). Alternatively, an actual loss of habitual control could result in a loss of action selection and/or initiation leaving the actor cognitively able to retrieve specific action outcomes and their value but unable to implement those actions. Cognitively simulating actions and thinking through their consequences sounds like planning. However, repeatedly engaging in cognitive simulation because of an attenuated capacity to complete the retrieval–evaluation–execution sequence does not resemble adaptive planning but rather *obsessive* or *intrusive thinking*. It might, therefore, be argued that whereas attenuated goal-directed control results in compulsive or intrusive actions, a loss of habitual control results in obsessive or intrusive rumination on the potential consequences of actions that are, then, unable (or less able) to be performed. Such failures could result from dysregulation in the competition between these control processes (resulting in a loss in the ability to inhibit habits) or in the cooperation between them (resulting in an inability to execute actions and so bring planning to an end).

Some years ago, we implemented this general scheme within an associative cybernetic model of instrumental performance (see Figure 3.1d). Briefly, a response tendency or urge is generated in a stimulus-response memory which then brings to mind the retrieval of a specific action and its consequences in an associative memory. Outcome retrieval in the associative memory activates an incentive memory of that outcome which, by marshalling specific motivational and emotional processes, determines its value, which acts to potentiate (or depotentiate) the motor effects of the stimulus-response urge and thus increase (or decrease) the probability that the action will be executed. It is this latter process that constitutes the cybernetic or feedback component of the model without which either response urges could run off unchecked or not run off at all (depending on their strength). Likewise, a significantly attenuated stimulus-response memory would result in the failure of activity in the associative memory to result in an actual response. One question that arises in this model is why there is any activity at all in the associative memory when the habit memory is attenuated. To understand this issue, it is necessary to introduce perhaps the single most important influence on the performance of goal-directed and habitual actions in the context of intrusive thoughts and actions: the effect of predictive stimuli on action selection.

Cognitive Control of Action

Although both prediction and control are necessary for adaptive behavior, discussions of the importance of predictive learning are usually confined to its role in determining conditioned reflexes of one kind or another. There are, however, multiple ways in which predictive learning can influence instrumental actions. Pavlovian predictors of important environmental events could elicit conditioned reflexes congruent or incongruent with our actions; for example, a stimulus paired with an aversive event could generate freezing, which would be incongruent with active avoidance responses. Furthermore, there is evidence that predictive learning can influence actions independently of the conditioned reflexes predictors produce. Indeed, there is good evidence that a stimulus that has become a reliable predictor of a valued outcome can enhance the performance of actions associated with that outcome while leaving actions associated with other rewarding outcomes unaffected. Furthermore, treatments that abolish the predictive validity of such stimuli can abolish these effects on instrumental performance without affecting their ability to evoke a conditioned reflex (Delamater 1995). Over and above conditioned reflexes, therefore, predictive learning clearly provides *information* regarding forthcoming rewards and punishers; if that information is important for adaptive behavior, its effects will then be mediated by changes in instrumental action, via its effects on actions sufficiently flexible to be modified in response to that information and to do so rapidly.

The influence of the information provided by predictive learning on instrumental action is typically studied in the laboratory using the *Pavlovian-instrumental transfer* paradigm (for a recent review, see Cartoni et al. 2016). In these experiments, subjects are exposed to a period of Pavlovian conditioning, during which cues of various kinds are paired with specific, usually rewarding, events (e.g., specific foods or fluids), after which they are trained to perform distinct actions to earn those same food or fluid outcomes (Figure 3.2a). In a typical rodent experiment, rats will be first given a period of predictive learning, during which they learn that stimuli S1 and S2 (e.g., tones or clickers) predict the delivery of distinct outcomes, O1 and O2 (e.g., dry food pellets or liquid sucrose): S1–O1 and S2–O2. Subsequently, they are trained to perform two novel instrumental actions (A1 and A2) for these same outcomes. They might be trained, for instance, to press one lever for the pellets and a second lever for the sucrose: A1 → O1 and A2 → O2. In a final test, the previous Pavlovian and instrumental phases are brought together to examine the influence of the former on the latter: rats must choose between A1 and A2 in the presence of S1 and S2 (i.e., S1: R1 vs. R2, and S2: R1 vs. R2). Importantly, no outcomes are delivered in this test phase, either after the stimuli or the actions. Thus, the test provides an opportunity to observe how predictive learning influences instrumental performance directly. Typically, the stimuli cause the rats to select and perform more vigorously the response previously associated with

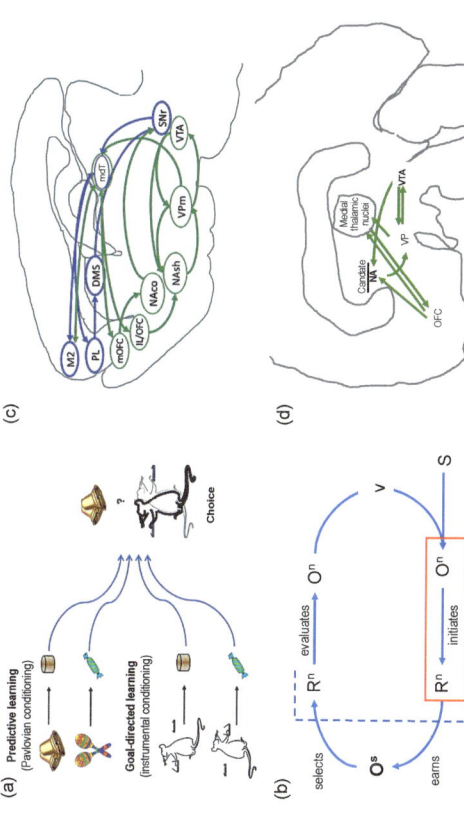

Figure 3.2 The influence of predictive learning on choice. (a) Design of a typical experiment used to study the influence of predictive learning (Pavlovian conditioning) on instrumental choice performance, here in rodents. Rats are first exposed to two stimuli paired with distinct outcomes and then trained to perform two actions for those outcomes before the influence of predictive learning on choice is assessed in an extinction test. (b) Evidence (see text) suggests that the influence of predictive learning is mediated by the ability of a stimulus, S, to retrieve a specific outcome, O ($S - O^n$), and so the specific action associated with that O ($O^n - R^n$). The outcome delivered as a consequence of R^n serves both as the goal of the action and as a stimulus that selects the next action. (c) Schematic of the cortical-striatal-pallidal-thalamo-cortical circuits underlying goal-directed action and the way predictive learning influences these actions. Note that the influence of stimuli on infralimbic and orbitofrontal cortex (IL/OFC) activity is driven by the ability of nucleus accumbens shell (NAsh) and medial ventral pallidum (VPm) to inhibit mediodorsal thalamus (mdT), and so disinhibit this input. The involvement of medial OFC and its connections with nucleus accumbens core (NAco) is crucial to the integration of this circuit with the larger goal-directed circuit and its influence on performance through medial agranular cortex (M2). (d) Dysfunction in the same cortical-striatal-pallidal-thalamo-cortical circuit in humans has been argued to underlie the compulsive actions associated with obsessive-compulsive disorder; caudate nucleus (CAUD) (after Modell et al. 1989).

the outcome predicted by the stimulus: given the scenarios described above, S1: R1 > R2 and S2: R1 < R2. Very similar effects are observed in mouse and human subjects to those observed in rats (Cartoni et al. 2016).

As intimated above, the influence of predictive stimuli on choice in the transfer paradigm depends on their predictive status; degrading the contingency between a stimulus and its specific outcome does not affect its ability to elicit a conditioned response but completely abolishes its effects on choice between actions. It is clear, therefore, that it is the information that the stimulus provides with respect to a predicted outcome that is critical to the ability of such stimuli to affect performance, rather than its ability to evoke the conditioned response. Nor, as it turns out, is this effect due to the ability of the stimulus to retrieve and activate goal-directed control generally via retrieval of the action–outcome relationship and, subsequently, the value of the outcome. One of the more interesting and telling effects in this literature is the finding that devaluing the outcome predicted by the stimulus does not affect the ability of that stimulus to influence choice (Rescorla 1994; Holland 2004). Although the predictive validity of the stimulus is critical, the value of the outcome predicted by that stimulus is not.

Finally, it is important to note that these effects of predictive learning on choice between actions are mediated by both excitatory and inhibitory action–outcome associations and their effects on performance. When a stimulus predicting a particular outcome elevates the performance of an action associated with that outcome, it does so without affecting the performance of other actions (Laurent and Balleine 2015). We have hypothesized that this is at least partly due to the fact that, in goal-directed learning situations, actions can become associated with the outcome that they deliver (e.g., A1 → O1, A2 → O2) as well as with the absence of outcomes that they do *not* deliver (i.e., A1 → no O2, A2 → no O1). Hence, a stimulus that predicts a particular outcome is likely to elevate the performance of actions associated with the predicted outcome, but not the responses associated with the absence of that outcome; information (e.g., S1 predicts O1) can elevate an action, R1, that delivers O1 but not another action, R2, that predicts no O1 (Laurent and Balleine 2015). In summary, Pavlovian-instrumental transfer provides insight into the way predictive learning affects instrumental performance. As a phenomenon, it reveals the following:

1. Presenting cues that predict specific outcomes elevates the performance of actions associated with those outcomes (without affecting those that are not) by selecting those actions and increasing (or inhibiting) their execution in an ongoing manner.
2. The ability of stimuli to produce these effects depends on how specifically they provide information about those outcomes.
3. The ability of these cues to provoke the performance of actions does not depend on the value of the outcome with which they are associated.

Of the theories advanced thus far to explain the effects of predictive learning on instrumental performance, the one best supported suggests that the association between the stimulus and outcome, acquired during Pavlovian conditioning, allows the stimulus to subsequently retrieve the outcome, thereby retrieving (or inhibiting) the action associated with that outcome (Figure 3.2b; Balleine and Ostlund 2007). Any failure to inhibit competing actions in the presence of a specific predictive stimulus would result in an unwanted, intrusive action. This is strongly reminiscent of the occurrence of intrusive, compulsive actions in various types of addiction as well as in multiple, severe neuropsychiatric disorders, including Tourette syndrome (Leckman et al. 2010), grooming disorders (e.g., skin picking, trichotillomania; Chamberlain et al. 2009), and obsessive-compulsive disorder (OCD) (Robbins et al. 2019). Furthermore, not only does the superficial similarity of the behavioral influence of predictive learning on instrumental performance suggest this, there is a very close similarity between the neural bases of Pavlovian-instrumental transfer and the circuitry previously implicated in the compulsive actions associated with these conditions.

Neural Bases of Pavlovian-Instrumental Transfer

A number of structures have been found to be involved in the way predictive learning affects action selection in transfer situations; for a summary, see Figure 3.2c. Many of these have also been implicated in either Pavlovian conditioning and/or instrumental conditioning, as would be anticipated. Treatments, for example, that lesion or block the activity of the basolateral amygdala (BLA) are particularly effective in blocking transfer effects, but they also abolish the ability of animals to encode the association of Pavlovian cues and instrumental actions with specific outcomes. Similarly, lesions to or inactivation of the mediodorsal thalamus (mdT) or the ventrolateral orbitofrontal cortex (vlOFC) abolish the selectivity of Pavlovian predictions with respect to specific outcomes, thereby abolishing the selectivity of transfer effects with respect to those outcomes (reviewed in Cartoni et al. 2016). However, the projection from the BLA to the nucleus accumbens shell is different in this regard: Disconnection of the BLA from the shell abolishes Pavlovian-instrumental transfer but has no detectable effect on either instrumental- or Pavlovian-conditioning processes. This suggests that this circuit is critical for the way information derived from predictive learning informs choice between instrumental actions (Laurent et al. 2015a).

In recent years, considerable detail has been added to this basic circuit in understanding how exactly the shell is involved in transfer. Briefly, what has become clear is that inputs to the shell from the ventral tegmental dopamine neurons are critically involved (Laurent et al. 2014) as are two other modulatory processes that are localized to the shell itself, involving the interaction of opioid and cholinergic processes (Laurent et al. 2012, 2014, 2015b). The latter process is controlled by the giant aspiny cholinergic interneurons that make up about 3% of the neurons in the shell. These neurons express delta opioid receptors on their membrane which bind enkephalin that inhibits the release of acetylcholine, thereby releasing its inhibitory effects on dopamine D1 receptor-expressing SPNs in the shell (Bertran-Gonzalez et al. 2012). Importantly, the degree to which delta opioid receptors are expressed on cholinergic interneurons is related to the effectiveness of predictive learning. Mice that show strong Pavlovian learning also show strong delta receptor expression, and this is true whether one assesses excitatory (Bertran-Gonzalez et al. 2012) or inhibitory associations (Laurent et al. 2015b).

The shell projects to the broader basal ganglia circuitry through several pathways, but particularly to the medial and lateral segments of the ventral pallidum. In a number of studies, we found that the more critical projection for transfer effects is to the ventral segment of the medial ventral pallidum (VPm) based on the effects of disconnecting the shell from the VPm versus the VP-l prior to the Pavlovian-instrumental transfer test (Leung and Balleine 2013). Shell inputs to the VPm drive the output neurons from this structure, which project to multiple efferent structures including the ventral tegmental area (VTA) and the mdT (Root et al. 2015). Again, using disconnection procedures, we found that the critical output for the transfer effect is to the mdT, whereas the projection to the VTA appears to be important for response vigor irrespective of the direction of that response (Leung and Balleine 2015). Thus, although damage to both VTA and mdT projections completely removes the transfer effect, damage to the VTA projection reduces the overall level of performance, whereas damage to the mdT projection blocks the bias in performance typically induced during the transfer test.

The VPm to mdT projection is GABAergic and its activity inhibits thalamic output neurons (Lavín and Grace 1994). These mdT output neurons inhibit the vlOFC (Alcaraz et al. 2016); thus, turning off this mdT projection disinhibits the vlOFC and alters the activity of this corticostriatal projection and its effects on performance. As such, this loop serves the function of maintaining the selective drive exerted by predictive learning on action selection, although precisely how it influences motor performance is currently unknown. One possibility is that, like other cortical areas that receive input from the thalamus, the OFC provides a return projection that, ultimately, excites motor output through premotor cortex (Hart et al. 2014). As this output is also likely to be engaged by the dorsomedial striatal goal-directed circuit, it is possible that the corticothalamic circuit is generally functioning to allow predictive cues to potentiate the performance of actions selected and evaluated by the goal-directed circuit (Hart et al. 2014). Whatever the merits of these speculations, it is important to recognize that any failure of the VPm to inhibit the mdT will result in a failure to disinhibit the OFC, and that the effect of this failure is not to block transfer effects per se but to render them nonselective; that is, the action typically inhibited in the transfer test via the inhibitory response-outcome association will be disinhibited and released as an intrusive, involuntary, or compulsive action.

Finally, one other structure implicated in the changes in instrumental performance induced by predictive learning is the medial OFC (mOFC). Damage to or inactivation of the mOFC also abolishes transfer effects without generally elevating performance; rather, Pavlovian cues have no influence on instrumental performance (Bradfield et al. 2015). One explanation for this effect is that the mOFC critically mediates instrumental performance in extinction; indeed we have recently argued that the mOFC functions to retrieve action outcomes when they are freely recalled from memory as opposed to when they are directly observable in the environment. We reached this conclusion from the finding that mOFC manipulations altered sensitivity to outcome devaluation, but only in tests conducted in extinction: when the devalued outcome was delivered contingent on performance, the response was adjusted appropriately. Importantly, these manipulations also altered the animals' sensitivity to Pavlovian-instrumental transfer (Bradfield et al. 2015). Thus, unlike the vlOFC, the mOFC is not just involved in influencing predictive stimuli on instrumental performance; it also affects experienced changes in reward on that performance. This suggests that its effects on transfer might be related to its involvement in retrieving the outcomes of goal-directed actions more generally when these outcomes are unobservable.

Neural Bases of Intrusive Thoughts and Actions

Over and above the relationship between the altered activity in various cortical-dorsal-striatal circuits and the disease states described above, theories related to the neural bases of addictive conditions and relapse as well as of OCD and related conditions have implicated changes in the OFC-shell-VPm-mdT-OFC loop mediating Pavlovian-instrumental transfer in those conditions (Fettes et al. 2017). The role of this circuitry in OCD is particularly well documented, especially the role of the vlOFC and mOFC and its reciprocal connections with the mdT in this condition. Thus, for example, it has long been known that the symptoms associated with OCD are ameliorated by surgical ablation of either the OFC or of the midline and intralaminar thalamus, including the mdT. Importantly, they are similarly ameliorated by thermal or gamma radio lesions of the ventral anterior limb of the internal capsule (also called anterior capsulotomy)—the region through which the white matter tracks pass containing the bidirectional fibers which connect the thalamus and the OFC (Pepper et al. 2015). More recently, deep brain stimulation of these fiber pathways has been reported to have a similar effect (reviewed in Greenberg et al. 2010).

Neuroimaging studies of OCD patients have implicated aspects of this circuitry in OCD, particularly the subregions of the OFC. Generally, the vlOFC has been reported to be hypoactive (Remijnse et al. 2006) and the mOFC to be hyperactive (Gillan et al. 2015) during tests that engage these regions; these findings are broadly consistent with other studies (e.g., Kahnt et al. 2012; Zald et al. 2014). Two important findings from the animal literature suggest that these differences in activity may be important for aspects of symptom formation. For example, a study by Burguiere et al. (2013), which uses the SAPAP3 mouse model of OCD, found that these mice show a persistent conditioned grooming response when it was elicited by a cue that had been paired with a water drop to the head (Burguiere et al. 2013). The authors hypothesized that this effect was generated by a hypoactive vlOFC and thus sought to activate this region in an attempt to reduce the excessive grooming response. They achieved this using optogenetic stimulation of the vlOFC and found that the persistent grooming was, indeed, inhibited by optogenetic stimulation of the lateral OFC neurons projecting to the striatum. A similar study by Ahmari et al. (2013), optically stimulated the terminals of mOFC cells expressing channelrhodopsin in the ventromedial striatum, centering on the junction between the nucleus accumbens core and shell. Stimulation of this pathway progressively increased grooming, which persisted after the stimulation (Ahmari et al. 2013). As such it appears that enhancing activation of accumbens projecting neurons in the vlOFC reduces compulsive grooming, whereas stimulating the mOFC projections to the ventral striatum enhances compulsive grooming in line with the hypo- and hyperactive phenotypes of the vlOFC and mOFC in OCD, respectively.

Sometime ago, based on similar kinds of observations in OCD patients, Modell et al. (1989) developed a circuitry model of OCD with striking similarities to the circuitry involved in Pavlovian-instrumental transfer (Figure 3.2d). They argued that "the primary pathogenetic mechanism of OCD lies in dysregulation of a basal ganglia/limbic striatal circuit that modulates neuronal activity in and between posterior portions of the orbitofrontal cortex and the associated medial thalamic nuclei" (Modell et al. 1989:32). Furthermore, they proposed that specifically the compulsive symptoms of OCD are associated with aberrant activity in a positive feedback loop in the reciprocally excitatory frontothalamic neurons, due to the loss of inhibitory input from the ventromedial portions of the striatum. Further, they postulated that: "the net result of ventral striatal activation is increased (inhibitory) pallidothalamic output: essentially the loop transduces excitatory input from the vlOFC into inhibitory output to the thalamus which serves to modulate activity of the frontothalamic circuit by means of an interposed negative feedback loop" (Modell et al. 1989:32).

The same circuit is crucial to normal Pavlovian-instrumental transfer. Damage to this circuit results in aberrant transfer; that is, it results in predictive learning eliciting *both* the performance of actions associated with the outcomes predicted by those stimuli as well as actions that predict the absence of those outcomes, which is the definition of an intrusive involuntary and maladaptive movement. This hypothesis suggests, therefore, that the transfer effect and its attendant circuitry, which provides such an important translational model of the cognitive control of actions, also provides a model of OCD in which abnormality in this circuit results in the inhibition of the vlOFC and, by extension, excitation of the mOFC.

Summary and Conclusions

Corticostriatal circuits play a central role in the induction of intrusive thoughts and actions that are associated with major psychiatric disorders and addictions of various kinds. Here, I point to two types of involvement based on an assessment of (a) the corticostriatal circuits engaged during the acquisition and performance of goal-directed and habitual actions and (b) the modulation of such actions by predictive learning. The first indicates deviations in the balance between the dual corticostriatal circuits that mediate goal-directed and habitual action control and the competitive and cooperative relationship between them that is required to generate adaptive behavior. Changes to the goal-directed circuit reduces its regulation of habits and allows the intrusion of stimulus-driven actions without regard to their consequences. Alternatively, changes to the habitual system could result in a loss of the cooperative processes through which stimulus-driven urges provide the basis for the retrieval and execution of goal-directed action. As a consequence, the operation of a planning process that retrieves action–outcome associations could result in the mental simulation of their performance without the ability to initiate those actions. This simulation has much in common with intrusive thinking.

One initiator of this planning process is predictive learning. Stimuli associated with the outcomes with which actions are associated can directly elicit those actions. Predictive learning can engage both excitatory and inhibitory aspects of action control: excitatory in the sense that actions associated with predicted outcomes are released, and inhibitory in the sense that such stimuli can block this influence from extending to actions associated with unpredicted outcomes. A circuit involving the orbitofrontal cortex, nucleus accumbens shell, ventral pallidum, and mediodorsal thalamus controls this balance between the excitatory and inhibitory effects of predictive learning. Interference with this circuit results in the release of inhibitory control, causing predictive cues to elicit actions associated with unpredicted outcomes, which, to a first approximation, is the definition of intrusive or compulsive action. Generally, therefore, the arguments presented here suggest that it is not a failure of habitual or goal-directed control processes per se, but rather the failure of the cooperative and competitive interactions between these processes that forms the basis of intrusive thoughts and actions.

Acknowledgments

Support for this work was provided by a Senior Principal Research Fellowship from the National Health and Medical Research Council of Australia (GNT#1079561). The author thanks members of the Decision Neuroscience Lab at UNSW and attendees at the ES Forum on Intrusive Thinking for the many discussions that have helped to shape this work.

4

What Is the State-of-the-Art Toolbox for How Circuits Mediate Behavior?

Michael R. Bruchas

Abstract

A critical challenge in the field of neuroscience as well as research into the neurobiological basis of behavior has been to establish links between the cellular and biochemical processes within the brain and nervous system that occur during the mediation of behavioral events. Developments over the last 10–15 years have provided several new means to accelerate, advance, and dissect the specific mechanisms for brain function. Developments across two realms of neuroscience and engineering have afforded researchers, clinicians, and biologists advanced abilities to facilitate the dissection, observation, control, and perturbation of neural systems within intact, behaving animals. These advances include electrical, optical, pharmacological, and specialized hardware which allow for closed-loop interfaces to monitor and manipulate neural function. This chapter explores how these recent developments have become integrated into our neurobiological tool chest. It describes current advanced approaches, and the limitations of each, and explores future pathways toward even better technologies needed to dissect the molecular, cellular, and circuit basis of behavior.

Introduction

The mammalian nervous system evolved over millions of years and contains a heterogenous composition of networks and cells, which send messages to one another to communicate information. This information is communicated by a variety of signals: electrical, chemical, and anatomical (i.e., architectural). These signals converge, amplify, or inhibit the flow of information to influence ultimately behavior in the organism. In many cases, specific central and peripheral nervous system diseases are caused by dysfunctions in these brain processes at molecular, cellular, and circuit-based levels. Some of these neuropsychiatric disorders include, but are not limited to, addiction, pain, and emotional disorders, such as anxiety, depression, or schizophrenia. For many generations, neuroscientists have sought to understand the neurobiological basis of behavior with the specific intent of resolving and better treating these types of disorders. Most current treatments rely on pharmacological or behavioral modifications from trained clinical practitioners. In many cases, treatments do not exist or individuals are resilient to any form of intervention.

A better understanding of the mechanistic underpinnings of behavioral processes is thus needed to develop new methods to engage and hopefully adjust the brain and spinal cord's function toward a more typical homeostatic state. The complexity of the brain and spinal cord's inner workings has limited the development of new therapies, but the technologies developed over the last 10–15 years offer promise in terms of uncovering these details, which in turn could identify new targets or methods.

Furthermore, there is a growing interest in the neurobiological basis of behavior as it relates to developing artificial intelligence, brain–machine interfaces, and alternative methods for generating complex processor-based systems and/or methods for adjusting human behavior in pathological states. New probes are being developed which allow for the delivery and recording of neuronal signatures in behaving animals. These technologies will provide enhanced functionality for researchers, but they also open up new avenues in clinical realms.

For neuroscience research to examine naturalistic and pathological behavioral states, it must address a key challenge: the full integration of minimally invasive, biological sensors, actuators, and pharmacological interventions. This technology is advancing at a rapid, exciting pace, and many new approaches have started to become available. Here, I focus on these advances (i.e., where we are in the field as well as the future pipeline of neurotechnologies) and discuss limits, ideas, and concepts for both biological and hardware-based neurotechnology.

Observing, Recording, and Manipulation of Neuronal Function *in Vivo*

The architecture of the central and peripheral nervous system is composed of a series of integrated modules that span the molecular, cellular, circuit, and system's levels. At the basic molecular level, neurons and glia in the brain express a selected series of proteins, which include ion channels (including iontropic receptors), peptides, pumps, and G-protein-coupled receptors (GPCRs). In the context of cellular homogeneity and heterogeneity, the brain is composed of millions of neurons, each of which tends to be enriched with various receptors, transmitters, or proteins. Over the last several decades, neuroscientists have generally worked to classify these types of cells and have begun to understand their diversity in response, release properties, and connectivity. The diversity of cell types in the brain and spinal cord, along with the various scales at which the nervous system operates, have made the design and implementation of various neurotechnologies and interfaces challenging. Here I will describe several technologies currently employed to observe and record neuronal activity in awake or head-fixed animals: how the technology is currently being used, as well as future iterations of related methodologies. Devices that interface with the nervous system must have the ability to decipher signals at multiple scales as well as sufficiently interface with biological tissue in a flexible, noninvasive manner.

Electrophysiological Methods

For nearly a century, the electrical properties of the brain and nervous system have been empirically measured and recorded. Investigators have used electrical probes to stimulate and mimic neuronal activity patterns in investigations of neural circuit functions during behavior. For many years, researchers were severely limited by the "channel count" of the electrophysiological probes for *in vivo* measures. This meant that we could not sufficiently sample large populations of neurons in behaving animals. More recently advanced materials engineering and manufacturing approaches have, however, brought forth new technologies, including large Utah arrays and Neuropixels (Jun et al. 2017). This latter technique integrates 960 recording locations within a 70×20 μm^2 area and allows for single unit action potentials to be isolated at very high resolution across multiple brain regions. In addition, it reportedly allows for stable tracking of single neuron activity over multiple days, thus allowing investigators to measure dynamic changes within neural circuits that change over time. This method permits significantly advanced throughput in neuronal recordings and provides the ability, given their small size, for simultaneous recordings of activity to be made over a wide range of brain regions.

Recent advances have also worked toward developing neural probes that are softer and more flexible in their ability to interface with the brain. Most neural probes are composed of hard surfaces, including metal, glass, and silicon semiconductors; these materials can activate the brain's immune response and lead to severe lesion, inflammation, and cell death. Recently, an injectable platform composed of mesh-based electronics, including 16 channels of platinum recording and stimulating electrodes, has become available (Zhou et al. 2017). This device was designed in a flexible, open manner to facilitate better integration with surrounding brain tissue and has the ability to isolate local field potentials as well as single units. It can also be operated chronically in the rodent brain for many months. This method provides additional surface and material structure that can interact with the brain in a noninvasive manner, while providing high-resolution actuation and observation data sets.

Current Limitations

Limitations of these new technological recording methods include the following:

- Neuropixel devices collect a large volume of data. Thus, due to the channel count and multiple-site recordings, data processing can be slow, time-consuming, and difficult.
- Integrating multiple streams of data over time while assuring for quality and meaningful data remains a significant challenge.
- Large teams of computational neuroscientists are required to ensure that the data is sufficiently and carefully managed, processed, and fit into a larger picture.
- These devices also do not include the means for optical coupling and isolating of selected optical approaches, which might help to identify the neuronal type being recorded.

Future iterations of the technique will need to employ streamlined data processing pipelines (now underway, as I understand it) as well as full integration with optical methods for applications (discussed below). In the case of flexible probes and the architecture therein, there are some limits to the technique in terms of channel count and integration with related optical approaches. Future iterations, however, may be able to incorporate some modifications to include these features.

Neuronal Actuators and Related Devices

Electrical recording offers precise, high-resolution information related to specific activity of regions, networks, and single cells in the behaving animal. However, one key limitation of the method is that investigators cannot simply rely on a unit's electrical signature waveform to classify the neuron as a certain type. Although some investigators do use this method, it is not considered definitive in most subdisciplines of the field. For over a decade, neuroscience has employed the use of genetics to specify particular cells and to manipulate these cells by incorporating light-sensitive proteins in the continually developing and evolving field of optogenetics (Lerner et al. 2016). The "workhorse" in the field has been channelrhodopsin-2 (ChR2), which acts to depolarize neurons upon activation. This tool is expressed in a genetically defined manner to allow for selected excitation of a given neuronal population, alongside experiments related to spatiotemporal sufficiency of a given neuronal subgroup in a particular behavioral output. In the field of neural circuits and behavior, along with *in vivo* neurophysiology, ChR2 has been used in numerous experiments to dissect the specific properties of various behaviors, including reward, decision making, addiction, pain, anxiety, depression, social interaction, feeding, and other homeostatic processes. In recent years, variations in ChR2 have been generated to provide powerful new abilities that further extend the function and utility of the optical approach. These include channels with faster kinetics, step-function properties, altered activation spectra (redshifted), and cellular localization (Yizhar et al. 2011; Klapoetke et al. 2014).

As these optogenetic stimulation tools became available, additional approaches were being developed to allow investigators to "silence neuronal activity." Here, the principal strategy was to develop photosensitive cation channels which could act to hyperpolarize a neuron and thus significantly minimize and prevent that neuron from generating an action potential and "firing." These inhibitory opsins include pumps for protons hydrogen (H^+) (called archaerhodopsin or bacteriopsin), sodium (Na^+), and chloride (Cl^-) (called halorhodopsin) (Yizhar et al. 2011). Like ChR2, these proteins have been modified to enhance function, expression, sensitivity, and wavelength, thus affording investigators more advanced methods to manipulate neural circuits. The key advantage to these inhibitory opsins is their ability to be harnessed for the determination of how a genetically defined neuronal population is necessary for a given behavioral event. The investigators can time lock activation of the optical silencing method within a given group of neurons and observe the behavioral consequences of that particular manipulation in real time.

While the optical tools described above offer spatiotemporal and optical control of neuronal activity through excitation or inhibition of given subsets of neurons, due to the constraints of their binary impact and the fact that their activation does not necessarily mimic naturalistic neuronal activity, additional optogenetic approaches have been developed. These include, for instance, methods for specifically manipulating cellular signaling, neuromodulation, and gene expression. In particular, a newer approach was developed whereby GPCR signaling (the primary mediator of neuromodulatory function in the nervous system) can be mimicked in a cell type- and neural circuit-specific manner. These chimeric optogenetic tools have been engineered using seven transmembrane-spanning opsins that contain the intracellular loops and C-terminal tail of GPCRs, which are typically expressed within mammalian neurons or glial cells. One of the first families of opto-XR receptors was the adrenergic receptor system, whereby optically active beta-2 and alpha-1 adrenergic receptors were generated (Airan et al. 2009; Siuda et al. 2015). Subsequently, a series of new opto-XRs, developed over the years, can activate a whole host of G-protein signaling pathways and neurons, including Gi, Gq, and Gs among others (Spangler and Bruchas 2017).

Finally, several other photoactivatable proteins have recently been employed in cellular studies and are beginning to be used *in vivo* in brain tissue. These optogenetic tools target the inhibition or activation of second messenger cascades (for a review, see Wiegert et al. 2017). These tools regulate downstream signaling using allosteric or proximity-based effects. They incorporate the use of flavoprotein domains, such as the light-oxygen-voltage domain and cryptochromes. These flavoproteins generally fit into one of three categories and act to initiate enzymatic activity, dimerize, or change conformation, all in response to light (Spangler and Bruchas 2017). The advantage of these newer optogenetic systems is that they can selectively and discretely target, in a spatiotemporally precise manner, specific intracellular signaling components within intact neurons or at the systems level; this allows investigators to probe how a specific cellular signaling, trafficking, or physiological event can be potentially directly linked to a behavioral outcome in real time.

Additional approaches, now widely used in behavioral neuroscience research, include "chemogenetic actuator" tools. These biological tools allow for selective modulation of GPCR signaling in specific tissues or cell types. Like their complimentary optogenetic partners, chemogenetic tools are very easily adapted to behavioral contexts. The most widely used of the chemogenetic tools are GPCRs, which have been designed to respond to specific ligands and couple to specific excitatory (Gq, Gs) and inhibitory G-protein linked neuronal pathways. In most cases, as with opsins, investigators use these DREADD (i.e., designer receptors exclusively activated by designer drugs) proteins by combining a particular genetic method (animal, viral, or both) to introduce the DREADD into a particular cell type. Through molecular evolution of the human muscarinic or kappa-opioid receptor, Roth and his group engineered a family of mutated receptors that are only activated by clozapine-N-oxide (CNO) or other ligands such as salvinorin B and compound 21, among others (e.g., Roth 2016). Currently, most investigators use the DREADD proteins hM3Gq and hM4Gi to excite (enhance probability) or silence neuronal or gliotransmission, respectively. These two proteins are typically introduced via a specific viral method (e.g., an adeno-associated virus, lentivirus, or herpes simplex virus) into particular cells in the nervous system, after which investigators inject a systemic ligand to activate each receptor protein for the desired effect. The activity of these receptors and the drug CNO typically peaks at 90 minutes. This approach, while providing cellular and network-level access, lacks spatiotemporal precision compared to optical methods. Thus, it is somewhat more challenging to incorporate into particular systems-level models of a given circuit's role in behavior.

Additional methods are now in the process of being engineered and tested in a variety of systems. These include the development of magnetically sensitive proteins as well as proteins that are sensitive to high-frequency vibrations, such as ultrasonic actuation (Stanley et al. 2015; Wheeler et al. 2016). These developments remain controversial (Meister 2016) yet could offer, if successful, a completely new noninvasive means to perturb and probe neural circuit function.

The use of various optogenetic and chemogenetic tools within neuroscience is continuing to expand on an almost daily basis. The key advances in protein engineering, crystal structure, cryogenic electron microscopy, and biochemistry are allowing for continued progress in this type of toolbox. The development of various sensitivities, color activation spectra, and functions coupled with novel hardware approaches will continue to advance the field and allow for more nuanced physiological neuronal activity patterns to be utilized. Furthermore, with the advances in biological substrates in optogenetics, additional hardware-based neuronal probes have been developed which offer advanced optogenetic targeting and function to decrease neuronal tissue damage or allow for tether-free animal movement in more naturalistic settings. These devices include new Michigan- and Utah-based optoelectronic probes (Deisseroth and Schnitzer 2013), along with printed flexible microLEDs and multifunctional polymer-based fibers. In some cases, devices can be powered using near-field or radio frequency communication parameters for completely untethered control of LED function, and these devices can also be used in most common behavioral setups in a closed-loop manner (Shin et al. 2017).

Key Limitations

Although optogenetic and chemogenetic approaches have provided unprecedented new knowledge about neural circuit function as it relates to behavior, they are constrained by several limiting functions. Of primary importance is the fact that sufficiency experiments that rely on using DREADD or ChR2 (and related opsins) utilize broad neuronal activation via light or chemical entity to activate a genetically defined neuronal population. In the case of chemogenetic hM3Gq-DREADD activation, neurons that express the receptor will respond in unison to the drug application; this increases their firing all at once, in a similar manner, as the drug is exposed to and binds the DREADD receptor onto a given neuron. In a similar manner, ChR2-based (or comparable) optical excitation results in photostimulation across the entire field of cells, whereby light can reach, with sufficient power, through the tissue to depolarize the population of neurons. The problem with this approach is that we know from *in vivo* electrophysiological studies and optical imaging approaches that neurons do not typically respond in a monolithic synchronous fashion. Indeed, neurons fire in patterned, stochastic ways to encode various behavioral responses. This is a severe limitation of current DREADD-, fiber- and LED-based optogenetic strategies utilized in behavioral neuroscience. The assumption is that by stimulating or inhibiting all the neurons in a given region at once, we can mimic the activity of the neural circuit as it relates to a particular behavior. This "binarization" of neuronal activation in circuit neuroscience should be questioned and resolved, so that we can better understand the discrete, heterogenous nature of circuit encoding and behavior in the nervous system. There is, however, some promise with respect to this limitation, and a few papers have recently utilized and highlighted this new method (Packer et al. 2015; Jennings et al. 2009). By using spatial light modulator-based two-photon microscopy, investigators have been able to image particular neuronal activity patterns and then "play back," in a finite manner, those patterns using optical stimulation across a single planer view. This approach is limited to a fixed neuronal number due to laser power and coupling restraints; however, the approach suggests that advanced optical holographic methods can be utilized to overcome the limits of monolithic, synchronous optical manipulations.

Detection and Visualization of Neural Circuit Function and Transmission

The complexity of neural circuits in the mammalian nervous system has been a daunting task to dissect. Canonical approaches have utilized *in vivo* electrophysiological methods to record activity of neurons and ensembles within discrete brain regions during behavioral tasks in awake, freely moving, or head-fixed animals. This method, used for many decades, has provided neuroscientists with a rich framework to understand how various networks in the brain respond under various behavioral states. While extracellular and multidimensional (high channel count) recordings have been instrumental to our understanding of neural circuits, they have been limited, for instance, by the following factors:

- They lack genetic or cell-type identification.
- They are unable to track neurons across days and trials with confidence, due to limitations of maintaining a single neuron during recording over multiple sessions.
- Channel and cell count are limited by region and array size, and for deep brain, significant issues arise due to lesioning of more dorsal structures in an attempt to reach limbic structures.

Fortunately, several modern approaches have advanced our ability to detect and visualize discrete neural circuits as well as to make claims about causality of various cell types, circuits, and networks in mediating a particular behavior. Although some of these approaches have only begun to be utilized widely in the community, they are at the forefront of neural technology and are likely to lead the field's efforts in the coming years.

In vivo Calcium and Voltage Imaging

Advances in genetically encoded calcium indicators (GECIs) represent the most recent avenue by which some of the limitations listed above are beginning to be resolved. Older versions of GECIs were limited in their capacity to resolve deep brain structures due to low signal, noise, and poor dynamic range. Newer variants have helped to resolve many of these limitations (Chen et al. 2013b; Odaka et al. 2014). The general principle of these GECIs is that calcium ions enter the cell following an action potential, and these sensors are then used to detect subtle changes in neuronal calcium by converting that signal into a fluorescent signal that can be measured using advanced microscopy approaches. This allows for a reliable proxy measure of action potential firing patterns, along with synaptic calcium dynamics in real time.

The recent development of ultrasensitive protein calcium sensors, GCaMP6.0 series, has been transformative for the field at large, because they allow for very reliable detection of calcium transient activity and circuit activity in deeper structures when coupled with the advent and use of both cranial window-based imaging and the gradient refractive index lens (GRIN), which provides optical advantages for researchers using single-cell imaging methods within deep brain limbic circuits. Cranial window-based imaging in multiphoton applications can provide larger fields of view, so that hundreds and even thousands of neurons can be imaged in a single animal over multiple behavioral sessions. New "mesoscope" microscopes are becoming available for this very purpose and will greatly expand the field. The utilization of these new biological and hardware-based imaging tools with fiber photometry as well as single- and two-photon imaging permits reliable terminal field and single-cell detection of calcium transient activity over single and multiple trials spanning days to weeks. Various hardware has been developed and optimized for maximizing the types of behavioral experiments in which GCaMPs can be used. Fiber photometry has gained substantial popularity in recent years due to its relative ease of use. This method employs a simple fiber optic and photometer-based detection system and provides a computationally simple data pipeline (Gunaydin et al. 2014) for measuring "bulk" calcium dynamics in a given neuronal population, and ease of use in freely moving behavioral studies. Further specificity of single-cell activity has been gained through the advent of GRINs lens-containing mini-endoscopes (Ghosh et al. 2011; Barretto and Schnitzer 2012); this allows for a less than 2 g microscope to be mounted to an animal's head, and a complementary metal oxide semiconductor image sensor for high-speed detection of calcium transients in single cells. This mini-scope approach allows for deep brain imaging during freely moving behavior, together with single-cell tracking over the course of multiple behavioral sessions (Mukamel et al. 2009; Xia et al. 2017). Finally, two-photon imaging in head-fixed, awake-behaving rodents allows for high-resolution long-term imaging with either cranial windows or GRIN lens implants for deep brain applications. This head-fixed two-photon imaging can be coupled with complex behavioral tasks, including virtual reality systems and spherical treadmills with simulated environments for increased complexity in both endoscopic and cranial window-based imaging platforms (Zhang et al. 2018; Jennings et al. 2019).

Additional tools are under development and being tested. These include an expansion of the color range for calcium indicator protein sensors, such as red fluorescent calcium indicators (RCaMP) and jRGECO (Akerboom et al. 2013). The advantage of these additional sensors is that they will improve the imaging depth within intact brain tissue, since near-infrared light scatters less through tissue. The other advantage of these red indicators is the future ability to allow for simultaneous multicolor imaging of diverse genetically or circuit-specific neuronal populations. These newly fashioned indicators could also allow for dual imaging of presynaptic axon terminals or coupling with other optogenetic actuator-based approaches. These indicators are currently limited by their signal-to-noise and dynamic range, but multiple groups are working to resolve and optimize these issues using forward genetic screens in high throughput testing scenarios.

One disadvantage of these calcium-based sensors is that they only sense calcium transient activity, not actual action potentials. That is, they resolve activity on a scale of seconds, not milliseconds. Compared to extracellular recordings and phototagging, this has caused some in the field to remain skeptical of the advantages of genetically encoded calcium imaging methods. However, recent advances in genetically encoded voltage indicators may hold some promise in resolving this issue. The recent developments of Archon, QuasAr2, and CheRiff allow for reliable voltage detection (Adam et al. 2019), which has been validated by electrophysiological studies to match kinetics directly with neuronal action potentials.

Imaging and Detecting Neurotransmitter and Neuromodulator Activity

A very recent and exciting development in biosensors for other biomolecules and neurotransmitters has come to the forefront. For many years it has been difficult to measure the actions of both fast neurotransmitters and neuromodulators in real time, using genetically encoded tools, during freely moving behavior. Classically, these molecules have been measured using microdialysis-based methods, which afford pristine detection with mass spectrometry but are limited by spatial and temporal resolution. Dialysis is also generally unable to detect larger molecules (e.g., neuropeptides), although some recent progress has been made with opioid detection (Al-Hasani et al. 2018). These include probes for glutamate (SF-iGluSnFR), acetylcholine (GACh), glycine (GlyFS), GABA (Chamelean), dopamine (dLight and GRAB-DA) (Patriarchi et al. 2018; Sun et al. 2018), as well as norepinephrine (GRAB-NE) (Feng et al. 2019). These sensors rely on technology based on fluorescent-resonance energy transfer or they take advantage of circularly permutated green fluorescent protein (cpGFP, also used in GCaMPs) fusion proteins within the third intracellular loops of specific GPCRs or related coupling domains. This allows the sensor to detect "binding" of the transmitter or modulator, and thus reveals the presence of a substance in a given response within a region or circuit. These new sensors are promising because coupled with modern viral and genetic tools, we can selectively express the sensors in various brain regions, cell type, and circuits, and record activity in discrete behaviors in real time.

While these protein-based sensor approaches offer a significant amount of promise in terms of our ability to dissect the role and function of specific neurotransmitters and modulators in real time during freely moving and head-fixed behavior, there are still several limitations, and advances will be forthcoming. Currently available sensors are mostly directed toward the detection of small molecules, including fast transmitters, cholines, and monoamines. While efforts are ongoing to use the same technology to detect neuropeptides, tool development has been limited due to the high-affinity nature by which neuropeptides bind to receptors and the inclusion of the cpGFP into the third intracellular loop of the GPCR. This is the same region of the receptor that dictates G-protein coupling and high affinity binding. There are some promising developing neuropeptide GPCR-based sensors on this front, none of which have yet been published or validated *in vivo*; however, if neuropeptides can be detected in a meaningful and genetically defined manner, it will open a host of very exciting possibilities in how neuropeptides coordinate neuromodulatory function within neural circuits that mediate a variety of behaviors (e.g., stress, anxiety, fear, addiction, and reward seeking).

These new transmitter sensors could open new avenues and enable us to address long-standing questions about neurons that co-release fast transmitters (e.g., GABA and glutamate) while simultaneously releasing monoamines and neuropeptides. We may be able to answer important questions about whether neuropeptides encode specific information on their own or in conjunction with specific fast transmitters under certain circumstances. Furthermore, these types of tools coupled with imaging would allow us to expand our understanding into the molecular and cellular basis of organization within neuromodulatory and neurotransmitter circuits. Are peptides active at specific locations, released in dendrodendritic form in some cases, or just released in mass in a volumetric manner? Although these types of biosensors hold great promise, further enhancement of their quantum yield (signal to noise) and sensors for other intracellular signaling molecules (including cAMP, kinases, and other cascades) will be needed for *in vivo* systems-level experiments in awake-behaving animal studies.

Conclusion

Optical tools provide unique methodologies in our quest to dissect neural circuits associated with behavior. Equipped with novel biological tools as well as new, less restrictive hardware and/or systems with higher resolution for imaging activity within neural circuits, investigators have been able to resolve specific spatiotemporal properties of discrete cell types, neurotransmitters, and neuromodulatory pathways in real time during discrete behavioral events. The challenges posed by these new methods include their invasiveness, their lack of temporal resolution (particularly with GECIs), as well as their dynamic range. Hardware limitations, in terms of the imaging window, pose limitations, for instance, on data stream management. As richer, high-resolution data becomes available through these new approaches, deciphering the data and utilizing new computational approaches becomes more imperative. Computational models, in turn, are needed to handle the new data, thus posing a future challenge.

The advent of these new technologies begs the question as to whether any of the tools described above might eventually pave the way to a better understanding of intrusive thoughts in animal models, and whether these could be applied in clinical translation. From the discussions at this Forum, it is clear that there are a variety of possible uses for novel tools in establishing causality and translation—provided that some of the limitations can be overcome. These include using closed-loop sensing of neuronal activity (GCaMP or other) and optogenetic (ChR2 or halorhodopsin equivalents) or pharmacological manipulations in a wireless setting. Real-time sensing during established behaviors defined to represent "intrusive thoughts" across species, alongside real-time feedback with optogenetic and pharmacological control, would establish causality and mechanisms for intrusive thoughts, at least in one sense. For example, deep brain stimulation has been widely used in clinical settings for a variety of neurological and psychiatric diseases, yet it has not been used in a closed-loop setting, whereby certain neuronal signatures, biomarkers, or other measures would be detected followed by a closed-loop infusion of a drug or optical/electrical stimulation.

The approaches outlined here, including optogenetic, chemogenetic, and electrical perturbation, could be amenable to these ideas if we could (a) measure signatures of intrusive thoughts that span particular brain regions with particular biomarkers and (b) overcome the limitations of expressing viruses in the human brain. Recent developments in retinal research and clinical trials with adeno-associated viruses, along with other viral delivery methods and many new hardware developments, could assist translational approaches in the future.

Collaboration between cross-disciplinary computational neuroscientists, biologists, psychologists, behaviorists, and clinical psychiatrists is of paramount importance and needs continual encouragement. Notwithstanding these challenges, the range of tools in the neuroscience toolbox continues to grow, offering innovative ways to resolve specific pathways, networks, and behaviors with increased granularity. Future efforts require specific focus on cross-species corroboration, computation, and analysis.

5

Convergent Experimental Systems for Dissecting the Neurobiology of Intrusive Thought

A Road Map

Shannon L. Gourley, Antonello Bonci,
Michael R. Bruchas, Shelly B. Flagel, Suzanne N. Haber,
Peter W. Kalivas, Amy L. Milton, Paul E. M. Phillips,
Marina R. Picciotto, and Jeremy K. Seamans

Abstract

Nonhuman experimental systems (also known as model organisms) are critical for understanding the neurobiology of intrusive thought. These model systems allow for the ability to manipulate specific neurocircuits, neurotransmitters, neuromodulators, and physiological and intracellular signaling events associated with behavioral markers that may be linked to intrusive thought. They permit unparalleled control over the external and genetic environments in ways and to degrees that are not possible in humans. Intrusive thought is an emergent property of multiple systems: emotional, cognitive, motor, and autonomic/somatic. In an animal model, one can ask specific questions about these systems and how they may be linked to, permit, or suppress intrusions. For example, how are specific connections, neuromodulators, or cell types involved in each of these systems, and how do they help form or maintain behaviors consistent with intrusive thought? Are positive versus negative valences unbalanced? Are common systems

---

Group photos (top left to bottom right) Shannon Gourley, Antonello Bonci, Suzanne Haber, Peter Kalivas, Amy Milton, Michael Bruchas, Shelly Flagel, Paul Phillips, Shannon Gourley, Jeremy Seamans, Antonello Bonci, Suzanne Haber, Marina Picciotto, Paul Phillips, Peter Kalivas, Amy Milton, Jeremy Seamans, Marina Picciotto, Shannon Gourley and Antonello Bonci, Shelly Flagel, Michael Bruchas

hijacked by intrusive thought, agnostic to the valence or content of the thought? Resolving these issues could be transformative for the treatment of several neuropsychiatric illnesses that are commonly characterized by intrusive thought. This chapter presents a road map for studying the neural mechanisms underlying intrusive thought using nonhuman experimental systems.

Introduction

Understanding the neurobiology of intrusive thought requires unfettered and unrestricted access to the brain. Thus, one turns to nonhuman experimental systems (also known as model organisms) because they allow admission to the brain as well as unparalleled control over the external and genetic environments, employing technical and experimental strategies that are not possible in humans. This access permits dissecting, quantifying, and manipulating specific neurocircuits, neurotransmitters, neuromodulators, and physiological and intracellular signaling events associated with behaviors. Once we develop strategies to infer intrusive thought in nonhuman experimental systems, several goals can be pursued, such as the identification of neurocircuits which are, or are not, associated with intrusive thought. Are distinct subcircuits, neuromodulators, or cell types involved in forming or maintaining intrusive thoughts with positive versus negative valence? Are common systems hijacked by intrusive thought, agnostic to the valence or content of the thought? Resolving these issues could be transformative in developing treatments for several neuropsychiatric illnesses that contain intrusive thinking as a pathogenic endophenotype. As discussed at greater lengths at other points in this volume, illnesses include common disorders such as drug addiction, posttraumatic stress disorder (PTSD), depression, and obsessive-compulsive disorder (OCD).

To develop a road map for studying intrusive thought in nonhuman experimental systems, our discussion begins by defining intrusive thought in the context of biological frameworks for the research laboratory. Next, we focus on conceptualizing intrusive thought as an emergent property of multiple systems. This leads us to formulate a road map for investigating intrusive thought in the future. Finally, we conclude by exploring the point from which we started and analyzing where we still need to go.

Defining Intrusive Thought in Biological Frameworks for the Research Laboratory

Our first goal is to set forth principles by which we can capture aspects of intrusions within the domain of experimental systems. Intrusive thought has been defined as *unwanted, unintended, conscious mental events* lacking control (Clark 2005). The aspects of this definition that we are best able to capture, operationally and quantifiably, in an experimental subject are neurobehavioral events that occur ectopically (e.g., out of their appropriate contexts), recurrence, resistance to change, and induction of arousal. Such events are insensitive to modification by external stimuli that would typically redirect neural or behavioral activity, and these intrusions often interrupt adaptive behaviors.

Our aim is to develop a framework for how we might understand the neurobiology of intrusive thought using nonhuman experimental systems. A primary challenge is the inability of our subjects to express their thoughts, so to speak, which forces us to interpret their behavior as a surrogate measure of thoughts. To address this issue, we describe in Table 5.1 key concepts that should optimize any given approach. These concepts include construct, predictive, and face validities. Construct validity refers to the degree to which a given experimental strategy accurately measures what it is meant to be measuring. Predictive validity refers to the extent a strategy can make accurate predictions about the human condition. For instance, if a drug has anxiolytic properties in humans, it should have anxiolytic properties in a valid task of anxiety-like behavior in a rodent or nonhuman primate. Finally, face validity refers to the degree to which a given strategy reflects what it is attempting to model. We might ask, "Does this approach seem like it will measure intrusive thought?" Of course, reliability and reproducibility are also key considerations. In addition, we highlight the notion of antecedents, which, in this chapter, refers to factors that predispose an organism to, or directly triggers, an intrusive thought.

Table 5.1   Validities and considerations in designing research strategies.

Construct validity	The interpretability, meaningfulness, or explanatory power of a given model; the degree to which a test measures what it claims to be measuring: How well does it capture the underlying constructs?
Predictive validity	The ability of a model to lead to accurate predictions about the human phenomenon: How well does a procedure identify pharmacological agents tested in model organisms that have therapeutic value in humans?
Face validity	The extent to which a test is subjectively viewed as reflecting the concept it intends to measure: Does it seem like it is really going to measure intrusive thought?
Reliability	Stability and consistency with which a variable of interest can be measured; phenomenon is readily reproduced under similar circumstances
Antecedents	The extent to which conditions in the model recapitulate factors that precede or trigger the phenomenon of interest (here, intrusive thought)

Conceptualizing Intrusive Thought as an
Emergent Property of Multiple Systems

Embedded in the argument that nonhuman experimental systems have utility in studying the etiology and neurobiology of intrusive thought are two fundamental notions:

1. Intrusive thought is, to some degree, conserved across rodent and primate species (and thus likely has some adaptive origins).
2. Corollaries and consequences of intrusive thought can be measured and quantified in the absence of speech.

To the first point, one can imagine instances in which multiple types of organisms would benefit from uninterrupted and intensive thought, such as when the goal is to escape from a predator. However, an individual must also be able to modify and shift focus when the situation changes (e.g., when the threat has been resolved) and engage in other behaviors that are more adaptive or otherwise suited to present and evolving contexts. A *failure* to inhibit intrusive thinking can distract from achieving adaptive goals.

As shown in Figure 5.1, we conceptualize intrusive thought to be an emergent property of multiple systems: emotional, cognitive, motor, and autonomic/somatic (for discussion of the origin of intrusive thought in these systems, see Roberts et al., this volume). We envision a world representation that contains these four coexisting elements, which homeostatically analyze and validate environmental and intrinsic (thoughts) stimuli to elicit adaptive behavior. Emotional and motivating content draw on circuitry in the central zone of Figure 5.1. Intrusive events trigger a deviation from the homeostatic condition; thoughts contain more excessive motivational and attentional relevance to the individual than is appropriate for the environment. In neuropsychiatric pathologies characterized in part by intrusive thoughts, this deviation is associated with a loss of proper regulation of the inner circuitry by the outer circuitry, as indicated in Figure 5.1 (for definitions of typical vs. intrusive thoughts envisioned by the model in Figure 5.1, see Table 5.2).

We envision that any given mental health disorder can coopt different domain hierarchies. Identifying these hierarchies could offer clues into the neurocircuits that one might explore in investigating etiologies and developing treatment strategies. For example, in disorders in which cognitive behavioral therapy can be effective, such as OCD, the cognitive domain plays a significant role in generating overall circuit feedback that restores homeostasis and control of the intrusions. We hypothesize that the distinct disorders or endophenotypes of disorders defined by DSM-5 have the order of domain dominance shown in Table 5.3.

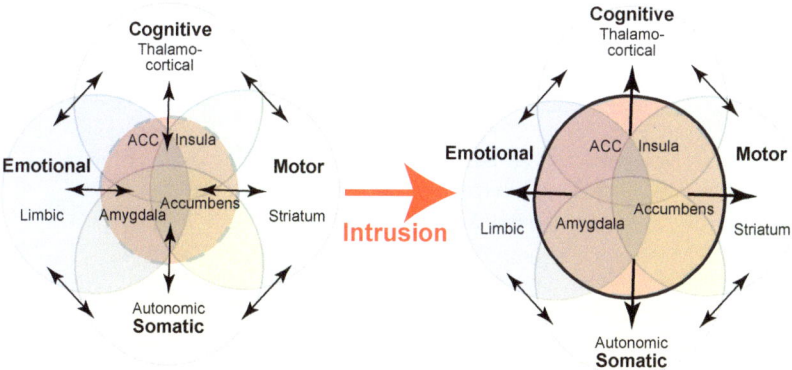

Figure 5.1   Intrusive thought is an emergent property of multiple systems. We propose that emotional, cognitive, motor, and somatic domains are recruited to interpret sensory or internal stimuli and generate adaptive responses. Navigating a complex world and adjusting our behaviors appropriately requires homeostatic involvement of these components. Depending on the arousal and motivation that emerges from this homeostatic interpretation, the central region is recruited (illustrated by the central red circle) to augment motivational value and attention. Each domain contains circuitry within the central region, as indicated by the brain nuclei listed (ACC, anterior cingulate cortex), that contributes salience to adaptive homeostatic interpretations and responses. Normally, we evaluate and modify our behavior, such as when in an aroused state that arises from a threat. In assessing the threat, we iterate between the external and internal circuits (bidirectional arrow), adjusting our appraisal of the stimuli through feedback between the circuits to generate the most appropriate responses. Though random thoughts occur, they are continuously appraised and only become problematic when the appraisal and motivation/arousal generated by the inner circuitry does not match the information received from the outer circuitry. Accordingly, the inner motivational circuitry becomes resistant to, or dominates, the outer cognitive circuitry (one-way black arrow from inner to outer circuitry). This leads to a loss of homeostatic response and manifests as excessive, maladaptive thoughts and possibly inappropriate behaviors. For instance, an intrusion producing a pronounced feeling of anxiety that cannot be regulated in a nonthreatening situation will produce autonomic, emotional hyperarousal and inaccurate conscious assessments of situations that could manifest as posttraumatic stress disorder.

Formulating a Road Map for Investigating Intrusive Thought

Keeping in mind the definitions, considerations, and concepts laid out above, we can begin to develop a meaningful list of behavioral and other factors that could be tractably measured in experimental systems (Table 5.4). For instance, we could capitalize on the ability of an external stimulus to distract an experimental subject from engaging in goal-directed actions:

- Will a rat in an operant-conditioning testing chamber respond for food reinforcers even in the presence of an opioid-related cue?

Table 5.2 Definitions of a typical versus pathological thought intrusion.

Typical	• Regulated by homeostatic cross talk between cognitive, emotional, motor, and somatic domains (see Figure 5.1).
	• Motivationally relevant information initially involves many classic limbic circuits and brain regions (e.g., ventral prefrontal and orbital cortices, insula, nucleus accumbens, and amygdala), see Figure 5.1, which act in concert with the outer circuitry to validate and regulate the state of motivation and arousal.
	• Level of arousal and motivation is appropriately managed by the outer circuitry, through feedback with the environment, to create and modulate the behavioral response.
Pathological	• Contains all of the elements of a normal event, except that the limbic circuitry (purple area, Figure 5.1) is not properly managed by the cognitive circuitry (outer circuit, Figure 5.1), thus creating an imbalance.
	• The resulting state of motivated hyperarousal leads to stress that is perpetuated without access to adaptive feedback and/or regulation by the outer circuitry and environment.
	• This imbalance can develop initially from any domain, and various combinations may be more typical in different neuropsychiatric disorders.
	• The different domain hierarchies which create the intrusion predominant in a given disorder offer a potential focus for experimental exploration into the underlying neuropathology of homeostatic loss of control.

Table 5.3 Utility of model organisms: behavioral measures relevant to intrusive thought. Here we broadly summarize behaviors and additional considerations relevant to investigations of intrusive thought in nonhuman experimental systems.

What can we measure?

• Disruption of goal-directed behavior (e.g., by drugs of abuse or aversive stimuli)

• Persistent avoidance of aversive stimuli (e.g., despite extinction conditions)

• Persistence of a given behavior, despite adverse consequences or punishment

• Persistence of a behavior in the absence of a conditioned stimulus or instrumental contingency (e.g., conditioned freezing that generalizes or fails to extinguish)

• Cognitive domains that are known to be affected in human conditions

Additional factors:

• Vulnerability factors (e.g., early-life stress)

• Individual differences

• Behavioral comorbidities that model known comorbidities in humans

• Recurrence and potential worsening with time (akin to sensitization or kindling)

- Alternatively, will the rat instead attend to the cue at the expense of goal-directed food seeking?
- What differs, at a neurobiological level, between the rat who becomes distracted and the rat who stays on task?

Similar to food-seeking behavior, drug-seeking behavior can have goal-directed properties, but the important measure that one might wish to collect in the context of intrusive thought is the degree to which it competes with a presumably more adaptive behavior, such as food seeking in a calorie-restricted organism.

Some recently developed procedures (a) force organisms to arbitrate between food-seeking behaviors and the avoidance of aversive stimuli and then (b) track the extinction of avoidance behavior in the absence of the stimulus (Bravo-Rivera et al. 2015; Rodriguez-Romaguera et al. 2016). Others measure the reaction of organisms to uncertainty (d'Angelo et al. 2014, 2017; Eagle et al. 2014; Morein-Zamir et al. 2018). Both strategies could be used to investigate mechanisms of intrusive thought, particularly when considered with factors such as individual differences or antecedents to intrusive thought, many of which can be recapitulated in the laboratory (Table 5.5).

Another type of intrusion that could be recapitulated in nonhuman experimental systems is the sense of incompletion of a task that looms until the process is completed. One example in humans draws from obsessive handwashing in OCD: the notion that one's hands must be washed is all consuming, generating hyperarousal and stress until one washes their hands, resolving the intrusion. Procedures in nonhuman experimental systems, such as persistent

Table 5.4 Examples of potential domain hierarchies involved in disorders containing maladaptive intrusions. The interacting domains defined in Figure 5.1 may be associated with particular neuropsychiatric conditions to a greater or lesser degree.

Condition	Proposed Hierarchy of Domains
OCD	Cognitive = motor > Emotional = somatic
PTSD	Emotional = somatic > Cognitive = motor
Craving in substance use disorders	Emotional = motor = somatic > cognitive
Rumination in depression	Emotional = somatic = cognitive > motor

Table 5.5 Antecedents to intrusive thought can be recapitulated in the laboratory. For arousal, long-term events are historical events that give rise to vulnerabilities and resiliencies, whereas short-term events refer to triggering factors.

Long-Term Events	Short-Term Events
Early-life experiences (early-life stressors)	Autonomic responses and stressors
Genetic correlates	Conflict
Environmental insults (drugs of abuse, trauma)	Emotional representations of environmental stimuli

response to a drug-paired cue or the "observing response task" (see d'Angelo et al. 2017), have utility in capturing both motor- and circuit-level aspects of incompleteness. In addition, most instrumental behavioral tasks include a discrete signal, such as a light, which designates the completion of a response requirement. Omission of that signal generates extended responding, a behavior that is potentially motivated by a sense of incompletion.

A third approach to studying the concept of incompleteness, which is not mutually exclusive, would be to measure neural signals that demarcate the completion of a response. For example, during cocaine self-administration, a phasic dopamine signal is observed in the nucleus accumbens (NAc) of rats upon completion of a response requirement (Phillips et al. 2003; Willuhn et al. 2012). In some animals, this feedback signal becomes diminished and, as a consequence, animals keep repeating the action, resulting in higher drug consumption (Willuhn et al. 2014).

Five Strategies for Investigating Intrusive Thought

To study intrusive thought in nonhuman experimental organisms, we propose five different approaches, summarized in Table 5.6 and discussed below.

Back-Translation, Susceptibility, and Resilience

It should go without saying that nonhuman experimental systems will be of greatest value if used in conjunction with appropriate types of behavioral analyses. There is, of course, inherent difficulty in translating clinical behaviors directly into an animal "model" of a mental health disorder, and one can debate whether it is even possible to model a mental health disorder in its entirety in animals (Bale et al. 2019). Back-translation offers a complementary approach. Broadly, this term refers to the identification of components of mental health disorders in humans that can be measured in nonhuman experimental systems. The concept of back-translation informs our first two categories: experimental approaches are driven by (a) factors implicated in human behavior, and/or by (b) vulnerability and resiliency factors that have been documented in humans. Ultimately, of course, the aim is to have "simultaneous translation" with different research approaches in humans and nonhuman experimental systems converging on the same findings (Milton and Holmes 2018).

Notably, one specific form of back-translation refers to the deconstruction of mental health disorders into specific psychological processes that can be studied in nonhuman experimental systems. In doing so, one should avoid measures that are subjective and self-reported; instead, the research scientist should look to data that are readily quantifiable and free of confounding influences. One example is performance on a battery of psychological tasks, such as the Cambridge Neuropsychological Test Automated Battery. By identifying deficits exhibited by patients (e.g., in attentional set shifting or working memory), it should then be possible to identify populations of animals that manifest the same deficits. Population variation could be induced through genetic manipulations, early-life experiences, more proximal life experiences (e.g., exposure to drugs of abuse), or perturbation of neural circuits or neurochemical systems. In other words, one could expose the nonhuman experimental organism to putative vulnerability factors that would be expected to exacerbate intrusive thought. The readily quantifiable nature of the deficits in psychological processing should allow for identification of potential neural and neurochemical etiologies as well as hypothesis testing using the full range of tools available to animal researchers.

Treatment Mechanisms

Another strategy, which we call "treatment mechanisms," refers to our understanding of why, biologically, certain treatments are effective. Cingulotomies improve intrusive thought in chronic pain management by mitigating the distressing nature of pain, but not the pain itself. The mechanisms by which this phenomenon occurs remain elusive and could hypothetically be examined in nonhuman experimental systems, opening a window for identification and deep interrogation of cells that are excited, inhibited, or otherwise modulated by cingulotomy. Such cell populations could then be manipulated pharmacologically, genetically, or through other strategies, such as those detailed by Bruchas (this volume). This approach would allow one to isolate neurobiological correlates of successful interventions whose identification could ultimately point in the direction of new and better treatment strategies.

Decrypting the Ensemble

The recurrent nature of intrusive thoughts suggests that some type(s) of similarly recurrent oscillatory or reverberating neural processes could be identified in effective models. These processes could then be exploited to expand understanding into the etiology of recurrent thought, a strategy that we refer to as "decrypting the ensemble." This strategy is inspired, in part, by evidence that in subcortical areas, the plasticity of conditioned fear-related behaviors is accompanied by transient, measurable changes in the expression of calcium-permeable and calcium-impermeable α-amino-3-hydroxy-5-methyl-4-isoxazolepropionic acid (AMPA) receptors (Clem and Huganir 2010; Rao-Ruiz et al. 2011). Additionally, changes in the ratio of subtypes of N-methyl-D-aspartate (NMDA) receptors can determine whether a fear memory ensemble is stable or susceptible to strengthening following plasticity (Holehonnur et al. 2016).

The specific content of intrusions can be common across a large number of patients and interacts with the environment (e.g., an increase in the number of patients reporting obsessions regarding contamination with acquired immune

Table 5.6 Approaches to empirically investigate aspects of intrusive thought in nonhuman experimental systems. These approaches are not mutually exclusive. Each approach is named and briefly described. Each strategy could be combined with stimuli meant to trigger or exacerbate an intrusive thought and/or procedures that introduce conflict. The motivation to pursue a specific strategy is described (i.e., the strengths of a given approach) and examples of each strategy provided. These lists are not exhaustive and some examples cross categories.

Approach	Motivation	Examples
1. *Back-translation* investigates genetic/molecular factors, circuits, and behaviors that are highly (often, causally) implicated in, or comorbid with, human conditions	• Informed directly by the human condition (i.e., high construct validity)	• Genetics: study the role of the *Fmr1* gene in repetitive behavior in fragile X syndrome • Circuits: investigate interactions between the prefrontal cortex and nucleus accumbens in rodent addiction models, an approach originally inspired by clinical research • Behavioral: investigate cognitive functions disrupted in neuropsychiatric illness (e.g., attentional function)
2. *Susceptibility and resilience* investigate known vulnerability and resilience factors in humans	• Potential for informing precision medicine • Understanding resilience could shed light on mechanisms of coping	• Genetics: study known genetic risk factors in Tourette syndrome and addiction • Behavioral: identify neurobiological effects of early-life stress or enriched environments
3. *Treatment mechanisms* investigates the neurobiology of effective treatments	• Informed directly by "what works and what does not" in treatment • Could shed light on disease mechanism or reveal new targets for treatment	• Surgical: deep brain stimulation, used to treat multiple conditions, can be studied in nonhuman experimental systems • Behavioral: conditioned fear extinction procedures could recapitulate aspects of cognitive behavioral therapy • Pharmacological: use of buprenorphine treatment for craving in opioid misuse

Table 5.6 (continued)

Approach	Motivation	Examples
4. *Decrypting the ensemble* decodes and recreates neural representations associated with events or stimuli not currently present.	• Causality and biomarkers can be simultaneously identified	• Use electrophysiology, imaging, or molecular labeling to identify cellular changes associated with a given behavior • Thereafter, attempt to *simulate* the same behavior by recapitulating neural or molecular changes
5. *Quantify natural behaviors* measures species-appropriate, ethologically relevant physiological events that occur at atypical time points or frequencies.	• Unbiased, exploratory, and comprehensive • Could benefit from within-subjects measures to capture individual- and population-level differences	• Ultrasonic vocalization, heart rate variability, physiological events (e.g., sleep, pupil dilation), clustering of naturally occurring behaviors (e.g., feeding, grooming) that manifest spontaneously in conjunction with an external stimulus (e.g., foot shock, aggressive encounter)

deficiency syndrome in the 1980s and 1990s). It is likely, therefore, that there are particular concepts and associated neural ensembles that are recruited for such intrusions. This phenomenon may reflect intrinsic differences in excitability in specific neuronal populations, as is the case for amygdala neurons which show high levels of cAMP response element binding protein (CREB) phosphorylation (i.e., activation), which are subsequently more likely to be recruited to fear-related neuronal ensembles following an aversive experience than other neurons (Josselyn et al. 2001; Frankland and Josselyn 2014; Josselyn and Frankland 2018). In other words, this subset of neurons with higher levels of phosphorylated CREB are primed to respond to input. They therefore could be an ensemble that could predispose an individual to an emergent (intrusive) event. Similar phenomena have been described in the context of immediate early gene expression (Suto et al. 2016; Whitaker et al. 2017), and these kinds of strategies could be deployed formally to study intrusive thought.

Quantification of Naturally Occurring Behaviors

One way to minimize anthropomorphic bias in interpreting animals' behavior is to utilize a final approach, which we term "quantification of naturally occurring behaviors." Here, the investigator records species-appropriate, ethologically relevant physiological events (e.g., ultrasonic vocalization or naturally occurring grooming or feeding behaviors) and identifies neural correlates when these behaviors deviate in quality or quantity from what is typical. In this and all of our categories, the introduction of triggers (e.g., drugs of abuse or stressors) will likely have utility in creating a situation that would cause intrusive thought, which then manifests in a behavior that can be studied.

Considerations

It is important to acknowledge multiple limitations inherent in nonhuman experimental systems as well as in the strategies that we propose. These include, but are not limited to, the following:

1. There is a risk that phenomena unrelated to intrusive thoughts are measured.
2. False negatives may arise, due to an inability to collect sufficient information (e.g., in cases of multiple interacting brain regions and/or neuromodulators).
3. Manipulations of molecules/cells using many current tools may not be physiologically relevant (see Bruchas, this volume).
4. In experimental organisms, the emotional content of any given manipulation is difficult to measure conclusively.

Utilizing multiple strategies should help us overcome these challenges and identify convergences.

Another challenge that we anticipate is resolving the complexity of many modern cellular/physiological data sets. A comprehensive brain map whereby affective circuits are defined by key features at a complete reductionist level is needed. Such a map would include:

1. RNA profiling of millions of neurons within the mammalian brain in naive versus "intrusive event-like states" within discrete brain structures (e.g., Campbell et al. 2017; Saunders et al. 2018; Gouwens et al. 2019; Mickelsen et al. 2019).
2. Mapping a functional architecture of these cell types alongside discrete behavior epochs or states using imaging and/or physiological processes.
3. Environmental, genetic, behavioral, and pharmacological dissociation of critical manipulations which impinge upon and alter the transcriptional state and functional responsivity of these maps.

Finally, a fruitful conceptual strategy may be to switch from assigning special functions to genes or neurons to computational perspectives of how ensembles work together and coordinate complex behavior.

Naturally, behavioral data sets can also be quite complex. A fertile strategy may be to differentiate learned association structures, based on the complexity of the information that is stored to support that association, using the "model-free" and "model-based" terminology from reinforcement learning. A model-free computation assigns a single dimensional value to a stimulus based upon the reliability of its association with a motivationally relevant outcome. In contrast, a model-based computation establishes a model of the environment that can be used to explore potential and inferred connections between stimuli and states. These concepts are discussed at greater length by Phillips and Milton (this volume).

Where Did We Start, and Where Are We Going?

A primary goal of this chapter is to offer a road map for future investigations of intrusive thought. In this final section, we describe a selection of relevant published and ongoing investigations and consider how these investigations may be further developed to interrogate intrusive thought in nonhuman experimental systems.

Global Modulation of Networks and Brain Regions versus Individual Parts of the Circuit: A Place In Between

Intrusive thought emerges from an interaction between several functional domains—emotional, cognitive, motor (see Figure 5.1)—and is often triggered by internal or external sensory stimuli. Typically, in response to stimuli, the interactions between these domains flow smoothly, with each taking a leadership role in the appropriate situation but then returning to the status quo. For example, an internal or external stimulus might invoke fear, overwhelming the individual until either an escape (movement) is executed or the cognitive control system takes over if the fear is unwarranted. Mediating between these functional domains requires the complex integration of information across them to resolve the situation. Intrusive thought can be considered to be a condition in which this mediation fails. Identifying the circuitry that underlies the integration of information processing across the different domains is a first step in understanding how and where these areas communicate, necessary to developing new therapeutic targets.

Karl Wernicke first recognized that connectivity of brain structures, rather than their locations, was the central feature of higher-order cognitive functions (Wernicke 1885/1994). Expanding on the idea, Geschwind suggested that this involves a combination of functional localization and connectivity, leading to the idea that the brain is comprised of complex, interrelated functional networks (Geschwind 1965a, b; Catani and Ffytche 2005). Functional imaging studies and graph theory techniques moved the field forward, demonstrating large-scale distributed networks and the existence of nodes and hubs (Sporns 2011). A node is an area that is connected locally or connected within a functional system. A hub is a node of a network that has unusually high connectivity to other nodes, or degree centrality, and high connectivity to other hubs, or eigenvalue centrality (van den Heuvel and Sporns 2013). Hubs are thought to represent regions for integrating and distributing information from multiple cortical regions. They likely play an important role in cross-functional computational tasks, such as integrating limbic, cognitive, and motor control calculations for decision making.

The rostral anterior cingulate cortex (rACC) is a good candidate for containing hubs, because it sits at the connectional intersection of the emotion, cognition, and executive control networks. Indeed, the entire rACC is considered a hub of the brain's global network (Buckner et al. 2009). However, the region is large, and inputs from different prefrontal cortical (PFC) functional domains vary across it. These connections could represent information processing sequentially across subregions (i.e., from valuation to cognition to action). Alternatively, a hub could be embedded within the rACC that integrates information across them. Consistent with the literature (Morecraft and Tanji 2009; Morecraft et al. 2012), mapping the distribution and relative strength of frontal cortical inputs across rACC in nonhuman primates reveals that PFC inputs to the rACC follow three general gradients:

1. Ventromedial PFC (vmPFC) and frontal pole inputs are strongest in the ventral and rostral parts of the rACC, with decreasing strengths in dorsal and caudal regions.

2. Frontal eye fields and premotor areas inputs are strongest in the dorsal and caudal regions, decreasing in rostral and ventral rACC regions.
3. Ventrolateral PFC (vlPFC) and dorsolateral PFC (dlPFC) inputs peak in a more central position.

One region embedded within these gradients, however, receives inputs from unexpected additional areas. In addition to inputs from expected connections from cognitive control areas, the dlPFC and vlPFC, this region is also connected with regions that are part of the emotional system, the orbitofrontal cortex (OFC) and vmPFC, as well as with the sensorimotor system, the frontal eye fields (Tang et al. 2019). Thus, this connectional hub within the rACC is in a position to *integrate information across emotional, cognitive, and sensorimotor systems*. It is perhaps unsurprising, then, that both PTSD and major depressive disorder show treatment response in an area in close proximity to the rACC hub (Mayberg et al. 1997; Pizzagalli 2011; Chakrabarty et al. 2016).

The striatum is also an important structure for integrating and distributing information. Although the striatum is classically divided into limbic, cognitive, and motor regions, embedded within this general topography, terminals from different frontal cortical areas interface in the rostral striatum, positioning it optimally to contain hubs (Haber et al. 2006; Averbeck et al. 2014). Indeed, in a specific location within the rostral caudate nucleus, terminal zones from the inferior parietal lobule, an area important for perception, converge not only with those from the dlPFC and vlPFC, as expected (Cavada and Goldman-Rakic 1991; Yeterian and Pandya 1993), but also with projections from the OFC and rACC. Thus, similar to the rACC, this hub combines inputs from several functional domains. The connections of the rACC and striatal hubs are examples of highly integrative, cross-functional regions with distinct combinational inputs that provide the anatomical substrate in which computations about motivation, internal states, cognition, perception, and motor control are linked to mediate adaptive behaviors based on the interaction of these functions. Disconnection of these hubs will likely result in an imbalance between goal-directed control, emotion, and higher cognition, and thus play a key role in maintaining intrusive thoughts.

Seeking the Source of Switching

One important aspect of countering or managing intrusive thought is the ability to *modify* thoughts and actions, a key ingredient in adaptive responding to external stimuli (Figure 5.1). Multiple structures discussed above are naturally involved in these processes and could be a focus for future investigations, particularly those that enjoy considerable homology between rodent and primate species. One such structure, the ventrolateral orbitofrontal cortex (vlOFC), has been intensively investigated using an instrumental contingency degradation procedure. In brief, the procedure requires nonhuman (or human) experimental systems to generate operant responses for rewards such as food or juice. Then, the experimenter modifies the likelihood that a given behavior will be reinforced, and organisms must update learned action–outcome associations to modify their responding optimally. In a series of investigations, inactivation of the vlOFC blocked the ability of mice to update response strategies (Gourley et al. 2013a; Zimmermann et al. 2017, 2018; Whyte et al. 2019), consistent with evidence that certain vlOFC neurons represent outcome-related memories (Namboodiri et al. 2019).

The vlOFC interacts with aspects of the dorsal striatum, a key constituent of goal-directed action, to coordinate action–outcome response flexibility (Gourley et al. 2013a; Gremel and Costa 2013). Meanwhile, the use of viral-mediated gene silencing and behavioral pharmacological strategies has revealed several essential molecular factors within the vlOFC that optimize its function. These factors include, but are likely not limited to, brain-derived neurotrophic factor (Gourley et al. 2013a; Zimmermann et al. 2017; Pitts et al. 2020) and its high-affinity receptor trkB (Pitts et al. 2018, 2020), Abl2 kinase (DePoy et al. 2017), GABAAα1 receptor subunits (Swanson et al. 2015), fragile X mental retardation protein (Whyte et al. 2019), and developmental expression of integrin receptors (DePoy et al. 2019). These investigations provide overwhelming evidence that the vlOFC is necessary for behavioral switching, and they potentially shed light on molecular factors that are disrupted when intrusive thoughts interfere with behavioral flexibility essential to day-to-day function.

One common factor linking all of these proteins is that they regulate the stability or turnover of dendritic spines, the primary sites of excitatory plasticity in the brain. Whyte et al. (2019) revealed that updating expectations regarding whether an action was likely to be rewarded reduced thin-type dendritic spines, considered immature, on excitatory vlOFC neurons in mice. Meanwhile, the proportion of mushroom-shaped spines, considered mature and likely containing synapses, increased, potentially solidifying newly modified action–outcome associations to optimize future decision making and behavioral flexibility.

One function ascribed to the OFC as a whole is the updating of expectations, particularly under ambiguous circumstances; by extension, unbalancing these connections via spine loss or inappropriate excitatory plasticity could render expectations ambiguous and thereby vulnerable to intrusion by competing impulses. Consistent with these notions, exposure to cocaine (Gourley et al. 2012a; DePoy et al. 2017; Pitts et al. 2020) and stress hormones eliminates dendritic spines in the vlOFC, and identical procedures cause failures in the action–outcome updating of stress hormones (Gourley et al. 2012b, 2013b; Barfield et al. 2017; Barfield and Gourley 2019). Further, drugs that improve behavioral updating appear to recruit local cytoskeletal regulatory systems (DePoy et al. 2017). As a final note, artificially stimulating excitatory neurons in this region also causes failures in action–outcome updating (Hinton et al.

2019), potentially by activating circuits associated with OCD. This idea is discussed at length by Balleine (this volume). Understanding the conditions under which specific vlOFC connections are stimulated or quiescent could shed light on how thoughts and actions can fail to be updated and become intrusive, and how one recovers from their intrusion by switching cognitive strategies or behaviors, maintaining adaptive flexibility between emotional, cognitive, motor, or somatic domains (Figure 5.1).

Unfortunately, cellular heterogeneity within several brain regions pertinent to this discussion remains undefined. For instance, even within the striatum, where cells subtypes can readily be distinguished based on dopamine receptor constituents, dopamine D1 and D2 receptor-mediated neuronal ensemble responses to cues and rewards over temporal dimensions, and as a function of experience, remain opaque (for review, see Castro and Bruchas 2019). Determining the contribution of each cell to learning, memory, stressor, and drug reactivity, for instance, has previously been studied using *in vivo* physiological approaches, yet the process of defining circuit- and cell-type specifics in time and space is still in its infancy. As such, comprehensively understanding the neurobiological bases of neuronal ensembles is a critical step if we are to dissect and understand the aberrant patterns (signatures) by which intrusive events occur.

Arousal Systems

Intrusive thinking includes arousal, and deviations from typical arousal states can predispose one to, or acutely trigger, intrusive thoughts, a notion emphasized in Table 5.5. As such, understanding the mechanisms of arousal presents a point of entry into understanding intrusive thought itself. Within the framework that arousal can decrease the threshold for permeation of intrusive thoughts, we might consider the ability of acetylcholine (ACh) release elicited by salient stimuli to alter the strength of signaling in thalamo-cortico-thalamic loops, both acutely and persistently, via synaptic potentiation (Aramakis et al. 2000; Kawai et al. 2007). This ACh-mediated elevation in activity may alter the threshold for transmission of sensory information from subcortical to cortical structures. The directionality of this signaling can vary across development, with different ACh receptors mediating increases or decreases in the transmission of sensory information (Aramakis et al. 2000; Heath and Picciotto 2009).

Although we recognize that hallucinations may not reach a formal definition of intrusive thoughts, it could be useful to evaluate the particular circuits for which we have direct evidence of a causal relationship with this perceived, maladaptive mental event. Pharmacological studies (Warburton et al. 1985; Fisher 1991), as well as evaluations of patients with loss of cholinergic neurons (Dauwan et al. 2018), reveal that blocking muscarinic ACh receptors or decreasing ACh levels in patients with Lewy body dementia (Tsunoda et al.

2018; Dudley et al. 2019) results in hallucinations. One proposal is that loss of either nicotinic or muscarinic activity in corticothalamic circuits may underlie these hallucinations (Esmaeeli et al. 2019). Another suggestion is that cortical ACh increases the signal-to-noise ratio of perceived events, and that "muscarinic receptor activation in the cortex is involved in confining the contents of the discrete self-reported conscious 'stream' " (Perry and Perry 1995:240). When cholinergic input to the cortex is lost, irrelevant sensory information normally confined to subcortical circuits enters conscious awareness (Perry and Perry 1995), and hallucinations result from "a failure of the metacognitive skills involved in discriminating between self-generated and external sources of information" (Kumar et al. 2009:119).

An additional set of studies suggests that there is a pervasive increase in ACh levels throughout the brain in patients who are actively depressed: for unipolar depression, see Saricicek et al. (2012); for bipolar depression, see Hannestad et al. (2013). Elevated ACh may be a risk factor for depression since remitted patients have intermediate ACh levels between actively depressed individuals and healthy controls (Saricicek et al. 2012), as measured by competition with a cholinergic ligand and validated by within-subject challenge with a cholinesterase blocker (Esterlis et al. 2013). Relatedly, ACh is a critical mediator of arousal and rapid eye movement sleep (Ma et al. 2018). At baseline, ACh input to the basolateral amygdala is very high, and tonic activity of the cholinergic system can thus control both the level of arousal of stress-related systems and the likelihood that a stressful event will activate the basolateral amygdala (Picciotto et al. 2012). Currently, there is no information on whether this increase in ACh levels is associated with intrusive thoughts (e.g., rumination in depression), but this topic could guide future experiments.

Manipulations and measurements of both cholinergic signaling and circuits modulated by its receptors may represent a cross-species neurobiological approach ripe for translational evaluation. One consideration is that intrusive thoughts can be represented in experimental settings by regulation of arousal states along a multidimensional continuum. A possible dimension along this continuum includes asynchronous, decoupled activity of the filter or gain domain which prohibits "normal" function in a given circuit, steering an organism toward a hyperaroused state. One particular example of this phenomenon is found in the locus coeruleus, which projects broadly throughout the brain, and its activity (tonic vs. phasic) is dictated by salience, context, and stress responsivity. The ability of the locus coeruleus noradrenergic system to dissociate attention signals from stressful ones depends on which inhibitory filters are engaged. Along the temporal dimension, intrusive thought may have the effect of dysregulating the inhibitory gain signal (typically regulated by neuromodulators such as neuropeptides, monoamines, and steroids), thereby producing an unwanted hyperaroused state.

Lessons from the Study of Cocaine

Nonhuman experimental systems of many species will self-administer drugs of abuse, including cocaine, thereby providing a strong measure of face validity to drug self-administration in animal models of human addiction. As in humans, drug seeking can intensify with time and experience, take on compulsive properties, and persist despite adverse consequences, allowing for the intense investigation of neurobiological etiologies.

Over the last 15 years, several research groups have modeled compulsive drug-seeking behavior in the face of aversive consequences in preclinical models, with the ultimate goal of mimicking, as closely as possible, the symptoms, diagnostic criteria, and features of addiction manifested by human patients (Deroche-Gamonet et al. 2004; Belin et al. 2008; Economidou et al. 2009; Marchant et al. 2014; Belin-Rauscent et al. 2016). In well-validated rodent models, voluntary drug-taking and drug-seeking behaviors coincide with mild foot shock punishments, which are used to create negative consequences following drug use. Exposure to foot shocks has revealed a divide in rodent phenotypes into two separate groups: (a) those which are shock sensitive, whereby the rat ceases to press a lever after receiving the foot shock and (b) those which are shock resistant, whereby the rat keeps pressing the lever despite receiving the foot shock. Much like human populations who develop compulsive drug abuse or addiction, approximately 30% of rodents exhibit the shock-resistant phenotype.

The utility of the punished model of compulsive drug use in identifying and developing translational therapeutics was demonstrated by Chen et al. (2013a). In this study, the authors discovered that shock-resistant rats self-administering cocaine show a marked reduction of activity in the prelimbic PFC, a subregion of the PFC that is important in mediating behavioral flexibility and decision making. When Chen et al. reversed hypoactivity of this brain region via optogenetic activation of the prelimbic PFC, rats significantly, and almost instantaneously, reduced their cocaine-seeking behaviors. These findings led to clinical trials using a well-known, noninvasive form of brain stimulation, repetitive transcranial magnetic stimulation (rTMS), previously used as a treatment for depression. These clinical trials revealed that rTMS reduces cocaine craving and cocaine intake, thus paving the way for larger double-blind clinical trials that are the first to offer a promising treatment against cocaine use disorder (Terraneo et al. 2016; Pettorruso et al. 2018). The results of these translational studies highlight the importance of continuing efforts in developing increasingly sophisticated rodent models of substance use disorders, which are fundamental in leading to the next generation of treatments against substance use and other addictions, and more broadly, intrusions on adaptive functioning.

The term "incubation" refers to progressive, time-dependent elevations in drug craving and sensitivity to drug-related cues. Incubation is thought to contribute to the maintenance and persistence of addiction and relapse (Lu et al. 2004; Venniro et al. 2016). By studying the incubation of cocaine-seeking behaviors in model organisms, we may also gain insight into certain forms of intrusive thought. Craving is considered a key factor in triggering relapse and can be triggered by drug-associated contexts or discrete drug cues. Although craving in humans is typically triggered through a combination of drug-associated contexts and cues, the two stimuli involve distinct and overlapping circuits. Thus, in animal models, it is useful to isolate them from each other during the incubation period; for instance, by extinguishing behavioral response to the drug-associated context with daily exposure to the context in the absence of drug availability. Accordingly, when the drug-paired cue is returned, it becomes possible to quantify behavior motivated only by the cue and to evaluate changes in circuitry and cell physiology produced by the cue-induced motivational state.

Within the context of dopamine-related pathologies (which presumably include substance use disorders), a change in the configuration of dopamine D3 receptors in the NAc has been observed, particularly when tonic dopamine levels are low. This change has not yet been fully characterized but may include a change in the ratio of the D3nf isoform; it also seems to enhance functional coupling with dopamine D1 receptors. This change has been associated with ticks in Tourette syndrome (Frau et al. 2016), L-DOPA induced dyskinesia (Fanni et al. 2019), and pathological gambling following dopamine agonist treatment in Parkinson disease (Pes et al. 2017). Treatment with the 5α-reductase inhibitor, finasteride, reverses associated molecular changes and ameliorates each of those pathological traits. Preliminary data indicate that finasteride also reverses incubation of cocaine craving and reduces escalated cocaine consumption (P. E. M. Phillips, pers. comm.), a finding that may be relevant to intrusive thought as well.

Distinct drugs of abuse (e.g., cocaine, heroin, alcohol) induce both similar and divergent neurobiological changes in brain regions like the NAc and frontal cortex. In light of the overlap in drug-seeking endophenotypes produced by, for example, the self-administration of opioids and psychostimulants, and the shared vulnerability to relapse in addiction, one argument has been that understanding the shared (rather than distinct) neurobiological factors might be particularly fruitful. One such example is elevated synaptic glutamate spillover from prelimbic PFC projections in the NAc during drug seeking for cocaine, heroin, alcohol, and nicotine. Under typical conditions, synaptic glutamate spillover is moderated by the glial glutamate transporter, GLT-1, located on astroglial end feet adjacent to the synaptic cleft. GLT-1 tightly controls basal extracellular glutamate, protecting against synaptic glutamate spillover. However, several drug classes downregulate GLT-1 and retract glial end feet from NAc synapses, modifications that have been directly linked to drug-seeking behavior (for a review, see Bobadilla et al. 2017).

Notably, acute stress has many overlapping effects, including decreased glial synapse coverage and reduced GLT-1 in the NAc (Garcia-Keller et al. 2016) and induction of glutamate spillover from unprotected glutamatergic synapses, which produces transient synaptic potentiation in NAc dopamine D1 receptor-expressing neurons (Scofield et al. 2016). Transient potentiation creates a situation that reinforces behavioral responses to stress- or drug-paired cues, making responding more persistent and more competitive with other stimuli (Kalivas and Kalivas 2016). Given that intrusive thought is similarly characterized by recurrence and resistance to outside stimuli that would typically redirect behavior, we should directly test whether these factors contribute to intrusive thought.

Sign Tracking, Goal Tracking, and the Interruption of Top-Down Control over Behavior

When rats are exposed to a Pavlovian-conditioned approach procedure, wherein an illuminated lever, a conditioned stimulus (CS), precedes the delivery of a food reward, an unconditioned stimulus (US), into an adjacent food cup, two distinct phenotypes may emerge (Flagel et al. 2009): *goal trackers* and *sign trackers*. Upon lever-CS presentation, goal trackers approach the location of reward delivery, whereas sign trackers approach and interact with the lever-CS itself.

For both sign trackers and goal trackers, the lever-CS is a predictor because it elicits a conditioned response. For sign trackers, however, the lever-CS also acquires incentive motivational value (incentive salience) and is transformed into a "motivational magnet" (Berridge and Robinson 2003). That is, for sign trackers, the lever-CS itself is attractive, elicits approach behavior, and acts as a conditioned reinforcer (Cardinal et al. 2002; Berridge and Robinson 2003; Flagel et al. 2009). Sign-tracking behavior can be considered compulsive because it will persist even if it results in omission of reward delivery, and is resistant to extinction (Tomie 1996; Flagel et al. 2009; Ahrens et al. 2016). Furthermore, relative to goal trackers, sign trackers are more impulsive (Lovic et al. 2011): they exhibit deficits in sustained attention (Paolone et al. 2013), show exaggerated responses to aversive stimuli (Morrow et al. 2011), and have an increased propensity for cue-induced reinstatement of drug-seeking behavior (Saunders and Robinson 2010). Recent evidence in rats (Eagle et al. 2014; Vousden et al., submitted) and humans (Albertella et al. 2019) indicates that sign trackers show greater levels of dysfunctional checking behavior of relevance to OCD.

The sign-tracker/goal-tracker animal model has been used to parse the neural mechanisms underlying two different learning strategies: predictive versus incentive learning. Sign tracking, or incentive learning, is dependent on dopamine in the NAc (Flagel et al. 2011b). In fact, using this model, it has been shown that the shift in dopamine in the NAc from the reward (US) to the cue

(CS) encodes the incentive value of the cue, not the predictive value (Flagel et al. 2011b). Relative to goal trackers, sign trackers show greater engagement of the cortico-thalamic-striatal "motive circuit" in response to a food cue (Flagel et al. 2011a). Within this circuit, the paraventricular nucleus of the thalamus (PVT) has emerged as a critical regulator of individual differences in cue-motivated behaviors (Haight and Flagel 2014; Haight et al. 2015; Kuhn et al. 2018).

The PVT is a midline thalamic nucleus ideally located to integrate cognitive, emotional, and arousal information from various areas of the brain and, in turn, to guide motivated behaviors (Kelley et al. 2005; Kirouac 2015). Specifically, the PVT receives dense input from the PFC, as well as subcortical areas, including brainstem nuclei such as the dorsal raphe and locus coeruleus and other areas such as the lateral hypothalamus and amygdala. The PVT then integrates this information and sends reciprocal output to some of the same regions, but also has dense glutamatergic projections to the shell of the NAc. In fact, the PVT can regulate dopamine release in the NAc, even in the absence of the ventral tegmental area (Parsons et al. 2007).

Within the context of the sign-tracker/goal-tracker animal model, neurons projecting from the prelimbic PFC to the PVT appear to encode the predictive value of reward cues, whereas subcortical systems surrounding the PVT encode the incentive value. Specifically, sign-tracking behavior is thought to result from hyperactivity of neurons projecting from the lateral hypothalamus to the PVT, and those projecting from the PVT to the NAc (Haight et al. 2017). The working hypothesis, therefore, is that cognitive representation, or the predictive value of the reward cue, is encoded in prelimbic PVC-PVT projecting neurons, and that this top-down process predominates in goal trackers. In sign trackers, however, where incentive learning prevails, subcortical processes are able to override this top-down mechanism. Thus, the PVT appears to act as a fulcrum between top-down cortical processes and bottom-up subcortical processes, and an imbalance between these processes may result in aberrant or psychopathological behavior. It is also intriguing, in light of our discussion of ACh systems above, that sign-tracking behavior in rats is associated with poor attentional control, mediated by an unresponsive basal forebrain cholinergic system (Kucinski et al. 2018). The neurobehavioral endophenotype of sign trackers may capture antecedents that predispose an individual to intrusive thoughts. For example, sign trackers appear to have an inherent imbalance between emotional and cognitive domains (with the PVT acting as a fulcrum between the two domains; see Figure 5.1), and this imbalance renders them more susceptible to behavioral control by intrusive experiences.

Avenues for Future Research

The concepts outlined above provide a number of avenues for future research. For example, the idea that we can capture and decode the neuronal ensembles associated with disruptions of behavior and intrusive events suggests that we can use those neuronal ensembles to test whether activation of neurons, pathways, brain areas, or interregional networks recapitulate or disrupt intrusive events. A prototypical experiment might be to provoke an initial adaptive behavior, such as grooming in a rodent as a result of a sticky substance on its fur (e.g., peanut butter). During the adaptive behavior, molecular techniques could be used to capture active neurons that drive both the sensory representation of the sticky substance and the motor output. Driving the activity of these neuronal ensembles repeatedly should recapitulate the behavior and the sensory representation, without the feedback of the cleaning of the fur or a clear outcome of the motor behavior. Measuring the increasing connection between the sensory and motor systems could be achieved with electrophysiology, or behaviorally, by measuring the likelihood of grooming to other, less intrusive stimuli (i.e., stimuli that would not generally elicit a robust response). Generalization to other stimuli could also be measured. Finally, recruitment of circuitry related to anxiety or emotional behavior could be measured, which might result from the mismatch between the environment and the sensory perception or motor outcome. This hypothetical "peanut butter test" could be generalized to other behaviors in which a specific initial stimulus and its adaptive motor outcome are initially paired and then dysregulated, by driving the neuronal correlates of the event and subsequent outcome in the absence of appropriate feedback (e.g., cued fear memories, drug-associated stimuli). Capturing aberrant outcomes of neuronal recapitulation may identify common neural mechanisms, whether of plasticity or systems-level generalization, which could then be probed in patients with intrusive thought.

Conclusions

Intrusive thoughts are a hallmark of several psychiatric conditions (e.g., OCD, panic disorder, major depressive disorder, and addiction). For individuals suffering from these disorders, intrusive thoughts gain inordinate control over their emotions and actions, interfering with daily activities and disrupting lives. Given that intrusive thoughts are common to multiple mental illnesses, it is surprising how little we know about the underlying brain mechanisms. This gap in knowledge stems from the fact that we have not yet, as a field, tried to capture explicitly the commonalities of multiple disorders, such as intrusive thoughts. Here, we have highlighted standard animal models, behavioral tests, and outcome measures that could be exploited to shed light on the neurobiological components of intrusive thought. We defined intrusive thought within a biological framework that can be probed in the research laboratory, with resultant models optimized to yield novel therapeutic targets. We proposed a conceptual model that captures intrusive thoughts as an emergent property of multiple systems (emotional, cognitive, motor, and autonomic/somatic) that are represented in hubs throughout the brain. When the neural choreography between these hubs and their corresponding nodes becomes disrupted, there is a loss of homeostatic and/or cognitive control which leads to maladaptive thoughts and inappropriate behaviors. In the laboratory, with careful experimental design and multidimensional levels of analyses, we can model this loss of control on both behavioral and neural levels. Here, we provided a road map that illustrates multiple routes by which different approaches can be used in combination to expand our understanding of intrusive thought. This road map is not proscriptive; rather, we hope that it will serve as a foundation for a novel avenue of preclinical research to advance our knowledge and ultimately lead to more effective therapies for a number of psychiatric illnesses that are characterized by intrusive thoughts.

Acknowledgments

The authors wish to acknowledge support from the United States National Institutes of Health (grant numbers DA014242, DA044297, MH117103, MH007681, MH100023, OD011132, DA039687, AA024599, MH106428-5877, DA033396, MH111520, MH112355, MH045573, and MH106435).

Psychology and Cognition

6

How Can Intrusive Thinking Be Measured *in Vivo* and Studied in the Context of Brain Mechanisms?

Marie T. Banich

Abstract

This chapter reviews different methods that can be used to examine and understand intrusive thought, beginning with behavioral methods. Common among these are self-report and diary measures of the experience, duration, and intensity of intrusive thoughts as well as self-reports of the difficulty in controlling such thoughts. These questionnaires, for the most part, have been tailored to the types of intrusions specific to a given psychiatric syndrome (e.g., flashbacks in posttraumatic stress disorder, thoughts of contamination in obsessive-compulsive disorder), which highlights the need to create a transdiagnostic self-report measure. Another common behavioral paradigm is to investigate intrusions after individuals are exposed to traumatic material, through a symptom provocation paradigm in individuals who have experienced trauma or an analog trauma (e.g., viewing a disturbing movie). Other behavioral paradigms, such as the Think/No-Think paradigm, specifically examine mechanisms of memory retrieval and suppression often thought to be disrupted in posttraumatic stress disorder.

Thereafter, it addresses paradigms for examining the neural mechanisms associated with intrusive thoughts. These approaches primarily couple behavioral techniques or paradigms with functional magnetic resonance imaging or electroencephalographic (EEG/ERP) methods. In addition to providing insights into the neural mechanisms that may underlie intrusive thoughts, these approaches may provide additional information regarding cognitive mechanisms, such as discerning whether memories are being suppressed or replaced. Discussion concludes by examining emerging approaches to the study of intrusive thinking. A main challenge is to find a method to verify that intrusive thoughts have indeed occurred. New paradigms that combine neuroimaging techniques with computational methods drawn from machine learning offer promise, as do techniques which allow intrusive thought processes to be examined as they occur during more naturalistic processing (e.g., watching a film).

Introduction

The question of how one can measure intrusive thoughts (i.e., thoughts that "pop up" into consciousness in a seemingly uncontrolled manner) is difficult to address, especially in a laboratory setting. It can be challenging to capture and to verify their occurrence by measures other than self-report. Moreover, traditional experimental measures used to explore mental processes (e.g., those that determine reaction time and/or errors) are not applicable. Despite these challenges, a number of approaches, both behavioral and biological, have been employed. In this chapter, I review those methods, discuss some new approaches, and suggest ways that the knowledge from other related arenas of inquiry might be used to inform potential novel approaches to this difficult question.

The need for scientists to have methodological approaches that will enable an understanding of the basic cognitive and neural processes underlying intrusive thinking is apparent when one considers that intrusive thoughts are ubiquitous across a large number of psychiatric disorders. Although intrusive thoughts are generally associated with posttraumatic stress disorder (PTSD) and are captured by the intrusions cluster of diagnostic criteria (American Psychiatric Association 2013), they occur in many psychiatric disorders: depression is characterized by intrusive negative thoughts and memory (Newby and Moulds 2011), anxious apprehension (i.e., worry) by repetitive intrusive concerns about future negative events (e.g., Fox et al. 2015), obsessive-compulsive disorder (OCD) by thoughts of contamination and/or harm to self or others (Bouvard et al. 2017), and schizophrenia by intrusions of semantic and sensory information (Elua et al. 2012). This commonality raises the possibility that intrusive thought across disorders may have a common underlying neural circuitry (Kalivas and Kalivas 2016). What tends to vary somewhat across disorders is the content of those intrusive thoughts (e.g., negative attributions of the self in depression vs. concerns about potential dangerous future outcomes in anxiety). In addition, intrusive thoughts in disorders such as PTSD are generally thought to be more sensorially based and of shorter duration, compared to ruminative thoughts associated with depression and worry, which tend to be more cognitive in nature, of longer duration, and recurrent (Speckens et al. 2007). Nonetheless, what all these types of intrusive thoughts share is that they appear to impinge upon consciousness in a somewhat uncontrolled manner. Hence, my focus here is on methods for examining the complete range of types of intrusive thoughts. To date, however, the majority of the work using methods to examine intrusive thought has focused mainly on individuals with PTSD, OCD, and/or nonclinical populations. Thus, there is a much larger range of individuals with psychiatric disorders to whom such techniques might be applied.

Here, I will consider a number of different ways in which intrusive thoughts have been examined. I begin with a discussion of behavioral paradigms, from questionnaires to experimental procedures, that examine how the frequency, nature, and control over intrusive thoughts can be measured. Thereafter, I consider how different brain-based techniques can be used to shed light on the underlying nature and neural bases of intrusive thought and conclude with a discussion of how intrusive thoughts might be verified in ways other than by self-report, specifically looking at emerging techniques that examine or track their representational content.

Behavioral Methods: Measuring the Frequency, Nature, and Control over Intrusive Thoughts

Self-Report and Questionnaires

Foundational work to understand intrusive thought comes from behavioral approaches that rely on self-report, of which there are two main types: The first tries to assess the content and nature of intrusive thoughts, while the second assesses the ability to control or manage thoughts, both those that are intrusive and other thoughts more generally. In addition, there is a third type of self-report that is a hybrid, assessing both the nature and controllability of thoughts. Questionnaires that examine thought content have generally been designed for specific psychiatric diagnoses. A number of these measures and their characteristics are outlined in Table 6.1.

For the most part, these measures appear to have good reliability and validity, so that scores on these scales can be used (or combined) in studies that take an individual differences approach in examining how scores on these measures vary with other metrics of interest. For example, scores on these measures can serve as covariates in brain-based approaches to determine whether the degree of brain activation in particular regions is associated with increasing scores on such measures. This approach is discussed in more detail below (see section, Examining the Neural Systems Related to Intrusive Thoughts).

Importantly, research using these measures indicates that most individuals experience intrusions and that intrusive thoughts occur across a continuum from nonclinical to clinical populations; for example, a similar factor structure is observed in both groups (e.g., Reynolds and Wells 1999). Factors that may distinguish clinical and nonclinical populations include the frequency, importance, and difficulty in managing those intrusions (e.g., Clark et al. 2014a). These findings are notable as they suggest that new and emerging methods used to examine intrusive thoughts in nonclinical populations could be fruitfully employed within psychiatric populations. For example, research with nonclinical populations has attempted to isolate the basic dimensions that underlie repetitive (although not necessarily intrusive) thoughts by asking individuals to rate the nature of their most common thoughts on a number of dimensions,

Table 6.1 Examples of typical self-report measures of intrusive thoughts and the ability to control thought. Typical self-report measures of intrusive thoughts and the ability to control thought. Example items from these measures are shown in italics.

Name	Construct Measured with Sample Question	Reference
Measures of intrusive thought content:		
OII	Thoughts related to contamination, harm to self and others, and taboo behavior, including frequency and believability: rate frequency of "exposing myself" from 0 "never" to 4 "always"	Purdon and Clark (1993)
RRS-SF	Continuous thoughts typical of depression separate from depressive symptoms, subscales: Brooding—comparison between one's current state and some unachieved standard: *What am I doing to deserve this?* Reflection—repetitive thoughts about problem-solving that might ameliorate negative affect: *Analyze recent events to understand why you are depressed*	Treynor et al. (2003)
Measures of the ability to control thoughts:		
WBSI	Difficulty in controlling thoughts, subscales: Intrusions—the degree of intrusion experiences: *I have thoughts that I cannot stop.* Suppression—tendency/ability to rely on thought suppression as a strategy: *There are things I try not to think about.*	Wegner and Zanakos (1994); 2 factor structure, Schmidt et al. (2009)
PSWQ	The degree to which a person worries: *Once I start worrying, I cannot stop.*	Meyer et al. (1990)
TCQ	Assesses strategies used to control thoughts: *I think about something else.*	Wells and Davies (1994)
IES-R	Measures intrusions after stressful or traumatic events: *I thought about it when I didn't mean to* or *pictures about it popped into my mind.*	Weiss (1997)

including self-relevance, frequency, importance, orientation with regard to goals, orientation with regard to social factors, and level of detail. Using a hierarchical clustering analysis, four major dimensions of thought content emerged: (a) level of construal (degree of temporal and perceptual specificity), (b) degree of personal significance, (c) temporal orientation (future oriented vs. past oriented), and (d) valence (positive, negative). Of relevance to the current discussion, scores on these dimensions are associated with characteristics related to mental health. For instance, higher levels of thoughts characterized by negative valence and high levels of personal significance are associated with higher levels of depression (Andrews-Hanna et al. 2013).

Table 6.1 (continued)

Name	Construct Measured with Sample Question	Reference
Hybrid measures of both content and control:		
ROII	Includes additional measures of the degree and manner to which obsessive thought can be controlled: rate 0 (never) to 5 (always), e.g., *Say stop to myself*	Purdon and Clark (1994)
ITQ	Frequency, degree of distress, and degree of difficulty controlling intrusive thought: *How disturbing are these thoughts for you? How difficult is it for you to get rid of these disturbing thoughts when they occur?*	Dougall et al. (1999)
EIS	Frequency, unpredictability, unwantedness, interference, and distress caused by the intrusive thoughts after analog trauma: *How often have you found yourself thinking to any degree about the rape scene since seeing the film?*	Salters-Pedneault et al. (2009)
IITIS	Examines thought content and strategies for control via a semi-structured interview, e.g., identify unwanted religious or immoral intrusions	RCIF (2007)

Abbreviations:

ESI: Experience of Intrusions Scale	PSWQ: Penn State Worry Questionnaire
IES-R: Impact of Event Scale-Revised, intrusion subscale	ROII: Revised Obsessive Intrusions Inventory
IITIS: International Intrusive Thought Interview Schedule	RRS-SF: Ruminative Response Scale, short-form
ITQ: Intrusive Thoughts Questionnaire	TCQ: Thought Control Questionnaire
OII: Obsessive Intrusions Inventory	WBSI: White Bear Suppression Inventory

Advantages, Limitations, and Potential Extensions

These types of self-report measures have advantages: they are typically short, easily administered, generally well normed, can be used with both clinical and nonclinical populations, and scores derived from them can be used as covariates in adjunct analyses. In terms of limitations, they require metacognitive abilities related to self-awareness and self-evaluation on the part of the respondent, which may be compromised in individuals with more severe psychiatric disorders. Perhaps most glaringly, however, is the fact that they can be narrow in scope, as most were designed to address a specific psychiatric disorder. As such, they tend to examine the types of processes (e.g., punctate vs. continuous) and specific topics of intrusive thoughts that characterize a given psychiatric disorder.

Hence, a questionnaire on intrusive thoughts and their control is needed that could be used more generally across individuals with a variety of clinical disorders, as well as with individuals who do not meet clinical criteria. There are numerous ways to design such a questionnaire, and it might be useful to include the following capabilities:

1. Analyze thought content via the assessment of underlying common factors across psychiatric and nonpsychiatric populations, such as their valence and temporal orientation.
2. Assess the degree to which an individual has difficulty in controlling thoughts as well as those individual differences that protect against intrusive thoughts (e.g., mindfulness) and the mechanisms by which they act (e.g., Emerson et al. 2017).
3. Distinguish intrusive thoughts from mind wandering and task-unrelated thoughts (e.g., Maillet and Schacter 2016).
4. Contain optional subscales that could assess intrusive thoughts specific to a given disorder (e.g., thoughts of contamination in individuals with OCD) as well as differentiate those from thoughts that occur in other psychiatric disorders.

Not only would the creation of such a questionnaire enable a finer assessment of the nature of intrusive thoughts and their control, it might also enable aspects of intrusive thoughts to be linked to specific symptom clusters across psychiatric disorders (e.g., fear).

Next, we take a look at other behavioral and experimental methods for examining intrusive thoughts. They are divided roughly into two types: the first engenders intrusive thoughts whereas the second examines how thoughts (or memory retrieval) can be controlled.

Engendering Intrusive Thoughts

To examine intrusive thoughts in an experimental setting, one approach is to actually engender them. The purpose of this method, often referred to as the *symptom provocation paradigm*, is to provoke intrusive thoughts, often of a traumatic nature (e.g., Brewin and Saunders 2001). Generally, individuals identify specific traumatic events in their lives or categories of stimuli that engender intrusions in a pretest session. Then, within an experimental session, specific pictures or stimuli that are traumatic are shown or individuals are exposed to a category of stimuli associated with trauma, such as combat noise or pictures for veterans (e.g., Daniels and Vermetten 2016).

A related approach, often referred to as *analog trauma*, is designed to induce and engender trauma-related responses (e.g., intrusive memories). This procedure typically involves having individuals watch a film that contains graphic depictions of traumatic events, such as physical or sexual violence (for a review, see Holmes and Bourne 2008). In an extension of this approach, virtual reality can be used to induce an analog trauma (Dibbets 2019). Across both symptom provocation and analog trauma, the degree and nature of thought intrusions can then be examined within the context of the laboratory or via a diary of intrusions for some specified time (e.g., one week) after exposure.

As discussed by Visser et al. (2018), a variety of points in memory processing could be disrupted by or associated with intrusive thoughts: from the original attention to and encoding of information to the access and retrieval of memories. Procedures designed to engender intrusive thoughts can be combined (a) to analyze the effects of other variables or manipulations at distinct time points (before, during, or after exposure) and (b) to examine how different processes (e.g., attention/focus, encoding, and recall) influence intrusions. Manipulations implemented before exposure include having individuals recall memories that induce high levels of self-efficacy (Krans et al. 2018) to put the focus on themselves, or playing a distracting video game to place the focus on something else (James et al. 2016b). Manipulations implemented during viewing of the material include varying the cognitive load during the task or instructions that lead to hyperarousal via hyperventilation (Nixon et al. 2007). Manipulations after exposure have included varying instructions on how to deal with the intrusions, such as rumination ("How can I drive again without thinking about what could happen?"), integration so as to distinguish the video from the person's own nontraumatic experiences ("Think about your own driving experiences"), or distraction ("Try to recall as many African countries as you can think of") (e.g., Zetsche et al. 2009; Horsch et al. 2017).

Advantages, Limitations, and Potential Extensions

The advantage of the symptom provocation and analog trauma methods is that they can be implemented relatively easily. They also have face validity, especially for syndromes such as PTSD, which is generally characterized by intrusions linked to a specific event or trauma. These methods can also be used with provocation for particular classes of items that might engender intrusive thoughts, such as those associated with OCD. Although individuals are typically asked to report intrusions, there are other possible means of assessing intrusions. For example, virtual reality approaches have the ability to provide additional information: upon reimmersion into the environment, one can examine the degree to which an individual avoids that portion of the virtual reality space associated with the traumatic scene (e.g., Dibbets 2019). With regard to still photos, one could redisplay portions of a traumatic scene (e.g., a car next to an overpass with a concrete barrier) without the specifically traumatic context (e.g., a bloody person lying in the road by the car) and determine the nature and duration of eye movements to the location of the trauma content. Relatedly, eye movements might be employed to determine when individuals are likely to be more inwardly focused, as has been used for lapses of attention (or mind wandering) during reading (e.g., Reichle et al. 2010). Finally, diary methods for recording intrusions after exposure could be expanded to use digital queries at random times via mobile apps and other smartphone technologies.

The vast majority of studies that examine intrusive thoughts do so from the perspective of long-term memory formation and retrieval, which is particularly appropriate when intrusive thought is driven by a specific event or circumstance, such as occurs in PTSD. Far less work has focused, however, on mechanisms related to intrusive thoughts, especially those not linked to a specific event, in terms of how they get "stuck" in working memory and current consciousness. Symptom provocation and analog trauma are not well suited to examining the nature of recurrent, ruminative, and cognitively based intrusive patterns of thinking, which are typical in depression or worry (anxious apprehension) but cannot be specifically linked to a particular point in time nor to a particular set of provoking stimuli. For example, depressive intrusive thoughts often focus on how one could "solve" the issues that lead to distress and negative affect. From thinking about social interactions with others to self-reflection on actions taken to an analysis of one's internal mood states, the topical range tends to be larger than, for example, thoughts of contamination in OCD.

Engaging and Examining Thought Suppression Mechanisms

Another method of examining intrusive thoughts is to determine the effects of formally trying to suppress an intrusive thought. In a classic version of this task, individuals are given a period of time (e.g., five minutes) where they are allowed to think of anything that comes to mind (often referred to as the free-thinking condition). Afterward, they are placed in either an expression condition, in which participants are told that they should think about a specific item (e.g., a white bear) for a given period of time (e.g., five minutes), or in a suppression condition, in which they are told to suppress thinking about that item (i.e., the white bear) for an equal amount of time. The participant then indicates by some means, such as ringing a bell or pressing a button, whenever the item comes to mind (Wegner et al. 1987). This results in a paradoxical effect: trying to suppress a thought at first leads to greater subsequent expression than if the idea had initially been expressed and then later suppressed (Wenzlaff and Wegner 2000). Meta-analyses find a small to medium effect of the rebound of thoughts after suppression in both clinical and nonclinical groups (Abramowitz et al. 2001).

From this initial approach, several variations have been employed. In one extension, individuals are asked to identify thoughts, images, or impulses that pop into their mind unexpectedly and in an intrusive manner. The number of intrusions during this free-thinking condition can be compared to a suppression condition as well as other potential manipulations, such as distraction involving thinking about something else (e.g., a past or future weekend with friends) or accepting a thought (e.g., think about the intrusive thoughts coming out of your ears on little signs held by soldiers, who walk them in front and then away from you) (e.g., Najmi et al. 2009).

Such measures have been used to examine individual differences in clinical symptoms, cognitive abilities, or age. Researchers have examined, for example, whether higher or lower levels of executive function and cognitive control influence the ability to suppress intrusive thoughts successfully in specific populations, such as those with PTSD (Bomyea and Lang 2016), or whether the suppression versus the expression of thoughts is linked to working memory ability (Brewin and Smart 2005). Using a different approach, others have examined whether intrusions vary according to the content (more specific to certain life periods) and age of the participants. In younger individuals, for example, career success or failure might constitute the focus, whereas for older individuals, memory loss (in particular, fear of forgetting friends and family) might be more prevalent (Beadel et al. 2013).

Another experimental method that has been used widely in the laboratory to examine control over thoughts, specifically memory retrieval, is the Think/No-Think paradigm (Anderson and Green 2001). In this classic paradigm, individuals are taught associations between a cue word (e.g., "ordeal") and a target word (e.g., "roach") to a given level of accuracy (e.g., 95%) that will ensure a solid memory trace. In the experimental phase, some cues are presented so that the participant must think about the associated target (Think condition), while other cues are presented so that the participant should *not* think about or allow the associated target into consciousness. Each cue is shown multiple times so that there are numerous opportunities to exert cognitive control over the memory of the target. Then in the test phase, the individual is shown cues for each of the initial pairs, and memory for the associated target is assessed. Memory is typically increased for Think trials and decreased for No-Think trials relative to a baseline of items whose cues were not shown during the experimental phase (which provides an index of forgetting since initial training). Hence, this paradigm is well suited to examine control over retrieval of information from long-term memory, which is highly relevant to disorders in which there is intrusive memory retrieval. Although initial studies used verbal stimuli, similar effects have been observed for visual and emotional stimuli, such as face–scene pairs (Depue et al. 2006). This may be more suited for studying populations where intrusive thoughts take the form of images (e.g., object–scene pairs), such as in PTSD (Catarino et al. 2015).

Advantages, Limitations, and Potential Extensions

These approaches provide methods for examining control over intrusive thoughts as well as the mechanisms (e.g., suppression, distraction) by which such control may be exerted. However, they rely on participant self-report of the occurrence of those intrusive thoughts and a certain amount of metacognitive awareness (i.e., internal monitoring of when those thoughts have occurred). Although not specific to intrusive thoughts, the Think/No-Think paradigm provides a robust and tractable experimental paradigm to examine the control over thoughts, especially with regard to memory retrieval. In clinical populations, its use is limited due to the length of the procedure (e.g., 30 minutes to 1 hour) and requirement that individuals learn and retain the pairs. Participants must be able to perform the initial learning and sustain an adequate level of attention and motivation to perform the task. As with the symptom provocation and analog trauma approaches, these methods have clearer linkages to disorders like PTSD and OCD than to the recurrent intrusive thoughts that are characteristic of depression and anxiety.

Neural Systems Related to Intrusive Thoughts

A substantial body of research has focused on examining neural processes that are associated with intrusive thoughts. In general, the main techniques used are magnetic resonance imaging (MRI), typically used to localize brain systems involved with intrusive thoughts, or electroencephalography (EEG), including event-related potentials (ERPs) which provide information about the timing of processes associated with intrusive thoughts. In general, these methods tend to be used in combination with one of the behavioral approaches discussed above.

The utility of such approaches is that they can provide insight into the mechanisms that may be generating intrusive thoughts. For example, although one must be cautious in making reverse inferences from patterns of brain activation (Poldrack 2011), evidence of prefrontal activity when attempting to limit the intrusiveness of thoughts is suggestive of an active control process, whereas evidence of activity in subcortical regions, such as the basal ganglia, would be suggestive of a more automatized process. Likewise, alterations in early ERP components (e.g., P1, N1) are more suggestive of attempts at control over sensory aspects of an intrusive thought, where alterations in later ERPs (e.g., P3 and N4) would be more suggestive of control over information in working memory or of a semantic nature, respectively.

Brain Processes Associated with Intrusive Thoughts

Measures Used to Engender Intrusive Thoughts

The symptom provocation paradigm, especially as it relates to individuals with PTSD, has been migrated into a neuroimaging environment. As with all functional MRI (fMRI) studies, the condition of interest must be contrasted with a baseline of some sort that does not engage the behavioral construct of interest. Often in these studies, brain activation in a symptom provocation condition is compared to baseline condition; this may involve processing information from a nontraumatic memory, emotionally neutral pictures (e.g., civilian or noncombat scenes), or non-emotional information (e.g., white noise or rest). Meta-analyses across such studies of individuals with PTSD (e.g., Sartory et al. 2013) have found that these paradigms reliably isolate a set of brain regions that differentially activate during the symptom provocation as compared to comparison conditions. This set of regions includes portions of the default mode network, considered to be involved in internal thought, self-referential processing, and autobiographical memory (Andrews-Hanna et al. 2014), as well as areas that process the emotional significance and valence of information, such as pregenual portions of the anterior cingulate and the amygdala. The involvement of these regions in autobiographical memory and emotion processing provide a piece of converging evidence, not provided by behavioral paradigms alone, that the symptom provocation technique is effective at inducing reexperiencing.

With regard to EEG/ERP methods, some studies (Roh et al. 2017) have examined differences in specific ERP components (e.g., error-related negativity) under symptom provocation as compared to other conditions in individuals with disorders characterized by intrusive thoughts (e.g., in individuals with OCD). Although relatively rare, other studies have examined EEG metrics, such as the hemispheric asymmetry of frontal alpha rhythms (as an index of approach and avoidance behaviors), to symptom provocation (for a review, see Meyer et al. 2015).

Measures Used to Examine the Experience of and Control over Intrusive Thoughts

One can also utilize neuroimaging techniques, in conjunction with behavioral methods that index when an intrusive thought occurs or when control systems are engaged, to limit or otherwise attempt to suppress such thoughts. One approach is to measure brain activation during the time periods in which intrusions occur and compare that to activation during time periods without intrusions. In some paradigms, the participant notes in real time when the intrusion occurs by pressing a button. Brain activation during the intrusive thought is then compared to some baseline, such as the time period right afterward when re-suppression occurs (e.g., Carew et al. 2013). Similarly, in the Think/No-Think task, one can examine neural activation on unsuccessful No-Think trials in which the item to be suppressed intrudes upon consciousness. This activation can be compared to Think trials, in which controlled (rather than intruded) retrieval has occurred (Hellerstedt et al. 2016), or to No-Think trials, in which the item is successfully suppressed (Levy and Anderson 2012). Examining brain activation during intrusions has been done with fMRI, although the fine temporal resolution of EEG/ERPs may be better suited. For example, EEG/ERPs can be used to provide a putative index of how long the intrusion remains in working memory (Hellerstedt et al. 2016) or to identify the onset of the process that is engaged to keep it from coming into working memory (Castiglione et al. 2019).

While some studies look at real-time intrusions, which provide a "state" perspective, other approaches examine this issue from a "trait" perspective.

Here, individuals are characterized as to the degree to which intrusions are experienced during their daily lives or over longer time periods. For instance, using daily diary entries, Kuhn et al. (2013) examined the average degree to which intrusive thoughts are experienced by an individual over a six-month period and then linked the intrusion rate to patterns of brain activation at rest.

Rather than focusing on intrusions in particular, another approach compares brain activation assessed by fMRI across various conditions, such as free thought, suppression of a given thought, or suppression of all thoughts (e.g., Wyland et al. 2003). Another way is to examine mechanisms that are involved in suppressing thoughts versus replacing thought (e.g., Benoit and Anderson 2012). As detailed below for aspects of working memory, mechanisms of item replacement, specific item suppression, and suppression of all thoughts appear to have partially overlapping but distinct neural mechanisms (Banich et al. 2015). These findings support separate consideration of these potential mechanisms of controlling intrusions.

Another issue that can be fruitfully examined using neural investigations is the degree to which the processes involved in suppression of thoughts are similar to or distinct from other categories of suppression. While a backward inference from brain activation to cognitive processes must be performed with caution, neuroimaging studies nonetheless can provide insights into the specificity of control over memory versus other processes. For example, in the same individuals, Depue et al. (2016) examined the degree to which activation during a memory suppression task (measured by the Think/No-Think task) engendered similar or separate neural mechanisms than either the suppression of emotion or the suppression of motoric responses (all compared to a domain-appropriate nonsuppression baseline). While all three tasks produced activation in right dorsolateral prefrontal cortex, it was the connectivity of this region to domain-specific processing regions (e.g., the amygdala in the case of emotion regulation) that differentiated these three types of suppression. Other studies suggest somewhat overlapping mechanisms of memory and emotional suppression (Gagnepain et al. 2017) as well as memory and motoric suppression (Castiglione et al. 2019).

Advantages, Limitations, and Potential Extensions

The advantage of capturing brain processes associated with both the engendering and controlling of intrusive thoughts is that they can provide more information than a simple behavioral reaction time (i.e., button press) or retrospective report. One must be cautious in inferring the engagement of cognitive processes from patterns of brain activation, even within the context of the broad set of knowledge regarding the neural circuitry underlying memory processes. Nonetheless, these patterns provide insight into what aspects of memory processing are disrupted during intrusive disorders, or whether control mechanisms are intact but mainly engaged at inappropriate times. A disadvantage of such approaches is that these neural metrics often require multiple trials for signal averaging; thus, the frequency of intrusions poses a potential limitation. Although they may readily occur, with increasing practice at suppression, they tend to become less frequent (Hellerstedt et al. 2016), which may limit the amount of data that can be collected.

The Multiplicity of Brain Metrics Available

The studies discussed above that used fMRI focused primarily on brain regions that become active during an intrusion or during the attempt to control an intrusion. Additional metrics, however, should be examined to see whether they can provide a distinct window into these processes. For instance, activation within cognitive control (e.g., dorsolateral prefrontal cortex) and memory-related regions (e.g., the hippocampus) has been implicated in suppressing memory retrieval, as has the connectivity between these regions (e.g., Depue et al. 2007; Benoit et al. 2015). Connectivity patterns could be examined using independent component analysis, which reveals groupings of brain regions whose activity follows a similar temporal time course during the suppression of a thought, as compared to other processes, such as visual imagery (Aso et al. 2016).

A variety of electromagnetic techniques can be applied to studying intrusive thoughts. These may focus on specific ERP components, such as the parietal old/new component, which occurs approximately 50–80 ms after stimulus presentation and is thought to be an index of memory retrieval (Rugg and Curran 2007). Such measures could be combined with measures of neural oscillations, recorded from the scalp (e.g., Depue et al. 2013) or via intracranial recordings in patients undergoing surgery for epilepsy (Oehrn et al. 2018), to provide information on the control of such retrieval. Magnetoencephaology has been used to examine downregulation of sensory aspects of long-term memory in the gamma band (70–120 Hz) in traumatized refugees (Waldhauser et al. 2018). Optical imaging methods (e.g., functional near infrared spectroscopy, which provides information on both the location and time course of activation) have been used to examine brain activation in individuals with PTSD during symptom provocation (Gramlich et al. 2017) as well as in individuals high in rumination during stress (Rosenbaum et al. 2018).

All of these methods record or otherwise observe the nature of brain activation associated with memory retrieval or control processes. In contrast, current work that focuses on using brain stimulation techniques (e.g., transcranial magnetic stimulation or transcranial direct current stimulation) to alter intrusive thoughts (i.e., to induce or disrupt them) is still preliminary. In one study, brain stimulation of prefrontal regions and the underlying white matter in three patients about to undergo surgery for epilepsy was found to induce intrusive thoughts. For example, when stimulated one patient reported: "The stimulation induces the disappearance of the word in my mind and replaces it with something else" (Popa et al. 2016:3). Another reported that he had "a thought that seems to come from nowhere" (Popa et al. 2016:4). To the best of my knowledge, brain stimulation techniques have not been used to disrupt thought. In addition, such methods may provide other insights into intrusive thoughts. For example, aspects of intrusive memories in PTSD tend to be over-generalized (Brewin 2011); that is, memories are not clearly differentiated. Transcranial direct current stimulation over lateral occipital cortex during the encoding of a memory leads to interference between memory representations, presumably because of coactivation and less differentiation between those representations (Koolschijn et al. 2019). Hence, such stimulation might potentially be used as a system to model aspects of intrusive thoughts in PTSD.

Advantages, Limitations, and Potential Extensions

Brain-based methods offer a wide variety of tools and a number of different metrics (e.g., brain activity, brain connectivity) that can be used to explore the mechanisms that underlie intrusive thinking. They provide converging evidence for purported mechanisms of intrusive thought and can be used to distinguish potential mechanisms involved in memory control and retrieval of intrusive memories. In addition, brain-based measures offer unique insights. For example, brain-imaging techniques have indicated that memory retrieval can be actively suppressed, as evidenced by a reduction below baseline in activation of hippocampal regions during attempts not to think about specific items (Depue et al. 2007). Recent advances in brain-imaging techniques allow information about intrusive thoughts to be gleaned from nonstructured and more naturalistic stimuli (e.g., a movie) without requiring a specific contrast between conditions (Huk et al. 2018). This opens the possibility for sophisticated computational algorithms to extract over time those critical patterns or signatures of brain activity that are associated with the formation or retrieval of intrusive thoughts.

Individual Differences: Approaches to Brain Anatomy and Function Associated with Intrusive Thought

Another approach is to examine how the functioning of neural systems varies, depending on differences among individuals in the degree of intrusive thoughts and/or the degree to which they can control such thoughts. Some studies examine aspects of the brain that are relatively static (e.g., brain anatomy or the organization of intrinsic resting-state networks) in individuals who experience high levels of intrusive thoughts: PTSD or OCD patients (e.g., Chen et al. 2018a; Gürsel et al. 2018) or individual self-reports from single or extended time periods (e.g., Kühn et al. 2013). While such studies may provide information about variation in potential brain structures involved in intrusive thought (e.g., the hippocampus), they may not provide information specific to intrusive thinking. For example, a stress response associated with traumatic events (e.g., increased cortisol and neurotoxicity of hippocampal cells) could cause reduction in hippocampal volume or shape. Moreover, examining brain anatomy and resting-state connectivity may not be ideal for studying intrusive thought, because both are relatively static, whereas intrusive thoughts, by nature, are time limited and dynamic.

Other approaches characterize individuals according to their level and/or controllability of intrusive thought to determine how these factors might affect brain activation. For example, during suppression of recently experienced items, individuals with a higher degree of self-reported difficulty in removing current thoughts from consciousness had higher levels of activation in Broca area; this presumably represents an inclination toward inner speech (Banich et al. 2015). Another recent and potentially profitable approach is to use magnetic resonance spectroscopy (MRS) to examine potential neurochemical mechanisms that may enable certain individuals to block memory retrieval. For example, individuals with higher levels of GABA in the hippocampus, as assessed by MRS, have a greater ability to suppress memory retrieval (e.g., Schmitz et al. 2017). The disadvantage of using MRS methods is that they are quite time consuming (e.g., 25 minutes). In addition, only a few brain regions can be interrogated during a scanning session, and the brain region interrogated is much larger (e.g., at least 3–4 times greater) relative to functional neuroimaging methods.

Advantages, Limitations, and Potential Extensions

Exploring individual differences in the psychological processes involved in intrusive thought has a long and fruitful history, and can be equally well applied for use with neural markers. However, for neuroimaging, an individual differences approach generally requires a larger sample size to detect covariation than is required in studies designed to detect group average patterns of activation (Cremers et al. 2017). Thus, utilizing an individual differences approach requires more time and money.

Future Frontiers for Brain-Imaging Methods

Using Brain-Imaging Methods in a Predictive Manner

There is much interest in determining whether an individual will experience intrusive memories after a distressing event. Prior work has examined whether certain characteristics of an individual and/or the way in which a distressing event is processed are robust predictors of subsequent intrusive thoughts (e.g., Marks et al. 2018). This work has been extended to examine whether measures derived from brain metrics might predict subsequent intrusions. For example, ERP measures of the effectiveness of suppression (greater amplitude of a fronto-centrally distributed N2) from the Think/No-Think task can predict subsequent intrusions after analog trauma (Streb et al. 2016). Relatedly, patterns of brain activation derived from machine-learning techniques, during the encoding of material in an analog trauma paradigm, can be used to predict the degree to which an individual will have subsequent intrusive thoughts (Clark et al. 2014b). Although in its infancy, such approaches may have much potential.

Neural Markers versus Self-Report

One important limitation of many of the methods described above is that they rely on self-report to verify an intrusive memory or control over thoughts, and thus do not provide insight into the nature of the representation of that memory. Neurally based measures have been used to try to address this issue.

Autonomic Measures

Autonomic measures have been used in conjunction with the behavioral methods discussed above. The idea is that reexperiencing traumatic events should induce physiological changes and that successful suppression of such thoughts should be associated with reductions in such physiological responses (e.g., May and Johnson 1973). Such measures have also been used in conjunction with an individual differences approach. For example, greater heart rate variability is associated with a better ability to inhibit thoughts, either in a structured thought suppression situation, the Think/No-Think paradigm (Gillie et al. 2014), or in the self-report of intrusive thoughts over specific periods of time (Gillie et al. 2015). Such physiological measures, however, are mainly nonspecific in nature and could reflect arousal, emotional distress, or anxiety, either in a state or trait manner that is unrelated to intrusive thoughts.

Neuroimaging Approaches

Although much work on intrusive thinking has focused on the retrieval of information from long-term memory, by nature, intrusive thoughts involve access to and active representation in working memory (for further discussion, see Visser et al., this volume). Understanding whether a thought is currently in the focus of attention in working memory is an important issue. Initial work suggests that brain-imaging techniques can be applied quite fruitfully to verify that individuals are indeed experiencing specific thoughts and/or manipulating them. In one study (Banich et al. 2015), individuals were shown a picture or heard a brief snippet of a familiar tune (e.g., "Happy Birthday") for four seconds; immediately afterward they then had to manipulate the item in one of four manners: maintain it, replace it with a different image/tune, specifically suppress the item, or clear their mind of all thought. Providing at least some evidence that participants were indeed manipulating their thoughts as instructed, a significant reduction of activity in primary sensory areas (e.g., visual cortex, auditory cortex) was observed averaged across all trials for the two conditions in which a thought needed to be cleared and removed (suppress the item or clear their mind of all thought), compared to when there was an active thought (maintain or replace). In addition, results indicated that while some neural mechanisms were commonly engaged across these operations (e.g., across replacing, suppressing, or clearing an item as compared to maintaining it), there are also specific neural mechanisms that differentiate the suppression of an item from clearing one's mind of all thought (Banich et al. 2015). Thus, neuroimaging may provide a neural marker and confirm specific mental operations performed on a given thought. This may lead us to differentiate the ways in which thoughts can be removed from working memory.

In follow-up work, machine learning was incorporated to expand the questions and issues that can be examined (Kim et al., submitted). Specifically, via a localizer task, a machine-learning classifier was used to distinguish specific categories of items (e.g., places, faces, fruit). These classifiers were then used on a trial-by-trial basis to determine the degree to which removing the thought was successful. If a thought is successfully removed, then the classifier fit should be poor. If the thought is maintained, then the classifier fit should be high. This approach enables us to examine the nature of the mental representation on a trial-by-trial basis and provides a means to determine the level at which such representations are maintained and/or cleared via specific category and subcategory classifiers (e.g., fruit: apples, grapes, pears). For example, one can examine whether clearing the thought of an apple generalizes to other fruit (e.g., grapes and pears). In addition, patterns of brain activation can be examined as a function of classifier fit to determine which brain regions are highly active when the classifier fit is low (indicative of clearing the thought), compared to when the classifier fit is high (indicative of the thought remaining). This, in turn, may provide insight into which brain regions are involved in exerting control over thoughts. While this study did not explicitly track intrusions, such methods could be extended to track the content of intrusive thoughts on a real-time basis by identifying multivariate neural patterns of distinct forms of intrusive content and evaluating the degree to which these patterns manifest on a moment-to-moment basis.

In summary, there are a variety of interesting new directions that might be used to provide novel insights into the nature of intrusive thought. These range from adapting and using paradigms from other areas of cognitive and affective psychology to investigate intrusive thinking, to new approaches and applications from brain-imaging methods that might be used to verify or predict intrusive thought.

7

Conceptualizing Intrusive Thinking at the Level of Psychological Mechanisms

Marie-Hélène Monfils and David M. Buss

Abstract

Intrusive thinking, the sudden occurrence of unwanted thoughts, images, or impulses, is a frequent and natural occurrence within our stream of consciousness (Clark and Purdon 1995). Present in both clinical and nonclinical samples, the high incidence of intrusive thoughts across the population renders challenging the task to identify meaning behind their occurrence. Their presence, frequency, and content do not appear, however, to be random. Intrusive thinking manifests differently in clinical versus nonclinical populations. They may be associated with certain emotions, thus offering a glimpse into their potential adaptive nature. This chapter examines what intrusive thoughts are and what they are not. It explores how they manifest differently in clinical versus nonclinical populations and asks whether these different presentations can provide insights into their origin. It evaluates intrusions as possible manifestations of adaptations and examines intrusions linked to evolved emotions (e.g., fear, rage, jealousy, and love). Identifying the possible reasons behind intrusive thinking may help guide future treatment.

Introduction

A commuter experiences a sudden urge to jump off the subway platform as the train arrives at the station. An individual engaged in cleaning up after dinner suddenly has a vivid image of throwing a plate against the wall, as a fit of rage intrudes their thoughts. Someone looking through their wardrobe to find something to wear that day suddenly hears a voice in their head saying, "You're such a loser." As another person walks their dog, a sudden image of repeatedly stabbing a passerby jarringly interrupts their train of thought. Another is driving to work when a voice interrupts their thoughts and suddenly proclaims: "You need to get on stage with a guitar! You're a rock star!"

Intrusive thinking, the sudden occurrence of unwanted thoughts, images, or impulses, is a frequent and natural occurrence within our stream of conscious-ness (Clark and Purdon 1995). Intrusions occur in both clinical and nonclinical samples, and their high incidence (80–90%) across the population renders chal-lenging the task to identify a meaning behind their occurrence (Clark 2005). It does appear, however, that their presence might not simply be random. The manifestation of intrusions differs in clinical versus nonclinical populations, and they may be associated with the presence of certain emotions that offer glimpses into their potential adaptive nature.

Here, we examine what intrusive thoughts are and what they are not. We explore *how* intrusive thoughts manifest differently in clinical versus nonclin-ical populations. We ask whether the different presentations in clinical and nonclinical populations might provide insights into their *origin*, and evaluate intrusions as adaptations.

What Are Intrusions?

According to Clark (2005:4), an intrusive thought is "any distinct, identifiable cognitive event that is unwanted, unintended, and recurrent. It interrupts the flow of thought, interferes in task performance, is associated with negative affect, and is difficult to control." This definition is generally consistent with others used to describe the phenomenon (Beck 1967; Horowitz 1975; Klinger 1978; Rachman 1981) and appears to include the following characteristics (Clark 2005):

- It is a distinct thought, image, or impulse that enters conscious awareness.
- It is attributed to an internal origin.
- It is considered unacceptable or unwanted.
- It interferes with ongoing activity.
- It is unintended.
- It tends to be recurrent.
- It easily captures attentional resources.
- It is difficult to control.

Clark also suggests that intrusive thoughts are negative, although it is unclear whether this criterion is universally accepted, a point to which we return below (e.g., Gregory et al. 2010).

Images versus Thoughts

Intrusive thoughts are acknowledged to manifest in different ways. Sometimes they present as images or scenarios, and other times as an internal voice devoid of imagery.

Memories versus Nonmemories

Though there is evidence that intrusive thoughts may be born from the recall of previous experiences, this is not always the case. Intrusions can sometimes consist of memories and other times not. There are instances in which the intrusion is autobiographical, but it is uncertain if that is always the case.

Spontaneous versus Triggered

Some intrusive thoughts are triggered by a cue. For example, an individual may experience an intrusive thought compelling them to stab themselves in the chest upon seeing a kitchen knife on the table. Seeing the knife might have served as a cue that prompted the intrusion. In other cases, the intrusion may seem to appear out of nowhere, without an obvious cue having acted as a trigger. The fact that a triggering cue is not identified does not necessarily mean that it was not present. Rather, it could be that the cue was subtler and not explicitly observed.

Valence of Intrusive Thoughts

Many definitions of intrusive thoughts imply that they are negative in valence (Clark 2005); however, Gregory et al. (2010) propose that intrusions may actually present as highly positive in individuals experiencing a hypomanic state. These could be similar in content to thoughts related to delusions of grandeur. Other situations suggest that positive intrusions exist, including in nonclinical samples. One example is that of infatuation in which an individual experiences intrusive thoughts about a loved one, and many such intrusions carry a positive valence (e.g., fantasies of union or sexual consummation). Even situations that appear to have a negative connotation on the surface could carry positive valence for the individual experiencing the intrusion. For example, imagining the suffering of an enemy could be quite positive in a scenario of homicidal ideation.

What Predicts Intrusive Thoughts?

Intrusive memories can be triggered by rumination, a phenomenon that is often present in individuals suffering from anxiety, depression, or both (Birrer et al. 2007). A number of mental health disorders are associated with the presence of intrusive thoughts (discussed further below); however, they also manifest in nonclinical samples. The overall incidence appears quite high: 80–90% of individuals in nonclinical samples report experiencing intrusions (Clark 2005). Below, we discuss possible origins of intrusions. There is evidence suggesting that attaching meaning or importance to intrusions can impact their frequency and controllability (Freeston et al. 1991).

What Intrusive Thoughts Predict

Intrusive thoughts are quite prevalent following trauma, although findings suggest that their frequency and severity is not predictive of posttraumatic stress disorder (PTSD) symptomatology (McFarlane 1988; Shalev 1992). Interestingly, Brewin et al. (1998) found that while the presence of intrusive memories either at baseline or prior to follow-up made an additional significant contribution to anxiety, it did not affect depression at follow-up (Brewin et al. 1998). Having high-frequency involuntary intrusive memories at baseline, however, significantly predicted later depression, even when controlling for the severity of symptoms at baseline (Brewin et al. 1999).

One interesting feature of intrusive thoughts is that they are a common feature across multiple psychiatric disorders.

Intrusions in Clinical Populations

The incidence of intrusions is quite high across multiple mental health disorders, where they are known to occur in individuals with obsessive-compulsive disorder (OCD), generalized anxiety disorder (GAD), PTSD, body dysmorphic disorder, eating disorders, depression, bipolar disorder, and others. The manifestation of intrusions appears to be partly affected by specific diagnoses. For example, an individual with OCD who engages in extreme handwashing might experience germ intrusions, whereas an individual with body dysmorphic disorder might get intrusions related to food items. For a more extensive discussion of intrusions in clinical populations, see Schlagenhauf et al. and Visser et al. (this volume).

Interestingly, intrusions appear to be dissociable from other characteristics that might be more specific to only one or two disorders. Whereas obsessions are thought to be characteristic of OCD, worry is a central feature of GAD (although not exclusive to it), and negative thoughts and rumination may typically be present in individuals with depression; intrusions are often present in all of these conditions. Let us now compare and contrast intrusions with worry, rumination, obsessions, and negative thoughts.

Intrusions versus Worry

Defined as "a chain of thoughts and images, negatively affect laden and relatively uncontrollable" (Borkovec et al. 1983), *worry* is a central feature in GAD, but it also occurs with high incidence in nonclinical individuals. In definition and in practice, worry and intrusive thinking are quite similar: They both interrupt ongoing thoughts and activities, and they can both present as thoughts or images, although worry occurs more frequently as verbal and intrusions more frequently as images (Clark 2005). Intrusions are thought to be less voluntary than worry; that is, worry can be brought on volitionally, whereas intrusions are by definition involuntary and disruptive. Another distinctive feature that disambiguates intrusions from worry is that intrusions are generally discrete and brief, whereas worry need not be.

Intrusions versus Rumination

Defined by Nolen-Hoeksema and Morrow (1991) as repetitive and passive thinking about one's symptoms of depression, *rumination* is to depression what worry is to GAD (Borkovec et al. 1998). To our knowledge, no study has directly compared the differences between intrusions and rumination; however, respective reports for each provide clues as to what disambiguates the two. Whereas intrusions are thought to be brief, sudden, and to involve generally unwanted thoughts or images, rumination involves a train of thought that is longer, repetitive, and recurrent (Clark 2005). It is possible that intrusions may trigger rumination, which in turn may precipitate a depressive or anxious episode. As such, the same content may be at the source of intrusions and rumination. In thinking about the distinction between intrusions and rumination, one might imagine that an intrusion could occur during rumination.

Intrusions versus Obsessions

Obsessions and intrusive thoughts are very similar, where the former appears to be an extreme version of the latter. Another characteristic that helps dissociate the two is that intrusions may sometimes be irrelevant to the self, whereas obsessions are relevant. Obsessions may often prompt behaviors such as compulsions that are intended to diminish the associated thoughts and manifest as OCD.

Intrusions versus Negative Thoughts

Intrusions can be dissociated from general negative thoughts in that the former is more likely to be irrational, whereas negative thoughts are more likely to be rational. In this context, rational refers to thoughts that are not at odds with the present context. An individual might be experiencing negative thoughts about their promotion prospect during a recession, for example. If, during a positive economy and after receiving a positive evaluation, they jarringly internally hear the words "you're about to get fired" just prior to giving an important presentation, their experience was an intrusion. Intrusive thoughts are more disruptive of day-to-day activity than general negative thoughts. Negative thoughts are a core characteristic of individuals with depression, and generally manifest as "thoughts" or in a verbal way rather than images. Intrusions may present either as verbal or as images, most commonly the latter. Unlike other forms of negative "processing," intrusions seem to be relatively common in nonclinical populations.

Manifestation of Intrusions in Nonclinical Populations and the Origins of Intrusive Thoughts

Little is known about the etiology of intrusive thoughts. Different theories have been proposed, but as yet we do not have a practical understanding of intrusions' origins. Here, we briefly review various theories on intrusions, examine their manifestation in nonclinical samples, and discuss the parallels between intrusions and memory retrieval.

Theories on the Etiology of Intrusions

Salkovskis (1988) suggests that intrusive thoughts might be an inherent aspect of problem solving. He proposes that despite being disruptive to thinking in the moment, intrusions may be useful, and that the very reason they appear suddenly and are intrusive and compelling could be that they are meant to be noticed. In other words, if intrusions appeared as a simply nondisturbing thought, we might not pay attention to them.

Rachman's view on intrusive thoughts is predominantly based on the etiology of obsessions (Rachman 1981). He also believes that an important contributing factor to intrusions is the development of a mood state that sets the tone for intrusions to occur. Rachman proposes, for instance, that individuals who are stressed and in a dysphoric mood state are more likely to experience intrusions. In such cases, individuals are also thought to have greater difficulty ignoring or suppressing the intrusive thoughts. He also suggests that certain personality characteristics (e.g., neuroticism, heightened anxiety) may make individuals more susceptible to experiencing intrusions.

Klinger (1978) proposes that intrusive thoughts are associated with "current concerns." In other words, intrusions occur when thinking is interrupted and the thought process shifts toward addressing what was brought about by the intrusion (the current concern). The intrusions, then, can be external cues or nonverbal events (Klinger 1999).

Horowitz proposed a reformulation of intrusive thoughts based on psychoanalysis. In his account of intrusive thoughts, Horowitz posits that active memory storage is characterized by an intrinsic tendency to repeat its represented contents, which continues until the storage of contents in active memory is terminated. Appropriate cognitive processing of the memory content terminates the process. Horowitz proposes that stressful events may yield intrusions that stimulate an active memory of an experience. This memory activation occurs repeatedly until there is integration of old and new information, perhaps to reconcile representations of a memory with a person's inner view of the world (Horowitz and Wolfe 2003). Horowitz's formulation appears particularly relevant to traumatic memories.

The general overarching theme across the views held by Salkovskis, Rachman, Klinger, and Horowitz is that, disruptive and disturbing as they are,

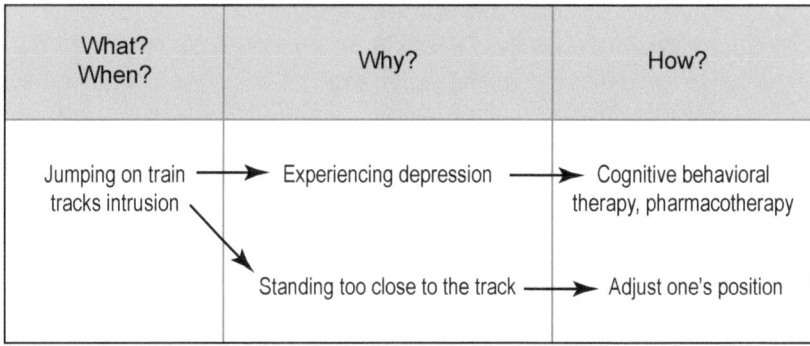

What? When?	Why?	How?

Figure 7.1 Identifying why intrusions arise may help guide how we can best treat them.

intrusions might actually serve an adaptive purpose. One difficulty in these interpretations of intrusive thoughts is that they are not exactly practical. That is, it is difficult to conceive how such theories might guide the development of future treatment. We believe there is value in examining the manifestation of intrusive thoughts in nonclinical populations and to extract the possible underlying adaptive basis for their presence. In short, if we are poised to identify *why* the brain produces intrusive thoughts, we should be better equipped to determine *how* to address them (Figure 7.1). What problem is the brain trying to solve? Are intrusive thoughts inherently harmful? Do they represent a beneficial mechanism gone awry?

Informed by these theories, we present an adaptationist perspective on intrusive thoughts. Thereafter, we examine the manifestation of intrusions in nonclinical samples through specific examples and extract two elements (content and process) that might provide useful insight into treatment avenues for mental health disorders in which intrusions are often present.

An Adaptationist Perspective on Intrusive Thoughts

From the perspective of modern evolutionary biology, adaptations are characteristics that evolved because they contributed in a specific way to solving a problem or challenge tributary to successful survival or reproduction (Williams 1966). Propensities to become fearful of snakes and spiders, for example, evolved because they led their bearers to avoid these dangers to survival (Öhman and Mineka 2001). Evidentiary criteria for invoking adaptation include economy, efficiency, and, importantly, improbable precision of functional design.

From this perspective, we ask: Do intrusive thoughts in nonclinical populations show evidence of functional design? Any sensible answer is reliant on further conceptual and empirical work. Guided by an adaptationist perspective, we offer a few preliminary suggestions or heuristics which rely on the following metatheoretical premises:

1. Organisms have finite time and resource budgets.
2. Organisms have evolved decision rules to prioritize effort allocated to some adaptive problems at the expense of others in temporal sequences: when faced simultaneously with tasty-looking ripe fruit, an attractive mate, and a dangerous snake, for example, humans prioritize effort allocated to avoiding a lethal snake bite, postponing effort devoted to the other adaptive problems.
3. Emotions such as fear, rage, disgust, and jealousy mobilize attention and effort to specific threats or challenges, orchestrating an organism's cognition, physiology, and behavior to address those challenges (Tooby and Cosmides 2008; Al-Shawaf et al. 2016).

One novel hypothesis that we are proposing here is that intrusive thinking may be one important design feature of evolved emotions that have this prioritization function, directing attention and allocating effort to solving some adaptive challenges at the expense of others.

An important feature of these adaptations is their probabilistic nature, guided by error management logic. Probabilistic nature simply means that adaptations only succeed in solving adaptive challenges with some likelihood, not invariantly. Although there is compelling evidence for evolved fears of snakes and spiders (e.g., Öhman and Mineka 2001), and these adaptations have undoubtedly saved many lives of their bearers, these adaptations do not invariably prevent life-threatening bites: more than 81,000 people worldwide die each year from snake bites. Evolved fears function probabilistically.

Error management theory is a metatheory of decision rules, combining signal detection theory with evolutionary theory (Haselton and Buss 2000). At an abstract level, when confronted with uncertain environments, there are two possible ways to err inferentially: making false positives and making false negatives. When recurrent cost asymmetries of making these two types of errors exist over evolutionary time, selection will favor decision rules to avoid the costlier error, even if they result statistically in more frequent errors. When perceiving a rustle afoot in a thick grassy wooded area, for example, one can err by inferring that a snake is absent when it actually is present (false negative) or by inferring that a snake is present when it is not (false alarm). In this example, failing to detect an actual dangerous snake in the grass is a costlier error than falsely inferring a snake's existence when there is, in fact, no real threat. Error management theory, in this case, predicts that fears of this sort have evolved to avoid the costlier error, generating avoidance of probabilistic threats, some or many of which will turn out to be false alarms. Although these evolved systems are biased, they are adaptively biased. Error management theory has garnered much empirical support, leading to the discovery of phenomena ranging from the auditory looming bias and the vertical descent illusion in the perceptual domain to the sexual overperception bias and infidelity overinference bias in the social domain (Haselton and Nettle 2006).

A final element in this framework, as applied to intrusive thoughts, is the *mismatch principle*, which states that evolved traits that were adaptive in the ancestral environments in which they evolved may misfire and become maladaptive in modern evolutionarily novel contexts (e.g., Spinella 2003; Li et al. 2018a). A prime example is eating disorders. Humans evolved in food-scarce environments and have evolved feeding adaptations to consume calorie-rich substances—those high in fat and sugar—when encountered and to easily store metabolic surpluses in the form of fat deposition. In modern environments that contain an abundance of these resources, easily obtainable with minimal effort in concentrated forms (e.g., fast-food restaurants or grocery stores), humans tend to overeat. Obesity and type 2 diabetes, absent in traditional hunter-gatherer cultures, are largely the result of these evolutionary mismatches, along with other factors, such as more sedentary living.

The elements of these principles lead to the hypothesis that intrusive thoughts are functional parts of evolved emotion systems. They are designed (in part) to mobilize attention and effort toward specific adaptive challenges. They are often adaptively biased, designed to avoid costly errors at the expense of more frequent errors. Some are maladaptive in modern mismatched environments that are widely discrepant from the ancestral environments in which they evolved.

Possible Adaptive Purpose of Intrusions in Nonclinical Samples

Individuals must constantly make decisions over choices and prioritize task importance. It is conceivable that intrusions act as a means to emphasize what should be worked through as soon as possible, in line with Salkovskis, Rachman, and Klinger. The fact that intrusions appear suddenly, are brief, and are often disturbing may emphasize the urgency of solving a potential problem. It is also conceivable that intrusions themselves present as a mechanism to *process* information, in line with Horowitz. Here, we begin by examining, in nonclinical populations, possible feelings that may be associated with intrusive thoughts, and we address their potential adaptive mechanism (intrusion content). Thereafter, we approach the possibility that intrusive thoughts in and of themselves are a mechanism that enables working through unaddressed but identifiable problems by viewing intrusions through the lens of memory (intrusion process).

Intrusion Content: Conducive Nonclinical Instances

Though they are not necessarily centered around a disorder, intrusions in nonclinical samples appear to occur along common themes across individuals. We contend that these themes may provide insight into the potential adaptive nature of intrusions, in line with the adaptationist framework proposed above. The following list is not exhaustive, but provides a starting point to examine different forms of emotions that can be at the source of intrusions. Their common characteristic is that they might serve an adaptive purpose; however, depending on the context during which they occur, intrusions can also reflect an adaptation gone awry.

Anger, Rage, and Revenge.   Anger has been hypothesized to be an evolved emotion, the expression of which functions (in part) to recalibrate someone else's welfare trade-off ratio with respect to you (Sell 2011). When a behavior affects two or more individuals, it can be selfishly skewed or altruistically skewed. Consider a roommate who has left dirty dishes strewn about the shared kitchen, expecting you to clean them. Expressions of anger to the roommate communicate that they have insufficiently taken your welfare into account and should adjust it in the future. In this simple example, intrusive thoughts and prolonged rumination function to stoke the emotion of anger until the roommate arrives back home and the rage can be expressed, ideally causing the roommate to recalibrate their welfare trade-off ratio with respect to you.

Now consider road rage: When someone cuts you off in traffic, it sometimes activates intense anger. In the modern environment, road rage sometimes produces violent car accidents when ramming the violator. The underlying emotion evolved presumably in small-group contexts in which its expression would cause the violator to recalibrate, taking your welfare more into consideration in the future. In a modern environment marked by dense urban living patterns, in which the handling of severe social violations has been outsourced to professional police, expressions of road rage can lead to disastrous and maladaptive outcomes (e.g., car crashes, personal injury, and death). The design feature of intrusive thinking that prolongs rumination about the violator was presumably adaptive in small-group contexts of the past, where social reputations mattered greatly and the failure to respond to violations could lead to a catastrophic loss of status. In the modern mismatched environment containing lethal 4,000-pound vehicles, traffic congestion, and swarms of anonymous strangers, intrusive rumination about someone who cut you off can lead to road rage and a maladaptive misfiring of this ancient emotion.

Jealousy and Infidelity.   Jealousy is an emotion that evolved to combat threats to a valued social relationship. If the relationship is a mateship, jealousy can be activated by cues to sexual or emotional infidelity, to signs of a partner's defection, or to threats posed by potential mate poachers or even by mate value discrepancies (Buss 2000; Buss and Haselton 2005). Infidelity is typically cloaked in secrecy, creating a signal detection problem for the partner. Once jealousy is activated, it can produce intrusive thoughts, prolonged rumination, and motivate vigilance to discern the nature and magnitude of the relationship threat. Intrusive thinking in this context can be functional, leading a person to gather relevant information and to allocate effort to warding off the threat, devoting resources to mate retention or to repelling the genuine threat posed by a potential mate poacher.

Intrusive thinking is a design feature of the jealousy adaptation; it leads its bearers to uncover and attempt to solve real threats to romantic relationships. Nonetheless, it can also misfire, leading to maladaptive outcomes. If the psychological detection of infidelity cues is set too sensitively, it can produce false accusations of infidelity, undermining the very relationship that jealousy was designed to protect. It can produce delusions of a partner's infidelity and pathological jealousy, leading to extreme violence toward a partner (Buss 2000). Because infidelities are typically concealed, cues to infidelity are inherently probabilistic. Based on error management theory, there is evidence that people overinfer infidelity to avoid the costly error of losing a partner to a romantic rival, even at the cost of making more frequent errors of inference (Goetz and Causey 2009). Moreover, many individuals who have been diagnosed by psychiatrists as having delusional or pathological jealousy turn out to have partners who, upon deeper investigation, have actually been unfaithful (Buss 2000). In short, it is difficult in any particular case to determine unambiguously whether jealous intrusive thinking is functioning as it was designed to function, or if it is misfiring and causing pathological outcomes.

Love and Romantic Infatuation. Intrusive thinking is a common feature of the infatuation stage of love, markedly present when separated from a loved one (Fisher 2016). It can interfere with work, cause other relationships to lapse, and even create a metabolic deficit when someone forgets to eat. Intrusive thinking often creates an idealization of the loved one, imputing maximal values to desirable qualities that have not yet been observed. Preoccupation presumably leads to efforts to woo a loved one or to become reunited with them after separation. After the infatuation stage fades and is replaced by a more subdued warmth and attachment, intrusive thinking subsides, allowing a reallocation of effort to other adaptive challenges, such as obtaining food, negotiating status hierarchies, or solidifying coalitional alliances. Intrusive thinking in the context of the infatuation stage of love is temporally delimited.

Like all adaptations, this one can go awry, misfiring in the modern environment. People develop romantic infatuations with movie stars, for example, when there is no possibility of meeting them, much less successful consummation. In the extreme, these can lead to criminal stalking, as in the case of John Hinckley Jr. who developed an intense infatuation with the actress Jodie Foster. He sent her numerous love letters, stalked her, and when his efforts failed to produce reciprocation, he attempted to assassinate President Ronald Reagan in a last-ditch desperate attempt to get her attention and demonstrate the intensity of his love and commitment to her. He now resides in a prison cell. In short, intrusive thinking can lead to disastrous outcomes, both for the individual and for others who become victims. When properly functioning, however, intrusive thinking leads to successful consummation of love.

This illustrates that intrusive thinking is not solely a design feature of the so-called "negative emotions." It is likely an evolved design feature of many emotions, including fear, rage, jealousy, shame, and guilt, as well as more positively valenced emotions such as love and sexual arousal. Nor is it always dysfunctional. Intrusive thinking is often a key design feature, motivating attention to pressing adaptive problems while postponing effort allocated to less important ones. Error management analysis highlights the difficulty of distinguishing functional from dysfunctional outcomes in any specific case, rendering the theoretical analysis of intrusive thinking more complicated than previously considered. Another example that generally involves positive intrusions pertains to the pursuit of goals or aspirations. It is important to note that while some emotional contexts are conducive to either positive or negative intrusions, others are likely to be more complex, as in the case of grief. Intrusions can often be autobiographical, that is, they relate to an individual's firsthand experience. Cast in this light, intrusions can actually be interpreted as a memory retrieval in some instances. We briefly consider intrusions occurring in the context of goals or aspirations and grief, and then further examine intrusions as memory retrieval to extract their potential adaptive process.

Goals or Aspirations. Intrusions can occur in scenario building of means to achieve. In such a case, the intrusions would likely be positive and inspire someone to pursue achievements. Much like the cases described above, the adaptive nature of goal and aspiration intrusions can go awry. For instance, an individual may experience fantasy-like intrusions that reach far beyond their abilities. In this case, the originally positive intrusions could grow to be a reminder of one's failures and hinder one's potential to succeed in a more achievable realm.

Grief. Grief is almost always triggered by the loss of a key social partner—a close friend, a romantic partner, or a family member. Research on intrusions during grief is limited but suggests the presence of both positive and negative intrusions in individuals experiencing the loss of a loved one (Boelen and Huntjens 2008). In the positive realm, mourners may experience intrusive memories of the loved one that died or fantasy reenactment. Through the lens of memory (discussed below), the purpose of grief intrusions could be that of strengthening a neurobiological trace, to keep the memory alive. Negative intrusions of grief can include memories of the death event or negative images or thoughts about the future. Early during grief, the intrusions, both positive and negative, may be helpful to the individual who experienced a loss. However, if persistent and enduring, they could interfere with a person's ability to move forward.

Evolutionary scholars have advanced two competing explanations of grief. One is that grief is an unfortunate nonadaptive by-product of love and attachment (Archer 2003), both of which are profoundly important adaptations in the evolved social suite of humans (Christakis 2019). The second is that grief serves several adaptive functions, such as identifying actions that might have led to the loss, motivating actions to prevent future losses and signaling to significant others (friends, family, mates) the need for help due to the loss (Nesse 2005). Other possible functions include signaling to others that you are a loyal coalitional ally and ruminating about the implications of the loss for replacing the lost one with an alternative mate or coalitional ally. Which of these competing hypotheses, or which combination, will bear fruit rests with future empirical research on individuals who experience loss and grief.

Intrusion Process: Intrusions as Memory Retrieval

There are useful parallels to be drawn between intrusions and other forms of memory retrieval. Memory retrieval can be broadly defined as recalling a prior experience, either following the presentation of an external or internal cue or through volitional control. Conceptualizing intrusions as memory retrieval appears very much in line with Horowitz's definition of intrusive thoughts and enables approaching the concept with an adaptive mechanistic view.

From this perspective, we can think of intrusions as potentially serving the adaptive functions described below. We can also conceive of intrusions as providing an opportunistic window or intervention. The latter can best be understood through the process of reconsolidation and memory updating.

Reminder That Certain Information Needs to Be Further Processed. By virtue of being interruptive and often irrational, intrusions are noticed. In this case, intrusions would likely reflect an event that has passed, which an individual may need to prioritize or address.

Warning to Allow Preparedness. Intrusions draw attention. Their purpose here is to enable an individual to react in the presence of a looming situation. As such, the intrusions could include content related to an individual's past experiences, but would certainly pertain to an individual's future.

Mechanism to Initiate Extinguishing or Exerting Another Form of Inhibitory Control over a Negative Memory. Retrieval of a previously consolidated memory (i.e., a memory that has been stored into long-term storage for longer than ~six hours) engages two seemingly opposing mechanisms: reconsolidation and extinction. Reconsolidation refers to a putative process which proposes that after retrieval, previously consolidated memories become destabilized and require renewed protein synthesis for long-term storage. Reconsolidation also offers an opportunistic window during which memories can be updated. In extinction, the repeated presentation of the conditioned stimulus in the absence of the unconditioned stimulus leads to a progressive decrease in the behavioral expression to the stimulus. Extinction can refer to both a process (the progressive decrease in fear throughout a session) and an outcome (e.g., resultant decrease in fear responding).

The concepts of reconsolidation and extinction have been extensively studied, and each provides an important avenue of improving psychotherapeutic outcome, particularly in anxiety-related disorders and addiction (Monfils and Holmes 2018). The two approaches have also been successfully combined to improve upon long-term therapeutic outcomes in the retrieval-extinction paradigm (Monfils et al. 2009). Approaching intrusions as memory retrieval could potentially enable the optimization of therapeutic approaches. If an intrusion is mechanistically akin to memory retrieval, it could provide an opportunistic window to intervene and attenuate their potency. Effectively, research suggests that behavioral or pharmacological interventions shortly after memory retrieval improves outcome above and beyond standard extinction-based approaches (Monfils et al. 2009; Schiller et al. 2010; James et al. 2016b; Telch et al. 2017).

Another way to handle upsetting memories, once recalled, is to exert a form of inhibitory control over them (other than extinction). A number of such approaches are discussed in detail by Visser et al. (this volume).

Memory Strengthening or Maintenance.   Once they are retrieved, memories generally strengthen if left untargeted (Inda et al. 2011). As such, while an intrusion may present a window of opportunity for treatment of a traumatic memory, if untreated, the intrusion could actually lead to memory strengthening as well. Such a mechanism could provide an adaptive explanation for certain intrusions (e.g., those that manifest during grief, or during goals and aspirations). In other cases (e.g., following trauma), intrusions could exacerbate a negative memory and render treatment more challenging.

Means of Escaping Boredom.   While there are often specific circumstances or conditions that appear to prime the presence of an intrusion, others may be more random. In such a case, an intrusion could conceivably serve the adaptive purpose of escaping boredom or monotony in a safe way by engaging in a daydream experience. In this context, intrusions could, for example, promote an individual to engage in mind wandering. An extensive discussion of mind wandering can be found in Visser et al. (this volume).

Summary

Intrusions are what we make of them. Although sometimes acutely distressing when they occur, intrusions can, in and of themselves, actually be innocuous or even positively valenced. What appears to be at the source of most of the distress experienced by intrusions is often the thought process that follows. Consider the following scenario:

A person is standing on the subway platform listening to their favorite song with headphones. Suddenly, a vivid intrusion appears in their mind: they see themselves jumping on the tracks, just as a train passes through. Thus far, the intrusion has not per se caused any harm. The person's reaction to the intrusion, however, can vary. One individual could simply think: "Whoa! That was crazy! Of course, I would never do that." At the same time, this person might experience a sense of comical relief: "Boy, if I had jumped, I wouldn't have to sit through all of those blankety-blank-blank meetings scheduled today." Another person might process the intrusion differently. For instance, they might think: "Why am I having these images? Should I jump? I am worthless. I don't want to die. Or maybe I do want to die. I don't know what to do. I'm worried about what I might do."

In the first response scenario, the person may not be bothered further (or at all) by the intrusion. In the second case, a person might perseverate on the experience or engage in behaviors to try to minimize the impact of the intrusion. This could potentially result in worry, rumination, and/or obsessions and associated compulsions.

In a sense, intrusions themselves may not be as distressing as what we make of them, and what we make of them is likely to be largely influenced by our state of mind (including, in clinical manifestations, the underlying pathology). In approaching treatment for individuals who experience intrusions, it is important to consider their possible adaptive nature. In doing so, it may be helpful to identify the intrusions' possible underlying content as well as the psychological process that a person's brain has determined should be engaged via the intrusive thought. Ultimately, identifying the possible "why" of intrusions may help guide "how" we can best treat them (Figure 7.1).

8

Current Psychiatric Perspectives on Intrusive Thinking

Florian Schlagenhauf, Andreas Heinz, and Martin Voss

Abstract

Various psychopathological symptoms share characteristics of intrusive thinking. Intrusive thoughts are part of the diagnostic criteria for posttraumatic stress disorder and obsessive-compulsive disorder but are also relevant in other psychiatric conditions, such as drug craving in addiction or rumination in depressive disorders. Intrusive thoughts must be differentiated from thought insertion observed in schizophrenia and related psychotic disorders. This chapter reviews the typical characteristics and content of intrusive thinking in the context of different psychiatric conditions and outlines current theories regarding the mechanisms of intrusive thinking.

Introduction

Intrusive thoughts can be characterized as repetitive, uncontrollable, distressing thoughts that enter conscious awareness unwantedly (Clark 2005). They are an important aspect of different psychiatric disorders, but they also manifest in the nonclinical, general population (Clark 2005; Garcia-Soriano et al. 2011). Prominent psychiatric examples include intrusions related to traumatic events in patients suffering from posttraumatic stress disorder (PTSD) as well as aggressive obsessions experienced by obsessive-compulsive disorder (OCD) patients (Heinz 2017). In schizophrenia or related psychotic disorders, the delusions, hallucinations, or thought insertions experienced by patients are repetitive, uncontrollable, and distressing (Heinz et al. 2016). In each of these conditions, the content of intrusive thoughts can be very different: in PTSD, the content may refer to a real autobiographic event whereas in OCD it may relate to an obsessive thought about contamination. Certain similarities do, however, exist: characteristic thoughts appear repeatedly (to a degree) and can interfere with normal functioning. A person perceives the thoughts to be unwanted and reports having no control over these thoughts; that is, they do not result from a deliberate or even effortful process, but appear involuntarily and automatically. Finally, intrusive thoughts create distress due to their content and/or characteristics.

In this chapter, we discuss intrusive thinking in the context of PTSD, OCD, addiction, and schizophrenia. For each of these psychiatric conditions, we outline the typical characteristics and content of intrusive thinking, together with the diagnostic criteria of the respective psychiatric condition. Discussion follows on current theories that aim to explain the mechanisms behind intrusive thinking, and we conclude by reviewing problems and open questions that require future attention.

Intrusive Thoughts in Psychiatric Conditions

Posttraumatic Stress Disorder

One prominent and required diagnostic criterion for PTSD is the presence of intrusion symptoms: a traumatic event is persistently reexperienced. According to DSM-5, Criterion B (American Psychiatric Association 2013), these intrusion symptoms encompass the following characteristics:

- Recurrent, involuntary, and intrusive memories of the traumatic event(s)
- Traumatic nightmares
- Dissociative reactions (e.g., flashbacks)
- Intense distress after being exposed to traumatic reminders
- Heightened physiologic reaction to trauma-related stimuli

To meet the full diagnostic criteria, a person must have been exposed to a life-threatening event and thereafter has to avoid trauma-related stimuli. Further, the person exhibits negative alterations in cognition and mood, which began or worsened after the traumatic event, and displays alterations in arousal and reactivity, such as hyperarousal.

The clinical characteristic of intrusive memory is that it "springs to mind unbidden—that is, against the person's will" (Visser et al. 2018). Such intrusive memories are forms of episodic memories of actually experienced autobiographical events, which are retrieved involuntarily. In its extreme form, the person intensely and vividly relives the traumatic event in the present. Such flashbacks involve the retrieval of detailed sensory features and are highly emotional. Typically, fragments and several distinct moments of the trauma are recalled, the so-called "hot spots," in a predominantly visual form (Visser et al. 2018).

Disturbance in memory seems to be the prominent feature of PTSD. It is widely agreed that multiple memory systems exist and that these rely, in part, on distinct neurobiological substrates (Henke 2010). "Declarative" memories are events and facts, which we can explicitly remember, and seem to depend on medial temporal lobe structures. "Nondeclarative" memories are implicit and not consciously accessible. They encompass procedural memory (e.g., aversive conditioning, motor skills, and habits) and are thought to be subserved by subcortical areas such as the amygdala (aversive conditioning) and striatum (skills and habit formation).

To what degree do different aspects of intrusive aversive memories (e.g., visual imagery or physiological responses) relate to different memory systems? Physiological reactions to trauma-related cues, also listed as intrusive symptoms in DSM-5, Criterion B (American Psychiatric Association 2013), are nondeclarative memories triggered by stimuli. Conversely, "unwanted emotion-laden memories that spring to mind unbidden in the form of sensory imagery" (Visser et al. 2018; see also Figure 14.1, Holmes et al., this volume) belong to the declarative system. Therefore, different neurobehavioral mechanisms most likely underpin the heterogeneous intrusive symptoms that are observed in PTSD. It has been suggested that intervention strategies should specifically target involuntary rather than voluntary retrieval (Visser et al. 2018), both intrusive memory fragments of a trauma as well as the conditioned responses to trauma-related cues are experienced involuntarily. Thus, a successful therapeutic intervention should aim to modify the specific underlying memory traces, while preserving an individual's ability to deliberately recall episodes and facts about the trauma (e.g., for legal reasons).

Obsessive-Compulsive Disorder

The defining features of OCD are repetitive, distressing, and inappropriate thoughts (obsessions) and/or actions (compulsions). According to DSM-5, obsessions are "recurrent and persistent thoughts, urges, or images that are experienced…as intrusive and unwanted, and that in most individuals cause marked anxiety or distress" (American Psychiatric Association 2013). Common foci of obsessions include contamination, pathological doubt, need for symmetry, and aggressive or sexual content. Patients fail to ignore or suppress obsessions and instead attempt to neutralize them through other thoughts or actions, such as by compulsively performing ritualistic behavior to undo the alleged harm of the obsessive thought or intention. Compulsions are thus repetitive behavioral or mental acts, such as checking, washing, or counting. These acts aim to reduce anxiety or distress, but they lack a realistic connection between the act and the goal that a person should achieve. Patients are aware that their compulsions and obsessions are unreasonable and inappropriate. The symptoms are experienced as alien and disturbing (i.e., ego-dystonic); still, they are recognized as being caused by the afflicted patient and not by an external agent (Heinz 1999).

Cognitive theories state that dysfunctional beliefs are the core of OCD and that compulsions develop to reduce anxiety. Appraisal models, for instance, posit that subjects who appraise intrusions as significant and meaningful (based on their dysfunctional beliefs), can develop OCD by escalating intrusions into obsessions (Julien et al. 2007). This hypothesis is based on the observation that nonclinical individuals experience intrusive thoughts, images, or impulses similar in content to individuals with OCD. Other recent theories conceptualize OCD as a disorder of habitual control (Robbins et al. 2019), where post hoc rationalizations of habitual actions contribute to obsessions (Gillan and Robbins 2014). Dual-system theories state that human learning and adaptive behavior are governed by two interacting control systems: one mediates goal-directed actions while the other supports habits (Balleine and O'Doherty 2010; Dolan and Dayan 2013). Habits are performed autonomously and largely independent of their consequences and are, therefore, inflexible to changes in reward contingency. In contrast, goal-directed actions are performed because of anticipated outcomes; thus, they rely on forward planning and allow greater flexibility to changes in contingencies (Friedel et al. 2014). There is a striking similarity between compulsions and habits (Gillan and Robbins 2014): Compulsions are automatic behaviors experienced as irrational and not in line with current goals (i.e., they are ego-dystonic). Behaviors characterized as habits are insensitive to action–outcome contingency and outcome value. Intrusive thoughts are perceived as unintended and not as deliberate and instrumental mental acts; this suggests that a more implicit and nondeliberative way of information processing is involved in intrusive thinking, more akin to a habitual system than a goal-directed control system. Thoughts can be accompanied by sensory qualities (e.g., intrusions in PTSD), or they can be heard aloud, which traditionally has been distinguished from thought insertion and classified as acoustic hallucinations (Jaspers 1946).

The temporal sequence of OCD symptoms seems relevant for an etiological understanding of OCD: Do obsessions come first and compulsive acts follow to reduce negative emotional states, as put forth by the cognitive theories of OCD? Alternatively, do compulsions develop first such that obsessions are secondary rationalizations of the compulsions? There seems to be limited longitudinal data to answer these questions reliably. However, a large proportion of children with OCD deny that their compulsions are driven by obsessive thoughts (Robbins et al. 2019). On the other hand, adult OCD patients report that intrusive images of dirtiness or contamination evoke the urge to wash or neutralize (Coughtrey et al. 2012).

Intrusive thoughts and compulsions present in OCD have been linked to dysfunction of frontostriatal circuits (orbitofrontal/anterior cingulated cortex, dorsolateral striatum/caudate, thalamus) (Robbins et al. 2019). Several neuroimaging studies of OCD revealed hyperactivity in these brain areas during rest (e.g., Baxter et al. 1987) and cognitive performance (e.g., van den Heuvel et al. 2005). Similar neurocircuits were activated during symptom provocation, mainly using visual stimuli to trigger OCD symptoms (Breiter and Rauch 1996). Moreover, tonic overactivity of this circuit seems to be associated with symptom severity (Adler et al. 2000) and predicts treatment response.

Frontostriatal dysfunction normalizes after successful treatment with psychotropic medication (e.g., Swedo et al. 1992) or cognitive behavioral therapy (e.g., Nakao et al. 2005).

In light of these neurobiological findings, Heinz (1999) suggested that dopamine dysfunction in the dorsal striatum is associated with motor tics, whereas more complex compulsive behavior patterns are triggered by cognitive concerns processed in the orbitofrontal cortex, which persist obsessively due to impaired feedback processed in dorsal striatal-thalamic-frontocortical loops. This neurocircuit model may help unify current theories that focus on cognitive versus habitual aspects of OCD.

Depressive Disorders and Rumination

Rumination is a maladaptive form of self-reflection and recursive self-focused thinking. Like obsessions, rumination involves recursive thinking about particularly self-centered negative information. However, whereas obsessive thoughts are usually experienced as aggressive or otherwise inappropriate (and are hence "unwanted" by the afflicted person), rumination often focuses on threatening environmental conditions as well as inappropriate or unfavorable character traits of the person (and thus involve self-blame for not being able to cope with the situation). Rumination has been hypothesized as an important factor in developing depressive symptoms and has been shown to exacerbate depression, enhance negative thinking, and impair problem solving (Nolen-Hoeksema et al. 2008). Self-focused rumination in depressive disorders has been linked to self-referential processes and the brain's default-mode network, particularly involving the subgenual prefrontal cortex, a brain area implicated in the processing of aversive information and the modulation of negative mood states (Kühn et al. 2013; Hamilton et al. 2015). Accordingly, positron emission tomography studies have revealed increased metabolic activity in this brain area among subjects with major depression (Drevets et al. 2008).

Major depression and OCD appear to differ with respect to their neurobiological correlates, which may reflect differences in relevant pathological mechanisms: In major depression, rumination may result from a failure to regulate aversive information input and associated personal concerns. In OCD, compulsive behavior appears to be triggered by obsessive thoughts aimed at compensating for the aggressive or otherwise unwanted content of these obsessions, yet fails to dampen concerns, which triggers repetitive action, resulting in reverberating circular interactions. In depressive disorders, other instances of intrusive thinking include suicidal ideations; mental images of killing oneself ("flash-forward" thoughts) have been shown to be associated with suicidal behavior (for further examples of mental imagery, see Holmes and Mathews 2010).

Addiction: Craving and Compulsivity

Intrusive thinking in addiction disorders includes thoughts related to drug consumption associated with a strong desire (craving) to consume a drug despite having reached a conscious decision to abstain from drug use (Heinz 2017). In this context, craving is defined as a strong desire or urge to use a drug or to engage in harmful activity, such as gambling for monetary reward. Further criteria for addiction include impaired control regarding substance intake, neglect of unrelated activities, tolerance, and withdrawal. Despite conscious decisions to do otherwise, recurrent substance use is a key characteristic of addiction, and it has been suggested that drug use becomes compulsive when subjects lose control over the powerful urge to consume a drug of abuse despite of aversive consequences (Everitt and Robbins 2016).

Indeed, a gradual shift from outcome-sensitive, goal-directed behavior to habitual behavior can contribute to automatic, habitual, or even compulsive drug intake, despite foregone positive outcomes and devastating (future) negative consequences. Drug-associated stimuli may acquire enhanced salience and act as appetitive Pavlovian cues that trigger automatic approach behavior (Robinson and Berridge 1993). Moreover, these environmental cues can impact (goal-directed) choice selection and behavioral adaptation through Pavlovian-instrumental transfer mechanisms, where affectively positive Pavlovian cues bias (unrelated) goal-directed behavior toward approach even when this is not useful in the instrumental context (Garbusow et al. 2016).

While healthy controls are able to arbitrate control between the habitual and the goal-directed system, a loss of control over certain behaviors (e.g., drug intake) might be due to a shift from goal-directed toward habitual control (Balleine and O'Doherty 2010; Dolan and Dayan 2013; Voon et al. 2015). Computational neuroscience uses "model-based" and "model-free" algorithms to explain goal-directed and habitual learning during, for example, sequential decision-making tasks (e.g., Friedel et al. 2014). A model-based algorithm views the environmental (or task) structure used for deliberative forward planning as a hallmark of goal-directed behavior, which in the case of sequential decision making refers to the transition from one environmental state to another. Model-free reinforcement learning algorithms reflect a retrospective and more rigid strategy that neglects environmental structures and relies solely on repeating previously rewarded actions. Initial studies in patients with different addictive disorders, including dependence on psychostimulants and alcohol, suggest impaired model-based control, thus shifting the behavior toward a model-free response (Sebold et al. 2014; Voon et al. 2015). However, in a more recent study, Sebold et al. (2017) found neither a general bias toward model-free (supposedly habitual) decision making in patients with alcohol dependence nor a poor treatment outcome associated with impaired model-based decision making. Instead, the balance between habitual (supposedly model-free) and goal-directed decision making differentiated alcohol-dependent patients (who later relapsed) from abstainers and controls only when individual alcohol expectancies were considered. This suggests that habitual behavior does not generally increase in addicted patients during choice behavior tests for non-drug-related rewards in a laboratory context; addiction-related habits appear to be triggered by specific cues and contexts in conjunction with previous experiences. In line with this hypothesis, a patient suffering from OCD, pathological gambling, and drug addiction described constant urges to perform habits compulsively related to his obsessions, while craving for gambling and drug intake was triggered only during certain time periods by specific drug or gambling-related stimuli (Schoofs and Heinz 2013). In light of these findings, compulsions in OCD appear to differ significantly from "compulsive" urges to consume drugs of abuse, warranting further phenomenological and neurobiological specifications (Heinz 2017).

Schizophrenia and Related Psychotic Disorders and Thought Insertions

Thought insertion is a positive symptom of schizophrenia and is regarded as a "first rank symptom" of the disease (Schneider 1959; Heinz et al. 2016). Not only does thought insertion constitute one of the most astounding positive symptoms of schizophrenia, it is frequently expressed. It occurs in approximately half of all patients diagnosed with schizophrenia (Sartorius et al. 1977), but appears to be absent in organic psychoses (Marneros 1988; Heinz et al. 1995).

In psychosis, patients typically report that thoughts are being "inserted" (in verbal form) by another agent into their head. Patients thus lose the feeling of "mineness" for a given thought; this marks a distinct difference between obsessions in OCD, ruminations in major depression, or drug cravings in addiction, all of which are "unwanted" and uncontrolled by the afflicted subject but not experienced as "alien" and attributed to outside agents. Vosgerau and Voss (2014) highlight the distinction between control, ownership, and authorship of thoughts. They argue that it is a conceptual truth that introspected thoughts are necessarily owned by the introspector (therefore ownership of thoughts cannot be disturbed), whereas lack of authorship over thoughts can be experienced in everyday phenomena (thinking "communicated thoughts," i.e., thoughts clearly formulated by another person) as well as in pathological conditions such as psychosis. By introducing another factor (e.g., control over thoughts), Vosgerau and Voss (2014) argue that the phenomenon of thought insertion is caused by a combination of these two factors—lack of control and lack of authorship—and that there is a double dissociation between both factors.

In an attempt to reveal the neurocognitive mechanisms underlying thought insertion, Campbell (1999) drew an analogy between thoughts and motor control processes and explained thought insertion in relation to the comparator model, originally developed for motor control (Frith et al. 2000). Campbell assumed that thoughts are comparable to motor processes (similar views were expressed by Feinberg 1978 and Ito 2008) and that every thought is preceded by an intention to think this thought. The actual thought occurring in one's stream of consciousness is then compared with the intention to think. When these two processes match, a feeling of authorship results; when they do not, attribution to another agent may occur.

While this offers an appealing framework to explain passivity phenomena, such as delusions of control or hallucinations, several problems arise when the comparator model is used to explain thought insertion (for a detailed critique, see Vosgerau and Synofzik 2010). One problem is that the account does not distinguish thinking (as a process to generate thoughts) from thoughts (as a result of thinking), making it difficult to pinpoint the difference between "influenced" thinking and "inserted" thoughts. Furthermore, it remains unclear what an intention to think a specific thought could be and how it can be distinguished from the actual thoughts; that is, why the intention to think is not naturally conceived of as the thought itself. Indeed, if every thought were to be preceded by an intention to think, we would run into an infinite regress: for each thought, we need a thought to get the process started, which in turn presupposes another thought, and so on.

A more recent attempt to conceptualize the phenomenon of thought insertion treats inserted thoughts as sensory events rather than motor processes (Sterzer et al. 2016). Building on detailed phenomenological descriptions from the early Heidelberg school in the first half of the twentieth century, which described thoughts that "become sensory" as being experienced as inserted, more recent studies explain thought insertion within the framework of predictive coding and Bayesian inference (Sterzer et al. 2016). Here, thought insertion is viewed not as the failure of introspection in a comparator process (Campbell 1999) but rather as the failure (or imprecision) of prior beliefs in a Bayesian inference process thought to be at the core of thought insertion (Sterzer et al. 2016). In analogy with aberrant salience attribution to external events, which could lead to the emergence of delusional mood and fixed beliefs (Heinz 2002; Kapur 2003; Heinz and Schlagenhauf 2010), internal events (e.g., verbalized thoughts) may also be experienced as overly salient, and therefore unusual, as well as surprising due to a lack of context and unusual structure (with a possible link to formal thought disorder). The individual's attempt to explain the aberrant salience and unusual character of such verbalized thoughts could result in their interpretation as being externally caused. Nonetheless, unintended or semantically inappropriate verbalizations can also be expressed in aphasia (e.g., due to a stroke) but are not accompanied by reports of "alien" involvement (Heinz 2017). Therefore, additional steps may be required to convince a person that a thought is "alien" and thus "must" be inserted by an external agent. In this context, it has been suggested that low precision of prior beliefs and/or increased sensory precision, both inaccessible to introspection, may render some thoughts so unpredictable that they are experienced as inserted (Heinz et al. 2019). Whereas beliefs and desires may have a role in prior beliefs regarding our own thoughts (Stephens and Graham 2000), what makes a thought feel alien is not that they are simply unwanted or "immoral" (in the sense of higher-level introspection), but rather that patients directly notice or perceive a thought to be "alien," thus pointing to unconscious mechanisms causing this experience. Such mechanisms may be usefully described by a Bayesian account of information processing in the central nervous system (Adams et al. 2013): during psychotic episodes, if prior beliefs are indeed imprecise in comparison to sensory input-driven posteriors, frequent prediction errors are made, which may trigger phasic dopamine release and cause salience to be attributed to otherwise irrelevant cues, including verbalized thoughts (Sterzer et al. 2016; Heinz et al. 2019).

If these considerations are correct, they may help to explain the difference between "alien" thought insertion and other forms of unwanted or intrusive thought content. In psychosis, impaired precision of prior beliefs may affect the whole experience of the world and self, which are experienced as unusual, alien, and often threatening. Thus, thought insertion in psychosis goes beyond the experiencing of unusual verbalizations (as in aphasia), unwanted cravings (in addiction), self-centered concerns (in major depression), intrusive memories (in PTSD), or obsessive thoughts (in OCD). The entire relationship between the individual and the real world is affected: that which was well known for a long time suddenly carries hidden meanings, harmless situations are imbued with a sense of danger, and longtime friends and family members become deeply alienated and may no longer be trusted. Since Bayesian accounts are supposed to reflect general functions of the central nervous system, such computational frameworks will have to explain how more restricted alterations in information processing, particularly with respect to verbalized thoughts, differ from the more fundamental alterations experienced in psychotic states.

Open Issues

To guide future enquiry, we conclude our discussion by highlighting unsolved problems that await clarification through future research. First, despite striking phenomenological differences between negative verbal thoughts, intrusive visual images, and memories, we need to know whether the underlying psychological and neurobiological mechanisms involved in intrusive thinking are similar across diagnostic categories. In support of shared transdiagnostic mechanisms, Gillan et al. (2016) has shown that reduced goal-directed control is associated with compulsive behaviors and intrusive thoughts.

Second, the relation between intrusive thinking and the concept of compulsivity needs to be elucidated, where compulsivity is defined as "a hypothetical trait in which actions are persistently repeated despite adverse consequences" (Robbins et al. 2012:82). Are intrusive thoughts one manifestation of compulsivity? If so, when is it expressed in maladaptive actions and when in intrusive thoughts? Can intrusions be understood as mental habits?

Finally, in psychosis, how can we mechanistically isolate disturbed authorship (or disturbed "mineness") from disturbed control over thoughts? Importantly, measurement instruments have to be harmonized on the clinical, psychopathological, behavioral, and neurobiological levels.

9

Neuropsychological Mechanisms of Intrusive Thinking

Renée M. Visser, Michael C. Anderson, Adam Aron,
Marie T. Banich, Kathleen T. Brady, Quentin J. M. Huys,
Marie-Hélène Monfils, Daniela Schiller, Florian Schlagenhauf,
Jonathan W. Schooler, and Trevor W. Robbins

Abstract

A classic definition of intrusive thinking is "any distinct, identifiable cognitive event that is unwanted, unintended, and recurrent. It interrupts the flow of thought, interferes in task performance, is associated with negative affect, and is difficult to control" (Clark 2005:4). While easy to understand and applicable to many cases, this definition does not seem to encompass the entire spectrum of intrusions. For example, intrusive thoughts may not always be experienced as unpleasant or unwanted, and may in some situations even be adaptive. This chapter revisits the definition of intrusive thinking, by systematically considering all the circumstances in which intrusions might occur, their manifestations across health and disorders, and develops an alternative, more inclusive definition of intrusions as being "interruptive, salient, experienced mental events." It proposes that clinical intrusive thinking differs from its nonclinical form with regard to frequency, intensity, and maladaptive reappraisal. Further, it discusses the neurocognitive processes underlying intrusive thinking and its control, including memory processes involved in action control, working memory and long-term memory encoding, retrieval, and suppression. As part of this, current methodologies used to study intrusive thinking are evaluated and areas are highlighted where more research and/or technical innovation is needed. It concludes with a discussion of the theoretical, therapeutic, and sociocultural implications of intrusive thinking and its control.

Group photos (top left to bottom right) Renée Visser, Michael Anderson, Marie-Hélène Monfils, Trevor Robbins, Daniela Schiller, Marie Banich, Adam Aron, Kathleen Brady, Jonathan Schooler, Florian Schlagenhauf, Quentin Huys, Michael Anderson, Daniela Schiller, Jonathan Schooler, Kathleen Brady, Quentin Huys, Florian Schlagenhauf, Marie-Hélène Monfils, Trevor Robbins, Renée Visser, Marie Banich

Introduction

While reading this chapter, you might be sitting in a coffee shop somewhere or traveling on a train. The sunlight comes in through the window and you envision how warm the planet might get. You are having a meeting in a few hours and various scenarios of it are running through your head. Perhaps these few examples have triggered some of your own thoughts, as you are no longer reading this article: your eyes are just glazing over these words. If this just happened and you rejoin us a few sentences below, you have just experienced an intrusive thought.

We all seem to have an intuitive understanding of what an intrusion is. Yet, attempts to define it immediately sets off debate. According to the one textbook that has been written on this is, the classic definition of an intrusive thought is "any distinct, identifiable cognitive event that is unwanted, unintended, and recurrent. It interrupts the flow of thought, interferes in task performance, is associated with negative affect, and is difficult to control" (Clark 2005:4). Central to this definition is the assumption that an intrusive thought always constitutes an unpleasant experience which negatively impacts functioning. Here, we revisit the definition of intrusive thinking, by considering all the circumstances in which intrusions might occur, their manifestations across health and disorders, and their neurocognitive basis. We start with a rather narrow, presumably more commonly accepted definition of intrusions being *conscious, involuntary, unwanted thoughts,* and arrive at an alternative, more inclusive definition of intrusions as *interruptive, salient, experienced mental events.* We discuss current methodologies used to study intrusive thinking, highlighting existing strengths as well as areas where more research and/or technical innovation are needed. We conclude with a discussion of the theoretical, therapeutic, and societal implications of intrusive thinking and its control.

What Are the Everyday Manifestations of Intrusions and Their Control?

Although intrusive thoughts are often associated with mental health disorders, they also occur to healthy individuals in everyday life (Purdon and Clark 1993; Berntsen 1996). Identifying broad circumstances under which these thoughts occur to people in general, irrespective of mental health status, may highlight their adaptive functions in healthy individuals and illuminate the processes that generate such intrusions. Moreover, considering the motivations that people may have for controlling intrusions in daily life can shed light on important psychological and social functions that the capacity for control helps to support.

Types of Content That Drive Recurring Intrusive Thoughts in Healthy Individuals

Although unintended thoughts can occur for any content, not all such thoughts are intrusive and bothersome. There are, however, certain contents that consistently trigger intrusive thoughts in otherwise healthy people. Here we discuss several examples and consider why this content is intrusive.

Emotionally Salient Events. Events that trigger intense emotions, such as psychological trauma, often lead to intrusive reminders, and the period of their intrusiveness extends for varying durations, depending on the event. Emotional events that trigger unwelcome intrusions are typically negative (e.g., anger, guilt, shame, sadness, fear, and embarrassment). However, positive events are also capable of triggering repetitive thoughts about pleasant memories, and this can be quite disruptive, especially when it interferes with focused attention that people need to perform certain tasks or activities (e.g., job responsibilities).

Incompletions. When people initiate a process and are then unable to complete it, due to an interruption or some other impasse, thoughts related to that incompletion tend to recur until the process is completed (Horowitz 1975). This hypothesis is reflected in the classical Zeigarnik effect proposed in the early twentieth century by Gestalt psychologists, who posited that people had superior memory for interrupted processes (Zeigarnik 1938). Incompletions can include both physical and mental tasks, as long as there is an unresolved problem. An interesting possibility is that the tendency for incompletions to precipitate intrusive thoughts may be amplified by the salience or emotional intensity of the incomplete process (Horowitz 1975). Intrusive thoughts related to the interrupted process (i.e., seeking comprehension, solving a problem) may persist until the situation is understood. This suggests that emotionally intense events may not intrude merely due to salience or encoding strength, but because they are accompanied by a compelling desire to comprehend them.

Relatedly, when people reach a significant impasse in solving a difficult problem, they continue to "work on the problem" in the background. For example, research on creativity and insight problem solving suggests that when people allow a period of incubation, insights may emerge spontaneously (the "aha" phenomenon), as though a process has occurred in the background (Gable et al. 2019). The commonly experienced "tip of the tongue" phenomenon in memory retrieval (described initially by W. James and S. Freud) is another example of when a temporarily forgotten item finds its way back into consciousness unpredictably. The content of people's mind wandering (i.e., when thoughts distract us from a task at hand) often includes unresolved problems salient to the individual, whether emotionally significant or not (Smallwood and Schooler 2006; Klinger and Cox 2011).

Intentions. A special case of "incompletions" worth distinguishing concerns incompletions arising from intended actions that must be deferred, an ability studied in research on prospective memory. This research indicates that although people often rely on the environment to remind them to perform intentions, or use intentional self-reminding strategies (e.g., lists), thoughts of the intended action can still pop into a person's mind, unbidden, particularly when intended actions have high importance or affect. Such cases suggest that intentions are maintained in an elevated state of accessibility, even without conscious rehearsal. Consistent with this possibility, deferred intentions often have a special active status in memory (outside of intentional rehearsal) that disrupts ongoing task performance, as illustrated by *intention interference* (Goschke and Kuhl 1993; Cohen et al. 2011; Bugg and Streeper 2019).

Anticipated Events. Although intrusive thoughts often involve past events or general ideas, they also concern events that have yet to happen. The associated anticipation can carry emotionally positive, negative, or even both consequences. Under some circumstances, anticipated events elicit conflicting feelings of excitement and apprehension in parallel. For example, an upcoming biopsy may return a diagnosis of cancer or good health.

Uncertain Events. When people are uncertain about a past or future event, this can promote intrusive thoughts for several reasons (Grupe and Nitschke 2013). For example, if someone is uncertain about whether they have already performed an action (e.g., locking the front door or taking one's pills), associated uncertainty may precipitate worry. Uncertainty about future outcomes can also result in persistently intruding thoughts, which in turn may prepare people for different outcomes, thus enabling them to be ready to respond appropriately.

Dissonant Facts, Events, or Beliefs. When a new fact, experience, thought, or impulse conflicts with one's beliefs or self-image, the resulting dissonance creates a tension that must be resolved. For instance, if a person commits an action that conflicts with their self-image, the implications for how they should revise their self-perceptions can be distressing. Are they the good person they think they are or not? Dissonance creates conflict that can trigger recurring automatic thoughts until the conflict is resolved. The desire to resolve the mental discomfort and stress induced by such discrepancies suggests that people strive for psychological consistency, an idea first proposed by Leon Festinger (1962) in his theory of cognitive dissonance.

Frequent Events, Stimuli, or Ideas. Not every intruding thought is emotional or related to incompletion or dissonance. Sometimes, the frequency of an event, image, or thought induces further repetitions. One striking example is the repeated hearing, in our imagination, of a song that we have recently heard on numerous occasions, known colloquially as "ear worms" (Hyman et al.

2013, 2015; Moeck et al. 2018). Research suggests that whether an ear worm develops is related to how often the song has recurred in recent experience.

Images.   As the foregoing examples illustrate, intrusive thoughts can occur in many representational formats. In some cases, intrusive thoughts can be verbal or propositional in nature, as might occur in persisting thoughts about incomplete processes, unsolved problems, deferred intentions, or incongruous, dissonant occurrences. In other cases, intrusive thoughts can be more sensory in character, including intrusions from any sensory modality. These mental images may play a special role in intrusive thinking, a possibility suggested by their prevalence in psychiatric disorders. Mental images are experiences of perception that occur in the absence of external sensory input (Kosslyn et al. 2006; Pearson et al. 2015). They are not limited to remembering the past, but can also include imagined future scenarios. While often benign, highly emotional images (e.g., of an upsetting event) can set off a cascade of other disruptive cognitive and emotional processes, including increased physiological responses and rumination (Lang 1977; Grey and Holmes 2008; Holmes and Mathews 2010; Ji et al. 2016; Holmes et al. 2017). Later, in discussions of clinical manifestations of intrusive thoughts, we revisit intrusive imagery in diverse psychiatric disorders.

**External and Internal Factors Triggering Intrusive
Thoughts in Healthy Individuals**

The previous discussion reviews content that is especially prone to generate intrusive thoughts, but it is worth separately considering the conditions that generate intrusions. In general, the likelihood of intrusions occurring in a given moment is related to both the presence of retrieval cues as well as the availability of control resources to resist unwanted intrusions. Here we discuss the critical role of cue-driven retrieval, matching physiological and mood state, monitoring processes that may cue thoughts, and the availability of inhibitory control.

Cue-Driven Retrieval.   Intrusive thoughts are often triggered by associations to environmental cues (Berntsen 1996). The most obvious example arises when a stimulus elicits an unwelcome reminder of a past event. For instance, after an argument, a friend's face may remind someone of the altercation, or a song may bring back memories of a loved one who passed away. Although intrusive memories provide clear cases of cue-driven retrieval, other forms of intrusive thoughts, such as feared future events, or unpleasant ideas or images, also seem to be cue driven. When *the same content intrudes more than once*, the content is reemerging from memory of previous thoughts. One can frequently identify specific cues in the environment, or in a person's patterns of thinking, that drive retrieval of perseverative content (e.g., Gagnepain et al. 2014; Benoit et al. 2016).

Matching Mood and Physiological State. Sometimes the cues that trigger retrieval are not specific stimuli in the world or concepts in memory, but rather are broad psychological or physiological states (Berntsen 1996). It is well established that information and experiences are often stored in memory in association with a representation of the context in which they took place (for a review, see Anderson and Hanslmayr 2014). This can include not only the spatiotemporal context (i.e., the environment), but also the mood, drug, or arousal state present at encoding. Later on, the chance of retrieving the content is higher when a person's state resembles the one at encoding. For example, experiences encoded in a sad or angry mood are more likely to be recalled at a later time, when a person is once again in a sad or angry mood than when they are in a happy mood.

Diminished Cognitive Control. In addition to the particular cues at retrieval and their match to encoding, other elements of the state of the person at retrieval may influence the frequency and persistence of intrusive thoughts. An important category of state-related variables that can influence intrusion frequency is whether a person is suffering from diminished cognitive control that might otherwise help the person to limit involuntary retrievals. If mechanisms such as inhibitory control prevent or reduce intrusions, then anything that compromises these abilities is a risk factor for intrusiveness, even in healthy individuals.

Several factors common in healthy samples may give rise to such deficits in cognitive control, including sleep deprivation (Nilsson et al. 2005; Drummond et al. 2006), general fatigue, stress (Shields et al. 2016), lack of exercise (Hillman et al. 2008), and intoxication. Although such changes in state are temporary for most healthy individuals, chronic depletion of cognitive control, especially with several of the above factors contributing, could put healthy individuals on a trajectory to develop persistent intrusions and changes in mood that are clinically significant. In addition, consistent, trait-related deficits in cognitive control may give rise to significant risks in controlling intrusive memories and thoughts, and may be a risk factor for psychiatric disorders (Levy and Anderson 2008).

Desirability of Control

In healthy individuals, when intrusive thoughts occur, they are usually perceived as unwanted, at least at that moment. As a result, people often try to exclude them from awareness in an effort to regain control over thoughts and emotions. When successful, such efforts enable a person to put unwelcome thoughts out of mind, thus diminishing their accessibility in memory and reducing their tendency to return. Such attempts to facilitate the forgetting process often serve important behavioral, emotional, and social functions. Here, we consider several contexts in everyday life in which people are motivated to forget thoughts for a functional reason.

Concentration during Tasks.   Sustained focus on a necessary task can be difficult when unrelated distracting thoughts intrude into awareness (Smallwood and Schooler 2006). Such thoughts can be of any valence. For example, receiving good news or bad news prior to leaving for work can undermine our focus. If distracting thoughts slow progress or increase the propensity for errors, it becomes essential and adaptive to put the distracting thoughts out of mind. Concentration is an extremely common selective attentional mechanism for trying to control intrusive thoughts, and successfully achieving concentration is clearly adaptive.

Executing High-Performance Cognitive and Motor Skills.   Professional athletes are often under extreme pressure to perform to a very high standard, and when they do not, there can be significant consequences for themselves or their teammates. In sports psychology, a literature on the causes of "choking" under pressure (Beilock and Gonso 2008) has emerged to address the processes that lead an athlete's performance to deteriorate under pressure. This often includes the inability to overcome intrusive thoughts that undermine the focus needed for top performance.

Regulating Pain.   Thoughts about pain, what it feels like, how uncomfortable it is, and worries about what it may mean often intrude persistently, even after the pain has resided. We consider people who can manage intrusive thoughts about their pain to be resilient and able to cope. In contrast, those who are overwhelmed by pain, who allow their discomforts to blossom into catastrophic thinking, not only pay a price for this distraction, their suffering becomes magnified and extended (Edwards et al. 2006). When confronted with a painful stimulus (e.g., the cold pressor task), people with a moderate history of adversity tolerate the pain longer and show less pain catastrophizing (the label given to excessive thoughts about the pain), relative to people who have no history of trauma (Seery et al. 2013). It appears, therefore, that in the face of physical discomfort, resilience emanates from the capacity to control intrusive thoughts about the painful stimulus, which not only reduces suffering but increases the ability to pursue other goals.

Regulating Affect.   When unwelcome thoughts intrude, they often evoke unwelcome changes in emotional state. For example, being reminded of an argument may trigger anger; images of an upcoming doctor's visit may trigger fear or anxiety; or seeing the same car that your ex-partner used to drive may evoke sadness. As a result, intruding thoughts trigger mechanisms that regulate emotion to return a person to a neutral or positive state. Although these endogenously triggered emotions are common and contribute substantially to psychological disorders, they have been neglected in research on emotion regulation, which has focused on regulating affect triggered by external emotion-eliciting stimuli. Understanding how intrusive thoughts can be downregulated in memory may provide critical insights into how people regulate the emotions that they elicit, contributing to successful *emotional homeostasis* (Engen and Anderson 2018).

Persisting in the Face of Failure. When memories or thoughts of past failures dominate someone's thoughts, they can be disruptive. Intrusive thoughts about failure are unpleasant and undermine feelings of competence and control; this may lead a person to abandon their goals earlier than they should or to fail to improve their performance. When someone can set aside past failures, this enables them to improve their skills or knowledge or to find creative solutions to the problems that led them to fail. Persistence, especially dogged persistence in the face of challenges and setbacks, requires successful regulation of thoughts. If persistence is an adaptive trait enabling major personal achievements, the inability to control thoughts of failure can limit personal growth.

Protecting Self-Image. People sometimes feel embarrassed or ashamed about things that they have done. Alternatively, others may say hurtful things that undermine self-confidence. Although people vary in how they respond to such events, they often try to put the unwelcome content out of mind. Correspondingly, they show greater forgetting for negative feedback, exhibiting a remarkable capacity to not remember their faults or misdeeds. Work on mnemic neglect (Sedikides and Green 2009; Sedikides et al. 2016) has established this pattern of forgetting and linked it to threats to a person's self-concept: recent threats to one's self-image are forgotten. Moreover, work on the positive self-illusion (Taylor and Brown 1988) has linked this tendency to improved mental and physical health: healthy individuals have a higher view of their capabilities than would be supported by observers. In contrast, depressed individuals, ironically, often have a more accurate view of their circumstances. Similarly, most individuals show a very powerful and replicable positivity bias in terms of the autobiographical memories that they ultimately retain (Walker et al. 2003), whereas depressed individuals very often show the reverse tendency.

Justifying Inappropriate Behavior. Sometimes people commit unethical, immoral, or hurtful acts about which they feel shame. Despite this, most people like to think of themselves as decent, creating dissonance between one's beliefs and deeds. Intrusive thoughts about these discrepancies often develop and must be addressed. Research on ethical amnesia indicates that people often forget their ethical lapses (Kouchaki and Gino 2016; Stanley and De Brigard 2019), suggesting that people suppress these uncomfortable thoughts.

Maintaining Attitudes and Beliefs. A person's political, religious, or personal beliefs are often resistant to contradictory evidence. When new facts contradict our beliefs, dissonance is created that must be resolved. This resolution often involves controlling intrusive thoughts to forget the inconvenient information. For example, Republicans and Democrats show enhanced directed forgetting for attitude statements that are incongruent with their beliefs, compared with congruent statements (Waldum and Sahakyan 2012). Moreover, a person's memory can be shaped by selectively recounting an event (Cuc et al. 2007; Stone et al. 2012, 2013), a form of thought substitution (Benoit and Anderson 2012; Anderson 2001). Intriguingly, this can undermine memories of omitted facts, a phenomenon known as *socially shared retrieval-induced forgetting* (Cuc et al. 2007; Stone et al. 2012, 2013). The selective forgetting that people exhibit for facts that are incompatible with their beliefs suggests that healthy individuals control their thoughts to avoid the discomfort of inconsistency.

Forgiving Others and Maintaining Attachment.   Sometimes friends or relationship partners commit offenses that provoke anger, giving rise to persistent intrusive thoughts. Indeed, intrusive thoughts about anger can be intense for some people and can take considerable time to "get over," placing strain on personal relationships and requiring the ability to override thoughts and impulses originating from the anger. This ability is predicted by individual differences in inhibitory control, suggesting that overcoming intrusive thoughts involves, in part, control processes that suppress them in service of forgiveness. Consistent with this possibility, when people decide that an offense may be forgiven, it is easier to suppress intrusive thoughts about the offense, ultimately leading to worse memory (Noreen et al. 2014). Thus, suppressing intrusive thoughts of anger ultimately contributes to a healthy capacity to forgive and forget in social relationships.

Similar considerations apply when there is a need to maintain an attachment relationship with a parent, guardian, or powerful authority figure (e.g., a boss), which may be essential to survive or thrive in an environment. In such cases, the capacity to control recurring intrusive thoughts enables a person to maintain normal relationships despite this conflict and, in some cases, to forget about the discrepant content.

Mind Wandering: An Everyday Manifestation of Intrusive Thinking?

One interesting example of everyday intrusive thinking is the widely studied experience of mind wandering. Returning to our example at the beginning of this chapter, we are all familiar with suddenly realizing that while our eyes have been moving across the page, our minds have been temporarily sidetracked by thoughts unrelated to the text. Such a situation is known as *mind wandering* (Smallwood and Schooler 2006). Using thought sampling techniques in which individuals are intermittently queried as to whether their thoughts are engaged in the task at hand, research suggests that people mind wander up to fifty percent of their waking hours (Killingsworth and Gilbert 2010). The surprising frequency with which people engage in mind wandering raises an intriguing question: Should mind wandering be viewed as everyday intrusive thinking? Let us consider the following aspects of this question.

Unwanted. Mind wandering is often unwanted because it disrupts task performance across many cognitively demanding domains (for a review, see Smallwood and Schooler 2015). It is, for instance, a major source of car accidents (Gil-Jardiné et al. 2017). Yet while mind wandering interferes with current tasks, it can be useful for completing more distal goals, such as planning (Baird et al. 2011) and creative problem solving. For example, when a vigilance task was interposed between two trials of a creativity task, mind wandering interfered with the vigilance task but enhanced performance on the second round of the creativity task (Baird et al. 2012). In a diary study, both creative writers and physicists indicated that twenty percent of their ideas occurred outside of work (Gable et al. 2019). Thus, although mind wandering can interfere with the task at hand, it often contributes to progress on other problems.

Unintended. Individuals frequently report mind wandering and are often not even aware that they are engaged in it, until they are caught, as in the experience sampling probe (Schooler et al. 2011). Nevertheless, individuals may sometimes deliberately abandon a task in favor of other thoughts (Seli et al. 2015). The lack of awareness that one is mind wandering (i.e., lack of meta-awareness; Schooler 2002) contributes to more disruptive mind wandering episodes (Schooler et al. 2011), which are neurologically distinct from on-task thinking (Christoff et al. 2009). Notably, lack of meta-awareness has been associated with mind wandering about unwanted thoughts (Baird et al. 2013a). This suggests that intrusive thoughts may similarly slip below the radar of meta-awareness (Takarangi et al. 2014).

Recurrent. Although up to fifty percent of people's waking hours is spent mind wandering (making mind wandering a recurring form of thought), the content of mind wandering routinely varies from one episode to the next. Hence, whether mind wandering is recurrent depends on the level of analysis.

Associated with Negative Affect. In their influential paper, Killingsworth and Gilbert (2010) hold that "a wandering mind is an unhappy mind." This characterization reflects the observation that when individuals mind wander, they are routinely less happy than when they are on task. Nevertheless, the degree to which episodes of mind wandering drive negative affect may be partly determined by content. For example, Franklin et al. (2013) found that participants were generally less happy when mind wandering than when on task, unless they were mind wandering about something interesting, in which case they were actually happier.

Difficulty of Control.   Individuals' perceptions regarding the controllability of mind wandering varies considerably (Zedelius and Schooler 2017), and these perceptions influence its frequency and impact. Individuals who view mind wandering as outside of their control mind wander more often, and their performance is more disrupted than those who view it as a mental state they can manage. Collectively, evaluation of the relationship between mind wandering and the definition of intrusive thoughts indicates that mind wandering is often, but not always, a form of everyday intrusive thought.

Summary and Commentary

As the above discussion illustrates, intrusive thinking occurs in a range of circumstances in healthy individuals. Such thoughts do not, by themselves, indicate a mental health disorder but are a normal feature of life. Moreover, attempts to suppress intrusive thoughts are often not only highly successful in healthy individuals, they are in fact essential to achieving numerous social, emotional, and cognitive goals, including the successful regulation of emotional state (Engen and Anderson 2018). Not all healthy people are equally effective at controlling intrusive thoughts, however, and vulnerability to persistent intrusions is critical to understand. These observations highlight the importance of understanding when and how intrusive thoughts become pathological and contribute to psychological disorders.

What Are the Main Manifestations of Intrusive Thinking in Mental Health Disorders?

As indicated in the preceding sections, occasional intrusive thoughts are normal occurrences and even commonplace in certain circumstances. However, there are a number of mental health disorders for which intrusive thinking is a core symptom, becomes problematic, and can interfere with function. Some common disorders in which intrusive thinking is a critical element are briefly described below; for further detail, see Schlagenhauf et al., Banich, as well as Brewer et al. (this volume).

Posttraumatic Stress Disorder

Symptoms of intrusions in posttraumatic stress disorder (PTSD) include recurrent, unwanted, involuntary, and distressing memories of a traumatic event (intrusive memories); reliving the traumatic event viscerally, as if it were happening again (flashbacks); upsetting dreams or nightmares about a traumatic event; and severe emotional distress or physical reactions to cues that remind an individual of the event (e.g., the smell of burned rubber reminding someone of a car accident or a ski mask that triggers memory of an armed robbery).

Intrusive memories or flashbacks are often precipitated by environmental cues that remind an individual of the traumatic event. Flashbacks are immersive, causing individuals to "dissociate" from their bodies and feel that they are back in the situation in which the trauma occurred. It is not clear whether intrusions and flashbacks are qualitatively different phenomena or fall on a spectrum of severity in which flashbacks represent a severe form of intrusions. The intensity and content may be different, yet both are multimodal sensory images that are interruptive, salient, experienced, unwanted, involuntary, and often recurrent.

Obsessive-Compulsive Disorder and Related Disorders

In obsessive-compulsive disorder (OCD), intrusive means the feeling of "being out of control." This feeling is experienced in different phenomenological domains since OCD develops over time. OCD is a process with different clinical stages rather than one single stage. These stages follow a dialectic interaction in which an intrusive event elicits a response and the response amplifies the intrusive event. Through different neurobiological adaptations, the course of OCD eventually worsens. Thus, OCD should be regarded as a disease process that develops through the amplifying interaction between (the reflection and resistance of) the person (mind) and the disorder (brain). The following example describes how intrusions may develop in OCD.

A young mother recently gave birth for the first time and now carries the responsibility of brand-new motherhood. Her partner leaves daily for work, and thus she is home alone with the child. Viewing her young baby in the crib, a thought appears in her mind: she imagines that she could strangle her baby in the crib and that no one is there to prevent her. The mere presence of the thought is intrusive because it occurs beyond her free will. She feels out of control and is unable to control her thinking. She is worried because the idea of strangling her baby does not fit her ideal of motherhood. The content of the thought is intrusive and ego-dystonic because it is not in line with her self-image. She wonders whether she could really strangle her baby and begins to feel out of control. How could she be certain that she would not strangle her baby, given the fact that humans are notoriously unpredictable? As the implication of the thought is intrusive, she feels anxious because the thought confronts her with being out of control.

The emotional value (anxiety) of the thought is intrusive. The presence, content, implication, and emotional value of the thought all have an intrusive quality. She actively resists the thought because it annoys her and feels intrusive. The process of reflecting or resisting, however, serves to reinforce the frequency and intrusive strength of the thought. The thought becomes obsessional. Her attention is completely drawn to that one single thought. Obsessionality is a dysfunction of intentionality: the incapacity to shift focus or attention to another topic due to a stronger and longer intentional relation with the mental act. The thought is intrusive because of its obsessive nature. She cannot suppress the thought. Moreover, she is compelled to think about her obsession. Compulsivity is a dysfunction of sense of agency: she is forced to think about the intrusion, contrary to her willpower. The thought is intrusive because of its compulsive nature. Gradually the thought becomes more present and repetitive; it loses its original meaning, but becomes intrusive because of its duration and repetition. The thought has become a full-blown obsession. Obsessions are answered with compulsions (e.g., obsessing about germs could lead to compulsive handwashing). Though initially successful in reducing anxiety, gradually they become intrusive since the acts have to be performed compulsively. Note that both obsessions and compulsions are intrusive, and both have an obsessional and compulsive quality. Eventually, the anticipatory power of the intrusion is so overwhelming that reality testing gets disturbed. She does not know anymore whether she has or has not strangled her baby. Thoughts may take on a delusion-like character, with psychotic features.

Substance Use Disorder

Intrusive thinking in individuals with substance use disorders (SUDs) may include planning to procure the drug, recall of the experience related to its use, as well as the anxiety related to lack of access to or possession of the drug. However, there is large variability between patients, dependent on disease stage. In the initial stages, the individual may perceive thoughts of the drug as non-intrusive and innocuous or even pleasurable, at least as long as they believe to be in control of drug intake, and that such thoughts do not significantly interfere with daily activities. Once the disease progresses toward more severe stages, the individual begins to perceive thoughts of the drug as intrusions. This usually happens when the individual realizes that they have lost the ability to make decisions regarding whether or not to obtain and/or take the drug.

In individuals with SUD, the onset of intrusive thinking varies widely. Intrusive thoughts may appear during acute withdrawal from the drug, or years after the last exposure to the drug. Furthermore, they can be elicited by an internal cue (e.g., a memory of an event related to the drug, or a particularly emotionally negative moment while the patient is alone at home) or external cues, such as other people taking the drug, movie scenes, or physical cues around the patient (e.g., restaurants, bars, or images related to the drug or addiction).

In conclusion, intrusive thinking is a major component of SUD as well as in other addictions, such as pathological gambling. It occurs throughout every step of the life cycle of SUDs and, in some cases, can be the *primus movens* of a series of actions and events leading to relapse and drug taking.

Mood Disorders

Clinically, intrusive thoughts are neither a core nor a necessary feature of mood disorders. However, they appear regularly in a manner that has important bearing on its treatment, severity, and outcome, specifically in the form of rumination, suicidal thoughts, negative automatic thoughts and "flight of ideas."

Rumination is a frequently observed cognitive feature of depression, involving the repetitive and persistent focusing on the causes of the current state of distress and its likely consequences. Depressive rumination is typically centered around personal shortcomings, faults, failings, and mistakes (Treynor et al. 2003). Rumination has intrusive features: it is highly interruptive and distracts individuals from engaging in other tasks. While individuals often state they do not want to ruminate, they also hold metacognitive beliefs that it is important to do so (Borkovec and Roemer 1995). Rumination is associated with the onset of depression and, when combined with negative cognitive styles, predicts the duration of depressive symptoms (Nolen-Hoeksema et al. 2008). The extent to which it is hence desired and indeed under volitional control is unclear.

Individuals suffering from suicidal thoughts report at times imagery related to potential ways of committing suicide (Holmes et al. 2007). This imagery can be experienced as unwanted, intrusive, and interruptive.

Negative automatic thoughts are regarded as a central feature of depression in cognitive models of depression (Beck 1976). In this view, events that are related in some form to core negative beliefs or schemata can trigger fast interpretations or "automatic thoughts," which, due to their negative nature, promote a depressed, negative mood. These automatic thoughts appear very rapidly and can profoundly influence behavior. However, as these automatic thoughts closely relate to core beliefs, they tend not to be experienced as intrusive or unwanted.

Individuals with manic episodes of bipolar disorder can display a "flight of ideas," whereby they exhibit a rapid sequence of unrelated thoughts. This state is often experienced in the context of mania and tends to be positively experienced as a phase of heightened creativity. In mixed states, however, it can also be perceived negatively and exhibit loss of control of one's thinking. Hence, although intrusive thoughts are not central to mood disorders, they do feature prominently and in areas that are thought to be closely related to the core mechanisms of the illnesses.

Anxiety

Two subtypes of anxiety have been distinguished: anxious arousal (e.g., panic) and anxious apprehension (e.g., worry). In cases of anxious arousal, the individual has fearful reactions and thoughts that lead to somatic symptoms and/or somatic symptoms are interpreted in a fearful manner. For example, in simple phobias, an individual will see a phobic object (e.g., a spider) and have intrusive thoughts regarding that object (e.g., "The spider is about to crawl on me and bite me") that are often accompanied by somatic symptoms (e.g., sweating, increased heart rate). In cases of panic disorder, the individual has somatic symptoms, such as increased heart rate and dizziness, associated with intrusive thoughts (e.g., "I am having a heart attack and am going to die"). This situation of a bodily state leading to intrusive thoughts is somewhat akin to SUDs, in which a somatic state (i.e., withdrawal) can lead to intrusive thoughts of craving.

In contrast, in cases of anxious apprehension, intrusive thoughts tend to be related to a future event: about the feared event itself, ways to avoid the fear, or the discomfort/harm that might be associated with the event. In some cases, intrusive thoughts can be more distressful when a person anticipates the event, than during the event itself. This appears to be associated with the fact that individuals with anxious apprehension have an intolerance of uncertainty (e.g., not being 100% sure that the flight will not involve severe turbulence).

Individuals that experience anxious arousal and/or anxious apprehension may know that their fears are likely unrealistic or overblown. This knowledge, occurring simultaneously with the experienced fear and intrusive thought, can cause distress and lead to additional intrusive thoughts related to poor self-evaluation similar to depressive rumination: "What is so wrong with me that I cannot get on an airplane like everyone else? Why am I such a baby?" However, intrusive thoughts associated with anxiety tend to be more future oriented whereas intrusive thoughts associated with depression tend to be more concerned with past events.

Psychosis

Psychosis is a state characterized by hallucinations and delusions, which is common in schizophrenia spectrum disorders but also observed transdiagnostically. Hallucinations are false perceptions, mostly experienced with similar sensory quality, as if originating from an outside stimulus but without an external stimulus being present. For example, a patient with schizophrenia might repeatedly experience hearing voices that are negatively commenting on what they are doing and perceive the voices to come from outside their head. Auditory verbal hallucinations of this type can repeatedly occur and severely interfere with daily life, although frequency and distress vary.

Delusions are false beliefs held with high subjective certainty and confidence despite contrary evidence. Delusional ideation is very salient for the patient and center on the person. For example, common topics are delusions of reference or prosecution. Deluded persons do not experience the delusional ideation as interruptive to one's thoughts. Delusional ideations are part of the person's belief system and thus are ego-syntonic; still, they can be disruptive for the person's relation to their interpersonal and social surrounding. A special case is delusion of control such as thought insertions, when (mostly verbal) thoughts are experienced as being "inserted" by another agent into one's head. Inserted thoughts can be salient, interruptive to the train of thought, and unwanted, but are attributed to an outside agent rather than to oneself, and therefore cannot be experienced as ego-syntonic nor ego-dystonic.

Attention Deficit Hyperactivity Disorder

Attention deficit hyperactivity disorder (ADHD) is a neurodevelopmental disorder characterized by a persistent pattern of inattention and/or hyperactivity-impulsivity that interferes with everyday functioning (American Psychiatric Association 2013). People with ADHD are highly distracted by things happening in the outside world as well as by internal thoughts. The issue is whether these distracting internal thoughts should be classified as intrusions. To an external observer, they are certainly disruptive and maladaptive. Compared to other disorders, however, the intrusions may, on average, be much less salient since an individual with ADHD can be distracted by any event, salient or trivial. The distraction can be relatively diffuse, and thoughts may not have specific, recurrent content. While certainly maladaptive and poorly controlled, these thoughts are also not necessarily unwanted; they may function as a welcome distraction (e.g., when feeling bored). An experience sampling study found that individuals with ADHD symptoms experienced excessive disruptive mind wandering, together with little meta-awareness on how to regulate this (Franklin et al. 2017). In general, one could ask whether having an exceptionally low threshold for internal and external distraction might render irrelevant the concept of intrusions; in such cases, the inability to focus and follow goals (i.e., stay on task) changes the goal hierarchy. Thought meandering becomes the default; there is simply no focused thought process upon which this meandering intrudes. This lower threshold could make people with ADHD more vulnerable to other disorders and the experience of intrusions, in general (Abramovitch and Schweiger 2009).

Revisiting the Definition of Intrusive Thinking: A Synthesis

From the description of all the possible manifestations of intrusive thinking in everyday life, as well as across clinical syndromes, a number of features can be distilled that best capture what intrusions are. Each of these features will be discussed below. From this, we select what we think are the key features of intrusive thinking to arrive at two definitions: one narrow, probably more typical definition of intrusions, and a broader, more inclusive one.

Consciousness.   To what degree are intrusions conscious? Is it possible to have unconscious intrusions? It is hard to conceive of an example where an intrusion is unconscious. However, the term consciousness is often confused with meta-consciousness (Schooler 2002); that is, being aware that something is an intrusion rather than merely experiencing it. Because of this ambiguity, experience may be a less controversial term, though arguably more abstract. In addition, while intrusions can encompass a range of unexpected events or actions, intrusive thinking explicitly refers to cognition. This may be incompatible with nonhuman animal models and may also overestimate how voluntary our usual thoughts are (see also Liu and Lau, this volume). Instead of conscious thought, we suggest the broader term experienced mental events.

Unwantedness/desirability.   Although intrusions are not necessarily negative in content, they may be unwanted in the moment, as they distract from the task at hand. The question then is: Does it matter what the task at hand is? Are intrusions during daydreaming unwanted, as long as they have no clear negative content? Can something even be identified as an intrusive thought if it occurs during daydreaming? One can even think of examples of clear pathological intrusions that are not necessarily unwanted. In depression, for instance, rumination is experienced as a strategy to solve a problem. In some cases, it reflects long-term aspirations or a society that makes something unwanted. It boils down to the question of what wanting is; different individuals (or even different voices within an individual) interpret what is wanted differently in the moment. Because unwantedness may not be a defining feature in every instance of intrusive thinking, it seems necessary to develop a narrower clinical definition that is more closely related to the patient's experience of being out of control.

Involuntary/Controllability.   Generally we think of intrusions as being involuntary, though this can mean different things. Involuntary can refer to the fact that a thought was unintended or uncontrollable once it appears. Are thoughts that are unbidden, but easily dismissed, intrusive? We suggest including involuntary in a narrow definition of intrusive thoughts, but not in a broader one (see below).

Disruptiveness.   While unwantedness and controllability may not be a defining feature of every instance of intrusive thinking, a necessary characteristic is that intrusive mental events interrupt and disrupt current cognition. By definition, they intrude upon ongoing processes (e.g., a task or a gentle, natural cadence of unconstrained thought). The disruptiveness may, however, not be experienced as such by the person with intrusive thoughts, though clearly it has maladaptive consequences to an external critic.

Salience.   Something can only interrupt if it has gravitational pull; that is, if it captures attention (see also Fedota and Stein, this volume). Such gravitational pull may lie in different, not mutually exclusive, aspects (e.g., its valence, highly negative or positive), its vividness, its novelty, or its incongruence with a current state.

Valence.   Inherent to Clark's definition (Clark 2005:4) is the assumption that intrusive thoughts are unpleasant, but this is not always the case. Examples of potentially positive intrusive thoughts may include being in love, or experiencing feelings of grandeur (as occurs during manic or psychotic episodes), or when one has an insight ("aha" moments). Therefore, we suggest characterizing intrusive events as salient to the person, but not necessarily as negative in valence.

Content and Shape.   Intrusions can take different shapes: they can present as verbal thoughts, slips of action, or mental images. Content varies in valence as well as in time frame, ranging from the past, present, or future. We suggest not including content or shape as a defining feature of intrusive mental events as it might vary widely between different instantiations of intrusive thinking in healthy and clinical states.

Punctate versus Extended.   Intrusions usually appear unexpectedly, with a sudden onset and a limited duration. They can, however, last for an extended period of time, as in the case of an uncontrolled flashback episodes in PTSD, or if we include delusions, mind wandering, worry, craving, and rumination as forms of intrusive thinking. In these latter cases, one could argue that only the brief, initial episode is the intrusion that sets in motion a cascade of secondary cognitive processes. Intrusive suggests that something intrudes upon something; if these secondary cognitive processes continue for an extended period of time, they become the primary cognitive process, perhaps even changing the goal hierarchy. Should such a primary cognitive process be labeled intrusive thinking?

Recurrence.   While an ongoing process should perhaps not be labeled intrusive thinking, intrusions can repeat themselves and still qualify as intrusions. In fact, a key feature of intrusive memories in PTSD and other clinical syndromes is their recurrent nature. It may be the frequency as well as the recurrent content that distinguishes clinical intrusions from nonclinical intrusions.

Trigger.   As described at the outset, intrusions can be triggered by external cues, such as certain sights or smells or through internal cues such as mood states. However, intrusions are more than reflexive orientation to salient stimuli: the stimulus-evoked mental event has to interrupt the ongoing process significantly.

Agency.   As described above, to be defined as intrusive, thoughts must be attributed to oneself rather than an external agency.

In conclusion, a commonly accepted definition of intrusions may be that they are *conscious, involuntary, unwanted thoughts.* While easy to understand and applicable to many cases, this definition may mean different things to different people and not encompass the entire spectrum of intrusions. An alternative, more inclusive definition may be that of intrusions being *interruptive, salient, experienced mental events* that generally recur, particularly in clinical syndromes. While the latter definition encompasses the clinical definition of intrusions, it does not differentiate nonclinical from clinical phenomena. Clinical intrusive thinking differs from its nonclinical form with regard to frequency, intensity, and maladaptive reappraisal. Intrusive thinking with clinical significance is often part of a clinical syndrome and results in distress and disability as a general identifier of mental illness. In Table 9.1, we evaluate to what degree each of these key features for both definitions is manifested across symptom clusters; for a slightly different conceptualization of intrusions in pathology, see Monfils and Buss (this volume).

From Table 9.1, it is clear that symptoms occurring in PTSD and OCD best exemplify the construct of intrusive thinking. Less prototypical examples of intrusive thinking are flash-forward events, drug craving, suicidal ideation, rumination, and worry. In these instances, intrusions may not always be involuntary, unwanted, or clearly interrupt a task at hand. Rather, patients may voluntarily engage in these thoughts, for example, as a way to dampen present agony (suicidal ideation) or to mentally expose oneself to a feared situation in the future (worry), thereby reducing the risk of being overtaken by panic unexpectedly. In drug craving, these thoughts are only unwanted and interruptive to the degree that an individual intends to abstain from using. It is important to note that while *worry* is part of the DSM-5 criteria for generalized anxiety disorder and as such is a real symptom, rumination is considered a risk factor for depression, but not a diagnostic criterion. However, both constitute repetitive negative thoughts, with common underlying features as well as unique features (e.g., Hur et al. 2017), and seem to qualify as intrusive thinking.

It is less clear to what degree symptoms in ADHD qualify as intrusive thinking and, relatedly, mind wandering in healthy individuals. Typical daily thought meandering certainly does not elicit the derailing effects that clinical intrusions do and may not even be interruptive. Impulsive actions and tics do not seem to qualify as they are not clear experienced mental events (i.e., they do not necessarily involve awareness). Finally, it is unclear to what degree hallucinations, thought insertions, or delusions should be considered examples of intrusive thinking. These thoughts may not generate the same feeling of uncontrollability and intrusiveness, and an individual may not even realize that these are intrusive thoughts, such as when an individual truly believes that the police or secret service is monitoring their thoughts, or when someone is getting commands via imagined voices. Hallucinations and thought insertions

Table 9.1 Features of intrusive thinking across symptom clusters.

Symptom	Related disorder	Thought			Mental event			Further specifier
		Conscious	Involuntary	Unwanted	Experienced	Salient	Interruptive	
Intrusive memories	PTSD	x	x	x	x	x	x	Recurrent, past oriented
Flashback	PTSD	x	x	x	x	x	x	Recurrent, past oriented
Flash-forward	Bipolar disorder	x	?	?	x	x	?	Future oriented
Obsessions	OCD	x	x	x	x	x	x	Recurrent
Compulsions	OCD	x	x	x	x	x	x	Recurrent
Drug craving	SUD	x	x	?	x	x	?	Depends on motivation for abstinence; recurrent, future oriented
Pathological worry	Anxiety	x	?	?	x	x	?	Future oriented
Depressive rumination	Depression	x	?	?	x	x	?	Past oriented
Suicidal thought	Depression	x	?	?	x	x	?	Future oriented
Thought insertions	Psychosis	x	?	?	x	x	?	No agency, attributed to outside agent
Auditory hallucinations	Psychosis	x	?	?	x	x	?	No agency, attributed to outside agent
Delusions	Psychosis, depression, mania	x	–	–	x	x	–	No agency
Inattention	ADHD	x	x	?	x	–	x	Relatively diffuse
Impulsive action	ADHD	x	x	?	–	?	x	
Mind wandering	ADHD	x	?	?	x	?	?	
Tic	Tourette syndrome	–	x	x	–	?	x	Recurrent
Urge	Tourette syndrome	x	x	?	x	x	x	Recurrent

are usually attributed to an outside agent. In our definition of an "experienced mental event," a sense of agency is implied.

How to Best Study the Processes Underlying Intrusive Thinking and Its Control

In this section, we consider a number of general mechanisms that underlie the likelihood for a person to experience intrusive thoughts, and how they are assessed. As suggested above, situations that trigger intrusions all tap into mechanisms related to the ability to control thoughts, actions, and mnemonic processes. Most of these processes come into play at the time of (involuntary) memory retrieval or when habits or compulsions interfere with goal-directed behavior. Here, we elaborate on what is known about these mechanisms and the methods to investigate them.

Cognitive Control Processes

Action Control

Goal-directed actions are those that we acquire to bring about a change in the world that accords with our basic desires. In this view, I perform an action, *A*, because I want a certain outcome or goal, *O*, and believe that *A* is the way to achieve *O*. Action *A* may start life under goal-directed control; however, with extended training performance, it can become more automatic and invariant, elicited by environmental stimuli, instead of to achieve specific consequences, thus becoming habitual. Both actions and habits are acquired and represented in parallel rather than serially. Therefore, it is possible for certain errors to emerge as a consequence of two controllers attempting to control action simultaneously: suddenly switching from goal-directed to habitual control can, for instance, result in the intrusive performance of an action that is adaptive in another situation, but not in the present situation where it is unwanted (e.g., continually turning on the windscreen wipers rather than the blinkers in a rental car). Conversely, intrusive goal-directed control can disrupt the smooth operation of a well-learned habit. It is well known, for instance, that suddenly exerting goal-directed control while playing a complex piece of music on the piano can result in reduced capacity to perform the piece accurately. The same holds for well-trained athletes (see Balleine, this volume).

Both actions and habits require top-down control to prevent excessive and hence maladaptive impulsive behavior. Clues about neural systems supporting the regulation of intrusive thoughts come from the field of action control, capitalizing on putative similarities between inhibition of motor responses and thought control. In the simple Stop-Signal paradigm, for each trial, subjects get ready to respond when a Go signal occurs; then, on a minority of trials, they try to stop in response to a subsequent Stop signal (Lappin and Eriksen 1966; Verbruggen and Logan 2008). Anderson and Green (2001) have long suggested that stopping a memory or thought from intruding is like stopping a motor response. When the stopping process fails, an intrusion occurs. Several studies have provided evidence consistent with this, although it is not yet certain whether there is a common process, common principle, or common circuitry (e.g., Morein-Zamir et al. 2010; Depue et al. 2016; Guo et al. 2018; Castiglione et al. 2019). Assuming there *is* a commonality between the stopping of a motor response and the termination of a memory or thought, it may be useful to adopt a distinction from the motor stopping literature between global suppression and selective suppression (Aron 2011). Stopping a memory or thought can be achieved by a global systemic suppression mechanism, which suppresses all memories (Hulbert and Anderson 2018), or through a selective stopping process in which the target is suppressed by retrieving an alternative memory (Hertel and Calcaterra 2005; Bergström et al. 2009; Benoit and Anderson 2012).

Pavlovian-instrumental transfer (PIT) provides another experimental paradigm relevant to intrusive thinking and may provide evidence of intrusive thinking in nonhuman animals as well as humans. In the case of positive PIT, experimental animals (or humans; Freeman et al. 2014) are trained on two actions to earn distinct reward outcomes: A1 → O1 and A2 → O2. They are then exposed to two stimuli paired with the same outcomes but without the opportunity to perform the actions: S1 – O1, S2 – O2. They are then allowed to perform the two actions in extinction (without reward) but in the presence of the two stimuli. In this situation, the animals will perform the two actions at a low baseline rate. When one of the other stimuli is presented, however, this leads to an interruption in the animals' performance and causes them to immediately shift performance to the action which, in training, earned the goal or outcome predicted by the stimulus (specific PIT). General PIT occurs when the stimulus reminds the subject instead of the valence of the goal and similarly may enhance responding nonspecifically. Probably even more relevant to intrusive thinking is negative or aversive PIT in which the occurrence of a negative Pavlovian conditioned stimulus (associated, e.g., with painful electric shock or loud white noise) enhances avoidance or withdrawal behavior or disrupts appetitive behavior, as in conditioned suppression.

These findings suggest that the presentation of the stimulus during the task might *intrude* on the animals' ability to perform their continued action. In this sense, the stimulus might be regarded as an intrusion, bringing to mind an outcome. The occurrence of the interruptive stimulus reminds the animal of its associated action, the performance of which provides a readout of that intrusion. PIT is not influenced by extinction of the predictive learning nor by the devaluation of the instrumental outcome. In these cases, the stimuli continue to drive the performance of a specific action, demonstrating continued stimulus-mediated retrieval of the specific, now unwanted, outcome.

Working Memory

While much research relevant to intrusive thoughts, such as those that occur in PTSD, has focused on long-term memory formation and retrieval (the topic of the next sections), significantly less work has been done on the role that working memory mechanisms may play in intrusive thoughts. Yet, by their very nature, intrusive thoughts can only be *intrusive* to the degree that they can gain access to working memory and hijack an individual's attention. Working memory is both a capacity-limited and time-limited buffer that allows information to be held online in service of current goals or objectives. As with long-term memory, there are distinct general processes that may be relevant to intrusive thoughts. Each of these processes is discussed in more detail below as well as by Banich (this volume).

The first process allows an intrusive thought to gain access to working memory or to be gated into working memory. Thoughts may be intrusive when they have priority for access to working memory. Information that enters working memory may be drawn from the external or the internal environment (e.g., current thoughts, long-term semantic or episodic memory, habitual actions), and may do so via its salience. If the information is drawn from the external world, it may be salient due to perceptual characteristics; if drawn from either the external world or internal milieu, it may be salient due to conceptual or abstract characteristics. For example, negative information may be quite salient for individuals who are depressed, and drug-related information may be quite salient for those with SUDs. Information may also gain access to working memory through a more controlled selection process (Feldmann-Wüstefeld and Vogel 2019). Individual differences in the efficiency of such selection processes have been observed (Vogel et al. 2005), raising the possibility that processes which allow for selection of information into working memory are altered in individuals who experience intrusive thoughts.

In the second process, an intrusive thought needs to be given priority within working memory so that it dominates current thought and actions. Thoughts may be intrusive when they are stronger and "out compete" other items within working memory, so being at the focus of attention even when they are not relevant for the goals or processes at hand. Generally, it is assumed that this selection process requires cognitive control, and one that can be examined, for example, using the emotional Stroop paradigm in which individuals are shown a bivalent stimulus: one feature is task relevant and should guide responding, the other is not. For instance, an emotional word (e.g., "joy") is overlaid on a face with an emotional expression, and individuals must respond based on the emotional valence of the word (positive, negative). When the word and the face are incongruent (e.g., the word "joy" on an angry face), reaction time and error rates are higher than when the items are congruent (e.g., the word "joy" on a smiling face). This effect is thought to arise because the task-irrelevant information, the face, is salient—so much so that it interferes with processing of the task-relevant information (in this case, the word). At least some part of the interference observed for incongruent as compared to congruent trials is thought to arise at the level of working memory representations (e.g., Banich 2009). In this paradigm, the salient information is provided by its stimulus characteristics (i.e., being a word not a face). In other cases, such as the "recent negatives," task information can be selected based on its temporal tag. Thus, individuals are shown a series of four items on a screen and, after a brief delay, a probe prompts them to decide if that item appeared in the immediately preceding set of four. On certain trials, the probe is one of the four items from the prior trial, but not the current one. In such cases, individuals show reductions in performance compared to a less recent probe. This paradigm demonstrates that time can be a marker for controlling what information is within the current focus on attention. Future research might explore the degree to which selection (either on the basis of perceptual, topical, or temporal characteristics) is affected in individuals who experience intrusive thoughts. To the degree that these control mechanisms are defective, they may not allow for intrusive thoughts to be moved from the focus of attention, or manipulated or reintegrated in such a way as to reduce their saliency (e.g., via reappraisal).

In the third process, information must be removed at some point from working memory to allow more relevant information to be placed in this limited buffer. Theoretically it has been argued that removal may occur via three potential mechanisms: (a) passive decay, (b) being replaced by something else, or (c) being actively removed (Lewis-Peacock et al. 2018). Thoughts may be intrusive to the degree that they are particularly resistant to such processes. These processes may be especially important for disorders of intrusive thoughts which are more abstractly cognitive (i.e., not sensory based) and more repetitive in nature. Moreover, in certain cases it may just be specific categories of information that are specifically resistant to removal. For example, the degree of rumination observed in depressed individuals is associated with difficulty in removing negatively biased information from working memory (Joormann and Gotlib 2008). This difficulty may occur because of an entrenched overriding schema for thought that focuses attention to negative material.

Long-Term Memory

Many intrusive thoughts concern contents that have been encoded into long-term memory. This content includes not only personal autobiographical memories of past events, but also other contents that clearly rely on declarative memory but are not autobiographical memories per se (e.g., images of feared future scenarios or memories of prior thoughts). As noted earlier in this chapter, these intrusive thoughts are often elicited by retrieval cues in the environment or, in some cases, cues that are internally generated by the person. While sometimes positive or neutral, intrusions often have a negative tone. In their extreme form, intrusive memories involve vivid, multisensory images from highly aversive events, constituting the hallmark symptom of disorders such as PTSD. When looking at the role of mnemonic processes underlying intrusive memories and opportunities to intervene, there are different stages of memory that need to be considered. A distinction can be made between the encoding of a memory and its retrieval: within the context of encoding, processes of encoding suppression can limit the impact of an event; within the context of retrieval, one can focus on opportunities for memory modification as well as mechanisms for suppression. Critically, intrusive thoughts involving content encoded into long-term memory (whether autobiographical, future scenarios, or memories of prior thoughts) can be conceptualized as involuntary retrieval episodes.

Memory Encoding

Some evidence suggests that the later intrusiveness of an emotional memory is influenced by processes at the time of encoding and shortly thereafter. Although emotional intensity contributes to recurring intrusive thoughts for a variety of reasons, the impact of emotion on both the strength and nature of the memory that is encoded may be especially important. As reviewed later in the chapter (see section, What Are the Neural Systems Relevant for Intrusions and Their Control?), abundant evidence has documented interactions between the amygdala and the hippocampus during emotional events, putatively contributing to a strong encoding that makes a memory more prone to later retrieval. Indeed, one view is that intrusive emotional memories are simply stronger memories, making them more accessible and easily triggered (Berntsen and Rubin 2008, 2013; Rubin et al. 2016b). In line with this, emotional memories are typically vivid and are often particularly resistant to the passive processes that usually lead memories to be forgotten over time (Hamann 2001); active forgetting processes, however, may behave differently, as discussed next. Another view is that memories for the specific episodes of which elements intrude may actually be worse; that is, they are less contextualized and therefore more fragmented (Brewin 2016; for a challenge to this view, see Rubin et al. 2016a; Bisby and Burgess 2017). This has led to the proposal of a "dual representation" of traumatic memories, with a hippocampus-based system underlying the neutral, declarative aspects of a memory, and a system involving the amygdala and sensory areas representing sensory, emotional aspects of a memory (Brewin 2014; Bisby and Burgess 2017). Although this may be an oversimplification, there is evidence that the intrusiveness of a memory can be selectively reduced while leaving the voluntary recall of a memory intact (Holmes et al. 2009, 2010; James et al. 2015; Gagnepain et al. 2017; Lau-Zhu et al. 2019) and vice versa (Bourne et al. 2010). This finding is compatible with the possibility that these types of memories may rely on distinct systems (Visser et al. 2018), or at least distinct representations (Gagnepain et al. 2017).

Whereas emotional intensity at encoding can amplify a memory's intrusiveness, this can often be mitigated by control processes that limit encoding and consolidation. For example, studies using experimental models for trauma show that engaging in a competing visuospatial task, during or shortly after experiencing an event, can reduce the frequency of intrusions of that event over the subsequent week, as evidenced through self-reports in a subject's daily diary (Holmes et al. 2004; 2010; Lau-Zhu et al. 2019). In addition, a sizeable body of work on item-method directed forgetting shows that when people are instructed to forget the immediately preceding memory item in a list, the instruction has a large impact on later retention of the memory, rendering to-be-forgotten items significantly less accessible, even for emotionally unpleasant scenes (Anderson and Hanslmayr 2014). Such directed forgetting effects arise not only on recall tests, but also on recognition and implicit memory tasks, indicating that participants can successfully block the encoding of recent events into long-term memory when motivated to do so. This encoding suppression mechanism is likely one coping process that healthy individuals use to limit the impact of upsetting events, reducing their intrusiveness (Anderson and Hanslmayr 2014). It may also be the key process underlying the previously discussed phenomenon of mnemic neglect (Sedikides et al. 2016).

Memory Retrieval: Modification

Sometimes the encoding of unwanted content occurs despite a person's efforts to prevent it. When this happens, people are vulnerable to later involuntary retrievals when reminders occur in the environment. During these retrievals, several processes become relevant that could modify the memory and play a critical role in whether intrusions develop or are mitigated. Understanding these memory modification processes presents opportunities for intervention, even for memories that have been successfully consolidated (see Figure 9.1). Much of the research on memory modification has been conducted within the tradition of classical conditioning in experimental animals, although memory modification work also extends to declarative memory.

In research on memory modification, retrieval or recall of a previously consolidated memory has been proposed to engage two possible mechanisms: extinction and reconsolidation (Suzuki et al. 2004; Lee et al. 2006; Clem and Schiller 2016). In extinction, the repeated presentation of a conditioned stimulus leads to the progressive decrease in, and a resultant reduction in, the behavioral expression of a memory. In reconsolidation, the presentation of an isolated retrieval is thought to initiate a molecular cascade that can be bidirectionally modulated to either strengthen or weaken a memory trace (Dudai 2004; Suzuki et al. 2004; Lee 2009). For example, blocking *de novo* protein synthesis or other cellular and molecular processes critical for memory destabilization and reconsolidation prevents subsequent memory expression (Nader and Hardt 2009; Flavell et al. 2011; Elsey et al. 2018; Orederu and Schiller 2018).

Studies utilizing a noninvasive behavioral approach combined principles of reconsolidation and extinction to update a conditioned fear memory. Specifically, one day after rats acquired associative memories through fear conditioning, they were presented with an isolated retrieval trial; then, within the reconsolidation window, they received an extinction session. This paradigm led to an enduring decrease in conditioned responding, which unlike standard extinction (i.e., extinction not preceded by an isolated retrieval trial) did not result in the return of fear as assessed via renewal, reinstatement, and

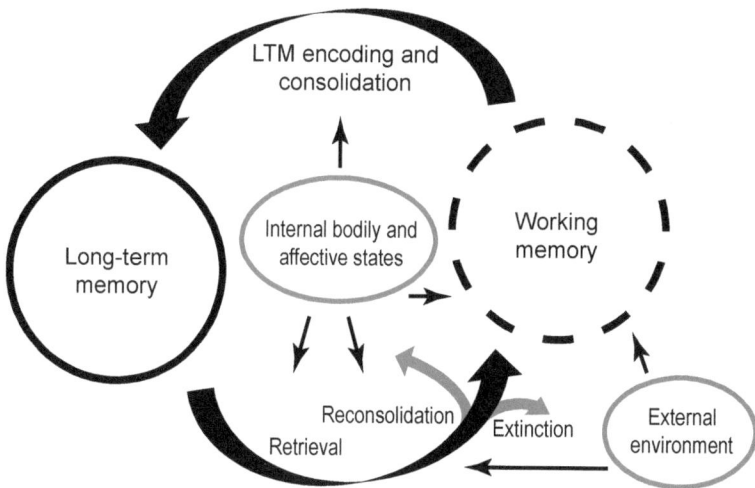

Figure 9.1   Interactive interplay in memory formation, maintenance, and modification. When initially acquired, memories are fragile and short lived. Some items or events may only be required for a brief moment and thus may be only temporarily maintained in working memory. Others may enter the process of consolidation and be stored in long-term memory (LTM). Upon retrieval, a consolidated LTM may reenter working memory and be temporarily put to use or trigger mechanisms of extinction or reconsolidation.

spontaneous recovery, and led to a retardation of fear reacquisition (Monfils et al. 2009). The findings provide evidence that retrieval followed by extinction promotes an updating of the original emotional memory rather than the formation of a competing memory trace (Cahill and Milton 2019). This idea was extended to fear conditioning in healthy humans (Schiller and Delgado 2010), individuals with phobias that received *in vivo* exposure (Telch et al. 2017), as well as a number of other forms of learning in rodents and humans (Xue et al. 2012; Sartor and Aston-Jones 2014; Luo et al. 2015; Björkstrand et al. 2016; Germeroth et al. 2017; Lee et al. 2017). In a clever twist, researchers have started to employ a procedure involving a distracting task (the computer game *Tetris*) following retrieval to update a traumatic memory (James et al. 2015; Iyadurai et al. 2018; Kessler et al. 2020; see also Holmes et al., this volume).

Memory retrievals serve as an opportunity to modify memories and indeed, other aforementioned protocols could actually be understood through the lens of reconsolidation updating (i.e., a change in the memory during a period of destabilization/restabilization). If intrusions are viewed as involuntary retrievals, they may provide a window of opportunity to modify an unwanted memory (see Figure 14.3 in Holmes et al., this volume), using either extinction mechanisms with extended exposures (Cassini et al. 2017; Hu et al. 2018), reconsolidation updating (with extinction, cognitive restructuring, a distracting task or pharmacological disruption), or suppression mechanisms (described below). It may be important to consider that failure to act in the face of pathological intrusions could lead to their reinforcing a memory trace. Alternatively, the unique nature of intrusions may represent a form of memory that is resistant to modification. For instance, it may not destabilize upon retrieval or fail to promote extinction or reconsolidation processes. In this case, more targeted behavioral or pharmacological interventions (or a combination of the two) may help.

Memory Retrieval: Suppression

As discussed, intrusions can be conceptualized as involuntary retrievals, whether they are autobiographical memories, images of feared future events, or persistent thoughts. Efforts to control intrusions therefore involve controlling retrieval. Here, we discuss two processes that contribute to controlling such thoughts: retrieval suppression and retrieval-induced forgetting.

Retrieval suppression refers to the act of trying to stop an ongoing retrieval process, usually triggered by a reminder (Anderson and Green 2001). For instance, a loved one's face may elicit an intrusion of a recent argument that caused upset or a dog barking may trigger memories of the night a person got news of a friend's death. Other retrievals may be of a feared future: a medical bill may remind someone of an upcoming medical test. Because people do not enjoy the aversive emotional states that such reminders can create, they often exclude the offending content from awareness to regain their focus and composure. The aim is not merely to stop the intrusive thought at the moment, but to diminish its recurrence. Before describing retrieval suppression, we first discuss differing views on the value of suppression and whether it is desirable to encourage.

The Utility of Suppression. The clinical psychology literature contains conflicting observations on the value of thought suppression. Many clinicians, particularly those concerned with trauma or anxiety, maintain that suppression is an intrinsically unhealthy response to intrusive thoughts or memories. Indeed, it is common to view suppression as not only unhelpful, but as a significant risk factor for developing psychiatric disorders. Some studies have found that questionnaires putatively measuring thought suppression (Wegner and Zanakos 1994) are related to worse symptom severity in PTSD and other disorders characterized by intrusions. Given these facts, one can readily understand why clinicians would conclude that suppression is maladaptive. As a result, many therapists and therapies (e.g., acceptance and commitment therapy) discourage suppression and encourage people to accept and interact with their intrusive thoughts.

Several observations, however, are incompatible with the view that suppression is maladaptive. First, suppressing unwelcome thoughts is a widespread behavior that serves many different functional purposes, as outlined above. Simply concentrating on an important task, by its nature, requires that an individual excludes distracting content. Second, achieving emotional balance after an upsetting event involves "getting over" unwelcome thoughts and feelings, a process that requires active regulation of thoughts. People who have difficulty with upsetting occurrences (e.g., getting over arguments or perceived slights) are viewed as coping less well than people who quickly recover. Third, after a trauma, most people initially experience intrusive thoughts and memories that usually diminish over time. People who cannot reduce their intrusions are classified as having mental health concerns; those who are capable of reducing intrusive thoughts effectively are considered resilient, a clearly positive attribute. Fourth, people with anxious thoughts about a feared event that they cannot discard are considered as less healthy, compared to those who are able to set aside fears and cope well. Finally, the proposal that trying to stop unwanted thoughts is intrinsically ineffective is oddly discordant with the massive literature on inhibitory control and, more broadly, attentional control, in which controlling cognition and behavior is not only desirable but an essential capacity of intelligence. Indeed, the same clinicians that classify suppression as maladaptive often maintain beliefs that are contradictory to that stance upon closer inspection. For example, the belief that attention bias modification (i.e., training people to ignore unpleasant interpretations) is a desirable therapy amounts to a belief that it is desirable to become skilled at ignoring unpleasant contents, exactly what people do through suppression. The belief that people with intrusive thoughts suffer from cognitive control deficits that make them vulnerable to such thoughts assumes that resilient people can set aside upsetting thoughts, which must therefore be beneficial.

What do we make of these contradictions? What accounts for the view that suppression is maladaptive, given its ubiquity and utility? One reason derives from work on thought suppression using the white bear paradigm (Wegner et al. 1987). In this paradigm, participants are instructed not to think about a particular thought (e.g., a white bear) over a five-minute period. During that time, if they happen to think about the thought, they were asked to report this. Afterward, they are given a five-minute free expression period. Wegner found that participants asked to suppress thoughts of a white bear ironically experienced more white bear thoughts than participants who were not asked to suppress. This enhancement of the unwanted thought after suppression was termed the *rebound effect* (Wenzlaff and Wegner 2000). Across nonclinical samples, meta-analyses show that people exhibit a small to medium rebound effect compared to control instructions (Abramowitz et al. 2001). Wegner and Zanakos (1994) also introduced the White Bear Thought Suppression Inventory, a scale intended to identify people prone to thought suppression. This instrument became widely used in clinical research and was found to correlate with both rebound effects and clinical symptomatology. Coupled with this question-naire, the white bear paradigm provided a compelling narrative that explained why psychopathologies were associated with intrusive thinking: unpleasant thoughts led to suppression, which led, in turn, to a paradoxical rebound effect, yielding a vicious cycle that worsened psychological symptoms (Wegner and Zanakos 1994).

There are significant difficulties with treating the white bear task as a model for controlling intrusive thoughts. First, the task requires the integration of the very thought to be suppressed with the task goal; that is, to know what task one is doing and to monitor whether one is achieving it, one reactivates the thought by asking: "Am I thinking about white bears?" The task is self-defeating be-cause it requires one to violate the task's goals to check whether the goal is being accomplished (Anderson and Huddleston 2012; Engen and Anderson 2018). Although the situation modeled by this task may model a slice of clini-cal reality, work using this paradigm has been overgeneralized to all instances of suppression (as will be seen shortly). Second, the White Bear Suppression Inventory—thought to quantify thought suppression frequency—has now been shown to have two factors opposed to one another: one measures thought sup-pression frequency while the other, the experience of intrusiveness (Blumberg 2000; Höping and de Jong-Meyer 2003; Rassin 2003). Research indicates that only the latter correlates with clinical symptoms, which is not surprising. Third, meta-analytic treatments demonstrate that suppression, as measured with the white bear task (and its rebound effects), shows few reliable differences across control and psychiatric populations, raising questions about the validity of the relationship of the process being measured and disordered control over intru-sions in everyday life (Magee et al. 2012). Finally, the clinical observations that led clinicians to be attracted to the claim that thought suppression is counterpro-ductive—reduced suppression in patients—can be explained more simply by positing that patients have difficulties applying inhibitory control rather than by suppression being intrinsically maladaptive. In line with the former view, sup-pression is an otherwise healthy coping response which, when disordered, poses a risk factor in developing intrusive symptoms—a very different interpretation with different clinical implications. For example, if suppression is not intrinsi-cally maladaptive, but simply not functioning properly in some populations, it is reasonable to expect that interventions can be developed to improve it. This possibility would be ruled out by the view that suppression is maladaptive.

Another reason why some clinicians have been opposed to suppression in-volves confusions in terminology. Researchers have not distinguished retrieval suppression from other constructs that may well be maladaptive and that might appear equivalent to retrieval suppression. For example, researchers have falsely equated retrieval suppression with the emotion regulation construct of expressive suppression and other seemingly related terms, such as avoid-ance, cognitive avoidance, and distraction. Careful analysis of the situations captured by these terms reveals that retrieval suppression is cognitively distinct from these other concepts:

- White bear suppression (Wegner et al. 1987) refers to the explicit intention to not think about a specific unwanted thought for a specific period of time (e.g., 5 min). The unwanted content is incorporated as part of the goal and integrated with it, making it, by definition, not possible to avoid periodic attention to the thought. In contrast, retrieval suppression involves the attempt to stop or cancel the retrieval of content associated with a reminder while sustaining attention to the reminder. A person's goal does not specify the content to be inhibited but rather concerns the desire to shut down "whatever is associated with the reminder."

- Expressive suppression (Gross and John 2003) refers to the strategy of inhibiting behaviors associated with emotional states, such as adopting a poker face to hide emotions. Expressive suppression involves motor control, not control over thoughts or feelings. Retrieval suppression, on the other hand, involves suppressing mnemonic processes and content.

- (Cognitive) avoidance (Ehlers and Steil 1995; Williams and Moulds 2007) refers to entirely avoiding reminders that could otherwise trigger unwanted thoughts or emotions. It fundamentally involves not confronting reminders and not adapting one's internal response to them so that unwanted contents are preserved without alteration and may continue to intrude at the slightest provocation. It is widely considered a maladaptive coping process, a "not dealing with the problem." In contrast, retrieval suppression requires a person to confront reminders directly, attend to them, and adjust their internal retrieval and affective, conditioned responses to them, features shared with cognitive behavioral therapy and exposure therapy in particular.

- Distraction refers to the removal of attention from emotions or thoughts and refocusing it onto other innocuous stimuli in the world or to unrelated thoughts. It is worth distinguishing general distraction (where attention shifts to stimuli, topics, or activities unrelated to the intruding thought) from specific distraction, which involves interacting with reminders. If general distraction leads one to remove attention from reminders to intrusive thoughts or emotions, it functions in a manner similar to avoidance and fails to engage suppressive processes that might adjust one's internal mnemonic and emotional responses. If distraction takes the form of generating alternative associations, thoughts, or memories in response to a reminder, this form of specific distraction would constitute thought substitution, which retrains the response to the cue to elicit different content, potentially leading to altered patterns of thought in the future. Sometimes, because retrieval suppression involves attending to reminders and not the memory, or instead, attention to thought substitutes, researchers falsely equate retrieval suppression with general distraction, from which it clearly differs.

Because clinical research has often not attended to these distinctions, it is easy to see how, with the paradoxical rebound effects reported in the white bear paradigm, clinicians would have taken a dim view of suppression. In recent years, retrieval suppression has been studied intensively. As we look at next, this work integrates models of the control over intrusive thinking within the bedrock of cognitive neuroscience research on inhibitory control.

Retrieval Suppression: Methods of Study and Basic Findings.   Work on retrieval stopping focuses on a situation that is similar to motor response inhibition. Participants are confronted with a cue that reminds them of a memory and are asked to focus on that cue for trials of several seconds, while preventing the memory from entering awareness. This situation parallels motor response inhibition tasks, such as the Go/No-Go and Stop-Signal tasks, in which participants are presented with stimuli designed to elicit a motor response, yet are required to stop the action. To study retrieval suppression, Anderson and Green (2001) introduced the Think/No-Think paradigm. Participants are trained on cue-target pairs until they can recall the target items when given the cues. These pairs can be unrelated words (e.g., ordeal roach), picture pairs, or even cues related to autobiographical memories. Participants then enter the Think/No-Think task. On each trial, participants receive a reminder from one of the pairs, colored either in green or red. Green reminders signal the participant to retrieve the associated item and keep it in awareness for the trial's duration; red reminders signal that they should attend to the stimulus but prevent the associated item from entering awareness at all. Usually, a given item is only suppressed or retrieved (not both), and will be repeated many times (usually, 8–16 times). A final memory test follows, giving participants all studied cues to determine the impact of retrieval suppression on the suppressed content. Performance on the final memory test is computed not only for the Think items, which were repeatedly retrieved, and No-Think items, which were suppressed, but also for baseline pairs studied in the training phase but which did not appear in the Think/No-Think phase.

A key finding that emerged from this work is that if people consistently suppress the retrieval process, memory for the suppressed No-Think items declines compared to Think items as well as to baseline items receiving the same initial training as No-Think items. Because suppressing retrieval impairs retention of the suppressed content, this phenomenon is known as suppression-induced forgetting (for a review, see Anderson and Hanslmayr 2014). Suppression-induced forgetting increases with the number of times that people suppress the unwanted content and resists monetary incentives on the final test for successful recall (Anderson and Green 2001; Hulbert and Anderson 2018). Forgetting generalizes to novel test cues for the suppressed items, exhibiting a property known as cue independence: suppression-induced forgetting occurs not only for simple word pairs, but also for objects or complex scenes, irrespective of whether they are neutral or negatively valenced (Depue et al. 2007; Anderson and Hanslmayr 2014). Suppression-induced forgetting has even been observed with autobiographical memories, although the effect emerges more in memory for details (Noreen and MacLeod 2013). Importantly, a growing body of work shows that suppression-induced forgetting does not merely influence explicit memory, it also affects implicit memory regardless of whether perceptually driven (Kim and Yi 2013; Gagnepain et al. 2014) or conceptually driven tasks are used (Hertel et al. 2012, 2018; Taubenfeld et al. 2019). Of clinical relevance, retrieval suppression also reduces affective responses to the suppressed content when participants view it later, whether measured with ratings (Gagnepain et al. 2017) or psychophysiology (Legrand et al. 2019; Harrington et al. 2020). These findings suggest that retrieval suppression, for healthy participants at least, regulates the negative affect associated with intrusive memories (Gagnepain et al. 2017; Engen and Anderson 2018).

Although most research on retrieval suppression focuses on final test performance, recent studies have increasingly focused on online measurements of intrusive memories during the Think/No-Think phase (Levy and Anderson 2012; Benoit et al. 2015; Hellerstedt et al. 2016; van Schie and Anderson 2017; Legrand et al. 2019; Harrington et al. 2020). In these studies, the Think/No-Think task is unchanged except that after every trial, participants provide a phenomenological report of their experience. Specifically, they judge whether the associated memory entered awareness. On Think trials, awareness of the memory is the intended outcome, and people usually report awareness on nearly all trials (averages of 90–100% of trials). On No-Think trials, however, because participants are instructed to prevent awareness of the memory, any awareness that does occur constitutes *clear evidence for an intrusive thought*. Importantly, when intrusions occur in this task context (e.g., on 30% of No-Think trials), the inference of involuntariness is particularly clear. Indeed, one of the hallmark features of an involuntary, automatic behavior is its tendency to occur, despite efforts to stop it. Intrusions during No-Think trials therefore provide a robust and theoretically clean measure of a vulnerability to involuntary retrieval. Other paradigms hoping to measure intrusions (e.g., diary studies) cannot make a clear attribution of involuntariness (for discussion, see van Schie and Anderson 2017).

Contrary to what might be inferred from the white bear paradigm, people can suppress the retrieval process and reduce intrusive thoughts elicited by reminders. For example, across repeated suppressions one observes a dramatic downregulation of intrusions between the initial trials (e.g., 60% intrusions) and final trials (e.g., 25% by the tenth trial). Moreover, the slope of the downregulation in intrusions is often correlated to suppression-induced forgetting (Levy and Anderson 2012; Hellerstedt et al. 2016), and suppression effects persist to produce memory deficits on both direct and indirect tests, contrary to the notion of rebound. Work on retrieval suppression aligns the control of intrusive thoughts with work on inhibitory control in general, especially work on action stopping. Indeed, given that nobody questions that actions can be stopped and that a giant literature has developed which documents action-stopping mechanisms in humans and other animals, it would be more surprising to find that internal actions, such as retrieval, were not also subject to control (Anderson and Green 2001). This conclusion fits the observation that psychiatric populations with intrusive thinking also have deficits in cognitive control (Catarino et al. 2015; Waldhauser et al. 2018). Suppression-induced forgetting has also been related to self-reports of rumination (Fawcett et al. 2015), perceived success at thought control in real-life settings (Küpper et al. 2014), and self-reports of intrusive memories collected in the week following exposure to a trauma film (Streb et al. 2016). These findings support the view that suppression is a helpful coping process that is compromised in clinical populations, rather than an intrinsically maladaptive approach to intrusive thoughts.

Retrieval-Induced Forgetting: Methods and Basic Findings.   Thus far we have focused on the role of inhibitory control in intentional retrieval stopping. Inhibitory control, however, can also be engaged in another context to forget distracting memories: selective memory retrieval. During selective memory retrieval, a person seeks to retrieve a particular event, idea, or fact. Reminders, however, are often associated to many traces, creating interference known as retrieval competition. Retrieval interference is ubiquitous, whether one is retrieving an episodic memory, a fact, a general idea, or simply an object's location (Anderson and Neely 1996). Isolating the desired trace despite this interference engages prefrontally mediated inhibitory control mechanisms to suppress competing memories. This suppression induces aftereffects on the inhibited traces causing forgetting. Thus, ironically, the very act of remembering something can cause forgetting of competing memories; this is known as *retrieval-induced forgetting* (Anderson et al. 1994; Anderson and Spellman 1995; Levy and Anderson 2002; Anderson 2003; Bäuml et al. 2010; Storm and Levy 2012). Although retrieval-induced forgetting is not intentional, this process can be exploited in a motivated manner to deliberately forget unwelcome memories (Benoit and Anderson 2012). The procedure by which this phenomenon is studied will be described below.

Retrieval-induced forgetting becomes particularly relevant to controlling intrusive thoughts when it is deployed in a motivated way; that is, as a key strategy for deflecting unwelcome thoughts, refocusing on benign knowledge. For example, after an argument with a friend or partner, one might (upon seeing them again) try to recall more pleasant thoughts about them (Anderson 2001). Optimists routinely seek the positive in any event, even if unpleasant things happened, and in so doing, may reshape their memories to fit a more pleasant existence. Indeed, the widely demonstrated positivity bias in autobiographical memory retrieval (the disproportionate accessibility of positive over negative autobiographical memories) significantly correlates with the amount of retrieval-induced forgetting that people show on laboratory tests with neutral materials (Storm and Jobe 2012). Thus, the capacity to exhibit inhibitory control over memory predicts a sunnier view of one's past than might be expected. Critically, retrieval-induced forgetting may be a mechanism that explains why reappraisal is effective as an emotion regulation strategy, given that reappraising an unpleasant memory involves selectively retrieving the contents of an event or thought (Engen and Anderson 2018). If conventional therapies for addressing intrusive thinking rely on reframing, rescripting, or reappraisal, retrieval-induced forgetting may be central to reducing the accessibility of negative interpretations and intrusions.

Selective retrieval's role in motivated forgetting has been studied with the retrieval-practice paradigm as well as with a variant of the Think/No-Think paradigm, known as thought substitution (Hertel and Calcaterra 2005; Bergstrom et al. 2009; Joormann et al. 2009; Benoit and Anderson 2012). In thought substitution, participants are instructed to stop retrieval of the unwanted memory by retrieving diversionary content, either self-generated or supplied, related to the cue (Hertel and Calcaterra 2005; Bergstrom et al. 2009; Benoit and Anderson 2012). Retrieval of such thought substitutes significantly impairs retention of the intrusive thoughts, although the neural mechanism underlying this process is dissociable from that involved in retrieval suppression (Benoit and Anderson 2012). Thought substitution as a strategy for controlling intrusive memories has been studied extensively using the Think/No-Think task, especially in dysphoria and depression (Joormann et al. 2009; Stramaccia et al., unpublished). Given people's natural pre-disposition to replace intrusive contents with distracting thoughts, how thought substitution affects memory is important to study.

Summary and Commentary

Most research on intrusive thoughts has focused on long-term memory formation, retrieval, and their control. The substrate for intrusive thoughts is laid down when memories are encoded. Many intrusive thoughts are related to past experiences, such as those associated with substance use or specific events (e.g., feeling rejected, being traumatized). Other intrusive thoughts may not refer to specific episodic memories (e.g., images of feared future events or ruminations) but are nonetheless stored in memory and governed by its mechanisms. If a memory is encoded in a particular way, it may be more accessible to reactivation at a later time. Highly salient memories may, in particular, be easily triggered by external and internal cues, many of which do not reach conscious awareness. Intrusions, therefore, appear to pop spontaneously into a person's mind. Such potent memories may be related to specific types of emotional information (e.g., negative information as in depression and anxiety) or to a particular topic (e.g., danger and threat in PTSD and anxiety; substance-related information in substance use disorders).

Substantial research has focused on cognitive control processes, including the role of inhibitory control mechanisms in intrusive thinking. Memory control mechanisms related to action control are known to suppress retrieval of specific, episodic long-term memories and can work more globally to block all episodic memory retrieval in response to a cue. Alternatively, reminder cues can be used to selectively retrieve alternative distracting content, suppressing the intruding thought via the mechanisms of retrieval-induced forgetting. Moreover, retrieving information may trigger destabilization and/or restabilization of a thought. As such, retrieval provides an opportunity not only to increase the strength of a memory, but also to reduce it. In addition, working memory processes are critically linked to the encoding and retrieval of long-term memory and are important to guiding the selection of information for consolidation into, and retrieval from, long-term memory (see Figure 9.1). It seems likely that for an intrusive thought to be experienced, it would enter working memory, making this system central to having an experienced (conscious) mental event.

Summary of Commonly Used Methods and Desiderata for Investigating Intrusive Thinking

Methods and Paradigms

The previous section described examples of research into the mechanisms underlying intrusive thinking and their control using a variety of methods. Here we present an overview of the methods and paradigms that are most commonly used to study intrusive thinking in humans, nonhuman animals, or both, each with its own strengths and weaknesses.

- *Experience/thought sampling* is a form of self-report that is often used to assess mind wandering in the present moment (Larson and Csikszentmihalyi 1983; Bolger et al. 2003; McVay et al. 2009; Schooler et al. 2011; Baird et al. 2014; Fraley and Hudson 2014). One form has people pressing buttons when they catch themselves mind wandering or experiencing other intrusions. This can be done in the lab or in real life (e.g., using a diary or smartphone). Alternatively, one can also probe people at irregular intervals to report whether they were on task or not. A comparison of probed versus unprobed self-report shows that having an experience and knowing that you are having an experience are different things (Schooler et al. 2011; Baird et al. 2014). The first requires metacognition and this is imperfect, and possibly worse in clinical populations.

- In the *emotional Stroop task*, individuals view items with two overlapping dimensions, such as a word superimposed on a face (Williams et al. 1996). The participant bases their response on a task-relevant dimension (e.g., valence of a word) while ignoring the task-irrelevant dimension (e.g., valence of a face). Proneness to intrusions is defined as the degree to which reaction times or error rates are increased on incongruent trials (e.g., the word "suicide" on a happy face) as compared to congruent trials (e.g., the word "joy" on a happy face).

- The *trauma film paradigm* is used to model intrusive memories of distressing events (Horowitz 1969; James et al. 2016a). It uses film stimuli in the laboratory, which contain traumatic content that can bring about intrusive memories subsequently in daily life. These memories are typically recorded in a diary, allowing for a frequency count of intrusive memories, or via button presses during a provocation task (both are forms of experience sampling).

- *Pavlovian conditioning* is the classic model by which simple associative learning and memory are studied, and it is well suited for research across species (Pavlov 1927; Rescorla and Holland 1982; LeDoux 2003; Nader and Hardt 2009). In this paradigm, an initially neutral stimulus (conditioned stimulus, CS+), such as a triangle, is repeatedly paired with an intrinsically aversive stimulus (unconditioned stimulus, UCS), such as an electric shock, while another conditioned stimulus (CS–), such as a circle, is never paired with the UCS. With sufficient CS+/UCS pairings, the CS+ acquires the same aversive qualities as the UCS and will elicit a conditioned defensive response that can be measured, for example, by the amount of freezing or avoidance behavior in nonhuman animals, or skin conductance, acoustic startle response, heart rate, pupil dilation, action tendencies, UCS expectancies, and subjective distress in humans. After repeated presentations of the CS+ without the UCS, the defensive response usually diminishes, a process referred to as *extinction learning*. Intrusions may often be conceptualized as conditioned responses to external or internal reminders.

- *Stop-Signal and Go/No-Go* are tasks that require inhibition of already initiated responses. Error rates in response to Stop-Signal or No-Go trials may reflect the outcome of diminished control of actions in a similar way as intrusions reflect a diminished control of thoughts.

- In the *Think/No-Think paradigm*, participants learn a set of cue-target pairs, either word pairs, picture pairs, or autobiographical memory cues (Anderson and Green 2001). Next, trials appear, each presenting a reminder cue and a task cue: a green box around the reminder cue signals that the associated memory item needs to be brought to mind whereas a red box signals that retrieval of the associated memory item needs to be suppressed. This method allows one to measure, during the suppression task itself, whether participants experienced intrusions (i.e., intrusion reports during the No-Think trials) in response to the reminders, despite their best efforts to stop retrieval.

- Retrieval-induced forgetting is studied using the *retrieval-practice paradigm* in which participants learn to associate a cue with multiple associates; participants are then set the task of retrieving a subset of these items, but not the rest (Anderson et al. 2000; Bekinschtein et al. 2018). Rather than providing explicit instructions to forget, this paradigm capitalizes on a natural process to select relevant information. It leads to forgetting of nonselected, competing memory items through a process of inhibition. Retrieval-induced forgetting is related to a variant of the Think/No-Think paradigm that uses thought substitution rather than retrieval suppression, and when deployed in a motivated way, may be a model for how individuals control and diminish intrusive thoughts. Retrieval-induced forgetting is now also being studied in rodents, capitalizing on rats' intrinsic curiosity (spontaneous object recognition paradigm). After encoding multiple objects in an environment, selective retrieval of one object leads to forgetting of other objects in a manner directly analogous to that seen in humans and that depends on the prefrontal cortex. An advantage that this paradigm enjoys over retrieval suppression is that it allows us to understand memory inhibition mechanisms, not only at the level of neural systems (through imaging in humans) but also through foundational neurobiological work in nonhumans using electrophysiological, lesion, optogenetic, and molecular biological methods.

- With *Pavlovian-instrumental transfer*, a subject undergoes Pavlovian threat conditioning and is also trained on an instrumental reinforcement task (reward or avoidance) in a different context (Campese et al. 2013; Cartoni et al. 2016; Watson et al. 2018). In the test phase, during extinction (or baseline performance) of the instrumental task, the CS+ is presented and its possible rate-altering effects measured as evidence of negative Pavlovian-instrumental transfer. There are several variations of this basic procedure, which can also be appetitive in nature and have general and specific aspects of transfer in relation to the overall goal. The degree to which the presentation of the CS+ interrupts an animal's (human or nonhuman) goal-directed behavior is taken as a proxy for the experience of an intrusion.

These paradigms should continue to be developed. In addition, we wish to stress that promising new approaches are emerging from computational neuroscience, as we illustrate next.

Computational Approaches to Intrusive Thinking

We turn now to a few considerations from a computational perspective. A common feature of intrusive thoughts is that they occur at the "wrong time." They are remembered or imagined contents that are not appropriate for the situation or the particular problem our minds are focused on at that time. But what constitutes the "wrong" time? How do we, our minds, or our brains choose the "right" thoughts or the "right" time for particular thoughts? Are there any general principles that guide the selection of thoughts, and what might they be? Computational modeling of decision making provides a framework for this problem and suggests a number of potential answers that might apply to different mental illnesses (Huys et al. 2016; Redish and Gordon 2016). At their core, these accounts rely on the notion that thoughts are expensive and that organisms have much to gain from deploying them well.

Our starting point is the acknowledgment that most decision problems faced by humans on a daily basis are so complex that they radically outstrip our computational abilities. The mismatch between computational demands and resources requires us to invest our computational resources with wisdom. Unfortunately, exerting any such wisdom further complicates the problem: the optimal deployment of limited computational resources is itself a decision problem that has the same form as the original decision problem, but which is vastly more demanding (Russell and Wefald 1991). This problem is displayed in Figure 9.2a. Hence, we are faced with the double challenge consisting of hard problems and the even harder problem of apportioning our cognitive resources.

There is, however, a silver lining for the study of thoughts. As the problem of optimally apportioning computational resources is essentially of the same nature as the original decision-making problem, the same formalisms used to study choice can potentially be brought to bear on the problem of thought selection (Lieder et al. 2018a, b). Errors in these processes might, in turn, link the emergence of intrusive thoughts to well-defined optimal approaches to solving decision-making problems under resource constraints (Huys and Renz 2017). Below, we review several computational accounts of thought choice and discuss how intrusive thoughts could arise.

Affective Biases

A bias toward thinking about negative events can arise through utility-weighted sampling, a process whereby the samples (in this case, what we choose to think about) are not just proportional to the probability of the event but also to its importance. Humans routinely overestimate the probability of extreme negative events (Lichtenstein et al. 1978), and such biases have long been viewed as a signature of irrationality. However, they may alternatively reflect a rational use of limited resources. When many different outcomes are

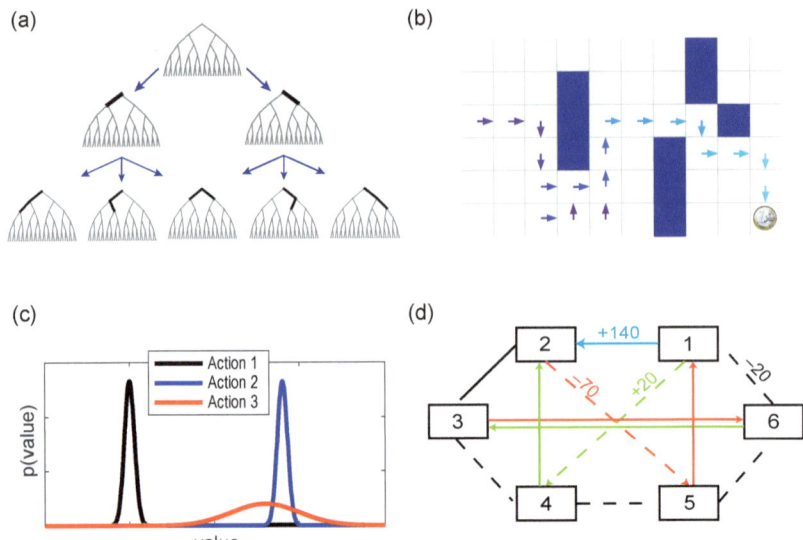

Figure 9.2 Computational approaches to thought guidance: (a) The meta-reasoning problem of optimally apportioning limited resources can be formalized as a decision-making problem over decision-making problems. While the standard decision-making problem is the tree at the top without thick lines, the meta-reasoning problem is a tree of such trees, where each branch corresponds to choosing to evaluate one branch (thick lines) in the original decision problem. Intrusions could relate to a tendency to only sample one set of options (arrows). (b) Prioritized replay: In this state-space, if the reward (Euro coin) is received in the bottom right corner, then standard model-free learning updates only the state immediately adjacent to it. Prioritized replay allows memories to be reused multiple times to update more distant states; here, propagating the information about the reward all the way back to the starting state. If the need or gain functions determining what memories to replay are altered, then this could result in repetitive replay of the same memory (arrows) as well as a failure to update distance states (i.e., to integrate the memory with other memories). Adapted after Mattar and Daw (2018). (c) Guidance of thoughts by habits: Consider the situation where a valuation system provides distributions for likely values of actions. Here, action 1 is clearly inferior to the other two, and it appears that action 2 is the best. If the agent were to invest computational effort into refining the estimates of the values of the actions, it would be best to examine action 3, as it may be even better than action 2. In this manner, a goal-directed or model-based system could elaborate on approximations provided by a simpler valuation system. This would also mean that the goal-directed thought choice could be (mis)guided by habits. (d) Stopping aversive thoughts: In this task, participants are extensively trained to learn to navigate a maze where each transition incurs some gains or losses. When given the opportunity to plan freely, they stop internal simulations when they encounter one of the salient losses in red. Adapted after Huys et al. (2015).

possible, it becomes difficult to consider them all, and humans are thought to simulate instead a few outcomes in their mind and average the results of these few simulations. If these simulations are too few in number, they are likely to miss very important outcomes. Simulating in proportion not only to the probability, but also in proportion to the utility, allows the resulting estimates to be more robust. That is, the apparent overestimation of extreme events may help mitigate cognitive limitations and explains a number of apparent irrationalities (Lieder et al. 2018b). Applied to the setting of intrusive thoughts, it provides one argument for why salient negative events should be simulated even when they are very unlikely. In fact, the more aversive the event, the higher the likelihood of simulation, suggesting one path by which the perceived negative valence of a simulated event might increase the frequency with which it is thought about.

Trauma Replay

A computational process that may be related to intrusive thoughts in PTSD is that of prioritized replay. Briefly, learning from experience is often slow. A prominent way of learning from experience through iterative updates of expectations with prediction errors is particularly slow. Indeed, this is one reason why it has been considered to be a computational account of habitual learning (Daw et al. 2005). Figure 9.2b shows that this is because experience at any one state or stimulus, *s* only leads to learning at that particular state or stimulus and is only propagated to adjacent states upon the next transition from those adjacent states back into state *s*. Such experience influences "knowledge" only very locally and does not generalize. One solution to this is to store episode-like chunks of autobiographical memory and replay them multiple times so as to spread the effect of any one experience to other states. This can substantially accelerate learning (Schaul et al. 2016). However, replaying memories is also costly, and hence the key here is to again deploy the resources (in this case, which memory to replay) efficiently. Computational models of this process identify two terms that determine which memory to replay: a need term and a gain term (Mattar and Daw 2018). The need term captures how likely state *s* is to be encountered in the future (Russek et al. 2017). Clearly, using computational resources to learn about states that will never be visited is not useful. The gain term captures how much the memory is likely to change behavior.

As this gives a normative account of when to optimally replay particular memories, it should reflect the tendency to replay both intrusive and non-intrusive memories experienced after laboratory induction procedures. It also provides an interesting window on features of intrusive memories in PTSD. The gain term depends on the implied change in behavior and, as such, a memory should only be replayed if it implies a change in behavior. This suggests an important modulatory role of generalization processes seen across anxiety disorders (Laufer et al. 2016). For instance, replay tendency after traumatic abuse by a trusted person should be influenced by the perceived importance of this event for other relationships. The more relevant it is judged for other relationships, the higher the replay tendency should be. More generally, it captures the notion that excessive negative appraisals of the sequelae of a trauma might relate to the emergence of PTSD symptoms (Ehlers and Clark 2000).

Obsessions

In the context of OCD, intrusive thoughts exhibit a different quality to those experienced in PTSD (see Monfils and Buss, this volume). Extensive evidence suggests that OCD involves a shift from goal-directed toward habitual behavior, with most evidence pointing to an impairment in goal-directed or model-free control rather than an explicit change to habitual or model-based processes (Robbins et al. 2012; Gillan et al. 2013, 2015; Voon et al. 2015). How could an impairment in model-based control give rise to the emergence of intrusive repetitive thoughts as seen in OCD? One avenue arises from the notion of value of information (Keramati et al. 2011). Consider the situation in Figure 9.2c, where one option is clearly good and a second one clearly bad. The third option appears slightly worse than the best one, but could be better. The optimal investment of limited cognitive resources in this case would be to examine this one option, again because this investment of cognitive resources has the potential of altering behavior. This could underlie intrusive thoughts in OCD if there was a drive provided by uncertain habitual or model-free evaluations, such as through distributional reinforcement learning (Dabney et al. 2017), coupled with impairments in goal-directed evaluations which fail to result in improved predictions.

Rumination

Frequently seen in depressive disorders, ruminations are usually described as a tendency to think repetitively about the causes of distress without engaging in active problem solving (Nolen-Hoeksema et al. 1993, 2008; Treynor et al. 2003). Viewed in the context of intrusive thinking, they appear to involve a prominent failure to inhibit or discontinue aversive sequences of thoughts. This raises a theoretical point not yet addressed above—that of thought inhibition. Thus far we have focused on which thoughts, evaluations, or memories should be chosen and have not yet addressed the monitoring of a thought that appears not to be fruitful nor the question of when it should be terminated.

This question has been examined in some detail using computational accounts of the task in Figure 9.2d. Here, individuals are trained to navigate a maze where each transition yields rewards or losses. They are then dropped into one state randomly and asked to search a route of a given length through the maze that maximizes their total earnings. The problem they face is a binary decision tree, as in Figure 9.2a. Individuals are much more likely to identify optimal routes that do not transition through salient losses than those which do, independently of the size of the large loss (Huys et al. 2012, 2015). Computational models of the choices in this task suggest that participants internally simulate potential routes through the maze and terminate simulation when they encounter a salient loss. Functional magnetic resonance imaging suggests that this inhibition recruits the subgenual anterior cingulate cortex (Lally et al. 2017), a region known to be important in depression (Drevets et al. 1997), its treatment (Mayberg 2009), and rumination (Hamilton et al. 2015). Indeed, individuals who score high on self-reported rumination show reduced pruning; that is, a reduced tendency to terminate thoughts when encountering large losses during their internal simulations (Q. Huys, pers. comm.), suggesting that rumination might directly relate to an inability to inhibit aversive thoughts, possibly via impairments involving the subgenual anterior cingulate.

Desiderata for More Sophisticated Behavioral Paradigms to Measure Intrusive Thinking and Its Control

Studying intrusive thinking is challenging for several reasons. Intrusive thoughts are typically spontaneous, making it hard to predict when they occur, and therefore, when to measure them. Although paradigms exist that can reliably induce intrusive thinking under controlled circumstances in the laboratory (as detailed above), additionally allowing for the investigation of its neural correlates, these paradigms may not capture all circumstances under which intrusions occur in real environments. In addition, while it is possible to infer the occurrence of intrusion-like events from nonverbal behavior, measuring an "experienced mental event" ultimately relies on self-report. Indeed, historically intrusive thinking has been studied with self-report questionnaires (for an overview, see Banich, this volume).

A general issue is how well the more objective, laboratory-based measures of intrusive thinking correlate with subjective measures. A common finding for other constructs (e.g., impulsivity) is that they do not intercorrelate particularly well (Nombela et al. 2014). Why might this be the case? One notion is that the subjective measures are in some sense "noisier" and more prone to error. It is well known, for instance, that people sometimes have great difficulty in expressing their conscious evaluations or descriptions of their thinking, and there may be considerable interindividual ability in this capacity. However, this notion can perhaps be dismissed, as there is also considerable evidence of superior test-retest reliability for questionnaires than objective measures, such as those based on reinforcement learning parameters (see Table 5 in Bland et al. 2016). It does appear likely that objective tests may capture much narrower aspects of the construct under study, thus resulting in a looser association with the more generic aspects captured by a composite measure of a psychometrically well-designed questionnaire. Another important conclusion, however, is that the objective measures and the subjective responses obtained from questionnaires are simply tapping into quite different processes, the latter most obviously addressing the contents of monitoring operations in meta-consciousness. This reliance on meta-awareness can be problematic and may be especially compromised in clinical populations. More importantly, in the field of intrusive thinking, there is a clear need for the development of better self-report instruments: at present, there does not appear to be a measure that captures all aspects of intrusive thinking in one instrument. In particular, there are four gaps in this field (for discussion, see Banich, this volume).

1. Most questionnaires were designed with a particular clinical syndrome in mind (e.g., PTSD, OCD, craving). As such, a general intrusive thinking scale does not exist.
2. For the most part, the questionnaires do not assess both the content of the thought as well as the ability to control those thoughts, broken down by content.
3. There are few, if any, questionnaires that specify the quality of intrusions across numerous dimensions, such as their form (verbal, images, urges), vividness, salience, and frequency.
4. There are no questionnaires that systematically assess how intrusions are triggered or the context in which people experience intrusions.

Looking forward, future research should focus on designing a new, theoretically motivated questionnaire, refining existing objective measures and capitalizing on the potential of computational approaches.

What Are the Neural Systems Relevant for Intrusions and Their Control?

The exact neural underpinnings of intrusions are currently unknown, but we could speculate about processes involved in intrusive thinking by combining what we know from the study of cognitive control, including action control, and the study of emotional memory processes.

Cognitive Control

Action Control

A critical prefrontal region for stopping movements is the right inferior frontal cortex (rIFC) as shown by lesion and transcranial magnetic stimulation studies (Aron et al. 2014). It is thought that the rIFC, in concert with the presupplementary motor area, implements stopping through a (fast) hyperdirect pathway to the subthalamic nucleus (STN) of the basal ganglia, and this suppresses thalamic drive back to cortex, leading to movement cancellation (reviewed by Bari and Robbins 2013; Jahanshahi et al. 2015; Wessel and Aron 2017). Importantly, in the standard case, subjects apparently stop using a global mechanism. For example, stopping the voice suppresses the hand (Badry et al. 2009; Cai et al. 2012). This broad skeletomotor suppression begins around 120 ms after the Stop signal and lasts for 100 ms or more. It is thought that the broad suppression relates to a wider putative impact of the STN on basal ganglia output (Aron 2011), just as the degree of global motor suppression in the Stop-Signal paradigm relates to the level of oscillatory power in the STN (Wessel et al. 2016).

Yet subjects can also stop selectively when they are *forced* to do so (Aron 2011). For example, they are able to stop one response (or hand) while continuing with another with minimal interference, and this does not result in broad motor suppression (Majid et al. 2012). It is possible that this form of selective response suppression relates not to a hyperdirect cortico-STN connection but rather to a frontal-striatal-pallidal, so-called, indirect pathway. For example, people with degeneration of striatum and pallidum, who are thought to have an indirect pathway disorder, could not stop as selectively in the paradigm and did not show typical physiological signatures of selective stopping (Majid et al. 2013).

As will be discussed below, this global versus selective picture in motor stopping may have relevance for global versus selective control over memory. Future work could develop behavioral paradigms to look at this selective control on the analogy of selective response suppression, taking into account the observation that selective stopping is best done when the subject proactively sets it up ahead of time (Cai et al. 2011; Majid et al. 2012); that is, it could be possible to go into a situation preparing to suppress a particular memory intrusion.

Working Memory

The three working memory operations that we discussed above may be particularly relevant to intrusions:

1. Gain access or gate information into working memory.
2. Select information within working memory to be given priority.
3. Remove information from working memory.

In general, the neural correlates of working memory mechanisms have been well described (D'Esposito and Postle 2015) and likely involve prefrontal areas that are involved in executive control processes (e.g., selecting among information in working memory), basal ganglia mechanisms that work to gate information into working memory, and posterior brain regions that help to provide sensory or abstract (e.g., semantic) representations of information to be activated and/or placed in working memory. Most of the paradigms utilized in cognitive neuroscience require the confluence of processes whereby information enters, is selected, and updated in working memory (e.g., the N-back task). Nonetheless, the neural bases of some of these three main processes have been distinguished.

With regard to gaining access to working memory, research suggests that the basal ganglia may play an important role in determining when the gate to working memory should be opened so as to let in new information (see Badre, this volume). It is proposed that dopaminergic reinforcement learning mechanisms help to provide information on when the gate to working memory should be opened. Hence, this process is not conscious and controlled. The basal ganglia form multiple loops with distinct regions of the cortex, which are semi-segregated based on the nature of information they carry (e.g., sensory, motoric). As such, these basal ganglia mechanisms might selectively or concurrently act to allow access to working memory. In the case of intrusive thoughts, especially when they are repetitive in nature, it may be that through habitual learning, the intrusive content can more easily open the gate and gain access to working memory. While alterations in the accessibility of information to working memory may affect many different syndromes in which intrusive thoughts are observed, the specific basal ganglia loop involved may differ according to the disorders (e.g., more motoric in substance use, more visual in anxiety), as discussed by Balleine (this volume).

With regard to selecting information within working memory to be prioritized (or buffered from interference), processes often considered to be executive aspects of working memory, meta-analyses suggest that selection relies on frontal and parietal regions across a variety of tasks (Nee et al. 2013). For example, using Stroop and emotional Stroop tasks, research suggests that these tasks tend to involve regions associated with the frontoparietal, cingulo-opercular as well as dorsal and ventral attention networks (e.g., a meta-analysis by Chen et al. 2018b), regions associated with cognitive control. Activation in these areas has been found to be altered by characteristics related to intrusive thought. For example, an individual's degree of worry influences the degree of activation (Engels et al. 2007) as well as the time-course of activation (Levin Stilton et al. 2011) across lateral and medial prefrontal regions. Portions of the superior parietal lobe may be particularly important in selecting or shifting what information is currently within the attention focus of working memory (Tamber-Rosenau et al. 2011). These regions are engaged in executive processes more generally, not just specifically with regard to working memory.

With regard to removing information out of working memory, the picture is less clear. As discussed by Banich (this volume), most methods examine removal from the perspective of replacement; that is, how new information is placed into working memory and/or how the current information is buffered from removal. These operations involve cognitive control regions, including both the frontoparietal network and the cingulo-opercular network. Furthermore, variance in activation in these areas has been observed in individuals, like those who are depressed, who behaviorally show difficulty in removing certain types of information (e.g., negative emotional information), as indexed by subsequent interference from those items (Foland-Ross et al. 2013).

Less research has examined how one can take information currently held within working memory and dispose of it completely. This has been a difficult question to address because it is hard to verify that an item has indeed been removed. Other than with behavioral methods such as self-report, which may not be reliable, few (if any) methods are able to confirm that a thought has been removed or inhibited. Yet understanding such processes could have important implications for how interventions for intrusive thoughts might be created.

While there is little research in this area, some work has demonstrated that brain imaging techniques can be used to help verify the removal of information and simultaneously to examine the control mechanisms by which such removal occurs. In one of the few studies of this nature, Banich et al. (2015) utilized activity in the sensory cortex as a rough proxy for whether information was currently active in working memory. With this approach, they verified the presence of such activity when individuals were maintaining information about a picture just viewed or a short tune just heard, or when they replaced the item with something else. They also observed a lack of such activity when the item was removed from current thought, both when participants specifically suppressed that item as well as when they did so by clearing their mind of all thought. With regard to control mechanisms, a hierarchy of function was observed. Common across the replace, suppress, and clear conditions—all of which require a shift in attention away from the original item to something else, compared to the maintain condition—was activation in superior parietal regions implicated in shifting attention among items in working memory (Koenigs et al. 2009). When information had to be removed from working memory (the suppress and clear conditions) as compared to when an item was present in working memory (the maintain and replace conditions), activation was observed over regions of lateral prefrontal cortex involved in executive processes that act on working memory. Finally, for the clear condition as compared to all the other conditions, there was increased activation in the insula, suggesting a shift of attention to bodily states (Craig 2011) as well as activation in inferior parietal cortex, which may represent a mechanism for altering the bottom-up salience of information (Cabeza et al. 2008), either by reducing the salience of visual information or by increasing salience of information derived from bodily states (see also Fedota and Stein, this volume). In relation to long-term memory, the global "clear" condition on the contents of working memory may be analogous to a global stopping process that inhibits retrieval of all information from the hippocampus, whereas the "suppress" condition may be analogous to suppression of a specific memory, as discussed in more detail below.

In summary, the brain regions relevant to working memory processes involved in intrusive thought likely involve (a) content-specific cortical regions needed to access the representation of information underlying the intrusive thought: visual areas for visual images, language- or semantically related regions for thoughts, limbic regions for emotional information, and hippocampus for episodic memory, (b), basal ganglia mechanisms that influence what gains access to working memory (see Badre, this volume), and (c) prefrontal mechanisms involved in control processes that select and manipulate information within working memory or act to engender its removal.

An unresolved issue concerns the extent to which the processes deployed in suppressing contents actively held in working memory are distinct from or overlapping with those involved in suppressing the long-term memory retrieval of unwanted thoughts, prompted by reminders (see section on retrieval suppression), and whether these different types of mental control tap into unique processes that may be differently affected in psychiatric disorders.

Long-Term Memory

Memory Encoding

For over a century, it has been recognized that memories are initially unstable, subject to interference, until they are stabilized through a process of consolidation (Ribot 1882; Muller and Pilzecker 1900; Burnham 1903). Once consolidated, long-term memories are largely protected from interference, save windows of destabilization that may occur upon retrieval (Sara 2000; Nader 2003). Reconsolidation involves a relatively brief (few hours) cascade of molecular and cellular processes enhancing synaptic efficacy via structural changes; a longer process (days and weeks) involves system-level connectivity changes between the hippocampus and cortical areas (McGaugh 2000; Dudai 2012). Emotionally significant experiences trigger the release of adrenal stress hormones, stimulating norepinephrine in the amygdala, which in turn modulates plasticity in the hippocampus, cortex, and other brain regions (Cahill et al. 1994; Southwick et al. 2002; Cahill and Alkire 2003; Strange and Dolan 2004; Hurlemann et al. 2005). Abundant evidence links dysfunction in these circuits to psychological disorders in which intrusions are a symptom. Figure 9.3 presents a schematic overview of the neural circuits underlying these major domains and their possible (dys)function in intrusive thinking.

To the degree that encoding of information is tied to a specific event, such as that often associated with PTSD, aberrant functioning in the emotion circuits as well as the hippocampal and medial temporal structures that support episodic memory are likely involved (Liberzon and Martis 2006; Pitman et al. 2012). Evidence points to hyperactivity in the amygdala and dorsal anterior cingulate cortex (the putative human homologue of the prelimbic medial prefrontal cortex), whereas the ventromedial prefrontal cortex and hippocampus evince hypofunction, accompanied by impaired extinction learning and recall (Milad et al. 2009; Milad and Quirk 2012; Logue et al. 2018). Most recently, specific computations of threat, including learning parameters such as value, prediction error, and learning rate, have been characterized in PTSD. It was found that PTSD severity was related to overweighing of prediction errors

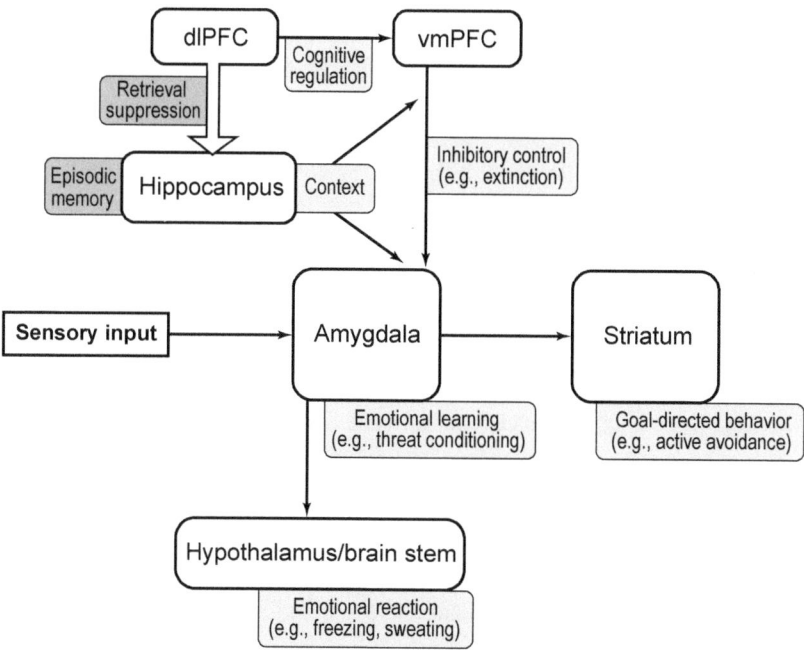

Figure 9.3   Most of what we know about the neural mechanisms of emotional learning and memory comes from animal studies that utilize threat conditioning and extinction as a model. Based on these studies, sensory information from neutral stimuli in the environment that reliably coincide with emotionally significant outcomes converge in the lateral nucleus of the amygdala where associative learning occurs, conferring emotional value on the conditioned cues. Projections from the lateral to the central nucleus engage descending projections to the hypothalamus and brain stem, which mediate the expression of the conditioned response (e.g., freezing). Projections from the lateral to the basal nucleus onto the striatum form a path that promotes active coping and goal-directed behavior. The prelimbic region of the medial prefrontal cortex (mPFC) connects with the amygdala to sustain the expression of emotional responses, while the infralimbic region acts to counteract amygdala output and diminish emotional responses. As such, the infralimbic mPFC (ventral mPFC in humans) is a critical region in the acquisition and recall of extinction learning. The hippocampus exerts contextual control over the expression of threat learning. The dorsolateral prefrontal cortex (dlPFC) is the region involved in top-down cognitive regulation of emotional responses by influencing amygdala via the vmPFC (Hartley and Phelps 2010; Schiller et al. 2010; Milad and Quirk 2012). The dlPFC also exerts top-down modulation of hippocampal activity to induce retrieval suppression (Anderson et al. 2016).

(akin to enhanced sensitivity to negative surprise) and to impaired amygdala and striatal tracking of the negative value of conditioned cues (Homan et al. 2019). Neuroimaging studies that examine the relation between viewing analog trauma and subsequent intrusions are sparse but do tentatively suggest that activation of regions in the salience network distinguishes distressing scenes that later intrude compared to equally distressing scenes that do not (Bourne et al. 2013; Battaglini et al. 2016; Clark et al. 2016), highlighting that consolidation processes are important in the formation of intrusive memories. Together, these studies invoke neural mechanisms for the learning and flexible modulation of emotional memories. Aberrant functioning of the salience and memory circuitries might induce emotional inflexibility, lack of proper contextual control, and impaired ability to diminish inappropriate emotional reactions, allowing strong negative memories to persist and intrude.

Similar mechanisms may come into play if a memory is associated with particular actions or procedures, such as may occur in substance use disorders. Here, the encoding likely involves basal ganglia systems and alterations in specific cortical processes (Balleine, this volume), such as visual regions that may become tuned to particular stimuli (e.g., paraphernalia associated with substance use). Across disorders, when information is associated with emotional salience or significance (e.g., salience of negative emotional information in depression, association of loud sounds with a traumatic event in PTSD), such encoding likely involves the amygdala, as well as the striatum, insula, and dorsal anterior cingulate cortex (Fedota and Stein, this volume). However, whether intrusive memories are stored in a qualitatively different way than other emotional memories, or are merely strong memories and therefore more accessible, is still a topic of debate (Berntsen and Rubin 2008, 2013; Brewin 2014; Bisby and Burgess 2017).

Retrieval Suppression

Although we know little about the neural mechanisms of intrusions, we know a great deal about the inhibitory control processes deployed to reduce their occurrence. For an in-depth review of the neural basis of memory control, see Banich et al. (2009) and Anderson and Hanslmayr (2014). For a review of memory control in relation to emotion regulation, see Engen and Anderson (2018). For a detailed consideration of memory control in relation to primate anatomical pathways and neural circuits, see Anderson et al. (2016).

Prefrontal-Hippocampal Interactions as a Basis for Retrieval Suppression. Research on retrieval suppression was initially premised on the parallel between stopping prepotent actions in response to triggering stimuli and stopping internal processes, such as memory retrieval (in response to reminders), to control intrusive memories and thoughts (Anderson and Green 2001). In both cases, a stimulus (or, sometimes in the case of retrieval stopping, an internally generated cue) initiates an automatic process (action preparation or retrieval) that the person wishes to stop, and both trigger a race between a Go and a Stop process. Given this analogy, one might expect both similarities and differences between retrieval and action stopping: similarities in the prefrontal control regions engaged in service of stopping, but differences in the target regions with which control processes interact to implement stopping. Because retrieval stopping involves stopping retrieval and not physical actions, a plausible suppression target would be the hippocampus, a brain structure involved not only in encoding new memories but also in their retrieval, at least for recently acquired events and thoughts (Anderson and Green 2001; Anderson et al. 2004; Anderson and Hanslmayr 2014).

Abundant evidence indicates that retrieval suppression, unlike motor response inhibition, relies on prefrontally mediated downregulation of activity in the hippocampus and other medial temporal lobe regions to stop retrieval, presumably by preventing pattern completion (Anderson et al. 2004, 2016; Depue et al. 2007; Banich et al. 2009; Levy and Anderson 2012; Gagnepain et al. 2014, 2017; Benoit et al. 2015; Schmitz et al. 2017). Suppression reduces hippocampal activation not merely relative to active retrieval in the Think condition, but also relative to passive baseline conditions, and this negative BOLD response arises from negative coupling between the right lateral prefrontal cortex and the hippocampus, established using effective connectivity analysis (Benoit and Anderson 2012; Gagnepain et al. 2014, 2017; Benoit et al. 2015; Schmitz et al. 2017). Indeed, within-subject comparisons of action and memory stopping establish a clear double dissociation, with retrieval suppression downregulating the hippocampus more than action stopping, but action stopping downregulating motor cortical areas (M1) more than retrieval suppression (Schmitz et al. 2017). Stopping of unwanted memories and thoughts thus involves a distinct frontohippocampal inhibitory control pathway that suppresses hippocampal activity. Hippocampal suppression appears to be a blunt instrument that acts globally on the hippocampal state. For example, when a person tries to suppress retrieval to depress awareness of a particular unwanted thought, the forgetting arising from that suppression is not limited to the item people intend to suppress; rather, any other recently encoded memories that occur either before or after the act of suppressing something are also forgotten, even if they are entirely unrelated to the content being suppressed (Hulbert et al. 2016). Thus, suppression of unwanted thoughts induces an *amnesic shadow* in the temporal surround of the suppression attempt, creating both anterograde and retrograde amnesia effects in healthy people. This finding has been linked to the global suppression of hippocampal processes that not only stop retrieval but also disrupt encoding and stabilization processes necessary to retain recent experiences (Hulbert et al. 2016). This global, systemic disruption of hippocampal activity is analogous to the global stopping identified in motor response inhibition.

What do we know about how the prefrontal cortex achieves this form of inhibitory control over hippocampal activity? Primate anatomical studies tell us that top-down suppression of hippocampal activity is unlikely to be direct, not only because there are no direct connections between the lateral prefrontal cortex and the hippocampus (Anderson et al. 2016), but also because long-range projections from the hippocampus are largely excitatory. To achieve an inhibitory effect in the hippocampus, if the negative BOLD response is truly inhibitory, the prefrontal cortex must drive local populations of inhibitory interneurons within the hippocampus to disrupt its function. Because all inter-neurons in the hippocampus are GABAergic, this observation suggests that individuals with higher concentrations of hippocampal GABA may show a superior ability to suppress hippocampal activity by prefrontal influence. Recently, Schmitz et al. (2017) found evidence of this, using a multimodal imaging study that combined fMRI with magnetic resonance spectroscopy. These findings indicate that people with higher concentrations of hippocam-pal GABA showed greater downregulation during retrieval suppression, more successful forgetting of intruding thoughts, and greater negative coupling be-tween the right prefrontal cortex and the hippocampus. As such, local concen-trations of hippocampal GABA may provide a pivotal function that enables the prefrontal cortex to implement long-range inhibitory influence necessary for control over intrusive thoughts. This discovery sheds new light on evidence of diminished hippocampal GABA in many disorders characterized by intrusive thoughts (Schmitz et al. 2017), which may be a heretofore unrecognized risk factor in the pathogenesis of disordered control over intrusive thoughts.

Despite the unique hippocampal targets involved in implementing re-trieval suppression, it is equally clear that both retrieval and action stopping processes engage overlapping regions in the dorsolateral (BA 9/46/10) and ventrolateral prefrontal (BA 44/45) cortex, suggesting the existence of do-main general supramodal inhibitory control regions that may dynamically recouple with task-specific target regions (Depue et al. 2016; Schmitz et al. 2017; Guo et al. 2018). These domain general regions are strikingly right lateralized, strongly consistent with the long-standing claim by Aron et al. (2004, 2014) that inhibitory control is right lateralized, although the regions clearly include both dorsolateral and ventrolateral regions, and not simply ventrolateral prefrontal cortex. These supramodal regions have been identi-fied both through within-subjects conjunction analyses of action and retrieval stopping (Depue et al. 2016; Schmitz et al. 2017) as well as conjunctions performed on quantitative meta-analyses of independent studies from many laboratories (Guo et al. 2018). Interestingly, action stopping and retrieval stopping also appear to engage highly colocalized regions within the basal ganglia (Anderson et al. 2016); for a detailed discussion of anatomical hy-potheses, see Depue (2012), Guo et al. (2018), Paz-Alonso et al. (2013), and Balleine (this volume).

Cortical Modulation and the Reinstatement Principle.   Although the discus-sion of direct retrieval suppression emphasized hippocampal suppression as the principal mechanism through which intrusive memories and thoughts are controlled, the hippocampus is not the only target of inhibitory control dur-ing suppression (see Banich as well as Balleine, this volume). For example, when people suppress visual objects or scenes, downregulation takes place in the hippocampus as well as in visual cortical regions, such as the fusiform cortex and the parahippocampal place area, respectively. This leads to the generalization that areas outside the hippocampus involved in reinstating the unwanted memory in awareness are suppressed in parallel with the hippocampus by the right lateral prefrontal cortex, an idea known as the reinstatement principle (Gagnepain et al. 2014; Gagnepain et al. 2017).

Intrusive Thoughts and the Triggering of Control.   Intrusions of unwanted contents into awareness during retrieval suppression are particularly important for triggering top-down inhibitory control processes that suppress the hippocampus and cortical regions. Using the intrusion judgment procedure outlined earlier, Levy and Anderson (2012) found that downregulation of activity in the hippocampus and other medial-temporal lobe regions was largely confined to trials in which participants reported that a memory had intruded into consciousness. Strikingly, the extent of hippocampal downregulation during intrusions predicted subsequent forgetting with a correlation of 0.7, whereas hippocampal activity during non-intrusions was unrelated to later forgetting. These findings suggest that intrusions play an important role in triggering top-down control by the prefrontal cortex to cancel the retrieval process, and that the critical inhibitory action that disrupts later retention of the intrusive thought arises in the purging of the intrusion from awareness. Consistent with this possibility, Gagnepain et al. (2017) replicated Levy and Anderson's (2012) evidence for intrusion-specific downregulation in the hippocampus with aversive scenes and showed that negative coupling between the right dorsolateral prefrontal cortex and the hippocampus was significantly greater during intrusions than during non-intrusions.

The findings by Gagnepain et al. (2017) are especially relevant to the current discussion, given that they involve the suppression of unpleasant and intrusive images, prevalent in psychiatric disorders. Gagnepain et al. (2017) also found intrusion-specific downregulation in both the amygdala and the parahippocampus. These downregulations robustly predicted both intrusion frequency and later reductions in negative affect for the suppressed content: the stronger the downregulation in the amygdala, the lower the number of intrusions and the greater the subsequent reduction in negative affect for the suppressed scene. These findings indicate that top-down inhibition during intrusive thoughts plays a critical role in modifying the representations that support both memory and emotion about the offending content. Consistent with this, Legrand et al. (2019) found that suppressing unpleasant images from awareness also significantly reduced later psychophysiological measures of emotion elicited by the scenes, such as heart rate deceleration. Similar findings have now been observed with skin conductance responses (Harrington et al. 2020).

The importance of purging intruding thoughts from awareness highlighted here points back to the discussion above of what relationship, if any, the processes involved in stopping retrieval have to those involved in regulating working memory. One assumes that when an intrusion is retrieved and enters awareness, there is a good chance that the intruding content has entered working memory, however briefly. Some evidence supports this. For example, Hellerstedt et al. (2016), using event-related potentials, found evidence that the frontal negative slow wave (NSW), observed over the right prefrontal cortex, is modulated during intrusions. This component has been linked in prior work to the storage of information in working memory. Consistent with this, participants during Think trials show a prolonged NSW that lasts throughout the full several seconds of the trial; in contrast, non-intrusions show little evidence of this, consistent with the exclusion of items from working memory. Intrusions, however, showed a brief increase in the NSW, which was rapidly eliminated within the first seconds of the trial, suggesting a brief penetration of the intruding item into working memory. Perhaps relatedly, Castiglione et al. (2019), using time frequency analysis, found evidence that during No-Think trials, there is a robust increase in frontal beta component during non-intrusions. Given the prior linkage of this component to motor response inhibition, these findings suggest that intrusions may reflect an initial failure of inhibitory control that allows the intruding content to penetrate working memory.

Related Phenomena Observed with Item-Method Directed Forgetting.   Although we have focused primarily on retrieval suppression, related work on item-method directed forgetting supports the hypothesis that, in parallel with retrieval suppression, people can also suppress encoding. By encoding suppression, we mean the possibility that the same inhibitory control processes which modulate hippocampal activity during memory retrieval to disrupt retention may also be deployed shortly after encoding to terminate stabilization processes in the hippocampus that might promote the formation of an enduring memory. For example, several studies indicate that when participants are instructed to forget an item that they just encoded into memory on the preceding trial, activation increases in the right dorsolateral prefrontal cortex and decreases in the hippocampus. As with retrieval suppression, connectivity analyses indicate that the prefrontal cortex couples with the hippocampus during Forget trials, especially on trials when the item is successfully forgotten (Rizio and Dennis 2013; Wierzba et al. 2018). In another compelling study, intracranial recordings indicated that lateral prefrontal cortex interacts with the hippocampus during instructions to forget to promote forgetting (Oehrn et al. 2018). These findings converge to suggest that suppressive processes are not limited to controlling intrusive retrievals, but may be also be used prophylactically shortly after an unpleasant experience to limit the footprint of that experience in memory. Moreover, this application of inhibitory control to suppress unwanted contents immediately after they are encountered seems related to inhibitory processes involved in purging the contents of working memory (e.g., Holmes et al. as well as Banich, this volume), although these two strands of research are not usually considered together.

Summary and Commentary

At present, little is known about the neural underpinnings of intrusive thinking, and the underpinnings presumably vary across its different manifestations. In contrast, research on cognitive control and emotional memory has identified key systems involved in generating and suppressing intrusions. Concerning intrusion generation, areas involved in signaling salience, such as the amygdala and striatum, not only play a role during the encoding of salient events or thoughts, but also respond to reminders of those events (external cues). Their salience signal may help bring back to mind memories of these events or thoughts, possibly via reinstatement of multimodal cortical representations of these events. In addition, these areas may themselves generate internal cues (mood states) that trigger intrusions. With regard to the suppression of intrusions, frontal areas implicated in inhibitory control over actions and thoughts play a clear role. In particular, the dorsolateral prefrontal cortex can suppress the reinstatement of a memory or of an imagined future event by downregulating hippocampal and neocortical activity, while also downregulating emotional responses via amygdala suppression. Suppression reduces the frequency of intrusions and impairs memory, also for new information presented around the time of retrieval suppression (i.e., creating an *amnesic shadow*). Some evidence indicates that it also reduces negative affect associated with suppressed content.

What Are the Implications of Intrusive Thinking?

In this chapter we have revisited the definition of intrusive thinking by systematically considering all the circumstances in which intrusions might occur and their manifestations across health and disorders. We define intrusions as being *interruptive, salient, experienced mental events* and propose that clinical intrusive thinking differs from its nonclinical form with regard to frequency, intensity, and maladaptive reappraisal. We have reviewed the neurocognitive processes underlying intrusive thinking and their control, including action control, working memory processes, long-term memory encoding, retrieval and suppression, and methods for studying them.

Functional Perspective: The Adaptive Nature of Intrusive Thinking and the Desirability of Suppression

Despite being commonly associated with mental health disorders, intrusive thinking commonly occurs in the absence of psychological problems and is thus by itself not indicative of pathology. In fact, there are many instances in which having a thought pop into mind to disrupt ongoing cognitive processes is actually beneficial, such as recalling an action that is required (e.g., paying a bill) or solving a problem (an "aha" moment) (see also Monfils and Buss, this volume). If mind wandering is a form of intrusive thinking, research shows that the capacity to "jump out of the present set" can be adaptive and relates to creativity. Its adaptiveness is likely reduced when mind wandering becomes excessive or too diffuse, such as in ADHD, or when the content is unpleasant.

The potential adaptive nature of intrusive thinking may be independent of the desirability to control it. In healthy individuals, frequent and exciting thoughts (e.g., love infatuations) might disrupt concentration on the task at hand and therefore not be adaptive, yet still wanted in the moment. Alternatively, intrusive thinking about mistakes that we have made may allow us to adjust our behavior and become a better person, yet unwanted in the moment. The discussion below on sociocultural implications provides compelling examples of this. In addition, as a clinical symptom, intrusive thinking is not always unwanted (e.g., exciting flashforward thoughts that occur during a manic episode), yet often maladaptive. Whether the opposite is also possible (i.e., unwanted intrusive thinking in mental health disorders may in some cases serve an adaptive purpose, such as in acute stress disorder as part of processing trauma), is an open question that requires more research.

Clinical and Therapeutic Implications

Could excessive intrusive thinking be an endophenotype (or neuroendophenotype) for mental disorders? According to the novel, more dimensional approach to understanding psychiatric nosology (Cuthbert and Insel 2013), the propensity toward intrusive thinking could underlie several otherwise distinct, categorically defined disorders, such as OCD and depression, the implications being comorbid, shared dysfunctions of common neural systems or networks, with obvious implications for treatment. Considering some of the dimensions already postulated, there are clear relationships with such constructs as inhibitory control and working memory. It remains unclear, however, whether intrusive thinking comprises a unitary construct itself or is a collection of phenomena that meet our definitions. For example, intrusive thinking could arise as an emergent feature of different neural networks processing perceptual inputs, on one hand, or neural networks underlying internal factors such as intentions and mood states, on the other. However, such malfunctions in different networks could depend, for example, on a common molecular or neurotransmitter deficit. This issue can perhaps best be resolved when we have firmer information about the neural substrates of intrusive thinking. There are some promising indications of this (see chapters by Philips, Fedota and Stein, Balleine, Badre, Gourley et al., and Roberts et al., this volume) as well as possible genetic relationships, which will depend on having precise definitions of the phenotype obtained, for example, through definitive objective tests and a standardized general instrument for assessing intrusive thinking.

The mechanisms and paradigms discussed have implications for therapeutic applications. An idea often discussed in the clinical literature is that suppressing intrusive thoughts in clinical disorders is counterproductive, an idea substantially influenced by Wegner's white bear thought suppression procedure (Wegner et al. 1987; Wegner 1997). However, substantial recent evidence with the Think/No-Think procedure suggests that findings from the white bear paradigm have been overgeneralized and that other processes, such as retrieval suppression, are effective in modulating thoughts, at least in healthy individuals. If so, there may be situations in which suppressing thoughts could be useful to improve an individual's ability to function, even in clinical disorders. For example, it may be that both trauma-focused interventions, such as eye movement desensitization reprocessing, and paradigms using cognitive interference, such as memory retrieval procedure and visuospatial task (Holmes et al., this volume), which appear to be effective in reducing the frequency of intrusive memories of trauma, work through a process similar to the retrieval-suppression mechanism described above. While it is clear that in some disorders (e.g., OCD), attempts to suppress intrusive thoughts can lead to rebound, clinical disorders involving intrusive thoughts are heterogeneous, as are patient populations who have these disorders. Thus, it is critical to explore the therapeutic implications of paradigms which have demonstrated efficacy in controlling, replacing, and suppressing intrusive thoughts in human laboratory settings (see also Brewer et al., this volume).

A particularly promising therapeutic approach for fostering the effective control of intrusive thoughts is mindfulness-based cognitive therapy, which integrates meditation techniques with cognitive behavioral strategies (Külz et al. 2014). One of the primary skills taught in such approaches is the ability to control one's thoughts to focus on breathing. The act of releasing thoughts in mindfulness practices is highly reminiscent of the control strategies invoked in the Think/No-Think paradigm, further illustrating their potential pertinence to the treatment of intrusive thoughts. Mindfulness practices may also be helpful in furthering individuals' meta-awareness of having intrusive thoughts in the first place (Baird et al. 2014). Specifically, such practices may enable individuals to identify episodes of unwanted thoughts that might have otherwise been experienced but evaded explicit acknowledgment (Baird et al. 2013a; Takarangi et al. 2014). Identifying intrusive thoughts in the light of meta-awareness may enable individuals to invoke the necessary mental control strategies required to release them.

Sociocultural Context

Intrusive thinking and its control (or lack thereof) take place within a sociocultural context. Examples of this include the AIDS-HIV epidemic of the 1980s. In his book, *The Man Who Couldn't Stop: OCD and the True Story of a Life Lost in Thought*, David Adam (2014) provides a vivid example of how frequent concerns about infection drove the induction of his own obsessive and compulsive symptoms. Certainly, there are many other examples of widespread general concern (e.g., nuclear war, Brexit, global pandemics) that intrude into our everyday consciousness and may lead to pathological consequences.

Although many of us successfully resist such concerns, this too can lead to societal consequences, as can be currently observed through the striking example of the climate crisis. Since the Industrial Revolution, global temperatures have increased by 1°C and humankind is well on track to experiencing an additional increase of 0.5°C by 2030 (Xu et al. 2018). In line with the worst-case scenario put forth by the Intergovernmental Panel on Climate Change (IPCC 2018), this half a degree will likely correspond to a 50% increase in extreme weather events worldwide (e.g., droughts, floods, snowstorms, hurricanes, cyclones) and exact devastating consequences, including mass migration, agricultural failure, and deadly heat events (Xu et al. 2018). The near total consensus of climate scientists, exemplified by the Paris Agreement, which almost all governments signed (and almost none are honoring), is that we have to reduce emissions soon. Failure to do so will result in continued temperature increases that will soon be beyond human control and ultimately lead to a "hothouse planet" by 2100, or perhaps even sooner (Wallace-Wells 2019).

The climate crisis is surely creating daily thought intrusions in hundreds of millions of people. Such intrusions are likely characterized by interrupting, salient, experienced events (imagery, emotion, moral feeling) that recur. This example illustrates the complexity of judging whether intrusive thinking is maladaptive or not, and the range of responses that people can have. For instance, against the unimpeachable backdrop of scientific knowledge, such intrusions appear highly adaptive: *they compel action* and yet only a very small minority of people *are* currently engaged in action—the great majority of global citizens are not taking action. Of these, some may deny the science or the predicted impacts, or they may accept the science and impacts but deny that any serious action is warranted (e.g., because they are ideologically committed to the current economic system). In Western societies, polling shows that most people fall in the latter category (accepting the science and the impacts), yet they are not acting beyond some minor adjustments in their personal lives, possibly because action would be inconvenient to one's career or lifestyle or because it would require confronting grief and fear in a way that one is not yet prepared to do. Since such people do understand and accept the terrifying imperative to act, but are not doing so, they could be characterized as exerting control over the intrusive thinking caused by the climate crisis. The form of control being used is probably based on reappraisal: people express degrees of fatalism ("the problem is too big or too hopeless," "it is too late"), nihilism ("humans deserve what is coming to them"), a deferral of responsibility to policy makers ("it is a government problem"), or presently unwarranted faith in technology alone to solve the problem (Hansen 2018). Nonetheless, as more extreme weather events, for instance, increase over time, so too will levels of anxiety, until the intrusion breaks through the threshold of control and people clamor for action.

In summary and in accordance with the definition of intrusive thinking as interrupting, salient, experienced events (imagery, emotion, moral feeling) that are also recurrent, the climate crisis is clearly generating intrusive thinking in a wide range of people around the world. This, in turn, creates conflicts which can inculcate attempts at control (e.g., at the reappraisal level). Such tensions exacerbate poor mental health as well as the feelings of instability and fear, which are already driving populist political regimes (Latour 2018). Meanwhile fake news disseminated within the broader system functions to degrade the salience signal that is currently essential to generate the type of population-level intrusive thinking that would compel quick action. This sociocultural example is interesting because it represents the flip side of many of the psychiatric-related examples discussed in this chapter. Whereas intrusions were often characterized as pathological, the climate crisis example demonstrates how intrusive thinking is a good thing for our biosphere and civilization, and attempts to control these intrusions are maladaptive. The survival of human civilization depends on more intrusive thinking right now.

Concluding Summary

Our discussion focused on the psychological bases of intrusive thinking and its control that may occur every day in healthy individuals, as well as in psychiatric disorders. We surveyed the range of phenomena that can be construed as examples of intrusive thinking and endeavored to reach satisfactory definitions and classifications of the different forms of intrusive thinking that will aid further research. The least constrained definition was one emphasizing the interruptive, salient, and experienced nature of intrusive mental events (as compared with a common definition which specifies unwanted and conscious, as well as interruptive criteria). Recurrence is a further property which may be more important for psychiatric manifestations. Agency, meta-consciousness, and (mal)adaptiveness or desirability are considerations that further define the boundaries of what can be considered as intrusive thinking. Based on this analysis, PTSD and OCD appear to be prototypical disorders of intrusive thinking, although its elements appear in a range of other psychiatric diagnoses, including addiction and depression, though probably not psychosis.

The main part of our discussion focused on neurocognitive mechanisms of intrusive thinking and its control, which included impairments particularly in the regulation of memory retrieval, as well as of affect and action control. We brainstormed future possible approaches to investigating intrusive thinking, with priorities for designing a new, theoretically motivated questionnaire, refining existing objective measures and capitalizing on the potential of computational approaches. Finally, we considered the importance of this in the context of broader social-cultural and clinical-therapeutic issues.

Systems and Models

10

Networks Relevant to Psychopathology and Intrusive Thought

John R. Fedota and Elliot A. Stein

Abstract

Intrusive thoughts are regular occurrences in healthy cognition. Across a variety of psychiatric conditions, however, such thoughts can become unconstructive and perseverative. Failures in computations to estimate the salience of the content of these thoughts are at least partly responsible for these clinically relevant disease symptoms. This chapter reviews neuroimaging results that show specific and related dysfunction in the calculation of salience at multiple neuroanatomically and functionally linked regions of interest, both cortically and subcortically. Transdiagnostic evidence for dysfunction in the striatum, thalamus, and prefrontal cortex is reviewed, as is a theoretical framework placing these regional findings in the context of large-scale brain networks. It is argued that changes in nodal function and network communication are signatures of a failure to properly shape predictions about the reliability and utility of external and internal stimuli, leading to maladaptive attentional capture and behavior, including intrusive thoughts.

Introduction

Spontaneous thought that is unrelated to current task demands is a normal part of healthy cognition. According to estimates from thought probe experiments, 30–50% of waking cognition is unrelated to specific tasks (Kane et al. 2007; Killingsworth and Gilbert 2010). However, even in normal cognition, this frequent mind wandering appears to bear an emotional cost. When prompted during mind wandering, participants reported being less happy than during task-focused cognition (Killingsworth and Gilbert 2010).

Spontaneous thoughts rarely occur only once. Instead, they are often recurrent. Across all types of repetitive thought, a common feature is an internal focus on "one's self and one's world" (Segerstrom et al. 2003). The content of these internally focused thoughts can be either positive or negative, such as daydreaming versus worry (Watkins 2008; Christoff et al. 2016). An extensive taxonomy of repetitive thoughts exists (Watkins 2008), which describes both constructive and unconstructive consequences of intrusive and repetitive thoughts that can be linked to psychiatric conditions.

The ubiquity and diversity of repetitive thoughts during cognition beg the question: What characteristics mark the transition from healthy, spontaneous thought to clinically relevant, repetitive intrusive thought? It appears that the valence of the repetitive thought is an important factor in determining clinical relevance. In Watkins's (2008) taxonomy, negatively valenced thoughts with unconstructive consequences include depressive rumination, worry, and perseverative cognition. These consequences can clearly be linked to psychiatric conditions such as anxiety disorders, major depressive disorder (MDD), obsessive-compulsive disorder (OCD), posttraumatic stress disorder (PTSD), substance use disorders (SUDs), and schizophrenia (SZ). Negative repetitive thoughts (NRTs) become clinically relevant when their magnitude or frequency increases or when they become perseverative and difficult to control or eliminate (Kalivas and Kalivas 2016).

Cognitive Constructs Relevant to Negative Repetitive Thoughts

Beyond a taxonomy of NRTs, a description of the cognitive construct(s) underlying these clinical phenomena is necessary. Here, the goal is to operationalize the definition of NRTs as an intermediate step to identifying specific brain regions, circuits, and large-scale networks where the identified constructs are neurobiologically instantiated.

Cognition, intrusive or not, can be conceived as the making of predictions about the environment, testing those hypotheses via sensory processing, and subsequently updating and refining these predictions based on experience. Thus, optimal interaction with the world requires accurate beliefs about the environment and the ability to update those beliefs as new *reliable* evidence is encountered. This process can be computationally formalized as Bayesian predictive coding, a general theory of brain function that specifies how goal-directed behavior is motivated through the integration of prior beliefs with sensory information (Rao and Ballard 1999; Doya et al. 2007; Itti and Baldi 2009; Friston et al. 2012; Aitchison and Lengyel 2017). This framework integrates disparate cognitive processes including learning, reward, executive control, attention, and sensory processing.

Briefly, Bayesian predictive coding involves the integration of prior beliefs and available sensory evidence to refine posterior estimates of beliefs. Any mismatch between these two distributions signals the need to update beliefs considering the evidence encountered. The computational weight given to probabilistic estimates of either prior beliefs or sensory evidence is governed by the relative precision of each (Mathys et al. 2011). That is, precisely estimated priors are less susceptible to modulation based on sensory evidence, while poorly estimated (imprecise) priors are readily adjusted by sensory evidence. As different sensory inputs can vary in their precision and information content, one cognitive imperative is to estimate which available sensory input will provide reliable and informative information to calculate better posterior estimates of the environment (Parr and Friston 2017, 2019). To this end, unambiguous sensory data should be amplified when present and sought when absent.

Cognitively, this predictive estimation of potentially reliable sensory sources has been described as the attribution of salience (Parr and Friston 2019). By salience, we mean a quality that is particularly noticeable or deemed important to the individual in a given context (Uddin 2014; Kahnt and Tobler 2017; Miyata 2019). As such, elements that provide unambiguous sensations should be ascribed higher salience, as they will provide reliable information for the adjustment of posterior estimates and iteratively improve subsequent decisions. The degree to which new information alters posterior estimates as compared to prior beliefs is termed *Bayesian surprise* (Itti and Baldi 2009; Barto et al. 2013).

Salience is closely related to value, but a key distinction is that value is a signed currency that varies monotonically from negative to positive whereas salience is an unsigned currency, where both negative and positive predicted outcomes can have equivalent salience (Kahnt and Tobler 2017). This is because both positively and negatively valent elements of the environment can improve posterior estimates; each can be informative in the refinement of posterior estimates.

When working properly, this iterative cycle of hypothesis (prior)–result (sensory evidence)–conclusion (posterior) allows for the flexibility and learning characteristics of human cognition. However, in the case of NRTs, improper predictions of the salience of elements in the environment will lead to suboptimal processing, including perseveration on elements with overly precise priors and/or a failure to guide attention to elements with weak priors.

For example, if previous experience creates an overly precise prior belief about the reliability of information to be gleaned from a stimulus (e.g., a drug cue or emotionally valent memory), its salience will be increased. Such an increase in anticipated salience will lead to the focus of greater attentional resources on the stimulus, potentially at the expense of alternatives in the environment. This dysregulated focus on one thought at the expense of others is a hallmark of intrusive thought, as described in the following sections. Indeed, the computational framework of Bayesian predictive coding has been shown to be useful in describing specific deficits across a variety of psychiatric conditions associated with NRTs: hallucinations in SZ (Sterzer et al. 2018), drug cravings in SUD (Gu and Filbey 2017), and perseverative focus in OCD (Levy 2018).

Below, we evaluate neuroimaging data across nodes and networks previously implicated in the cognitive processes related to the estimation of prior probabilities (salience) and attentional modulation within the framework of Bayesian predictive coding. Not surprisingly, the multiple points of failure in the information processing cascade requires discussion of a wide range of implicated brain regions. In addition, these individual brain regions can be supra-ordinately organized as nodes in large-scale networks, providing a systems-level perspective on dysfunctional communication associated with the estimation of salience. Each source of potential salience attribution dysfunction will be addressed in turn.

Ventral Striatum and Potentiated Response

A large body of literature has shown the striatum to be responsive not only to value and reward, but also to the salience of a given stimulus, independent of its positive or negative valence. In humans, these reward (magnitude * signed valence) and salience (magnitude * *absolute value* of valence) responses are separable: salience-evoked activation is seen in the ventral striatum (Zink et al. 2006; Jensen et al. 2007; Bartra et al. 2013), the insula, and the dorsal anterior cingulate cortex (dACC), while reward (positive valence) is also encoded in the striatum and across various brain regions, including the orbitofrontal cortex (OFC) (Litt et al. 2010; Bartra et al. 2013; Kahnt et al. 2014).

Dysfunction in striatal signaling related to the identification of salient environmental stimuli is observed across psychiatric conditions associated with NRTs. When compared to healthy individuals, those at risk for developing psychosis show heightened activation to task-irrelevant stimuli within the ventral striatum (Roiser et al. 2012; Schmidt et al. 2016). The observed increase in ventral striatal activation to irrelevant stimuli suggests an oversensitivity to uninformative cues in the environment. Within the Bayesian predictive coding framework described above, during normal cognition these irrelevant cues should be ascribed reduced salience, as they provide little to no sensory information with which to refine posterior estimates of the task environment.

Similar biases in ventral striatal activation are seen in other conditions such as OCD and bias in processing losses (Jung et al. 2011), MDD and anhedonia (Whitton et al. 2015), as well as SUD and cue reactivity (Volkow et al. 2010; Kühn and Gallinat 2011). Further, in a recent review of SUD, Moeller and Paulus (2018) observed that ventral striatal activation patterns are related to long-term abstinence outcomes: increased activation in response to drug cues is related to worse clinical outcomes (i.e., increased substance use and recidivism), whereas increased activation to monetary or nondrug reward cues are related to better clinical outcomes. In each case, ventral striatal activation is biased in its sensitivity to a variety of environmental cues. Thus, the dysregulated salience attribution does not appear to be a consistent, transdiagnostic insensitivity to reward or aversive stimuli. Instead, it appears there is an inability to distinguish properly between task-relevant and task-irrelevant (or intrusive) cues.

However, the question remains: How does such an increase in evoked activity precipitate a perseverative or intrusive thought? One well-articulated neurobiological mechanism, instantiated in the ventral striatum, for such a transition to salience misattribution and uncontrolled attentional capture is the glutamate-mediated transient synaptic potentiation model (reviewed by Kalivas and Kalivas 2016).

Briefly, this model suggests that upon presentation of a previously potentiated cue (e.g., a drug-related cue in SUD or an object of obsession in OCD), excessive glutamine release in the nucleus accumbens (NAc) core leads to transient synaptic potentiation that biases the attribution of salience to the potentiated cue being processed. In addition, this bias in the NAc core leaves the region less responsive to alternative cues that would normally be more fully encoded (Kalivas and Kalivas 2016). This dual mechanism likely increases the magnitude of the difference between potentiated and alternative cues and further instantiates salience of the potentiated cue.

The molecular mechanism for such a transient biasing of NAc core processing has been described in a rodent self-administration model: Drug-seeking behavior was related to transient potentiation of D1 receptors in the NAc core following presentation of drug cues. Moreover, following brief access to cocaine this potentiation rapidly extinguished, only to be reinstated following 45 minutes of forced abstinence (Spencer et al. 2017). These dynamics of potentiation are consistent with models of SUD that describe preoccupation with and craving for drugs, both of which can be viewed as NRTs (Koob and Volkow 2009).

Once a cue is ascribed a salience incommensurate with its task relevance, transient synaptic potentiation within the ventral striatum may sustain this bias by both increasing the coded salience of the potentiated (i.e., maladaptive) cue, while simultaneously decreasing the salience of any alternative, competing (i.e., goal-directed) cue. Computationally, this cascade is consistent with biased processing of specific subsets of previously encountered stimuli (e.g., drug cues in SUD), potentially leading to a reduced ability to modify posteriors based on experience. Maladaptive ventral striatal activation across a variety of psychiatric conditions is consistent with this interpretation.

Cortico-Striatal-Thalamo-Cortical Loops Convey Salience Signals throughout the Brain

Regardless of how the NAc core encoding of salience is biased in disease, neuroanatomical evidence clearly shows processing in the ventral striatum does not occur in isolation. Dense and reciprocal interconnections between striatum, thalamus, and cortex in the form of cortico-striatal-thalamo-cortical (CSTC) loops allow for complex communication among brain areas when processing salient and motivationally relevant information. These connections also provide an anatomical mechanism to convey reciprocally looping ascending (striatum–cortex) and descending (cortex–striatum) influences on salience computation. The physical connections along CSTC loops have been described in great detail (Haber and Knutson 2009; Haber 2016), while tract-tracing results in nonhuman primates can be clearly related to patterns of resting-state functional connectivity in humans (Parkes et al. 2016; Choi et al. 2017).

Classically, three loops linking the dorsal striatum with presupplementary motor area, frontal eye fields, and dorsolateral prefrontal cortex (dlPFC), and two loops linking the medial and ventral striatum with the OFC and ACC, respectively, have been defined (Alexander et al. 1986). These connections form a gradient in cortical projections to the striatum, whereby ventral striatal inputs are associated with PFC areas processing emotion, caudate inputs with cognition, and putamen inputs with sensorimotor processing (Haber 2016). These interconnections suggest that the biased signals processed in the ventral striatum are ultimately conveyed throughout the brain, including specific regions discussed below. Many of these are primary nodes of large-scale networks across the cortex relevant to attentional control (Uddin 2014; Heilbronner and Hayden 2016) and psychiatric disease (Menon 2011; Sutherland et al. 2012; Kaiser et al. 2015).

However, it is important to note there is no clear anatomical or functional boundary between the ventral and dorsal striatum, and the cortical projections to and from the striatum create a continuum of connectivity (Haber and Knutson 2009; Choi et al. 2017; Marquand et al. 2017). Thus, assigning a one-to-one connectivity relationship between striatal and cortical regions is not possible. In fact, it has been estimated that the terminal fields of dACC, ventromedial PFC, and OFC cover almost 25% of the striatum, which is an overrepresentation as compared to the cortical volume of the brain (Haber et al. 2006). This overrepresentation is especially germane to the current discussion due to the role of OFC and dACC in reward and salience processing, respectively (Bartra et al. 2013). Instead of a well-defined gradient of CSTC loop connections, clear zones of integration as well as convergence are observed across the striatum (Haber and Behrens 2014; Choi et al. 2017; Marquand et al. 2017).

Choi et al. (2017) describe a homology between tract tracing in nonhuman primates and resting-state functional connectivity in humans, illustrating clear delineations in the pattern of physical and functional connection between the ventral and dorsal striatum. This pattern of connectivity agrees with previous tract-tracing evidence in nonhuman primates (Chikama et al. 1997). Specifically, more ventral striatum is strongly connected to the dorsal anterior insula (aIns) (Chikama et al. 1997) and dACC (Kunishio and Haber 1994; Parkes et al. 2016). In contrast, the rostral dorsal caudate is a hub of integration, with projections to and from a variety of brain regions associated with attentional control, including caudal inferior parietal lobule (IPL), ventrolateral, dorsolateral, and dorsomedial PFC, dACC, and OFC (Choi et al. 2017; Marquand et al. 2017).

The observed dichotomy in striatal projections to cortex is consistent with circuits differentially impacted by chronic exposure to drugs. Our group has identified a *ventral* striatal–dACC circuit whose resting-state functional connectivity is reduced in chronic cocaine users and a *dorsal* striatal–dlPFC circuit whose connectivity is increased in the same group as compared to healthy controls (Hu et al. 2015). Further, the balance between the up- and downregulation of these circuits was correlated with DSM-IV-TR compulsivity symptoms in cocaine users. Similarly, in first-episode SZ, reductions in ventral striatal–dACC resting-state functional connectivity are observed and have been correlated with reported symptom severity (Lin et al. 2017). These findings show that ventral striatal–dACC connectivity modulations are relevant across multiple conditions associated with NRTs (in this example, SUD and SZ). Further, the relationship between connectivity strength and clinically diagnostic criteria suggests a role for these circuits as potential neuroimaging biomarkers of disease severity.

The CSTC loops between cortex and striatum are interspersed with connections from striatal regions to various thalamic nuclei, which in turn are connected back to the cortex. The cortical and thalamic connections to the striatum are coordinated, meaning interconnected cortical and thalamic regions both project to the same striatal area. For the purposes of this discussion, the central-medial (CM) and medial parafascicular (PF) nuclei of the thalamus, which are connected to medial PFC areas including dACC (Behrens et al. 2003), are also connected to the ventral striatum (Van der Werf et al. 2002).

An illustrative example of impairment across nodes of the described CSTC loops in the processing of highly salient stimuli is seen in an imaging paradigm that employs erotic pictures (Metzger et al. 2010). When the salience of an anticipated erotic picture is processed by healthy participants, the nodes of this CSTC loop linking dACC and ventral striatum via the CM/PF thalamus become activated, as identified via high field fMRI. Here, CM/PF thalamus, dACC, and aIns showed increased activation during the anticipation of a salient, erotic picture (the ventral striatum was outside of the field of view in this study; Metzger 2010). Gola et al. (2016), provide further supportive evidence by showing enhanced anticipatory processing of erotic pictures within the ventral striatum of men seeking treatment for problematic porn use. Treatment seekers in this later study displayed enhanced striatal activation to the cues predicting erotic pictures but not to cues predicting monetary gains.

Taken together, Metzger et al. (2010) and Gola et al. (2016) show that anticipation, not consumption, of highly salient stimuli increases activation in each node of the CSTC loop (i.e., ventral striatum, CM/PF thalamus, and dACC). The observed activation within these nodes during *anticipatory* processing is consistent with the conceptualization of salience as a predictive computational process for encoding anticipated relevance of a specific stimulus in the environment (Parr and Friston 2019), in this case an erotic image. These results provide an additional example of biased salience processing in a behavioral addiction (Potenza 2015; Kraus et al. 2016). This further broadens the scope of conditions associated with NRTs and dysfunction in core nodes of these CSTC loops.

The role of the thalamus in health and disease is an area of active inquiry. The traditional designation of the thalamus as a passive "relay station" for information processed elsewhere in the brain is being reconsidered. Recent evidence suggests thalamic influence on cortical connections, including the coordination of activation and direct control of synchronicity between cortical regions via gain control (Saalmann 2014), and active filtering of information (for a review, see Halassa and Kastner 2017). Consistent with this more active processing role, the mediodorsal thalamus is suggested to amplify signals in the PFC (Parnaudeau et al. 2018) and to extend representations in the PFC over longer time durations than those associated with cognitive processes, like working memory (Pergola et al. 2018). Advances in thalamic parcellation (Kumar et al. 2017) and quantification of its resting-state functional connectivity with the entire brain (e.g., via 7T fMRI) will only improve the resolution of these findings and further articulate the role of thalamic nuclei in salience attribution.

Prefrontal Cortex: Regions of Interest Implicated in Psychiatric Disease

With NAc core dysfunction now described along with neuroanatomical links between this region and thalamic and prefrontal regions, including the dACC and aIns, our focus shifts from striatal to cortical areas implicated in psychiatric disease associated with NRTs. A recent meta-analysis of structural (n > 15,000) (Goodkind et al. 2015) and functional (n > 5,000) (McTeague et al. 2017) data from psychiatric patients across a variety of disorders associated with intrusive thoughts (SZ, MDD, SUD, OCD) identified specific yet overlapping areas of dysfunction. Specifically, the pattern of gray matter loss in patients was circumscribed to dACC and bilateral aIns (Goodkind et al. 2015). In agreement with these structural findings, hypoactivation in cognitive control task-evoked activity was also observed in the dACC and right aIns as well as in the left dlPFC and right IPL (McTeague et al. 2017). The tasks employed in the meta-analysis did not probe intrusive thoughts per se, but rather top-down cognitive control more generally. That said, the pattern of hypoactivation across disease conditions showed reductions in regions associated with both attentional control (i.e, left dlPFC, right IPL) and the calculation of salience (i.e., dACC, aIns).

The aIns and dACC are coactivated across a wide variety of cognitive control and attention-related tasks. In fact, activation in these nodes are among the most observed results in the fMRI cognition literature (Behrens et al. 2013).

Given their ubiquity in the extant literature, aIns and dACC have been theorized to play a central role in broad cognitive constructs, as in the detection of relevant information from both the external and interoceptive worlds and the coordination of appropriate attentional capture and behavioral response to these salient signals (Uddin 2014; Heilbronner and Hayden 2016; Nour et al. 2018). The degree of task activation in these regions increases with the demand for attentional control or with an increase in ambiguity of stimuli during perceptual decision making (Lamichhane et al. 2016).

The insula is associated with the integration of interoceptive information into calculations of salience (Critchley et al. 2004; Craig 2009) via ascending pathways communicating visceral and allostatic information (Critchley and Harrison 2013; Kleckner et al. 2017). Within the insula (and the dACC, as discussed below), these signals are integrated to create a single subjective image of "our world" (Kurth et al. 2010); interoceptive representations in the insula have been theorized to provide the basis for a perception of self via the integration of interoceptive signals related to physical state (Seth 2013; Namkung et al. 2017). This self-focused processing localized to the insula is strikingly consistent with the description of the content of repetitive thought as being "related to one's self and one's world" (Segerstrom et al. 2003).

Returning to a transdiagnostic theme and the observation of insular hypoactivity (McTeague et al. 2017), impairments in interoception are implicated across psychiatric conditions associated with NRTs (Khalsa et al. 2018). Insular activation in response to cued recall of a previously interoceptive challenge was diminished in subjects with MDD as compared to healthy controls (DeVille et al. 2018). Similar impairment in interoceptive processing focused on insular hypoactivation (Naqvi and Bechara 2010) has also been associated with SUD (Goldstein et al. 2009; Sutherland et al. 2013; Paulus and Stewart 2014).

In addition to a central role in the integration of interoceptive signals, the insula is a primary adjudicator between external (exogenous) and internal (endogenous) focus as a function of salience attribution (Sridharan et al. 2008; Menon and Uddin 2010; Uddin 2014). Characterization of the causal interactions between insula and dACC, however, shows additional divergence in their patterns of activation. Specifically, aIns has been shown to amplify the detection of salience within the dACC (Chen et al. 2014; Cai et al. 2015). This timescale is consistent with EEG spectral analysis showing insular activation precedes that of dACC (Chand and Dhamala 2016).

In contrast to the association with interoceptive processing in the insula, the dACC is associated with cognitive monitoring and control processes along with economic decision making (Botvinick 2007; Kolling et al. 2016; Shenhav et al. 2016; Alexander and Brown 2017). Recent integrative accounts suggest that the primary function of the dACC is to process multiple facets of information about the context in which a decision is being made to enable the appropriate goal-directed strategy (Heilbronner and Hayden 2016; Li et al. 2018b). That is, dACC codes task-state information (originating either endogenously or exogenously) that is relevant to the current demands on the individual via adaptive coding to discount or ignore irrelevant details (Heilbronner and Hayden 2016). An important difference between these context-relevant calculations and those occurring in the OFC, which are more closely linked with reward, is that while the OFC encodes the value of the current choice, the dACC integrates across multiple dimensions and across a longer timescale (Kennerley et al. 2011).

The related calculations of interoceptive salience and context-relevant stimuli within the aIns and dACC can be more formally combined via recent studies of belief updating in healthy populations. The belief updating calculation incorporates the integration of new, relevant information with existing expectations, termed Bayesian surprise (Barto et al. 2013). Divergence between prior belief and updated posterior beliefs characterizes the level of this "surprise." Recent work in healthy individuals shows that surprising (salient) information *relevant to updating beliefs* is encoded in the ventral striatum, aIns, and dACC whereas merely surprising but uninformative information is not (Nour et al., 2018). This is an important distinction, as these regions appear to differentiate the information content of different sensory streams. This function is in line with the construct of salience, as defined by Parr and Friston (2019): the anticipated reduction in uncertainty is associated with a specific element in the environment.

Thus, the increase in activation across the striatum, aIns, and dACC can be putatively related to the accurate identification of relevant information (i.e., the salience definition described above) used to refine beliefs and guide goal-directed behavior. Thus, along with the ventral striatum, the aIns and dACC play a combined role in distinguishing novel (unexpected, uninformative) and surprising (unexpected, informative) sources of information. Informative information is used to alter response strategies (e.g., attentional modulations or motor output), whereas surprising but irrelevant information is not.

Applying this idea more directly to psychiatric conditions that are linked to NRTs, a failure to accurately encode Bayesian surprise is consistent with the hypoactivations in aIns and dACC described by McTeague et al. (2017). Hypoactivation in these regions points to an inability to update beliefs efficiently, which in turn may perpetuate a positive feedback loop whereby poor estimates of salience guide attention to maladaptive elements in the environments, leading to perseverative focus on uninformative stimuli, which may precipitate NRTs.

The Salience Network in Intrusive Thought

Within the described CSTC loop, dysfunction and/or maladaptive bias toward potentiated stimuli is observed at multiple levels of salience processing. Additionally, many of the implicated regions of interest are regularly activated in concert depending on task demands. For example, Nour et al. (2018) show ventral striatal, dACC, and aIns activation in the processing of informative, novel information, with the dACC and aIns regularly coactivated (Behrens et al. 2013). These commonalities suggest a benefit to examining the brain at a higher level of organization, moving from regions of interest to dyadic circuits to large-scale network organization.

To this end, dACC and aIns are the primary nodes of the salience network (SN). Originally described by Seeley et al. (2007), and subsequently replicated using a variety of methodologies (Dosenbach et al. 2007; Smith et al. 2009; Power et al. 2011), the SN is a centralized processor that ascribes salience to stimuli (Uddin 2014) while coordinating attentional resources between an internal and external focus in response to task demands (Sridharan et al. 2008; Menon and Uddin 2010). Such a function is clearly consistent with the roles ascribed individually to aIns and dACC detailed above.

While the dACC and aIns form the primary nodes in the SN, additional large-scale networks are broadly relevant to cognition (Smith et al. 2009; Ji et al. 2019). The interaction between the SN and these networks leads to a more global view of the dysfunction associated with intrusive thought; such SN interactions can be conceptualized within a tripartite network model. Briefly, this model describes a default mode network (DMN), which includes the rostral and posterior cingulate cortices, parahippocampal gyrus, and bilateral inferior parietal cortex (Raichle et al. 2001; Buckner and DiNicola 2019), and an anticorrelated executive control network (ECN), which includes the bilateral dlPFC and parietal cortices (Honey et al. 2007). During interoceptive processing, the DMN is relatively more active and the ECN is relatively deactivated (Fox et al. 2005; Keller et al. 2013). During exteroceptive processing the reverse is true. Toggling between these two networks is thought to be mediated by the SN (Menon and Uddin 2010).

Indeed, regardless of the cognitive function ascribed to the individual regions of the SN, transdiagnostic findings implicating nodes of the SN, both structurally (Goodkind et al. 2015) and functionally (McTeague et al. 2017), have led to conceptualizations of SN dysfunction as a core transdiagnostic symptom of psychiatric dysfunction (Menon 2011). Transdiagnostic differences in activation within nodes of the CSTC loops connecting striatum to SN nodes have been described in detail by Peters et al. (2016).

Tripartite Network Model and Aberrant Cognition

An influential theoretical model by Menon (2011) describes a transdiagnostic hypothesis of aberrant salience calculation in psychopathology that centers on dysfunctional network connections between SN, DMN, and ECN. The model suggests that the normal adjudication between internal (DMN) and external (ECN) focus mediated by the SN (Sridharan et al. 2008) is commonly disrupted across psychiatric diseases. While entirely consistent with biased calculations within the nodes of the SN (aIns and dACC) and ventral striatum described above, this network-level model provides a systems neuroscience description of brain dysfunction and explicitly relates changes in salience attribution to emotional and attentional processing distributed throughout the brain.

The dynamic interactions between large-scale networks are determined by a variety of factors. For example, the integrity of connections within a given network (Honey and Sporns 2008; Boes et al. 2015) is important to determine the functional capabilities of that network and the communication fidelity both within and between networks. In addition, high-fidelity messages can fail to be communicated successfully if hubs of interconnection between networks are dysfunctional in disease (Cole et al. 2013; Gratton et al. 2018). These communications also occur dynamically, and the engagement or disengagement of functional connectivity between networks, a process mediated by the SN, are potential points of failure at a systems level.

Menon's model of dysconnectivity has proven prescient (Menon 2011). Across a variety of diseases associated with NRTs, network-level dysconnectivity between SN–DMN, SN–ECN, and DMN–ECN have been observed. Using the tripartite network framework, a meta-analysis (n > 16,000 total patients, including n > 8,000 patients) across a variety of psychiatric diseases (including MDD, OCD, PTSD, SZ) associated with NRTs identified transdiagnostic patterns of seed-based functional dysconnectivity both within nodes of the SN, DMN, and ECN as well as between the three networks (Sha et al. 2019). Hypoconnectivity is observed at rest across diseases *within* nodes of the DMN and SN and *between* nodes of SN and both DMN and ECN. In contrast, hyperconnectivity is observed within the ECN and distinct nodes within DMN, between the ECN and DMN, and between DMN and subunits of the SN.

These results are broadly consistent with Menon's tripartite model: reduced connectivity between SN and DMN or ECN suggests a reduced complement of information to integrate into accurate and/or flexible estimates of prior and posterior beliefs about the environment. However, hyperconnectivity between DMN and nodes of the SN are also observed, suggesting the balance between connections, as opposed to a unitary increase or decrease between networks, may be a distinguishing feature (e.g., Hu et al. 2015).

A second recent result further broadens the scope of analysis: interrogating brain-wide connectivity patterns via a connectome-wide association study approach (Shehzad et al. 2014) to identify brain regions whose whole-brain connectivity pattern is associated with the p factor, a hypothesized common factor underlying psychopathology across disease conditions (Caspi et al. 2013). Using a data-driven approach and a large data set (n > 600), four regions in the occipital cortex were identified where whole-brain multivariate connectivity patterns were correlated with p factor scores (Elliott et al. 2018).

These results are not directly interpretable within the tripartite network hypothesis, as the regions identified fall outside of traditional SN, ECN, DMN boundaries, when the identified occipital regions are used as seeds in a resting-state functional connectivity analysis, similar to those analyzed by Sha et al. (2019). Nonetheless, hyperconnectivity with both the ECN and DMN was positively correlated with p factor score. Importantly, while DMN and ECN within the tripartite network model are usually considered oppositional (Fox et al. 2005; Sridharan et al. 2008; Menon and Uddin 2010), the connectivity of both networks was similarly enhanced with an independent, and psychiatrically relevant, region of the occipital cortex. Further, the degree of this shared increase in functional connectivity correlated positively with p factor score.

These results point to network interactions beyond the SN (Peters et al. 2016) that may indirectly influence the salience calculations instantiated within dACC and aIns. In both of these recent cases, the network structure separating ECN and DMN appears to be reduced, either through hyperconnectivity between these two normally oppositional networks (Sha et al. 2019), or via increased coherence with a mediating node outside of the tripartite network (Elliott et al. 2018). The coherence between ECN and DMN may further bias the information integrated in SN, though an empirical demonstration of this remains outstanding.

Conclusion

In psychiatric conditions that include NRTs as a symptom, a common set of biases in information processing and dysconnectivity between specific nodes as well as large-scale networks is observed. In each case, these dysfunctions appear to reflect a failure to tune or modulate the response or connection properly, as opposed to a broad deficit in either task-evoked activation or connectivity. Recent advances in gathering large data sets across a diversity of psychiatric conditions have aided in revealing these dysfunctions.

It is important to note that *healthy* cognition includes the regular experience of intrusive thoughts. It is an increase in the perseverative focus on these thoughts that leads to clinically relevant dysfunction. Thus, it is not the presence of a response within or a connection between brain regions that is indicative of a disease as much as it is the inability to discriminate properly among alternatives or determine the most relevant information to guide decision making. We suggest that processing biases at the level of the striatum, thalamus, insula, and dACC indicate computational dysfunctions during the Bayesian predictive coding of salience.

As a representative example, we extend the model by Kalivas and Kalivas (2016) to incorporate a more explicit role for thalamus, dACC, and aIns along with large-scale networks within the tripartite model of brain function. Biased processing of potentiated stimuli, at the expense of alternative stimuli, within the ventral striatum is strongly linked to glutamate-mediated transient synaptic potentiation (Spencer et al. 2017). These signals are conveyed along the well-described CSTC loops (Choi et al. 2017) to the thalamus, where recent evidence suggests their representations in PFC may be amplified or sustained (Parnaudeau et al. 2018; Pergola et al. 2018), potentially increasing bias.

Hypoactivation of nodes within the SN (Peters et al. 2016; McTeague et al. 2017) leads to suboptimal integration of interoceptive signals (Kleckner et al. 2017), which may be biased due to the allostatic load of psychiatric disease more generally (McEwen and Gianaros 2011). In addition, the integration of endogenous and exogenous information to identify and fully process contextually relevant information (Heilbronner and Hayden 2016) is likely biased by the observed hypoactivity within nodes of the SN. These processing biases within SN lead to an inability to identify relevant information in a given context (Parr and Friston 2019) or to update beliefs to modify attentional strategies or behavior more generally (Nour et al. 2018). Finally, the integrative calculations of the SN may be further impacted by alterations in large-scale network structure (Sha et al. 2019), which further bias salience calculations by reducing the fidelity of information communicated with the rest of the brain.

In summary, bias at each stage of salience attribution leads to an overrepresentation of potentiated stimuli as well as to an insensitivity to counterfactual evidence, which normally signals the need to alter behavior. A better understanding of the calculations instantiated within each of these regions, and more importantly, a more holistic, systems-based picture of their interactions, is likely to identify novel therapeutic interventions that will allow us to mitigate the unconstructive consequences of NRTs and to treat the underlying dysfunction.

Open Questions

To guide future enquiry, we conclude by highlighting three problem areas that await clarification through future research. First, are NRTs the cause of the psychiatric diseases described or only a symptom? While the current conceptualization of failures in Bayesian predictive coding computations is consistent with the neuroimaging evidence of dysfunction in these conditions, few direct links have been described between these regions and the subjective experience of NRTs in patients (for a notable exception in SZ, see Sterzer et al. 2018). This computational framework, however, provides testable hypotheses to determine how the estimation and updating of beliefs about the environment may be causally linked to the experience of NRTs across conditions.

Second, what are the limits of the neurobiological framework centered on CSTC loops? Any discussion of salience necessarily implicates the entire brain. Which key nodes have not yet been accounted for in the current conceptualization (hippocampus, amygdala, dlPFC)? Especially in the estimation of beliefs, the central role of memory processes is currently underspecified.

Finally, what potential treatments do these circuits suggest? Given the multiple levels of systems that are impacted—from D1 receptors in ventral striatum (Roberts-Wolfe et al. 2018) to large-scale brain networks (Sha et al. 2019)—are multipronged treatments, such as simultaneous pharmacotherapy (Kalivas and Kalivas 2016) and transcranial magnetic stimulation (Peters et al. 2016) more likely to succeed?

Acknowledgments

This work was supported by the National Institute on Drug Abuse, Intramural Research Program and Center for Tobacco Products (U.S. Food and Drug Administration) Grant No. NDA13001-001-00000 (to EAS).

11

Brain Networks for Cognitive Control

Four Unresolved Questions

David Badre

Abstract

The last decade has witnessed a marked shift of emphasis in cognitive neuroscience away from simple localization of function and toward the organization, coding, and dynamics of brain networks. This is surely a healthy evolution of our science, and the study of cognitive control has benefited from this shift, as much as any domain. However, the emphasis on brain-wide networks for cognitive control has reopened some older debates, once thought resolved, while also introducing some new ones. This chapter focuses on four questions viewed as unresolved and fundamental because one's particular answer to them commits to some basic theoretical differences regarding cognitive control function: Are there one, many, or any networks whose primary function is best described as cognitive control? Are the networks supporting cognitive control in the brain "hub-like" or "hierarchical" in their intrinsic and extrinsic organization? Are the networks for cognitive control modulatory or transmissive in the pathway from thought to action? Does controllability apply at the level of cognitive function or brain state? Each question is defined and relevant background is presented that could inform a resolution.

Introduction

A longstanding problem in cognitive science and neuroscience concerns how the brain supports cognitive control. In broad terms, cognitive control refers to the set of mechanisms needed to organize our thoughts or actions to achieve a goal, particularly when the behaviors involved are not well learned or habitual (Stuss and Benson 1987; Logan and Gordon 2001; Miller and Cohen 2001; Badre and Nee 2018; Badre 2020). Cognitive control allows us to strategically select responses appropriate to our circumstances, to adjust our behavior on the fly, and to adapt to open-ended problems and novel situations. It allows us to sustain goal-directed behavior over multiple timescales and to withhold inappropriate responses, even when those responses are prepotent, habitual, or stem from the prevailing urges of the moment. Cognitive control function lies close to the heart of human intelligence and ingenuity. It is also vulnerable to deficits across many, if not most, psychiatric and neurological disorders, being at the base of many of the behavioral problems arising in those conditions. Thus, understanding the mechanisms by which the brain supports cognitive control is a problem of fundamental importance.

Understanding cognitive control is of direct importance for intrusive thinking, the definition and scope of which is addressed in detail in other chapters of this volume. Most definitions, however, require that intrusions are unwanted and are unrelated to our goals or the task at hand. Thus, control mechanisms are an important means by which we both avoid intrusive thoughts and manage their impact. It follows that understanding the brain systems that support cognitive control function will have important implications for intrusive thinking, both in identifying its sources and seeking its potential remediation. In this chapter I review the brain networks that support cognitive control as a general background for more direct consideration of intrusive thinking.

As with most domains of cognitive neuroscience, the last decade of research into cognitive control in the brain has witnessed a shift away from a paradigm of functional localization toward one of functional networks. Among the most robust and important observations to emerge from the overall network approach has been that sets of brain areas tend to covary mostly with each other and not with other areas (Power et al. 2011; Yeo et al. 2011; Buckner et al. 2013). Further, the structure of this covariation is not entirely due to spatial proximity. Rather, affiliated areas can be distributed in each lobe of the brain, whereas other areas that are spatially contiguous may not affiliate. These basic properties have allowed for definition of brain networks or clusters of areas that covary with each other at different scales (Power et al. 2011; Yeo et al. 2011).

Here, I focus on four big questions that are provoked when one takes a network view of cognitive control seriously:

- Are there one, many, or any networks whose primary function is best described as cognitive control?
- Are the networks supporting cognitive control in the brain "hub-like" or "hierarchical" in their intrinsic and extrinsic organization?
- Are the networks for cognitive control modulatory or transmissive in the pathway from thought to action?
- Does controllability apply at the level of cognitive function or brain state?

Obviously, this is not intended as an exhaustive list of questions about control networks. Rather, these are the kinds of questions that I find myself asking routinely, whether in my own work or in reading about others'. No one has definitive answers, and so these questions also remain contentious or unresolved.

My goal is not to provide answers in this essay, though I will express my own view. Rather, I will define each question and present some relevant background in the hope of provoking further discussion.

Are There One, Many, or Any Networks Whose Function Is Cognitive Control?

One of the oldest questions in the study of cognitive control or executive function is whether there exists one or many executive controllers in the mind and brain, or if there are executive controllers at all. The majority view has mostly been that, while there exists cognitive control function, it is not simply one thing. Rather, what we call executive function or cognitive control actually refers to a variety of specific functions and capacities.

Two camps reject this basic view. First, there are those who contend that there is one central system for cognitive control or executive function and that little to no decisive evidence exists for strong dissociations among subtypes of cognitive control functioning. The second camp argues that control is emergent from network processing in the brain, but that no particular area or network of areas is best characterized as primarily supporting "control." Finally, even among those who agree that cognitive control exists and has many facets, there has been little agreement about the exact type and number of these facets.

This core debate has unfolded in almost every domain in which cognitive control has been studied: from behavior to individual differences to neuropsychology to neuroimaging. Currently, it is playing out again in network neuroscience. I will devote some more space to this first question than the other questions as it also provides an opportunity to summarize some background on the networks relevant to cognitive control.

The Multiple Demand System: One Network to Control Them All

One reason that the unitary hypothesis has been so hard to falsify conclusively is that it is often the null hypothesis (Aron et al. 2015). It predicts that in any setting in which one attempts to locate a difference based on a type of cognitive control, there will be no difference. Thus, any imprecision in design, logic, or measurement has the potential to find evidence consistent with this unitary view by virtue of being inconsistent with the alternative. As a consequence, the unitary view has been something of a "zombie hypothesis" over the years: falsified in experiments that show dissociations in the brain or behavior, only to rise again a few years later when the same distinction is not found to generalize to a new task or the methodology changes. However, it is important to acknowledge that a failure to locate a difference, even in direct replication, is not itself positive evidence for a unitary controller. Rather, unitary controllers need positive predictions and evidence of their own.

In this light, the definition of the multiple demand system put forth by John Duncan and colleagues is appealing as a unitary controller view of brain organization, because it is based on a positive prediction: the multiple demand system is engaged when you perform any challenging or difficult task (Duncan 2013). Under these difficult circumstances, one needs to sequence the set of attentional states required to perform the task. It is also in these "hard" settings where one should expect the unitary cognitive control system to be engaged. Importantly, however, the specifics of the task in question or the demands that made the task difficult are not important. This system should be fundamental and domain general, so that it participates across these different task settings.

To test this hypothesis, Fedorenko et al. (2013) conducted an fMRI experiment in which they contrasted difficult versus easy conditions in a wide range of tasks. Difficulty was simply defined as a condition that took longer and induced more errors behaviorally. The tasks differed in their specific demands and the domain of input, such as between verbal or spatial. Nonetheless, when one contrasted the hard with the easy conditions of these tasks, a consistent set of areas was activated in each participant, as shown in Figure 11.1a. Given its definition, this network was dubbed the "multiple demand system" or MD system (Fedorenko et al. 2013).

The MD system has been studied extensively. It includes premotor cortex, lateral prefrontal cortex (PFC) around the inferior frontal sulcus, the intraparietal sulcus, the anterior cingulate cortex (ACC), the frontal operculum, and subregions of the basal ganglia (Fedorenko et al. 2013). This network has been associated with a variety of measures of flexible behavior, including general intelligence (Woolgar et al. 2010) and novel rule following (Tschentscher et al. 2017). In addition, most recently, it has been found to line up with the Human Connectome Project parcellation that is defined based on a range of structural and functional anatomical features (Assem et al. 2020).

As a unitary system, the MD system is proposed to serve a very general control function needed across multiple complex tasks; namely, the assembly of attentional episodes that are the smallest unit chunks of a complex problem (Duncan 2013). When people seek to solve a new or difficult task, it has long been thought that they must break the problem into parts (Newell 1990). From the MD theory, each part is defined by a set of input-output relations that are coordinated by attentional systems. The MD system is proposed to manage these attentional episodes and the transitions from one to the next. Thus, neural coding within this network is thought to be highly dynamic, changing from moment to moment in a trajectory determined by the flow of attentional episodes. The consistent and widespread observation of flexible and dynamic coding of prefrontal neurons from electrophysiological recording in the nonhuman primate shares a qualitative correspondence to this view of multiple demand coding (Rainer et al. 1998; Stokes et al. 2013).

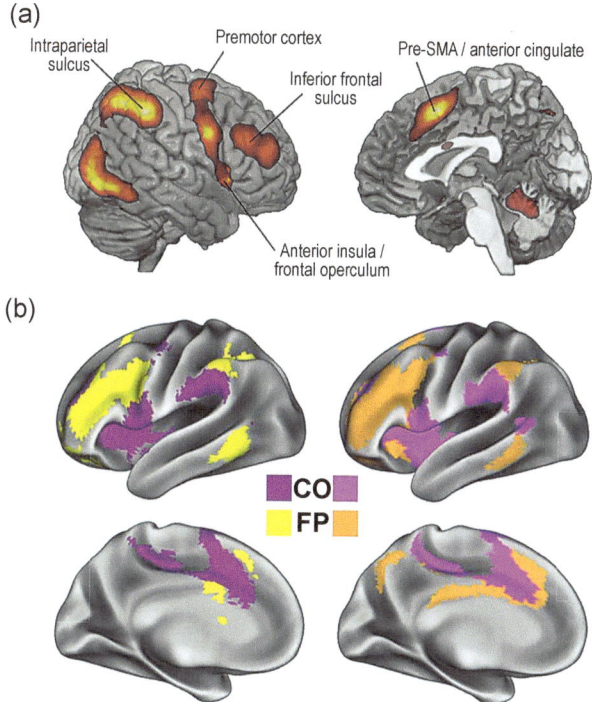

Figure 11.1   Networks activated across multiple task demands. (a) Activated regions of multiple demand systems: contrast of hard versus easy conditions in all tasks run (after Fedorenko et al. 2013). (b) Frontoparietal (FP) and cingulo-opercular (CO) networks defined through functional connectivity: different methods of network definition find convergent network definitions (after Gratton et al. 2018).

Frontoparietal Control System and the Cingulo-Opercular, Other Control System

As already noted, the functional definition of the MD system encompasses a wide and consistent set of frontal, parietal, and subcortical regions. However, evidence from analysis of functional connectivity in large resting-state data sets indicates that these areas are separable into at least two networks: a frontoparietal (FP) network and a cingulo-opercular (CO) network (Power et al. 2011; Gratton et al. 2018). Whole brain parcellations repeatedly locate differences in connectivity between these two networks across multiple methods, in large samples, and repeatedly in "deep sampled" fMRI subjects (Power et al. 2011; Yeo et al. 2011; Gordon et al. 2017; Gratton et al. 2018; Ji et al. 2019). Data from patients with brain damage to either regions of the FP or CO networks exhibit reduced functional connectivity at rest within that network but not across the networks, amounting to a double dissociation of functionally connected networks (Nomura et al. 2010). Further, this network distinction does not depend on studying connectivity at rest. A recent fMRI study reproduced this network difference in connectivity within the functionally defined MD system while participants performed a cognitive control task (Crittenden et al. 2016). Thus, the evidence is quite strong for a network distinction in functional connectivity among two major networks that make up the MD system.

Importantly, however, while the evidence for two networks within the MD system suggests distinct functions are served by these two networks, the evidence for what those functions might be is neither strong nor clear. A meta-analysis of executive function tasks proposed a functional distinction between the FP and CO networks based loosely on timescale of control (Dosenbach et al. 2006, 2007, 2008). This analysis noted that the FP network was activated in tasks involving task cueing or adjustments of a task from feedback. The CO network, by contrast, was activated for these features in addition to demands to sustain control over time. Based on these observations and follow-up work, Dosenbach, Petersen, and colleagues proposed a distinction between "control implementation" by the FP network and "task set maintenance" by the CO system (see Gratton et al. 2018). These functional designations are intuitive, but they are not specified in a concrete mechanistic or process-specific way. To date, no study has cleanly operationalized these processes and pitted them against each other. Thus, no evidence of a functional double dissociation between control implementation and task set maintenance presently exists for the FP and CO networks.

It is notable in this context that other prominent frameworks have attributed more mechanistic functional differences to the lateral PFC and dorsal ACC areas that overlap with the FP and CO networks, respectively. For example, Botvinick proposed that the dACC may be important for detecting conditions that require control, such as response conflict, and thereby signaling upregulation of control signals by lateral PFC (Botvinick et al. 2004). Recently, Shenhav et al. (2013) updated the conflict detection model to suggest that ACC computes the expected value of control, a signal that specifies the type and intensity of control carried out by lateral PFC. Others have proposed that dACC has access to stimulus-response policies which allows it to make predictions and detect errors in response outcomes (Alexander and Brown 2011). The predicted response-outcome model captures this mechanism and can account for a wide range of results from both electrophysiology and neuroimaging. Still other models have suggested that dACC plays a role in computing value of counterfactual plans to be executed in the future (Fouragnan et al. 2019). Thus, several models propose a functional distinction between dACC and lateral PFC, which might extend to the FP and CO networks, though there is presently little agreement about what these differences might be or consistent empirical evidence for these distinctions.

Hierarchical Control and Distinctions within the Frontoparietal System

Within the broadly defined FP network there is evidence for further functional distinctions and subnetworks (e.g., Dixon et al. 2018). These distinctions have been most consistently observed in the context of complex tasks that are designed to test hierarchical cognitive control (Badre and Nee 2018). Hierarchical cognitive control refers specifically to cases where we must control actions based on immediate contextual signals, while also being influenced by higher-order superordinate control signals that are either more abstract policy or extended in time.

In general, if a task requires tracking multiple contextual signals to keep overlapping behavioral policies separate, demands on hierarchical control grow. For example, in a recent experiment, children and adults were instructed with a set of mappings between cartoon characters and left or right button presses, the "Go" task (Verbruggen et al. 2018). Prior to performing the Go task, however, participants were asked to view all of the cartoon characters, pressing the right button to advance to the next character (the "Next" task). This meant that while performing the Next task, participants would occasionally press the rightward arrow to a character that required a left response later on during the Go task. Such an overlap of responses can cause conflict, evidenced in slowed response time during the Next task. However, this conflict, is reduced if one can impose a latent context that separates the episode of the Next task from the later episode of the Go task and their respective response sets.

Interestingly, when doing the Next task, children exhibited more conflict than adults; children had a harder time imposing this context episode on the task. Notably, this conflict was evident even though they had never performed the Go task and were only instructed on the response rules for this task. So, it was not rule following that was a problem for the children, perhaps counter to the widely held view. The conflict indicates they immediately implemented the rules just from the instruction. Rather, their slow response was a symptom of diminished hierarchical control capacity: they could not keep the latent task contexts separate.

Studies testing hierarchical control have consistently exposed differences within the FP control system (Figure 11.2a). Across a range of studies using fMRI, transcranial magnetic stimulation (TMS), and testing of patients with frontal lobe lesions, differences in policy abstraction (defined in terms of the number of conditions or branches in a decision tree between stimulus and response) yield differences along the caudal to rostral PFC, with the highest levels of abstraction associating with the rostral mid-dorsolateral PFC (Koechlin et al. 2003; Badre and D'Esposito 2007, 2009; Nee and D'Esposito 2016, 2017; Badre and Nee 2018). Further, manipulations of temporal abstraction, which refers to the degree to which a goal or task must be held pending over time, have found fMRI activation in the most rostral portion of the frontal cortex, the rostrolateral PFC (Koechlin and Hyafil

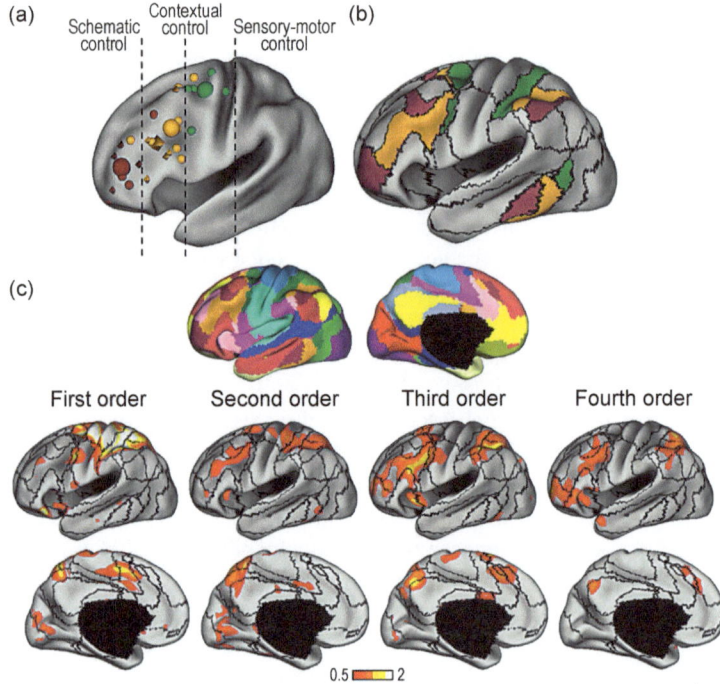

Figure 11.2 Relationship of functional studies of hierarchical control and brain networks defined from functional connectivity. (a) Results from a meta-analysis of hierarchical control studies (after Badre and Nee 2018). The colors distinguish three functional zones related to different hierarchical control demands related to using simple (sensorimotor control) or complex (contextual control) contexts to control responses. Schematic control refers to studies that manipulated temporal abstraction or subgoaling and branching demands. Small shapes are individual studies. Large shapes are average coordinates. Within the contextual control zone, spheres refer to second-order control and diamonds to third-order control and show a further separation rostral to caudal of these studies. (b) The 17-network parcellation from Yeo et al. (2011) with the three networks most overlapping the three zones highlighted (after Badre and Nee 2018). (c) The direct comparison of the Yeo et al. (2011) network parcellation with activation across four levels of hierarchical control, from Badre and D'Esposito (2007), shows the consistent network overlap in multiple lobes of the brain (after Choi et al. 2018).

2007; Desrochers et al. 2015; Nee and D'Esposito 2016; Badre and Nee 2018). Perhaps relatedly, the rostrolateral PFC has also been implicated in tasks requiring information from memory, future directed thought, counterfactual or alternative courses of action, or pending actions to act as control signals (reviewed in Badre and Nee 2018). For this reason, Nee and Badre gave this zone a general label of "schematic control" to emphasize its relationship with these types of computations.

Important to the present discussion, these distinctions along the lateral PFC are mostly encompassed within the broadly defined FP network. However, brain networks can be decomposed at multiple scales. Yeo et al. (2011) applied a clustering procedure to a functional connectome collected at rest in a large sample of participants. While one clustering solution, termed the 7-network parcellation, agreed with the coarse FP- and CO-network distinction, they also identified a finer-grained 17-network parcellation that broke up the FP network into more than one network (Figure 11.2b). This included, for example, a separate network for the rostrolateral PFC as distinct from the mid-dorsolateral PFC and from the premotor cortex.

By comparing this 17-network structure with the functional delineations identified by Badre and D'Esposito (2007) in the fMRI study of hierarchical control (Figure 11.2c), Choi et al. (2018) found that there was a correspondence between the functional bounds associated with task-based differences in levels of hierarchical control and distinctions within the 17-network structure. Further, there were also effects of hierarchical control in distinct regions of the parietal cortex and medial frontal cortex in accord with the network structure (Choi et al. 2018). In a direct comparison, it was found that network membership, rather than rostrocaudal location, best predicted the hierarchical level of a particular voxel (Badre and D'Esposito 2007). Thus, rather than a set of areas or a gradient going from front to back along the lateral frontal cortex, ranked by a factor like policy abstraction, Choi et al. (2018) found that there are a set of subnetworks within the FP network (or MD network) that are differentially activated, based on complex control demands such as policy or temporal abstraction.

Frontostriatal Circuits and Gating Interactions

A further network property of control that has been highlighted by the study of hierarchical control is the potential importance of corticostriatal loops in controlling interactions between separate frontal circuits (O'Reilly and Frank 2006; Collins and Frank 2014; Chatham and Badre 2015). It is well established that the basal ganglia form a series of loops with the frontal cortex via the thalamus (Alexander et al. 1986; Haber 2003). In motor control, these loops are thought to support a feedback-based gating function (Mink 1996). Specifically, candidate actions represented by cell populations in premotor cortex are initially too weak to fire, because thalamic drive is under tonic inhibition by the globus pallidus. However, these candidate actions in premotor cortex also send descending inputs to the striatum. The striatum, including putamen and caudate, receive broad inputs, not just from this premotor region but from cortex more broadly. Cells in the striatum are modulated by the presence of dopamine, which also induces plasticity so that these cells can learn which combinations of actions and context have been adaptive or not. Thus, the value of the actions considered in premotor cortex is computed as a function of what is being processed in cortex more broadly. If these actions have a history of being adaptive in this context,

"Go" cells in the striatum will elicit a cascade that ultimately disinhibits the thalamus and allows the action to be output (Mink 1996; Wickens, 1993; O'Reilly and Frank 2006).

One influential hypothesis is that these same corticostriatal feedback loops can operate over goal and context representations that are needed for cognitive control and that are maintained in working memory by the lateral PFC (O'Reilly and Frank 2006). This computation is described using the metaphor of a gate on working memory. When the gate is closed, information does not pass in or out of working memory. When it is open, working memory can be updated and top-down control signals deployed. The feedback loops of the basal ganglia could operate as these gates by controlling transmission from one cortical network to another through their disinhibitory action on the thalamus.

Consistent with this hypothesis, there is evidence from fMRI, patient, and pharmacology studies for these corticostriatal interactions during tasks that specifically manipulate input and output gating of working memory (Frank and O'Reilly 2006; McNab and Klingberg 2008; Baier et al. 2010; Chatham et al. 2014; Chatham and Badre 2015). Furthermore, the loops between the lateral PFC and the basal ganglia are ordered and topographic, such that there are both macro- and microlevel loops between cortex and striatum that are arrayed in an orderly fashion along the rostrocaudal dimension of the frontal lobes (Verstynen et al. 2012). Choi et al. (2018) reported convergent evidence of hierarchical ordering within the striatum in resting-state functional connectivity. Further, some evidence from fMRI and TMS provides functional support for separate loops that control context- and motor-level processing during rule learning and execution (Badre and Frank 2012; Jeon et al. 2014; Korb et al. 2017).

Interaction among multiple corticostriatal loops is a candidate mechanism for hierarchical control (see Figure 11.3; Frank and Badre 2012). Specifically, the gated output of superordinate contexts maintained in working memory by one network can act as a top-down influence on the corticostriatal gating loop controlling subordinate networks. In this way, multiple contingent contexts can interact hierarchically to control action.

Models of these multiple corticostriatal loop interactions have shown that they can efficiently learn abstract hierarchical rules, transfer these structures to new tasks, and exhibit the same quasi-parallel decision dynamics that humans employ when they perform hierarchical control tasks (Frank and Badre 2012; Collins and Frank 2013; Ranti et al. 2015). Further, gating of contextual representations is a means of controlling input and output through lateral PFC, thus breaking down hard problems into more manageable chunks. In this sense, these gating computations resemble Duncan's conception of an attentional episode (Duncan 2013). These computations emerge, however, from an interaction among separate, hierarchically ordered subnetworks.

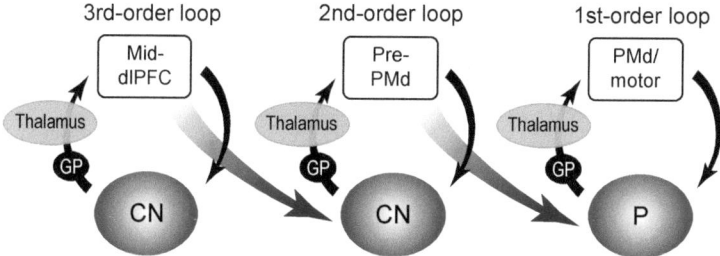

Figure 11.3 Schematic of a nested, interacting corticostriatal loop network for hierarchical control. The details of the corticobasal ganglia loops have been simplified in this diagram. Each loop is a feedback loop for one cortical network. However, the output of each network can act as a top-down influence on a lower-order loop. This nesting can provide a mechanism for multiple-contingent gating needed for complex, hierarchical control of behavior. Labeled areas are motor cortex (motor), dorsal premotor cortex (PMd), anterior dorsal premotor cortex (pre-PMd), mid-dorsolateral prefrontal cortex (mid-dlPFC), globus pallidus (GP), putamen (P), and caudate nucleus (CN). Reprinted with permission from Badre and Nee (2018).

The Stopping Network

A rigorous and compelling line of research has associated a separate corticobasal ganglia network with a distinct form of cognitive control from rule following and hierarchical control, namely stopping. Inhibitory control has long been a mainstay of cognitive control function. However, not all inhibitory behavior (e.g., slowing, stopping, or withholding what we are doing) is the consequence of an inhibitory process.

The distinction between inhibition as an outcome and inhibition as a countermanding or stopping process has caused considerable confusion in the literature (Macleod et al. 2003). For instance, the Go/No-Go task commonly used to study inhibitory control might tap into an inhibition mechanism that prevents an urge to respond on No-Go trials. Not responding to a No-Go cue, however, could simply reflect a decision not to go rather than an actual suppression of a Go response. This ambiguity clearly poses a challenge to the study of the systems underlying inhibitory control. Thus, to understand inhibitory control in the brain, it is important to test cases where an inhibitory process is required to stop an ongoing or initiated action or thought.

To test inhibitory control, the Stop-Signal task (SST) is the closest to a gold-standard paradigm that we have (Logan and Cowan 1984). An action must be selected on every trial of the task in response to a "Go" stimulus. However, these initiated responses must be occasionally stopped when a "Stop" stimulus onsets at a delay after the Go stimulus. Success on the SST will thus depend on the deployment and intensity of an inhibitory process, measured behaviorally as the Stop-Signal response time and correlated with individual differences in inhibitory control, including relating to real-world impulsive behaviors such drug addiction (Dalley and Robbins 2017). Not all impulsive behavior, however, is linked to inhibition tested by the SST.

Strong evidence from multiple sources has associated stopping in the SST with a brain network that includes the right inferior frontal cortex, the pre-supplementary motor area (preSMA), and the subthalamic nucleus (STN) (see Figure 11.4; Aron et al. 2004, 2007; Aron and Poldrack 2006). These regions are consistently activated in fMRI studies that employ the SST. Damage to the right ventrolateral PFC and preSMA causes deficits in stopping that are dissociable from other frontal regions, such as the dorsolateral PFC. Importantly, the right ventrolateral PFC, preSMA, and STN interact as a dynamic network to inhibit behavior (Aron et al. 2016; Wessel and Aron 2017). These regions are connected by direct white matter connections, the integrity of which correlates with the speed of stopping (Forstmann et al. 2012).

STN is a key node in this stopping network (Isoda and Hikosaka 2008; Li et al. 2008; Schmidt et al. 2013). It projects an excitatory influence onto the globus pallidus, thereby enhancing its inhibitory influence over the thalamus. This pathway can rapidly bypass the gating computations occurring in the corticostriatal loops and put the brakes on behavior. Recent evidence from an elegant optogenetic study in the mouse confirms these basic features in the context of the stopping that occurs during surprise (Fife et al. 2017). Specifically, excitatory stimulation of the STN cells that project to the globus pallidus caused cessation of licking responses in a mouse. Then, inhibition of the STN eliminated stopping due to a surprising stimulus.

The stopping network lies clearly distinct from the FP network involved in contextual control that was discussed above (Aron et al. 2015). Even subcortically, it appears most related to the distinct hyperdirect (rather than direct/indirect) pathways through the basal ganglia. Thus, motor inhibition may be another example of a dissociable form of control.

Further, there is growing evidence for a broader inhibitory role for this network beyond countermanding motor actions. For example, we observed increased theta band oscillations between preSMA and STN under conditions of greater uncertainty, and this coupling correlated with slowing of responses during the decision (Frank et al. 2015). Ostensibly through motor inhibition, the impact of control was functionally at the level of decision making. By stopping the output of a response, more evidence was allowed to accumulate before committing to a response; this is formally equivalent to setting a higher evidence threshold and making a more conservative decision. Finally, there is evidence that components of the stopping network, including the right inferior frontal cortex, may also inhibit cognitive actions, specifically the act of retrieval from long-term memory (Guo et al. 2018; Castiglione et al. 2019). In sum, there is a separate brain network for fast stopping, and there may also be further subnetwork distinctions within this domain.

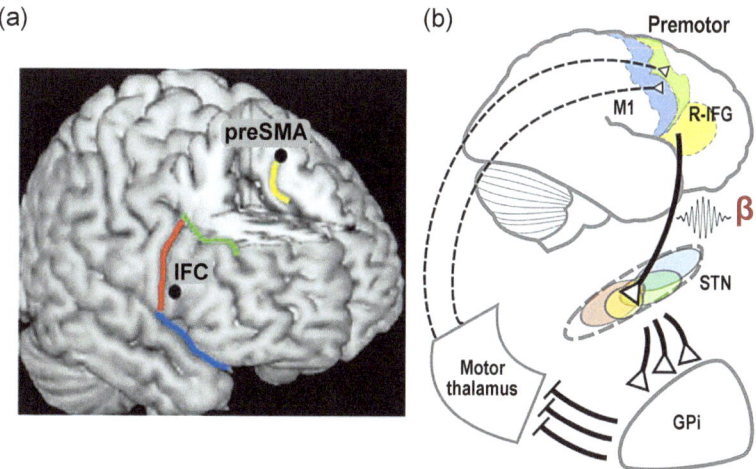

Figure 11.4   Networks critical for stopping. (a) Cortically, the right inferior frontal cortex (IFC), sometimes termed ventrolateral prefrontal cortex, and the presupplementary motor area (preSMA) have been consistently implicated as playing a causal role in stopping during the Stop-Signal task (after Aron et al. 2007). (b) Schematic of the pathways between cortex, the subthalamic nucleus (STN), internal globus pallidum (GPi), and thalamus that are thought to support fast stopping (after Aron et al. 2016). R-IFG: right inferior frontal gyrus.

Control without Controllers

The evidence presented above supports either one or several networks involved in control. However, a third perspective, most recently argued by Eisenreich et al. (2017), holds that none of these networks truly supports cognitive control as a unique function. Rather, since neurons are systems of distributed computation, they have emergent features of control that arise naturally in such systems. From this perspective, cognitive control is an emergent property of network computation, and there is no specific system devoted to cognitive control in the brain (Eisenreich et al. 2017).

There are many examples of distributed systems in nature that display controlled behavior without the presence of a central controller. Eisenreich et al. (2017) gives the example of a bee swarm searching for a good site to build a hive. Bees use dances to communicate to other bees that they have found, for instance, a good hive site. More bees will come to the site and do the dance if they agree with the location. Once enough bees are dancing at the site, the dance changes to a "build here" decision. At that point, a decision threshold has been passed, and the bees start to build. However, if there are multiple sites, there is conflict. The bee swarms at each location grow more slowly, and so more time is required to reach a decision. Importantly, this control adjustment is carried out at a "swarm level," not at the level of any individual bee, as no bee is aware of both locations. Thus, distributed systems exhibit dynamics that can be characterized as control.

To what degree is cognitive control similarly emergent? The strong version of this perspective proposes that there is no population of neurons in the brain that is devoted to representing a goal or that is directing actions toward it. Rather, goal-directed behavior emerges naturally from the systems devoted to action and perception and their local control dynamics. Control is distributed throughout the brain rather than being a function with a locus, whether that locus is a brain region or a network.

There is little positive evidence for the strong version of this viewpoint. Rather, the primary evidence is negative, questioning evidence for cognitive control networks in the brain and evidence of loss of function from brain lesions. For example, Eisenreich et al. (2017) focus on the complex and nonlinear mixed selectivity of neurons within the frontal and parietal association cortices in the networks discussed above to question whether these represent a goal or context as a top-down influence. In addition, Eisenreich et al. (2017) take a relatively dim view of human fMRI and neuropsychology, noting that there are debates in these literatures, or mixed results, regarding most proposed networks. They take these debates to suggest that the evidence in favor of regions or networks that are devoted to cognitive control function is inconclusive at best.

However, in my view, the evidence in favor of networks for control is not as ambivalent as Eisenreich et al. suggest. The mere presence of complex and nonlinear mixed selective cell coding constrains the mechanisms for control, but it is not, in and of itself, at odds with these neural representations serving a control function. Likewise, Eisenreich et al. overstate the inconsistencies found in the neuroimaging and neuropsychological literatures to some degree. As described in the preceding sections, many findings are now quite consistent and highly robust. The debates have boiled down to disagreements over functional interpretations of fairly consistent distinctions. Nonetheless, from their skeptical position, Eisenreich et al. make a crucial point not to underestimate the inherent controllability of any distributed system. In theorizing, it is important to distinguish this kind of distributed but local control from the centrally organized, goal-directed control we associate with cognitive control function.

Are Networks Supporting Control "Hub-Like" or "Hierarchical"?

To some degree, merely recognizing that a network rather than a specific brain area is important for a function like cognitive control commits the same shallow theoretical error as "blobology" did in the early days of cognitive neuroscience. Labeling a network merely assigns it a location without providing mechanistic insight or constraint on theory. However, the focus on networks for control, rather than individual areas, does offer an opportunity to consider new questions about the macro-level processing dynamics and functional organization of those networks. Note that this is a distinct question from that considered above, which is concerned simply with whether there are multiple control-related networks and what their function might be. The question considered here is what the nature of the interaction among these networks, and others, might be. In this respect, one question to emerge out of the study of networks for cognitive control is whether these networks are *hub-like* or *hierarchical* in organization.

Viewing cognitive control systems as hubs is an intuitive and appealing theoretical idea. In essence, cognitive control systems manage or modulate routing between other systems of perception and action to carry out tasks. Thus, these networks are central to the network dynamics of the brain, will be active across most tasks, and will exert broad influence. In other words, cognitive control networks are flexible hubs, with near proximity to all other networks, and with the ability to change their connectivity with multiple other systems as needed to coordinate their dynamics during a task (Figure 11.5a).

Evidence from fMRI functional connectivity has provided some support for the hub hypothesis. Cole et al. (2013) scanned participants while they performed sixty-four different mini-tasks in the scanner. This procedure allowed changes in connectivity to be assessed while people were shifting the rules and domains over which they performed the tasks. Cole et al. observed that the FP network showed the greatest variability in its connections with other networks across all the tasks relative to any other network, including the CO network. Furthermore, rather than just reflecting random variability in a small set of connections, FP also had the highest participation coefficient, which derives from how uniform its connections are across all networks. From these observations, Cole et al. (2013) concluded the FP network was acting as a flexible hub, changing its connectivity based on the task and thereby modulating the relevant network for a particular task. In subsequent work, this global cross-task connectivity of the FP network has been associated with fluid intelligence, a further clue to its potential importance for cognitive control, particularly during rapid task instruction and execution (Cole et al. 2015).

To some degree, the flexible hub model resembles a unitary central controller that is required to modulate all other dynamics in the brain. As already noted, however, there are likely distinctions among networks for control, even within the FP system itself. Indeed, a recent analysis of functional connectivity patterns of the FP network across multiple task conditions found that this hub-like network was decomposable into at least two networks with different patterns of connectivity, and that these patterns were similar to those identified by Yeo et al.'s (2011) 17-network parcellation (Dixon et al. 2018).

An important alternative hypothesis to a global hub is that the subnetworks for control relate to each other hierarchically, such that some networks exert higher-order influence over other subnetworks, which in turn exert control over more restricted domains (Figure 11.5b). The cascade model (Koechlin et al. 2003) essentially proposed such a dynamic along the rostrocaudal axis

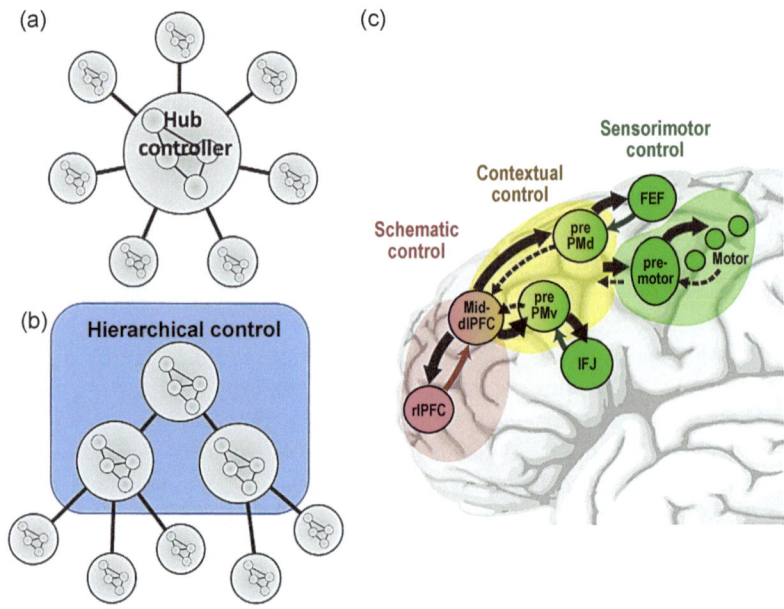

Figure 11.5 Hub versus hierarchical network organizations. (a) A hub network organization places a control network, like the frontoparietal network, at the center of coordinating other networks where it serves a general and fundamental role in organizing all other networks. (b) A hierarchical network organization allows for multiple controlling networks to share asymmetric influences with each other and to have differences in their domains of control and proximity to other networks. (c) A schematic summary of the results from Nee and D'Esposito (2016, 2017) showing hierarchical interactions among frontal lobe networks (after Badre and Nee 2018). Regions along lateral prefrontal cortex are shown within the three control zones referenced in Figure 11.2a. Heavy, unbroken arrows show strong directions of influence. Broken arrows depict weak influences. Colored arrows are domain- or task-specific influences. Abbreviations: mid-dorsolateral prefrontal cortex (mid-dlPFC), rostrolateral prefrontal cortex (rlPFC), ventral premotor cortex (prePMv), inferior frontal junction (IFJ) area, anterior dorsal premotor cortex (pre-PMd), frontal eye field (FEF).

of the frontal lobe, such that abstract temporally extended control signals in rostral frontal cortex influence more temporally proximate contextual signals in lateral PFC, which in turn influence action control by premotor and motor cortex. Other models of hierarchical control have shared similar dynamics, including among nested corticostriatal loops and through medial PFC (Frank and Badre 2012; Alexander and Brown 2015).

In a set of two fMRI experiments, Nee and D'Esposito (2016, 2017) provided evidence for a hierarchical structure within lateral PFC. These studies used estimates of effective or directional connectivity from dynamic causal modeling while subjects performed a set of complex tasks that engaged varying degrees of hierarchical control in verbal versus spatial input domains.

Hierarchical strength was defined in terms of greater outward than inward connectivity (i.e., a region has broader outputs than inputs, as defined in Badre et al. 2009).

The basic results from these experiments are summarized in Figure 11.5c: Mid-dorsolateral PFC was active in higher- (more abstract) but not lower-order tasks across both input domains. It exerted an influence on the more caudal dorsal premotor cortex and ventral premotor cortex regions that were active across both the simpler and more complex tasks, but only in the spatial or verbal domain, respectively. These caudal contextual control regions also received domain-specific input from sensorimotor regions. The mid-dorsolateral PFC received greater input from the rostrolateral PFC during conditions where temporal abstraction was required. In a follow-up TMS study, Nee and D'Esposito replicated these findings and showed that stimulation of nodes in this network produced behavioral effects that were broadly consistent with this information flow.

These findings from fMRI in humans converge with earlier anatomical studies in the macaque monkey. Goulas et al. (2014) performed an extensive meta-analysis of monkey anatomical projections using the CoCoMac database and focused on the connectional asymmetry that might drive hierarchy. They coded multiple sites in the PFC based on the same definition of hierarchy as above: any area higher in the hierarchy would have broader efferent connections to lower-order areas than the reverse. Consistent with Nee and D'Esposito, anterior mid-dorsolateral PFC (areas 45 and 46) showed the greatest asymmetry on this metric, relative to regions caudal to the mid-dorsolateral PFC or to the rostrolateral PFC which is anterior to it (Goulas et al. 2014). Notably, although Goulas et al. (2014) did find evidence that the mid-dorsolateral PFC was higher in terms of this network definition of hierarchy, it was not the most hub-like, based on a measure of betweenness centrality. This appears consistent with structural connectivity metrics in humans as well (van den Heuvel and Sporns 2013).

It remains open how one should characterize the dynamics among networks supporting cognitive control. The broadly defined FP system exhibits a hub-like character, with high participation and flexibility in connectivity across multiple tasks. There is also evidence that subnetworks within this overall system relate to each other hierarchically. In that system, there is no central domain general hub. Rather, the rostral mid-dorsolateral PFC is not active or necessary across all tasks; it is necessary during those complex tasks that require higher-order contextual control. Lower-order areas within the FP system are activated across more tasks, but they are domain specific. Thus, a hierarchical control architecture assumes that global control of the whole system emerges from limited, local, and hierarchical interactions among control networks. This contrasts with a hub network that manages interactions broadly and globally.

Finally, it should be noted that there are other hierarchical models we have not discussed that yield different organizations. For example, Barbas and Rempel-Clower (1997) proposed a laminar definition of hierarchy which distinguishes regions based on their output versus input layers of cortex. This laminar definition of hierarchy also predicts hierarchical interactions within the FP system, but places the more rostral areas, like rostrolateral PFC, higher (Goulas et al. 2014). Thus, the architecture and organization of networks for control remains a mostly open question at present, but at least one core distinction is between those proposing hub-based interactions and those proposing hierarchical ones.

Are the Networks for Control Modulatory or Transmissive?

A core assumption in most brain and network theories of cognitive control is that their function is modulatory rather than transmissive. This dichotomy was highlighted by Miller and Cohen (2001) in their seminal review on the PFC and cognitive control. Their claim was that the PFC does not lie along the pathway from stimulus to action. Rather, a series of pathways from input to output exist in the brain that differ in their various strengths of connection. Collectively, these pathways represent the full action repertoire of the system. What PFC contributes is a system set apart for maintaining contexts in working memory and deploying them as control signals that can bias competition among these pathways for behavior. From this perspective, then, PFC is modulatory, not transmissive. As such, one could remove the PFC and this would not prohibit actions from occurring in response to inputs. However, as PFC maintains high-level goals and contexts, its loss would prevent the system from selecting action pathways based on abstract, temporally remote, or task set information that is not available in the immediate stimulus. The result is primarily automatic behavior based on the strongest stimulus-to-response mapping.

As already noted, most current perspectives on the brain networks for control take this modulatory view. These networks serve control by maintaining contexts; then through gated hierarchical interactions or flexible hubs, they bias the right organization of the system to carry out the task that will achieve the desired goal or fits with that context.

There is renewed reason, however, to reconsider this accepted view, or at least the strong version that the PFC selectively maintains a context representation required for modulating other systems that route inputs to outputs. To see why, consider the problem of control as a route driven between two locations in a town. A good control system is set up such that any start point can reach any end point. This is often done by building some main roads through town that everybody uses. This is a generalizable system because the right combination of these roads can assemble any route. But, it also causes problems. As they are general and everyone uses them, such roads are susceptible to traffic. Thus, we have to add gates (traffic lights) and monitor where we are going. This is analogous to the interference or competition among stimulus-response (SR) pathways that we experience in a task that overlaps with other tasks because we are using general rules. Now consider that you have a particular route that is being used a lot. You could build an express road between those two locations in town: at some cost of time and asphalt, you then add some dimensions to your road system and gain a low traffic route. Increasing your dimensionality is costly but highly efficient, if you know you need a particular set of routes.

From this analogy, one way to think about the transition from controlled to automatic behavior is to view it as a transition from a reliance on generalizable, low-dimensional neural representations that are subject to interference to high-dimensional neural representations that take time to build but which directly map a combination of inputs to a response. These transformations could occur as transformations within frontal systems themselves. Early on, coordination among more networks, using gates and so forth, is necessary because a new task has to be assembled from low-dimensional components. However, over time, it is efficiently supported by a high-dimensional representation that allows a more direct route from input to output. It is still routed through the PFC, where multiple contexts and goals can affect it, but just differently in terms of the format of the routing (e.g., from low to high dimensions). This is different from the modulatory view which requires that there are always the same separate tracks from input to output: control acts like a switch operator deciding which track gets to run and when. This is among the distinctions that Eisenreich et al. (2017) made in their argument about emergent control systems and is captured by their schematic representations of different control architectures (Figure 11.6).

The evidence for this transmissive rather than modulatory model of control is limited at present, but there are intriguing clues. First, the computational trade-off described above between generalizable low-dimensional representations versus parallel high-dimensional representations has been shown in theoretical work using neural networks (Fusi et al. 2016; Musslick et al. 2017). Second, there is evidence from physiology in the nonhuman primate and multivoxel pattern analyses of human fMRI data that the FP network does not encode single contextual features of tasks but rather large conjunctions of multiple task features (Woolgar et al. 2011, 2016; Rigotti et al. 2013; Pischedda et al. 2017). Presently, we lack clear evidence that separate areas or networks represent separate contexts or elements of a task. Third, maintenance in working memory may not be a fixed-point system, wherein information is maintained in a single stable form to be accessed at any point as an external control signal. Rather, evidence from electrophysiology in monkeys and EEG in humans suggests that neural ensembles undergo dynamic change over time (Stokes et al. 2013). Thus, these representations are themselves expressed in trajectories toward an end point. Finally, evidence from nonhuman primates has shown that the nonlinear mixed selectivity of PFC neural representations supports high-dimensional capacity during task performance (Rigotti et al. 2013). This is what allows these populations to encode multiple mixtures of their inputs in unique patterns that can be read out by downstream cells. Rigotti et al. also

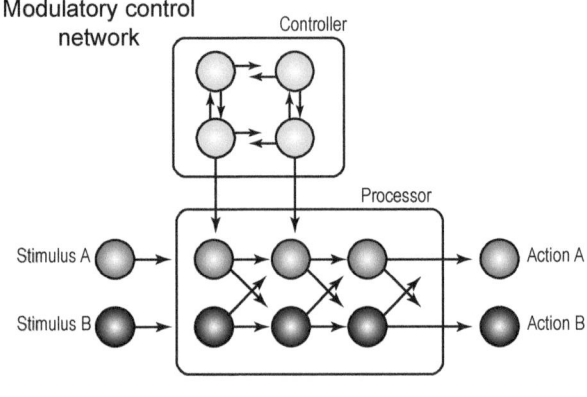

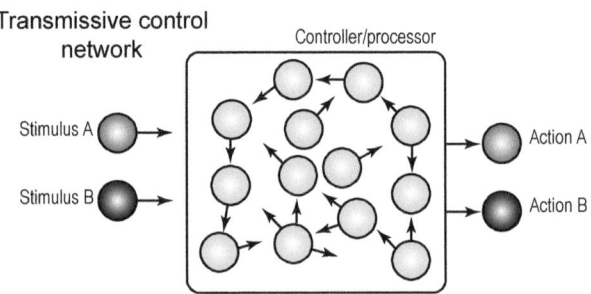

Figure 11.6 Schematic of the difference between modulatory versus transmissive control networks (after Eisenreich et al. 2017). In the modulatory control network (top panel), the contextual controller lies outside the pathways from stimulus to response. Its influence is like a switch operator, choosing which path from stimulus to response is enacted. Its removal removes controlled behavior, leaving behind only automatic behavior. In the transmissive control network (bottom), control networks are a part of the pathway from stimulus to response. The nature of these representations, however, changes over the course of experience with a task. Thus, transitions from controlled to automatic behavior are supported by features in the geometry of the population coding, such as high- versus low-dimensional coding.

provided evidence that this high-dimensional coding is behaviorally relevant, as trials in which a monkey committed an error were associated with a reduction in dimensionality.

It is important to emphasize that these results could be interpreted in several ways. It could be that high-dimensional representations are why the PFC can be flexible. In other words, high dimensionality allows multiple input states to be mapped to multiple output states. Alternatively, as monkeys in these experiments have been extensively trained on these tasks, they have automated the task and formed direct mappings from multiple input states to output states in a compact way. Regardless, both perspectives are largely transmissive in their view of the PFC rather than modulatory. PFC is part of the routing, but the *nature* of the routing is constrained by the format of its representation, either high or low dimensional. These observations reopen the question of whether control networks are best characterized as modulatory, sitting outside the basic flow of perception to action, versus being a direct part of it but changing their format and coding as a function of automaticity.

Does Controllability Apply at the Level of Cognitive Function or Brain State?

The conception of the mind and brain as a control system is one of executive function's most animating theoretical ideas, dating back at least to Norbert Wiener's mental-servo notions in *Cybernetics* (Wiener 1948) and the seminal studies on the human operator in motor control by Kenneth Craik (1948). These ideas and their descendants rely on the engineering formalism of optimal control theory. In a control system, there is a set point, which is the desired system state, and mechanisms of feedback or prediction that lead the system to adjust toward the set point either in response to or in anticipation of changes to the system state.

In the classic example of a thermostat, the state of the system that matters is the temperature in your home. The set point is the desired temperature. Feedback to the system in the form of temperature measurements can result in heating or cooling actions that will change the temperature of the environment until the set point is reached. This is feedback control. Fancier modern thermostat systems may also anticipate or learn about how ambient temperatures change over the course of a day. Such a system can engage proactive cooling or heating to maintain a stable set point, and so implement feedforward control. Regardless of its specifics, however, the efficiency with which a control system can reach its set point and the range of set points it can reach are a means of assessing its quality. Control systems can be evaluated, compared, and optimized on the basis of their ability to reach any particular desired state from any initial state (termed controllability) and the efficiency with which they do so.

In cognitive control, control theory concepts have been historically posed at the cognitive-functional level. The set point is defined with reference to some real-world defined goal or target, such as drinking coffee or making it to your connecting flight or naming the ink color in a Stroop task. An effective control system is one that allows us to reach the widest range of such goal states efficiently, either in the world or our cognitive system, given a similarly wide range of initial contexts and situations. In this conception, maximal controllability (i.e., being able to get to any output state given any input state) is presumably what cognitive neuroscientists intuitively mean when they use the term *flexible behavior*.

From a control theory perspective, psychological or neural mechanisms must gather feedback or make predictions about the distance to desired set points at this cognitive-functional level. Then, some mechanisms or processes are proposed that select and implement mental or physical actions to reach them (e.g., moving, remembering, thinking, naming, inhibiting). Learning is similarly based on feedback from the world about one's current functional-level state and the actions that were taken. In elaborating these models, neuroscience focuses on the neural mechanisms that implement these functional-level control operations.

The expected value of control model introduced above is an example of a theory of cognitive control that emphasizes optimal control theory (Shenhav et al. 2013). From this perspective, a control problem occurs when there is a disparity between a goal state and the current state, such as response conflict arising during a Stroop task. The control system, in this case the dACC, is able to compute not only this distance but the mental effort needed to resolve it, in terms of the type and intensity of the control signal needed. This is the expected value of control; namely, the value of achieving the goal discounted by the effort required to reach it. Decisions about what and when to engage control, then, can be made optimally by the brain as a function of these computations.

As with expected value of control, most models of cognitive control organize the control problem at this functional level and then draw links to neural mechanisms at different degrees of specificity. Recently, however, a new set of perspectives on control have emerged within a network connectivity framework and these emphasize a subtly different level of controllability. For example, a line of sophisticated work has applied advanced network analysis techniques of white matter connections to characterize the controllability of the brain (Laurent et al. 2015a; Betzel et al. 2016; Gu et al. 2017; Khambhati et al. 2018). In brief, these analyses have emphasized how the density and organization of connections affect transitions within the space of possible brain states. Analyses within this paradigm have emphasized two kinds of transitions (Laurent et al. 2015a): transitions to common, "easy to reach" states are associated with densely connected networks, like the default mode network, whereas transitions toward rare, "hard to reach" states are facilitated by networks with weak connections, such as the FP and CO networks (Figure 11.7). This suggests that these control networks are well positioned for maximal controllability, shunting the system into any desired brain state.

These exciting new ideas offer a powerful approach to understanding control in brain networks, and the search for translational, developmental, and clinical correlates of these metrics is ongoing (Cornblath et al. 2019). However, as described, the underlying model of control in these cases differs fundamentally from the traditional functional-level control systems described above. These systems define the control problem at the level of *brain state* rather than *cognitive function* or *real-world goal state*. In other words, the control problem is not how do you get that particular drink you want, but rather how do you get the brain into a particular state that corresponds to having that drink. The

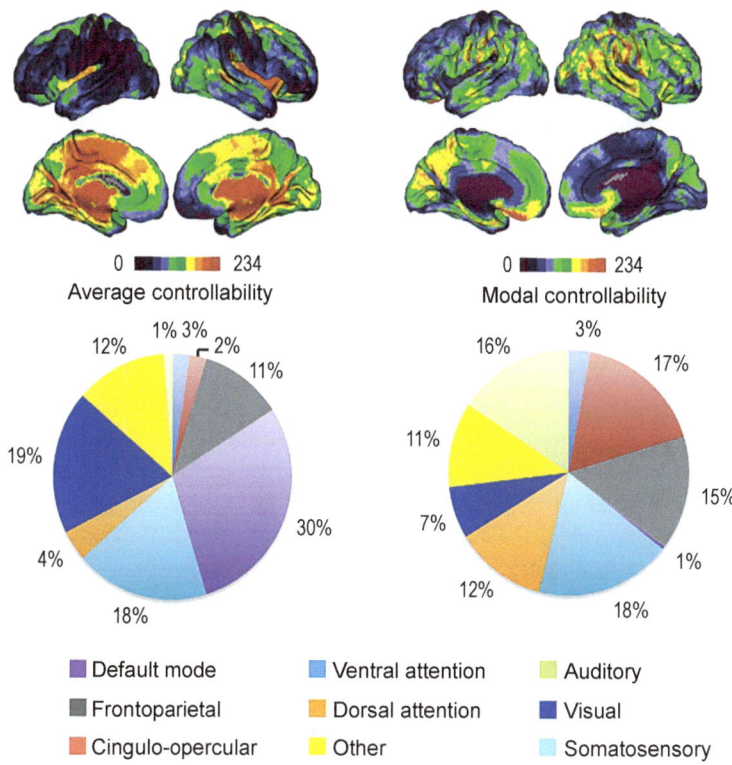

Figure 11.7 Analysis of network controllability in the structural connectome (after Gu et al. 2015). Left: Networks show the highest average controllability, which reflects how efficiently they can move to "easy to reach" brain states. The default network is the highest on this metric. Right: Networks show the highest modal controllability, which reflects the efficiency with which they can move into "hard to reach" states. Highlighted in this analysis are the frontoparietal and cingulo-opercular networks.

set point, then, is a target pattern of brain activity, not a real-world objective, and the control system must plot the distance between your starting pattern of brain activity and that goal pattern of brain activity through a functional connectome. Finally, controllability is not defined in terms of how you behave but rather how readily you can shift from the brain state you are in to any desired brain state.

So, why aren't these levels of theorizing about control the same? Isn't this network conception just a reductionist reframing of the original functional-level control problem in terms of brain states? This is certainly the way it is often posed and interpreted. However, a complication arises because of the classic philosophical problem of multiple realizability. This is exactly the case where this abstract philosopher's thought problem has some real implications for scientific theory.

Multiple realizability was famously raised by Hilary Putnam as an argument against identity theories in which each psychological state had one and only one implementation (Putnam 1967; Block and Fodor 1972). The argument is that multiple species or individuals within a species can realize the same psychological process, like pain or vision, with very different brains. Thus, there is a many-to-one mapping of brain states to psychological states. David Marr's idea of distinct levels of analysis relies on similar arguments about the asymmetry as one goes from computational to algorithmic to implementational levels (Marr 1982). Essentially, the functional or psychological level is abstraction over multiple possible realizations of algorithms and implementations within the brain.

The implications of this concept for cognitive control are important. For the reductionist reframing to correspond directly to the functional-level models, there must be a one-to-one correspondence between any one functional-level state (e.g., drinking) and a corresponding brain state (or a highly correlated class of such states). However, as the problem of multiple realizability highlights, this assumption is hardly guaranteed. Rather, the goal of having that drink may entail a wide range of activities to get there and a wide set of possible realizations of actually quenching one's thirst. Each of these is associated with a set of brain states. Some may not be more strongly correlated with each other than with other states, and so comprise a disjunct set. Thus, what ultimately connects this disjunct class of brain states is the functional-level outcome, drinking. As such, conducting control at the functional level would be the best way to ensure success, rather than making specific brain states a set point.

A further issue concerns feedback in a brain state control system. To work at the brain state level, the control system needs a means of detecting its distance from its set point. But, we don't have explicit access through the senses into our actual brain state, in terms of what neurons are firing and when. We do, however, have access to a functional-level description, like whether we are drinking or not. This feedback is also essential for the control system to learn and know what actions to take to reach a goal in the future. Without assuming a direct correspondence between the functional and brain state level that violates multiple realizability, feedback about the specific brain state target is not available to the control system in an obvious way. This is a problem if one's control system is operating primarily at a brain state level. We do, however, have perceptual systems that can assess the real-world outcome of our actions and can use these to assess our state.

What this discussion highlights is that computing the distance to a goal and seeking inputs that minimize that distance is different, depending on whether one is mapping the distance to a particular brain state or the distance to a disjoint set of such states defined by a functional outcome. Even if one allows for some correlation among that set of states, it is easy to see how the problem gets quite complicated in the latter case, if the control system has no access to the functional-level description and is instead optimizing control over brain states.

These complications notwithstanding, the issue of level of controllability remains unresolved. It is not clear to me that the conventional functional view is necessarily always the correct one or whether both options might not be true and influence matters under different circumstances. At the level of planning, awareness, and explicit control, the functional-level description of control might be the appropriate level at which to understand cognitive control for the reasons discussed above. As such, learning and feedback mechanisms must ultimately reference this level of analysis. Neural accounts must explain how the brain supports this functional-level control system, deploying neural mechanisms that interpret the state of the world with respect to goals, compute distances to real-world hypothetical and counterfactual outcomes, define the means to cross that space, and monitor progress as it goes. However, for other kinds of control, such as switching among well-learned tasks or adjusting on the fly to maintain a stable trajectory of behavior, things may be different. In these cases, the principles and constraints of brain-level network control may be the most relevant feature determining individual variability in success, even for functional-level outcomes.

In sum, whether control plays out at a primarily functional versus brain level is an important open question, particularly as we take more sophisticated approaches to understanding dynamics within the brain's connectome. These levels of controllability are not mutually exclusive, as both may influence controlled behavior. This discussion highlights the importance of being explicit about the level at which we assume control is occurring in our theorizing, and that it is not trivial to assume that brain state control is isomorphic with functional-level control.

Concluding Thoughts

The emergence of network neuroscience has brought with it an opportunity to reevaluate some older questions about control and to raise some new and exciting ones. Answers to these questions frame most of the major theories of cognitive control in the brain, with each of them staking out a position implicitly or explicitly along these lines. As I noted at the beginning, however, this is hardly an exhaustive list of the major questions facing control theorists. For example, neural dynamics clearly constitute a very important aspect of brain processing, and how dynamics among networks relate to control is only starting to be understood. One can easily think of other such questions. This essay is merely a starting point for considering the implications that network neuroscience holds for our understanding of cognitive control function and the level of neural and psychological mechanism. With relevance to this volume, these concerns will also constrain any conception of control of intrusive thought as well as the ways one might intervene.

Acknowledgments

This work was supported by the National Institute of Mental Health (MH111737) and National Institute of Neurological Disease and Stroke (NS108380) of the NIH, a MURI award from the Office of Naval Research (N00014-16-1-2832), and an award from the James S. McDonnell Foundation. I am grateful to Apoorva Bhandari, Olga Lositsky, and other Badre Lab members for helpful comments and discussions on these topics, and to Emily Chicklis for her help with figures and illustrations. Figure 11.7, from Gu et al. (2015), and Figure 11.4b, from Aron et al. (2016), are used under the terms of the Creative Commons Attribution 4.0 International License (CC-BY). Copyright for Figure 11.4a, from Aron et al. (2007), is held by the Society for Neuroscience

12

A Framework for Understanding Agency

Kayuet Liu and Hakwan Lau

Abstract

Why do some thoughts feel involuntary and intrusive? When should we hold someone responsible for their actions and thoughts when they all have some basis in the brain? Are we truly free agents when we are bounded by shared values and culture? This chapter presents a framework for how our consciousness of our own intentions and emotions allows us to form causal narratives about ourselves and the world. These narratives determine our sense of agency, and we ascribe responsibility correctly depending on the extent to which one is capable of forming culturally appropriate narratives. Different ways of characterizing consciousness are analyzed, with a focus on one that may prove most useful within the context of understanding individual agency. A variant of the higher-order view of consciousness is advocated that allows us to form causal, albeit imperfect, narratives about ourselves. However, it is because of these imperfect narratives that our understanding of agency and responsibility is formed. Thus, understanding how these narratives come about is an important first step to understanding agency and how some thoughts are considered involuntary and intrusive. Implications of this framework are discussed using examples from mental illnesses, addiction, suicide, and racism.

Do Our Brains Make Us Do It?

Aberrant acts committed by patients with severe mental illnesses (e.g., schizophrenia) are often considered not punishable. By law, if patients have lost their capacity to reason, society should not hold them criminally responsible (Mobbs et al. 2007). However, there seems to be a spectrum of controllability (Moscarello and Hartley 2017) within which we ascribe degrees of responsibility to patients suffering from different mental disorders. Take addiction as a contrasting example. Law aside, members of society often disagree on whether addicts are responsible for their own actions. Some hold that it is the addicts' own "decision" to go down the path of becoming who they are. Some have challenged the notion that addiction is a brain disease, because the neural correlates of addiction are not sufficient to cause addicts to do what they do in many environments (Levy 2013). The question, then, is: How "voluntary" are the behaviors of addicts (e.g., seeking out a drug) compared to those of patients with schizophrenia (e.g., talking to an imaginary friend in public), since both have bases in the brain? Indeed, when a healthy student is late for class due to procrastination rather than a traffic jam, this behavior is also based in the brain. In a sense, people's brains are the causes of their behaviors in all these instances. Yet we ascribe agency and responsibility differently in each case.

Perhaps this relates to how we see an individual's agency and responsibility in the context of culture. If someone is brought up in a society in which it is acceptable to take food off each other's plates without first asking, much as we may disapprove of such actions, we may accept that person's behavior more easily than had that person been brought up in a culture where such behavior was sternly forbidden. In this sense, are we truly free individuals, acting voluntarily out of our own desires or judgments? Or are we bounded very much by our shared values, so that our errors may reflect more on the failure of society rather than ourselves?

In this chapter, we do not attempt to solve these difficult questions, but rather hope to provide a useful framework within which they can be addressed. We will argue that the notion of consciousness is crucial to understanding these issues. However, there are many different ways to characterize consciousness. We evaluate a few accounts and focus on one that may prove useful within the context of understanding individual agency.

Classical Literature on Free Will

Traditionally, debates on free will concern whether our actions are predetermined; that is, whether our actions are genuinely *de novo*. The assumption is that, in principle, if we knew all current physical events of the world, together with a complete understanding of all physical laws, we should be able to predict the next events perfectly (Laplace 1951; Hoefer 2016). Within this context of determinism (Hoefer 2016), how can one be an "unmoved first mover" (Strawson 1994; Pereboom 2001)? Are our actions not already fully determined physically, before they actually take place? To the extent that some notion of freedom is possible within the deterministic framework (i.e., compatibilism; Fischer 2006; Nahmias 2016), it cannot be because these actions are random and thus unpredictable. Rather, we are responsible for our actions because we have some degree of autonomous control over our actions. To argue that our actions are genuinely free in this sense, one option is to argue that the physical world is not truly fully deterministic. Some have appealed to findings from quantum physics: that certain future events are not fully determined, even if all current physical events are known (Kane 1996). This debate is important and interesting, but many good reviews of the literature already exist (e.g., Sinnott-Armstrong 2008). Here we focus specifically on considering which cognitive architectures may allow some meaningful notion of control to happen, while assuming that classical Newtonian physics is sufficient for understanding cognitive systems such as ourselves. For an in-depth discussion of these issues, with regard to the implications of quantum physics, see Tse (2013).

The Consciousness Requirement

Shifting from the traditional focus on whether our actions are predetermined, it has been argued that for individuals to be held responsible for their actions, they need to be conscious of certain relevant events (Caruso 2012; Levy 2014). Intuitively, this makes sense: it seems unfair to hold one fully accountable for something that one isn't even aware of having done, nor having even remotely contemplated doing. Incidentally, there is experimental evidence that this "consciousness requirement" is in line with our folk psychological concepts of free will and responsibility (Shepherd 2017). Accordingly, many have focused on the question: To what extent do *conscious* mental states truly cause behavior (Pockett et al. 2006)? This diverges from the traditional question of determinism, because this new question is still meaningful even if consciousness itself were fully deterministic (Nahmias 2014). So long as the (deterministic) conscious processes in the brain causally influence our behaviors, there may still be an important sense in which we have control over our actions.

There are, of course, critics of the view that consciousness is relevant. For example, Smith (2005) claims that forgetting a close friend's birthday (i.e., something that one does not consciously choose to do) does not eliminate the responsibility of failing to call or send a card. We will not go into the details here (for further discussion, see Caruso 2012), but an important lesson to derive from these exchanges is that the arguments often depend on which specific notion of consciousness is at play. We will focus on this issue in the next few sections.

The Surprising Power of the Unconscious

The question of whether the conscious processes in the brain are causal to behaviors may seem trivial (Baumeister et al. 2011). Although we all feel that our conscious thoughts and decisions have causal efficacy, several lines of empirical studies seem to challenge this intuition. First, studies of unconscious priming, mostly coming from the area of social cognition, show that our actions and decisions may be influenced by unconscious cues (e.g., words or symbols irrelevant to the primary tasks at hand, or the gender or age of the experimenter), the meaning of which we are not fully aware (Bargh et al. 2012). If true, these findings may show that our actions are not as fully consciously determined as we thought. Some have called into question, however, whether these findings can be replicated (Chabris et al. 2019), and one could argue that the relevant cues are merely unattended but not truly subliminal.

Another line of studies focuses on the well-known Libet clock paradigm (Libet et al. 1983), in which subjective estimates of action onset and conscious intentions (i.e., the time span during which individuals feel that they are about to make an action) are reported. The actions concerned are typically simple motor movements, such as spontaneous flexing of the wrist or pressing of a button. Preceding these simple movements, it has been reported that the brain activity arises well before the onset of conscious intention (Deecke et al. 1969; Libet et al. 1983; Lau et al. 2006; Soon et al. 2008). These findings have stimulated many debates. Simple computational models have also been proposed, suggesting that the findings are exactly what we should expect since neuronal processing is noisy (Nikolov et al. 2010; Schurger et al. 2012).

The conclusion from these studies seems to be that our conscious intentions are preceded by unconscious brain activity. Some have taken this to mean that our conscious intentions are "determined" by the preceding unconscious activity, yet this interpretation is unwarranted. In most cases, what was shown is simply a weak statistical correspondence between the unconscious activity and the subsequent intention. In any case, whether conscious intentions are determined is beyond the scope of our current interests.

The Libet studies also led to another finding: the time between our conscious intention and actual action execution may be too small for meaningful causation to take place (Lau et al. 2006). Further, based on early neuroimaging studies (Lau et al. 2004), the corresponding putative "intention" areas of the brain can be targeted with magnetic stimulation (Lau et al. 2007): this showed that the reported onset of conscious intention can be influenced by stimulation even *after* the action is completed. Subsequently, other studies have also used psychophysical methods to produce similar results (Banks and Isham 2009).

Just because intentions may be subsequently revised, however, does not rule out the possibility that pre-revised versions of intention may occur prior to action. Ultimately, the actions involved in the Libet studies are simple and inconsequential, so that even if conscious intentions do not cause them immediately, this does not rule out that intentions may be important for subsequent and more complex behavior.

Importantly, Wegner and colleagues conducted studies outside of the context of the simple Libet paradigm, concluding that our conscious intentions may be generally illusory; that is, not causal to immediate actions (Wegner and Wheatley 1999; Wegner 2002). Likewise, it has been shown that unconscious information can influence more complex tasks, such as preparing to answer one type of question over another (Lau and Passingham 2007; Rahnev et al. 2012) or response inhibition (van Gaal and Lamme 2012).

In other studies, it was shown that unconscious information in the brain can facilitate different forms of associative learning (Taschereau-Dumouchel et al. 2018b), which in some cases revealed therapeutic potential. For instance, using a technique called multivoxel neuro-reinforcement (Watanabe et al. 2017), one can pair unconsciously the representations of a spider with monetary reward so that the subject will subsequently show reduced physiological threat-related responses to images of spiders (Taschereau-Dumouchel et al. 2018a). In other studies, using a similar experimental setup, powerful forms of reward-based learning have been shown to take place unconsciously (Cortese et al. 2019). In addition, subjective confidence regarding one's ability to discriminate certain stimuli can also be changed with this unconscious association method (Cortese et al. 2016).

In summary, unconscious cognitive processing seems very powerful indeed, especially regarding its ability to form statistical associations and to influence subsequent behavior (via priming). Does this mean that consciousness plays no causal role and has no function? The answer is not so straightforward. Lau (2009) argues that although unconscious processes are powerful, this does not mean that there is necessarily no room for consciousness to add further benefits. To discuss meaningfully the theoretical possibilities of the role for consciousness, we need to distinguish between different notions of consciousness.

Pure "Qualia" versus Sheer Cognitive Power

In the studies mentioned above, *unconsciousness* typically refers to stimuli or processes of which the subjects are unaware; that is, subjects do not know that they take place. So, in a sense, *consciousness* is just the opposite: subjects are aware of the relevant events. There is, however, a tradition in philosophy that analyzes consciousness as concerning pure qualitative experiences (Nagel 1974; Levine 1983; Block 1995). In some cases, philosophers invite us to consider that molecule-by-molecule functional duplicates of ourselves may lack such qualitative experiences altogether (Chalmers 1996). While such conceptual possibilities are intriguing, this notion of consciousness is not particularly interesting for our current purposes (Levy 2014). If we define consciousness as having no functional consequence, of course, it could play zero causal role.

On the other end of the spectrum, one could characterize consciousness as the capacity to access information and use it for purposeful behaviors. On one view, this form of consciousness, called access consciousness, is always supported by strong, stabilized signals broadcast throughout the different systems in the brain (Dehaene 2014; Dehaene et al. 2017). With this notion of consciousness, it is highly likely that consciousness will be causally relevant for important decisions in everyday life (Levy 2014). Once again this seems to depend, somewhat circularly, on the definition of consciousness we choose to adopt. Of course, if consciousness is *by definition* characterized by global, complex, and elaborate processes, it is likely functionally important. But does this ignore how much we seem to be able to accomplish unconsciously?

A Middle Ground?

Because of the above considerations, a notion of consciousness that is relevant for our current analysis should ideally allow for some functional consequences in principle, without *assuming* so from the outset (Rosenthal 2012). Is it possible that there is a view that mental event is conscious in the phenomenological sense while whether it contributes functionally to our rational thinking and agency ascriptions remains an empirical matter? One such possible account is the *higher-order view of consciousness*, which can be traced back at least to John Locke and Immanuel Kant (Lau and Rosenthal 2011). According to this higher-order view, mental representations of events in the world are by themselves unconscious. We can call these first-order representations. They only become conscious when they are *meta*-represented by higher-order representations. That is, higher-order representations *about* the first-order representations are necessary for making the content of the latter conscious. On this view, we can see why powerful forms of unconscious processing are possible: the same first-order representations with the same functional capacities can be conscious or unconscious, depending on the presence of the relevant higher-order representations (upon which one forms beliefs and more complicated narratives). Yet why do we need higher-order representations for first-order representations to become conscious? Traditionally, the arguments come from philosophical analysis (Rosenthal 2004): if one is aware *of* being in a certain mental state, one must represent oneself as being in that state (via the higher-order representations). Below we will elaborate further what this means in terms of cognitive architecture. Criticisms of this well-known theory, and their replies, have been extensively reviewed (Rosenthal 2005; Brown et al. 2019b).

Concordance with Current Science

Just because a philosophical notion of consciousness exists and can serve our purpose does not mean that we should adopt it. Fortunately, there is considerable empirical support for the higher-order view of consciousness (Lau and Rosenthal 2011). Neuroimaging studies have shown that subjective awareness of visual stimuli is associated with brain activity in high-level cognitive regions (e.g., in the prefrontal cortex), even under highly controlled experimental conditions in which the subjects are not merely processing the stimuli in simple (first-order) tasks (Lau and Passingham 2006). Also, neurological patients with selective damage of their visual cortex may lose the relevant subjective visual experiences but not their ability to correctly "guess" the identity of the relevant stimuli (Weiskrantz 1997); when visual stimuli are presented to intact parts of their visual cortex (as compared to the "blind" regions) leading to a conscious experience, higher activity in the prefrontal cortex was also found (Persaud et al. 2011). Disruption of activity in this prefrontal brain region selectively impairs one's ability to introspect whether one has successfully perceived the stimuli, without impairing (first-order) perception itself or memory (Rounis et al. 2010; Fleming et al. 2014). These findings are somewhat difficult to explain if we assume that consciousness is just strong, global information processing.

While these findings are reviewed in detail elsewhere (Lau and Rosenthal 2011), of particular interest to neuroscience is our emerging ability to assess, on a person-by-person basis, the efficiency of the relevant higher-order mechanisms (Baird et al. 2013b; McCurdy et al. 2013; Vaccaro and Fleming 2018). As we will see below, analysis of such individual differences is likely the key to understanding breakdowns of agency and responsibility.

Another line of support of the higher-order view comes from modern studies of artificial intelligence (Lau 2019). It is generally accepted that for neural network models to perform well, they can benefit from having the capacity for "predictive coding," in which a model can generate exemplar images (e.g., of cats or monkeys) top down. This improves the model's ability to classify images (e.g., of cats or monkeys). Training such a generative network, however, can be time consuming. To accelerate this process, another network called the discriminator, whose job is to detect forgeries (Creswell et al. 2018), can be trained to discern whether an image is genuinely from the world or created by the generative network (forgery). When we pit these two networks against each other, so that the discriminator wins a point for correctly identifying a forgery and the generative network wins a point by getting away with it, these two networks grow together not unlike rivaling siblings: they both become highly efficient in a relatively short time.

In the context of the human brain, it has been suggested that similar mechanisms of predictive coding occur. Neurons in the visual cortex may fire when a cat is presented to the subject, but similar neural representations are also involved when we imagine or remember a cat. The brain must be able to tell what causes these same neural representations to be activated in each instance. Given that these top-down generative mechanisms seem to be so efficient, it is likely that through development they are aided by the existence of a discriminator-like mechanism. Such a mechanism can determine whether a visual representation is generated by oneself or triggered externally by actual objects of perception. Plausibly, this mechanism can also tell when the same visual neurons may just be firing because of spontaneous noise. This conceptualization fits well with the higher-order view, in the sense that the output of this putative discriminator is akin to the higher-order representations necessary for conscious perception to occur. Normal conscious perception happens when the discriminator decides that a certain visual representation is truthfully representing the external world right now (Lau 2019).

Formation of Rational Beliefs

In part, based on the findings concerning the surprising power of unconscious processing, it has been argued that higher-order representations may have little additional utility besides making the relevant first-order content conscious (Rosenthal 2012). Using the computational interpretation above, however, we can identify some plausible key functions. In particular, one such function may be the formation of rational beliefs. By rational, we merely mean that these beliefs are *subjectively justified* in the sense that upon introspection, it should seem to the *subject* that the beliefs are *reasonable*. In general, it seems reasonable to believe in what we consciously see. Even when we have other reasons to believe that our perceptual system may be at fault—such as during lucid dreaming (Baird et al. 2019), at a live magic show, or after having knowingly ingested hallucinogens—there is still a strong temptation for us to believe what we see. According to the above interpretation of the higher-order view, this is because when we consciously see a cat, we have a first-order representation of a cat as well as a higher-order representation which states that the relevant first-order representation is a truthful representation of the world right now. The derivation from there to the belief that "there is a cat" is akin to a matter of syllogistic inference, so naturally such a belief may seem rational from the outset.

What good is having these subjectively justified beliefs? We do not deny that some beliefs may be unconscious,[1] but as human agents, we form narratives about the world and ourselves, and such narratives matter for our actions (Dweck 1999). In doing so, we tend to try to be coherent (Holyoak and Powell 2016). When we have many beliefs (including background, unconscious beliefs), this coherence is often difficult to achieve. One would therefore do well to include in this rational thinking system only beliefs of which we are reasonably certain. According to this perspective, beliefs that enter our narratives are mostly subjectively justified. In the case of perception, such beliefs can be as simple as "there is a cat." However, we also form beliefs about ourselves, our actions, and emotions, to which we now turn.

Self-Narratives in Agency and Emotions

As in perception, we know that motor representations in the brain also serve multiple purposes. Simple motor commands in the brain (in the primary and secondary motor cortices) are activated when we act as well as when we imagine performing the same action or when we observe others performing the same action. So presumably, as in the case of perception, some discriminator-like monitoring mechanism needs to decide when a motor command reflects what *oneself is intending to perform* (rather than just what one is imagining, or noise). This allows one to form the corresponding belief that "I intend this to happen."

---

[1] What we are claiming here is that beliefs based on conscious experiences are subjectively justified. Beliefs can, however, be based on unconscious experiences, but they will not be subjectively justified.

Taking an analogy from computers, the ability to form such beliefs seems useful. Let us say that a computer server in a large network sends a command to print ten pages to the printer. Another node in the network complains that the printer queue is now jammed and asks the first server to resolve the problem. It would be quite useful for the first server to know who contributed to the printing in the first place.

This self-directed nature of the corresponding representations is also relevant in emotions. Again, we know that for certain basic emotions, imagining them activates similar neural representations as experiencing them (Reddan et al. 2018). Likewise, when we emphasize and think about others' emotions, similar representations are involved. Therefore, when these first-order emotional states occur, the brain needs to know what the causes are. According to the view advocated here, one consciously experiences emotions when the relevant higher-order state points out that the first-order emotional representation reflects what *one is going through*. Thus, the corresponding belief may be simply that "I am angry" or "I am scared."[2] Having such beliefs may be quite useful in navigating social situations and in explaining to others why we behave a certain way.

Contrasting these with the relatively simple beliefs in the case of perception ("*there* is a cat"; see Figure 12.1), there is a sense that agency and emotions are intrinsically more self-involving. In fact, as Ledoux argues, without some minimal concept of the self, an animal may not experience basic emotions (e.g., fear) at all (LeDoux and Sorrentino 2019).

Why does one need to form these rational beliefs about oneself, which requires that the relevant first-order states be made conscious via meta-representations? The proposal is that first-order processes are powerful: we can use them to learn about statistical associations between events. However, mere associations are not coherent *narratives*. Moreover, narratives are stories in which events are *causally* related. When we say that Julius Caesar invaded a certain country *because* he was angry, we mean that his emotion *caused* certain behavior. When he decided to invade, presumably he saw himself as an agent who was causally *responsible* for the decision. Inferring about causality is, however, notoriously hard. It is technically a challenging problem from a statistical point of view (Pearl and Mackenzie 2018). Without controlled experiments in which we can manipulate the putative causes while holding all other things constant, assumptions need to be made and heuristics involving counterfactuals may need to be invoked (Bond et al. 2012; Chambon et al. 2018). As such, interpretations matter; there may be more than one way to tell a story based on the same facts. With these imperfect narratives, we form a quasi-rational understanding of why we behave a certain way, and we provide socially acceptable justifications of our actions, based on folk psychology.

---

[2] Forming such beliefs does not necessarily involve language ability.

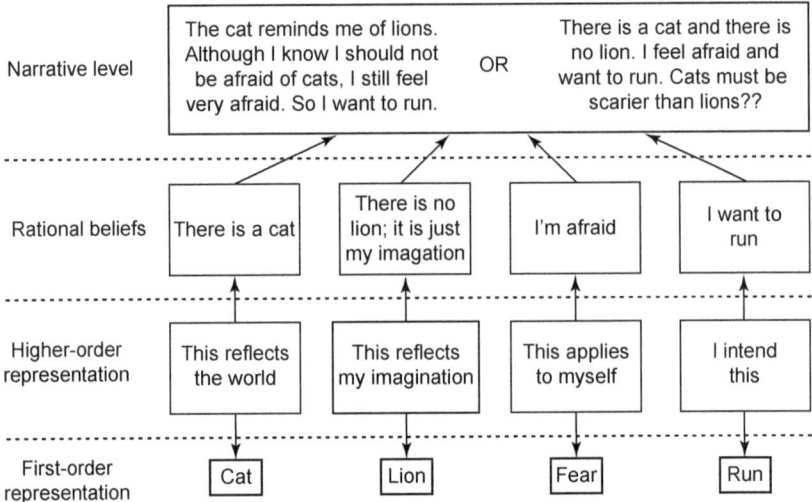

Figure 12.1 From first-order representation to self-narrative. Higher-order representations relate mental states to oneself (e.g., this reflects my imagination) upon which rational beliefs are formed. Different narratives, however, can be formed depending on how one relates the beliefs.

Some Related Views

The process of beliefs and subsequent narrative formation that we propose here is quite similar to the accounts of the post hoc explanations made by Gazzaniga's left-brain interpreter (Wolman 2019). To clarify, the above view does not mean that consciousness is the same as self-narrative (or self-consciousness). The view holds that the necessary and sufficient conditions for conscious experiences involve having the corresponding first- and higher-order representations. It is only by having these representations that we can form rational beliefs, based on which we form these causal narratives, while trying to be as internally coherent as we can.

This view links consciousness with rationality. Therefore, it is related to other models of rational decision making. In the Two Systems framework, championed by Daniel Kahneman (2011), first-order representations may roughly correspond to processes in System 1 (fast), with subsequent processes belonging to System 2 (slow). In reinforcement learning, there is a well-known distinction between model-based and model-free learning and decision making (Dayan and Berridge 2014). Roughly, higher-order mechanisms may relate better to model-based processes, whereas the first-order mechanisms may map to model-free statistical associations.

Our goal is not to replace or compete with these views. They are independently, empirically well supported, but they may serve different purposes. For example, although there may be a sense that the Two System approach or model-based versus model-free distinction may map onto conscious versus unconscious processes, such mapping is not intended to be clear cut. The higher-order view, on the other hand, is a theory of consciousness per se with direct empirical evidence. In the present context of understanding agency, the consciousness requirement is an important component of the overall argument. Accordingly, it is not clear if one should be absolved of the relevant responsibility just because decisions are made with System 1 (fast) thinking or model-free learning.

Another distinction is that one may conceive of the Two Systems framework as representing two parallel processes. The first- versus higher-order model here, however, stipulates that the two mechanisms are in a hierarchy. This hierarchical nature may have important consequences for treatment of mental disorders; intervention at the first level will causally impact on the higher level (Taschereau-Dumouchel et al. 2018b).

Understanding Mental Illnesses

This self-narrative account of agency may help us understand why some believe that patients suffering from severe mental illnesses may be less deserving of punishment than the unpunctual student, even though in both cases brain activity causes the relevant behavior. In the case of a patient with schizophrenia, the very basic mechanism of higher-order perceptual reality monitoring may be at fault. Thus, the patient may be unable to distinguish self-generated inner speech from externally triggered voices (e.g., from "God"), occurring from a breakdown between the bottom two levels in Figure 12.1. The patient may not be able to tell if an action is voluntarily produced or controlled by aliens. As such, the entire self-narrative system may well disintegrate. It is probably not fair to hold such patients accountable for their own behavior, if they are not correctly aware of who and what events caused these actions in the first place.

In the case of the unpunctual student, the higher-order system is presumably intact. What might have caused the (mildly) delinquent behavior may be a momentary overemphasis on the value of not having to rush or an attraction to another activity. These values represented in the first-order system are no less brain based and, in a sense, they too cause the resultant behavior. With an intact higher-order system, however, one should be able to appreciate that these first-order values are problematic and that one would be guilty all the same for acting a certain way.

What about cases of addiction and substance abuse? Between severe mental illness and everyday cases of delinquency, there likely lies a spectrum. In some cases of addiction, one may suffer first-order malfunctioning such that a substance may be associated with an unrealistic expected level of immediate reward, even when one is well aware that it cannot be good in the long run. In some cases, this higher-order mechanism may well be relatively intact, so one may be accountable for not recognizing the situation as such. However, there may also be cases where such a higher-order system is compromised, due to chronic abuse, which is known to impair the brain circuitry responsible for high-level control (Baler and Volkow 2006). Still, even when the higher-order system is intact, there are cases where alternative actions are not perceived as feasible or attractive, as suggested by the Rat Park studies (Alexander et al. 1978; Solinas et al. 2008); in such cases, the best solution may well lie in social policy rather than neurobiology (Hart 2013). Like others (e.g., Levy 2013), to make the correct judgment we think that we need to be able to adjudicate between the different cases in terms of both the specifics of the brain impairment as well as the environment. What we want to emphasize is that the distinctions between higher-order representations, beliefs, and self-narratives are crucial to the notion of agency.

A relevant intermediate condition to consider may be obsessive-compulsive disorders. Here, intrusive thoughts may primarily arise due to a malfunction at the lower levels; for instance, one may register the scene of an unclean bathroom as extremely threatening. Among these patients, some may genuinely believe that this is the case at the self-narrative level. Other patients, however, may recognize that the unclean bathroom is actually not that harmful, yet the visceral experience may be too much for them to overcome. In other words, patients may differ in terms of whether the conscious experience ultimately affects the healthy functioning of the entire higher-order system.

Because of these considerations, we call for more effort to subclassify the various disorders, including anxiety and depression. Is a certain patient suffering from malfunctioning at the first-order or higher-level, or both? Importantly, as LeDoux and Pine (2016) have argued, these different etiologies may need to be targeted independently. This is complicated by the fact that a disorder at one level may influence another level, as they are interconnected. To provide comprehensive treatment, one useful strategy may well be to target both the higher and lower levels (DeRubeis et al. 2008). Recognizing which level is the source of the problem for a particular individual will likely help finesse this process.

Thus, we argue that mental disorders are brain disorders and to understand agency and its breakdown we need to look carefully at the individual's condition, using the first- versus higher-order framework. Finding some correlates in the brain for certain misbehavior, however, should not absolve an individual of relevant responsibility. By understanding which brain correlates may be reflecting specific behavioral impairments, the framework advocated here provides a way to identify the theoretically relevant correlates at the different levels.

Understanding Responsibility of Individuals in Society

Mental illnesses can sometimes lead to one of the most devastating consequences: suicide. At times considered to be one of the most personal and existential decisions (Nietzsche 1955), suicide has also been analyzed as a social phenomenon (Durkheim 1951). In Emile Durkheim's classic work on the topic, he argued that many aspects of suicide depend on societal norms and values. If suicide were entirely a matter of personal decision or individual psychopathologies, it would be difficult to account for the stable cross-cultural differences in suicide rates (Durkheim 1951; Liu 2009). In what ways are we to understand suicide as being culturally and socially contextualized? Does the individual not make the decision after all?

The advocated view in this chapter can help us categorize the different ways in which social influences take place. At the first-order level, the statistical regularities of social events are picked up by the individual: How rarely does suicide occur? When the tragic event of suicide occurs, how do others react? As the individual becomes consciously aware of these events and their contingencies, the individual forms rational beliefs about suicide in social settings. Importantly, one also forms narratives about these events, in which one as an agent plays certain causal roles.

At the narrative level, interpretation matters. Different stories have different meanings. We interpret victims of suicide as causal agents. Why did so-and-so kill themselves? What drove them to such a desperate decision? Was it moral for them to do so? How does it causally affect their loved ones? These narratives apply to others as well as oneself and are naturally colored by our social understanding of the relevant concepts and implications. As such, societal norms and values influence suicidal behavior. However, we ultimately understand the decision is to be made by the individual concerned, within the given social context (Weber 1930).

Take racism as another example. One may perceive the presence of many youngsters of a certain ethnic group in a neighborhood as being statistically correlated with a higher occurrence of crime. Hypothetically, let us say that this association were statistically true in a certain context. Our unconscious first-order system may be truthfully picking up such an association, yet at the narrative level, where we ascribe causal relations to events, we do not necessarily have to interpret the ethnicity of the relevant youngsters as the *cause* of the prevalence of crime. More plausibly, both the ethnic makeup and occurrence of crimes in the neighborhood could be commonly caused by poverty and other forms of social injustice. Accordingly, the individual is still responsible for making the correct and appropriate interpretation and forming the correct narrative, given the same statistical facts picked up by the first-order system.

In this context, it is worth noting that in overcoming racism, much effort is focused on addressing our various implicit biases. What we have argued for here (excluding any possible statistical biases) is that there is also a level of narration to consider, where cause and effect is up for debate. Just because narratives are subject to interpretation does not mean they are too elusive to be worth studying. They can be changed and clarified through education and social discourse. At times, focusing on this higher level may well be more effective than methods focusing on changing our first-order implicit biases.

Closing Remarks on the Role of Higher-Order Mechanisms in Intrusive Thinking and Treatment

When considering therapeutic intervention in higher- or first-order frameworks, it may be useful to consider B. F. Skinner's perspective on whether the concept of human consciousness plays any meaningful role in science (Blanshard and Skinner 1967:325): "No major behaviorist has ever argued that science must limit itself to public events....As a behaviorist, however, I question the nature of such events and their role in the prediction and *control* of behavior" (italicized emphasis added). Skinner famously advocated studies in psychology based on the Pavlovian tradition, in which we focus on how an individual learns about the statistical associations between observable events, rewards, and behavioral responses. In our terms, this concerns the unconscious first-order level. Skinner (1971) went so far as to speculate that to engineer a better society, we should focus precisely on these basic mechanisms. Indeed, in health and disease, methods of intervention based on these Pavlovian principles have often been proven effective. However, as we see from the quote above, even a stern behaviorist as Skinner did not completely rule out the possibility that consciousness could ever be studied. The problem is that at this higher narrative level, things are complex and often depend on social factors that affect the various interpretations. Unlike Skinner, however, we are not so pessimistic about the possibility that we can understand this system. We have laid out how higher-order level narratives may be causal and may also inform ourselves about our role as causal agents. In the context of intrusive thinking, this means that whether a thought is considered intrusive or voluntary may ultimately depend on these imperfect narratives. The narratives themselves may be complex, depending on perspectives, but at least we can try to understand the underlying general mechanisms.

To conclude, we have provided some limited evidence, but no proof, that this framework is correct. The issues at hand are of immense historical significance. They concern whether we can understand individuals as rational agents. Prior to World War I and II, many great scholars from disciplines as diverse as sociology, economics, political science, psychology, and neuroscience opined on this topic. However, over the last half century, discussions on consciousness and free will have shifted toward physics. We suspect this may be due to contingent sociohistorical factors, some unfortunate and not necessarily productive. The important question of freedom of the individual, in the context of health and disease, in isolation as well as within social networks, may not benefit as much from insights from physics as from careful analysis of the neurocognitive structure (outlined throughout this volume) that underlies our narratives about ourselves. That psychotherapy can be just as effective as Pavlovian-based methods in the clinic, at least in some cases, suggests that we should not write off the intriguing possibility that our higher-order mechanisms can also be systematically understood, and modified, as needed.

13

Systems Approach to Intrusive Experiences

Angela C. Roberts, Rita Z. Goldstein, David Badre,
Bernard W. Balleine, Hugo D. Critchley,
Aikaterini Fotopoulou, Sophia Frangou,
Karl J. Friston, Tiago V. Maia, and Elliot A. Stein

Introduction

This chapter explores how intrusive experiences may occur at a systems level from psychological, computational, neurobiological, and physiological perspectives. A general scheme is proposed of the essential elements of an intrusive experience, and where in this scheme dysregulation could occur to increase the likelihood of an intrusive experience. It also considers a range of psychological and mathematical models that have been applied to explain how intrusions may ultimately happen, some of which are more closely integrated into neurobiological systems than others. These include a Bayesian model of active inference, integrated psychological and physiological models of interoception, and psychological and neurobiological models of working memory and associative learning and their relevance to concepts of flexibility and stability.

Phenomenology of Intrusive Experiences

Human mental operations (e.g., perception, emotion, cognition, metacognition, and action planning) are both complex and diverse. It is therefore important that we clearly define the phenomenological properties of intrusive thinking, particularly because it can encompass a wide array of forms, topics, and themes (see Visser et al., this volume). Here, we employ the term *intrusive experience* instead of intrusive thinking to denote that our deliberations apply to intrusive verbal thoughts, intrusive nonverbal thoughts (e.g., images,

---

Group photos (top left to bottom right)   Angela Roberts, Rita Goldstein, Bernard Balleine, Sophia Frangou, Hugo Critchley, Elliot Stein, Tiago Maia, Katerina Fotopoulou, Karl Friston, Elliot Stein, Bernard Balleine, Rita Goldstein, Hugo Critchley, Karl Friston, David Badre, Katerina Fotopoulou, Angela Roberts, Tiago Maia, Sophia Frangou, David Badre, Angela Roberts

music), intrusive impulses (e.g., motor actions), as well as intrusive bodily sensations.

Intrusive experiences have been conceptualized in varying ways (Rachman and Hodgson 1980; Parkinson and Rachman 1981; Salkovskis and Harrison 1984; Edwards and Dickerson 1987b; Freeston et al. 1991; Yao et al. 1999), and an overall consensus definition is currently lacking. The most common features across definitions involve the involuntary and disruptive nature and internal attribution of intrusive experiences; with regard to valence and controllability, there is greater variation (Rachman and Hodgson 1980; Parkinson and Rachman 1981; Salkovskis and Harrison 1984; Edwards and Dickerson 1987b; Moulding et al. 2014). Rather than attempting to provide a general definition of intrusive experiences—a goal that has tended to elude the field and has been tackled in more detail by Visser et al. (this volume)—we focus on three stages inherent to intrusive experiences (Figure 13.1). This deconstruction allows for the empirical probing of the processes and neural systems that underlie intrusive experiences, with the ultimate goal of identifying the most appropriate targets for intervention when such experiences become pathological. Consequent upon this model are the following parameters:

- The intrusion itself is inherently neutral. It is conceptualized here as a neural event (or cascade of events), the origins of which are likely to be relatively localized within specific brain circuits or networks.
- Intrusions undergo appraisal. During appraisal, attributes are assigned to the intrusion. By definition, intrusions are unintended and thus they will be appraised as involuntary. Assignment of other attributes and emotional responses to the intrusion will depend on its nature, content, and context (situational and personal) in which it occurs.
- Post-appraisal cognitive control mechanisms (Braver 2012) determine the response strategy to the intrusion.
- The resulting intrusive experience is not inherently pathological but rather a common universal human experience (Salkovskis and Harrison 1984; Freeston et al. 1991; Corcoran and Woody 2008; Bouvard et al. 2017). Some intrusive experiences, however, can be pathological, depending on their nature, content attributes, recurrence, controllability, and behavioral consequences (e.g., Julien et al. 2007; May et al. 2015).

Figure 13.1 Components of the intrusion experience.

The Intrusion

An intrusive experience has a neural "locus of origin," is sufficiently strong so that it spreads to brain regions with which it is closely linked, and propagates beyond a critical threshold which allows it to interrupt other processes and enter awareness.

The locus of origin can provide an intuitive account for the nature and content of the intrusive experience (Figure 13.2). Intrusive experiences of a sensory nature (e.g., images, music) are likely to originate within sensory systems. Intrusive experiences that involve movement are likely to originate in motor systems. Intrusive experiences that involve somatic sensations (e.g., thirst) are likely to originate in homeostatic systems (Figure 13.2; for a discussion of different neurological intrusion domains, see Gourley et al., this volume).

Tourette syndrome is a good example of where a locus of origin for the intrusive experiences can be identified. In Tourette syndrome, intrusive premonitory sensations and movements (i.e., tics) are associated with abnormal activation in somatosensory and motor cortical regions (Conceição et al. 2017). A locus of origin formulation is more challenging for intrusive experiences involving verbal thoughts. Recent advances in cognitive neuroscience, however, suggest that cognition in everyday life is dominated by thoughts that are not directly linked to sensory processing or task-directed behavior (Kane et al. 2007). Several terms (e.g., spontaneous cognition, unconstrained cognition, or mind wandering) are currently used to refer to these stimulus- and task-independent processes. In parallel, emerging neuroimaging findings have associated spontaneous cognition with connectivity within the default mode network, a functional brain network that is more active during stimulus- and task-independent periods (Andrews-Hanna et al. 2010; Dixon et al. 2014). However, it is important to note that our model postulates that regardless of the initial locus, the originating signals spread to additional brain regions following connectivity pathways so that intrusive experiences acquire multisystem associations once they reach a certain threshold (see below). In other words, they enter the global workspace (Dehaene et al. 1998) or form part of the winning coalition (Maia and Cleeremans 2005).

To account for how intrusive experiences occur, we propose two heuristic mechanisms: a breach and a permissive mechanism (Figure 13.3). These mechanisms are described separately although they may coexist. They are conceptually embedded within theories that view experience as the outcome of selective signal propagation in the face of competition (Dehaene and Changeux 2004; Beck and Kastner 2009; Graziano and Webb 2015) or global constraint satisfaction (Maia and Cleeremans 2005). The mechanisms for signal selection are currently unclear and have been described with various terms, including signal biasing or weighting (Sergent and Dehaene 2004; Beck and Kastner 2009), signal enhancement (Graziano and Webb 2015), biased competition (Desimone 1998; Deco and Rolls 2005), and gating (as we

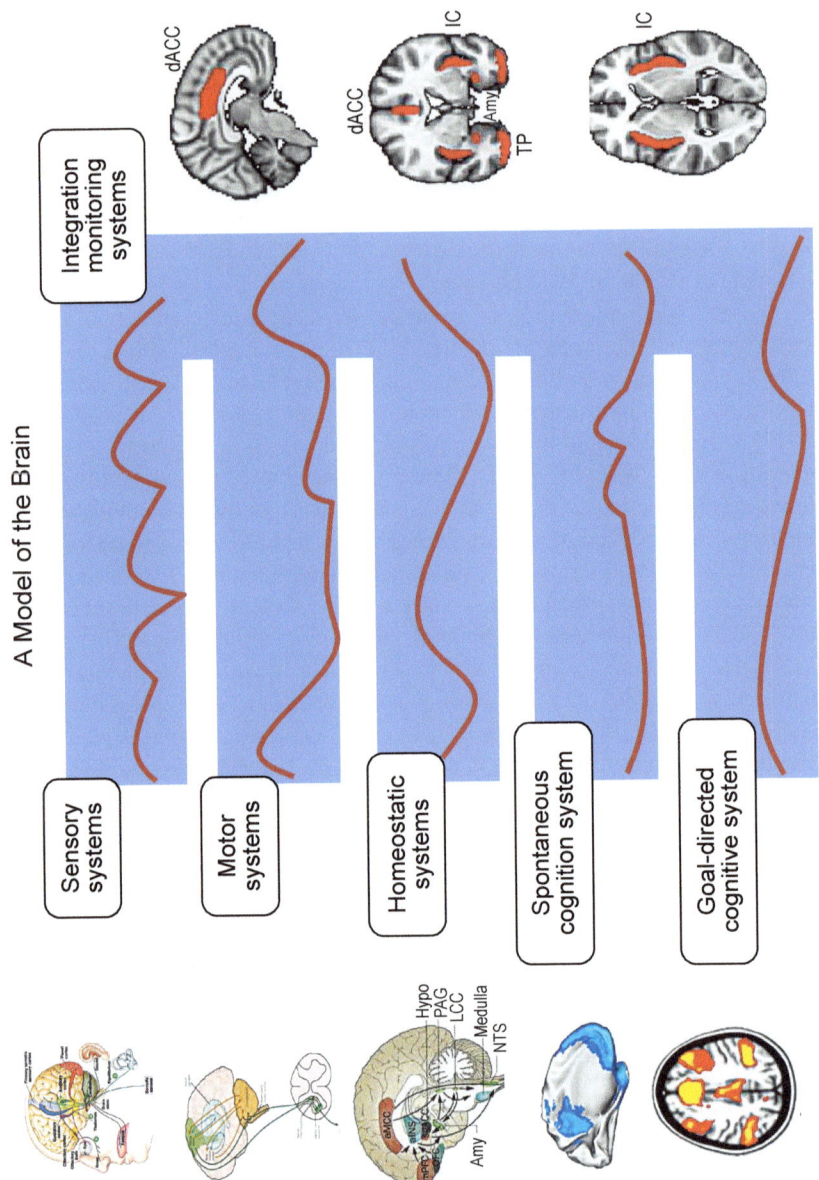

A Model of the Brain

Sensory systems

Motor systems

Homeostatic systems

Spontaneous cognition system

Goal-directed cognitive system

Integration monitoring systems

Figure 13.2 The brain is functionally organized into cognitive systems supported by spatially defined networks (Power et al. 2011). This simplified model illustrates brain activity within brain systems (left) that could host a potential "locus of origin" of an intrusion: sensory and motor systems, homeostatic systems involved in the integration of exteroceptive and interoceptive signals (Salvato et al. 2020), a system involved in spontaneous cognition supported by the default mode network (Raichle et al. 2001), and a system for goal-directed behavior supported by frontoparietal regions (Fox et al. 2005). The model posits that intrusions are experienced as such when they enter "awareness," which is likely to occur in the presence of additional recruitment of networks involved in monitoring (shown on the right), such as the salience network (Seeley et al. 2007). The blue areas (middle) represent the network space, with oscillatory activity therein denoted by red lines. Amy (amygdala), dACC (dorsal anterior cingulate cortex), Hypo (hypothalamus), IC (insula cortex), PAG (periaqueductal gray), LCC (lateral cerebral cortex), NTS (nucleus tractus solitarius), TP (temporal pole).

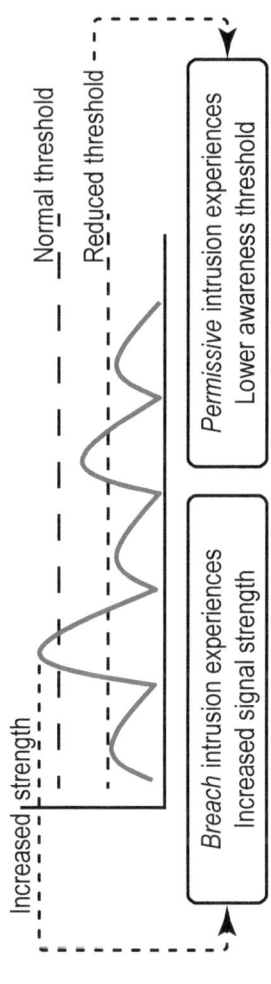

Figure 13.3 Breach and permissive intrusive experiences.

discuss in more detail below). Here we use the analogy of *awareness threshold* (borrowed from sensory perception) to visualize the moment a signal gains sufficient *biological momentum* to breach the threshold of awareness. Accordingly, breach intrusive experiences can occur because the strength, features, or contextual significance of the originating signal enables its selective enhancement. In contrast, permissive intrusive experiences occur when the threshold is transiently or persistently lowered and thus permits the propagation of weaker signals. The timing of intrusive experiences (i.e., when they occur) can be influenced at any point by the external environment as well as by internal states, which can be referred to as "motivational states" in that they combine representations of somatic states and overall general behavioral drives (discussed in more detail below). It also follows that intrusive experiences are influenced by genetic and molecular factors, including neurotransmitters (e.g., Bonvicini et al. 2016; Sinopoli et al. 2017), that define healthy within-individual variation (i.e., the likelihood of an intrusion within an individual) and interindividual differences (i.e., differences between individuals in the likelihood of experiencing intrusions), and that these may be associated with pathological conditions affecting brain integrity at multiple organizational levels (e.g., Keelan et al. 2019).

Generally, signals relating to survival (e.g., hypoglycemia) will generate breach intrusive experiences. The same could apply to abnormally generated signals, as in the case of Tourette syndrome, where abnormal sensorimotor activation spreads to other brain regions (e.g., the insula) and eventually breaches the threshold of awareness (Conceição et al. 2017). Signals relating to significant prior (e.g., childhood abuse, traumatic event) or immediate circumstances (e.g., negative thoughts about the self) may also be selectively enhanced and thus breach the awareness threshold. In such cases, the content of the intrusive experiences is more likely to be "personal" to the individual. The personal nature of the intrusion is also likely to constrain the range of its content; thus, such intrusive experiences are likely to be stereotypical. The intrusion experiences observed in posttraumatic stress disorder (PTSD) are prime examples as their content is repetitious and of personal significance (American Psychiatric Association 2013). Our model also predicts that permissive intrusive experiences are likely to have a more variable and circumstantial content because the lowering of the awareness threshold will permit the propagation of a variety of signals. Attention deficit hyperactivity disorder (ADHD) would be a prototypical example of a condition in which permissive intrusive events might occur. Currently, intrusive experiences in ADHD are considered in terms of abnormalities in attentional brain systems that gate awareness (Castellanos and Proal 2012; Bozhilova et al. 2018). As already mentioned, the dichotomization of intrusive experiences as breach or permissive does not imply that they are mutually exclusive. For example, up to 50% of patients with Tourette syndrome have ADHD, suggesting that breach and permissive intrusions may co-occur and determine clinical severity and complexity.

The Appraisal

During the appraisal stage, we postulate that the intrusive experience will attract unconditional and conditional attributes and emotional states. By definition, intrusive experiences will be unconditionally labeled as involuntary as they bypass processes of agency (see Liu and Lau, this volume; Gallagher 2012; Moore and Fletcher 2012; Braun et al. 2018). However, typical intrusive experiences retain the "sense of ownership"; that is, the sense of selfhood we attribute to our own bodily sensations, thoughts, and actions (Gallagher 2012). It is worth noting that they are distinct from psychotic experiences which, although often construed as intrusive (especially hallucinations and delusions), typically involve a loss of agency and self-ownership (Feinberg 1978; Moore and Fletcher 2012; Frith 2014).

The appraisal of intrusive experiences is a multisystem phenomenon that may, in some cases, rely on complex representations involving semantic/linguistic networks. During appraisal, the attributes assigned to intrusive experiences and the emotional responses they invoke will depend on their content, nature, and normative significance (i.e., alignment of personal beliefs and societal values) (Korsgaard 2009). We argue that the ultimate purpose of the appraisal is to determine the "likedness" of the intrusive experience; that is, the degree to which the experience is aligned with the individual's future plans (Figure 13.4). As used here, likedness aligns with notions of motivational relevance (Higgins 2011) and self-congruence (Rogers 1959; Higgins 1987) and, as mentioned above, the appraisal of the intrusive experience depends on the characteristics of the individual having the experience, including their exposures.

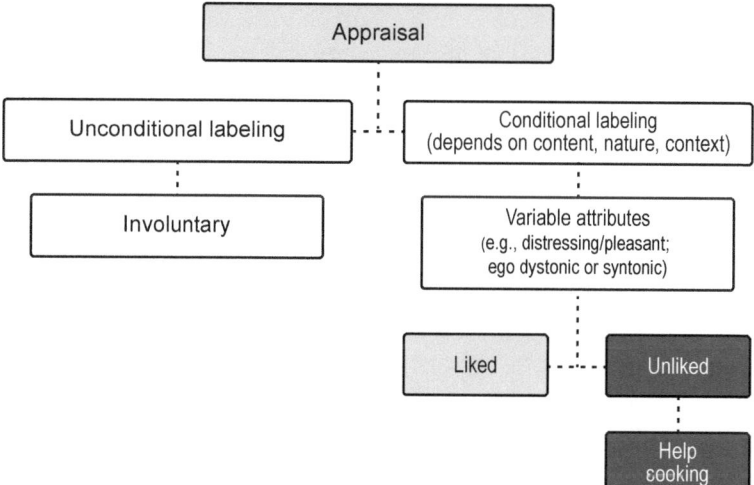

Figure 13.4 Appraisal of intrusive experiences.

Intrusive experiences that are appraised as distressing and "not liked" are more likely to be classified as clinically significant and to elicit help-seeking behavior. However, intrusive experiences can be of a positive nature and liked, as in thoughts associated with loved ones or that emerge from sudden insight or "eureka" moments (Kounios and Beeman 2014). Still, intrusive experiences that are deemed positive are not always adaptive and may contribute to further pathology by providing confirmation for maladaptive beliefs, as in hedonic hunger in individuals with restrictive eating disorders (Lowe et al. 2016).

The Outcome

The outcome of the appraisal will invoke mechanisms and networks that support selective attention, decision making, response inhibition, and response selection (Niendam et al. 2012; Langner and Eickhoff 2013; Zhang et al. 2017; Chen et al. 2018b). We assume that there will be no voluntary inhibition for liked intrusive experiences (Figure 13.5). The experience would either be allowed to decay or it could be maintained through attentional mechanisms. A liked intrusive experience may even act as a catalyst or starting point for another mental or motor plan. In such cases, the switch from the pre-intrusion state to a new one may be viewed as a positive outcome of the intrusive events. Eureka moments would fall under this category.

By contrast, "unliked" intrusive experiences will evoke attempts at voluntary inhibition. The success or failure of the experience will depend on the functional integrity of frontostriatal networks that are generally implicated

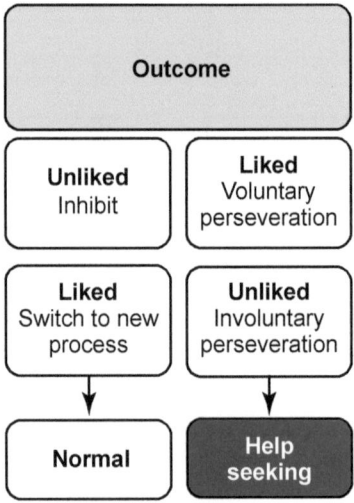

Figure 13.5 Outcome of intrusive experiences.

in inhibitory control (Niendam et al. 2012; see also chapters by Balleine and Badre, this volume; Bari and Robbins 2013). Of note, disorders characterized by intrusive experiences also present with a more general impairment of inhibitory control that affects multiple aspects of cognition and behavior (Gourley et al., this volume; Marsh et al. 2009; Shin et al. 2014; Morand-Beaulieu et al. 2017; Pievsky and McGrath 2018). Failure of inhibitory control is expected to give rise to perseveration and/or premature action (as exemplified in compulsivity and impulsivity, respectively), which may elicit secondary appraisals involving frustration, anger, and increased arousal directed at the failure to inhibit rather than the original intrusive experience. Such an outcome is likely to increase the allocation of attentional resources to the intrusive experience and the inhibitory failure; in some individuals, this may reinforce intrusion experiences, leading to a pathological loop.

Salience, Precision, and Value in Intrusive Experiences

Having considered the nature of intrusive experience in terms of definitions, phenomenology, and their implications in a clinical setting, this section provides a complementary perspective that takes its lead from systems neuroscience and, in particular, computational approaches that offer a formal and quantitative account of the phenomenology at hand. We introduce concepts of *precision*, *salience*, and (motivational) *value* that may help understand how and why intrusive experiences occur. We use two working examples that illustrate how dysregulation within these psychological domains may explain different sorts of intrusive experiences; namely, those associated with obsessive-compulsive disorder (OCD) and PTSD. This section concludes with a discussion of the conceptual implications in terms of computational architectures that underwrite intrusive experience, and how accompanying computational models and (neuronal) process theories can be used to characterize empirically observed behavioral and neuronal responses.

Brief Review of Active Inference with a Special Focus on the Nature of Precision, Salience, and Value

The treatment in this section considers cognition as a process of inference or belief updating in the brain. Specifically, we use an active inference framework to cast action and perception as solving an inference problem; namely, optimizing beliefs about states of affairs in the lived world and, crucially, beliefs about how the world should be sampled or navigated (i.e., beliefs about plans or actions). In short, we make a simplifying assumption that trains of thought can be associated with planning as inference (Attias 2003; Baker et al. 2009; Botvinick and Toussaint 2012; Baker and Tenenbaum 2014; Mirza et al. 2016).

Planning as inference rests on an internal model of how (unobserved) states of affairs causing (observed) sensations are generated. This is known as a *generative model*, usually expressed mathematically in terms of the *likelihood* of some observations, given latent or *hidden states* and *prior beliefs* about those states (for a more detailed account, see Appendix 13.1 and Figure 13.A1). In this setting, beliefs are nonpropositional (i.e., subpersonal) and simply refer to probability distributions encoded by synaptic activity or connectivity (in the sense of Bayesian belief updating or belief propagation). A simple example of active inference is the way that we forage for visual information. If I move my eyes from one position to another, the state of my oculomotor system will change, and this will have profound implications for the sensory impressions on my retina. A sequence of eye movements would then correspond to a particular *policy* or action strategy. My job is to infer the most likely policy that "something or someone like me" would engage, and then select a particular action (i.e., a next move) under that policy.

In selecting the most likely policy, I will necessarily refer to my prior beliefs about the policies I am likely to pursue; namely, those that provide the most evidence for my (generative) model of the world. This can be expressed formally in terms of a prior over policies, based on *expected free energy*. Free energy, in this instance, is known as an *evidence bound* in machine learning (Winn and Bishop 2005) and can be thought of as a measure of expected surprise or prediction error. Mathematically, expected surprise is also known as uncertainty. This means that I will select those policies (and implicit courses of action) that resolve uncertainty about the state of the world. This formulation of active inference emphasizes the two-way exchange between an agent and her world, where the implicit action-perception cycle means effectively that beliefs can change states of the world, which in turn change the sensations that update beliefs. For an illustration of this circular causality, see Figure 13.A2 in Appendix 13.1.

Motivational Value and Salience

Mathematically, expected free energy can be decomposed in a number of ways (see Figure 13.A2 for a decomposition into *risk* and *ambiguity*). For our purposes, the more prescient decomposition is in terms of *salience* and *value*. Heuristically, the (negative) expected free energy of a policy is equal to salience plus value (see Appendix 13.1 for details and how this decomposition relates to other disciplines in neuroscience):

$$\text{Expected free energy} = \text{Salience} + \text{Expected value}. \qquad (13.1)$$

In this setting, *salience* corresponds to the uncertainty resolving or intrinsic (epistemic) value of a policy. It is variously referred to as relative entropy, mutual information, information gain, Bayesian surprise, intrinsic motivation, or value of information (Barlow 1961; Howard 1966; Optican and Richmond 1987; Linsker 1990; Itti and Baldi 2009). Salience, therefore, reflects the information gain or resolution of uncertainty afforded by response to a cue: "How much will I learn, if I look over there?"

Salience can be contrasted with expected (extrinsic or instrumental) value, which is the motivational value of a policy defined in terms of outcomes that are preferred a priori. Expected value is an important construct in optimal control theory in engineering, reinforcement learning in psychology, and utility theory in economics. Expected value simply scores the expected returns (cf. rewards) following a particular policy, expressed in terms of the log probability of some prior preferences (i.e., preferred or expected outcomes). This is sometimes referred to as extrinsic value, as opposed to epistemic value, to make it clear that these are extrinsically supplied outcomes that provide a motivational value for the policies under consideration. The foregoing offers a definition of salience and motivational value in terms of active inference and the accompanying quantities or functionals of Bayesian beliefs encoded by neuronal activity and connectivity. So, what about *precision*?

Precision and Attention

Precision is an attribute of sensory outcomes or evidence at hand. Very precise data are informative, in the sense of having a high signal to noise. The important thing, from our perspective, is that precision has to be estimated or inferred in a context-sensitive fashion. For example, if I know that I am exploring an unfamiliar room in the dark, I know that the precision of visual sensations will be much lower than the precision of my somatosensory sensations. I would therefore assign a greater precision to the mapping between the hidden states of the world (e.g., "a chair in front of me") and the somatosensory outcomes (e.g., "I will feel this chair if palpated"). Conversely, if I know the light is on, I will adjust the precision of my visual mapping such that visual information is afforded much more precision and has a much greater influence on belief updating about the state of my room.

Psychologically, this is effectively the same as attention (Desimone et al. 1990; Desimone 1998; Womelsdorf et al. 2007; Parr and Friston 2019); in other words, a selective gating or attentional filtering affords one sort of sensory stream with more precision than another. When this deployment of attention is part of a policy (i.e., a covert action much like the premotor theory of attention), we have an attentional policy that is implemented through a selective gating of the sensory information at hand (Limanowski and Friston 2018). This will become an important concept later in our discussion of memory (see below), where policies that selectively afford precision to different sources of information correspond to *gating policies*. In hierarchical generative models, the level of the implicit gating or precision control may determine the nature of attention: exogenous versus endogenous (e.g., spatially directed attention vs. attention to a particular visual feature, respectively).

Applying Computational Models of Active Inference to Understand Intrusions in Obsessive-Compulsive Disorder and Posttraumatic Stress Disorder

According to the above formulation, precision is a ubiquitous attribute of the likelihood and prior beliefs that nuance or select the right kind of information for belief updating. This selection or gating rests on the excitability or lateral inhibition among competing representations at any level of a hierarchical generative model. As noted above, precision is context sensitive and must be inferred; this means that it depends on beliefs about hidden states (i.e., context) and, indeed, beliefs about policies (i.e., "what I am doing"). In contrast, salience and value are attributes of a particular plan or policy whose evaluation involves belief updates about the succession of states in the future, under a particular course of action. A salient act is one that resolves uncertainty or is likely to have the greatest epistemic affordance. The value of a plan is scored by the degree to which the outcomes are likely to be realized.

Let us now consider the computational pathology that might underwrite a typical intrusive experience in OCD: "checking behavior." Assume that there are two states of the world with which I am concerned: "the door is locked" versus "the door is unlocked." Any policies that resolve uncertainty about whether the door is locked will have a high salience. If I do not know a priori whether the door is locked or not, checking whether the door is locked has the greater salience and will, in the healthy course of things, resolve my uncertainty.

Imagine now that my generative model also predicts a state of physiological arousal due to the possibility that the door is unlocked, and all the catastrophic consequences that such a state of affairs could entail. If I can resolve my uncertainty and be 100% certain that the door is locked, then I predict that the associated interoceptive evidence for physiological arousal will also be attenuated. Now, imagine what would happen if I were unable to attenuate the precision of interoceptive signals:[1] I would check the door, expecting to find it locked and *expecting my arousal to subside*, but it does not.

I would now be in the curious situation of still being uncertain about when the door is locked because I have sensory (interoceptive) evidence at hand that I cannot have checked the door (because I am still physiologically aroused). This means that the epistemic affordance of door checking is still in play. In fact, unless I can attenuate my interoceptive signals, this uncertainty will

---

[1]   In active inference, a failure to attenuate the precision of proprioceptive or interoceptive signals is accompanied by a failure to engage motor or autonomic reflexes. In this example, a failure to engage autonomic reflexes means that a state of physiological (sympathetic) arousal would persist.

continue to be in play and induce successive checking behavior that may proceed indefinitely.

Notice in this example that checking behavior has been formulated in terms of aberrant salience because the action of rechecking the door does not lead to the resolution of uncertainty. This aberrant salience is suboptimal (i.e., pathological) because of a failure to attenuate or change interoceptive signals. In short, a failure of sensory attenuation led to aberrant salience and a persistent epistemic affordance that never resolves itself. In other words, no matter how many times I check the door, I never sense that my uncertainty has been resolved, which could further maintain a state of autonomic arousal. Expressed even more simply, this checking behavior is futile because there is an irreducible uncertainty about the state of the world due to a failure to attenuate interoceptive evidence from my body. This predictive processing, or active inference account of OCD, is based (and elaborates) on work by Kiverstein et al. (2019) and Rae et al. (2019a), and owes much to seminal accounts of why patients with OCD appear to be "stuck in a loop."[2] For example, Roger Pitman (1987:336) suggested that "the core problem in OCD is the persistence of high error signals, or mismatch, that cannot be reduced to zero through behavioral output," and that "the obsessive-compulsive's internal comparator mechanism is faulty. No matter what perceptual input it receives, it continues to register mismatch....It may be that in fact the action was well done, but the defective comparator cannot register it" (Pitman 1987:340).

In turn, Szechtman and Woody (2004:111) suggest that "the symptoms of obsessive-compulsive disorder...have what might be termed an epistemic origin—that is, they stem from an inability to generate the normal 'feeling of knowing' that would otherwise signal task completion." On the empirical side, Gentsch et al. (2012:656) found decreased sensory attenuation in OCD, which was suggested to "explain the tendency of individuals with OCD to continuously register error signals, and to experience dissatisfaction in outcome processing."

The somewhat contrived formulation of OCD, in terms of aberrant salience, focused on an account of intrusive experience that manifests in overt motor behavior. Does this explanation hold for intrusive thoughts, images, and experiences in PTSD? A plausible account could proceed along the following lines: Imagine that, at the point a traumatic event is experienced, there is some particular configuration of (interoceptive or exteroceptive) sensory inputs in play. The traumatic event can then induce a one-shot learning of the concomitant gating policy. When this pattern of sensations is encountered subsequently, it is extremely difficult to ignore, because sensory information is afforded great precision. These sensory cues will induce belief updating and the selection of

---

[2] This account is from the PhD thesis by Itzchak (Isaac) Fradkin: "Deficits in processing of prediction errors in obsessive compulsive disorder: Effects on action, thoughts, learning and agency," Hebrew University of Jerusalem, June 2019.

the traumatic narrative or policy that entails overt or covert action. In the latter setting, action is neither motoric (i.e., mediated by striated muscles) nor autonomic (mediated by smooth muscles) but *attentional* in nature. In other words, the gating policy is called up in an obligatory fashion, sometimes described in terms of modulating *sensory* and *prior* precision (Skewes et al. 2014; Ainley et al. 2016; Powers et al. 2017; Rae et al. 2019a).

This traumatic active inference or learning will induce a recapitulation of the internal policy or narrative that may enable posterior expectations all the way down to the sensory levels of perceptual hierarchies. In other words, a triggering event will *breach* attentional thresholds and induce a cascade of hierarchical and sequential processing that recapitulates the sequential narrative associated with the original trauma. The mechanisms behind such fictive (intrusive) experience are part and parcel of self-evidencing under a generative model. Common examples here include dreaming, imagination, and the generative or constructive perceptual processing associated with structure learning and eureka moments (Hinton et al. 1995; Botvinick et al. 2009; Gershman and Niv 2010; Tervo et al. 2016; Friston et al. 2017; Gershman 2017).

Based on this account, the intrusive experience induces a gating policy that prescribes covert (mental) actions that are manifest as internal scene construction and accompanying narratives (Peters et al. 2017; Wilkinson et al. 2017), as opposed to the mostly overt actions considered in the OCD example above. Clearly, the foregoing account does not offer a qualitative distinction between intrusive experiences that reflect an adaptive response to trauma and the psychopathology that results when intrusions are experienced (or manifest) as maladaptive and persistent. However, the computational account narrows down the field, in terms of where aberrant inference and learning may be operating in conditions like OCD and PTSD. Next, we consider the failure of sensory attenuation and subsequent failure to relearn the right sort of attentional response as a plausible candidate.

Summary

The two working examples of OCD and PTSD were introduced here to make a key point: the intrusive experience of OCD rests upon *aberrant salience* that is secondary to a *failure of sensory attenuation*; namely, an aberrant top-down modulation of sensory mappings. In contrast, the PTSD example appeals only to aberrant precision via a *breach* of sensory attenuation due to traumatic learning of a particular attentional set or gating policy. In other words, people with PTSD may lose the capacity to ignore the irrelevant and be plagued by breaches of attentional filtering or gating endowed by sensory attenuation. If one subscribes to these accounts, the conclusion is that the primary pathophysiology behind both kinds of intrusive experience is a failure of sensory attenuation that most likely involves interoceptive signals. Interestingly, a failure of sensory attenuation emerges in computational treatments of other psychiatric conditions (Skewes et al. 2014; Ainley et al. 2016; Powers et al. 2017; Rae et al. 2019a); in particular, schizophrenia and autism. [For a review of aberrant precision and sensory attenuation in psychiatry, see Stephan et al. (2016) and Friston (2017) for details and references.]

On this account, a minimal but sufficient explanation for intrusive experience is a failure of inhibitory control inherent in the sensory attenuation. The key thing that the active inference framework brings to the table is that this inhibitory control is not about the contents of perceptual experience, but the precision or attention afforded this content. From a physiological perspective, this is important because a failure of inhibition (i.e., a failure of sensory attenuation or attenuation of sensory precision) may be mediated not by hyper- (or de-) polarizing neuronal populations but by modulating their excitability or gain. In turn, this suggests the mechanisms that underwrite the pathophysiology of intrusive experiences are located either in classical modulatory neurotransmitter systems or the downstream effects on cortical excitability (as mediated by fast-spiking inhibitory interneuron coupling with pyramidal cells).

In summary, the emerging picture is of a deficit in the neuromodulatory mechanisms (and dynamics) that implement the top-down control of attention; namely, its sensory attenuation. A natural corollary is that there may be as many different forms of intrusive pathologies as there are neuromodulation mechanisms and projections. Irrespective of this diversity, and the accompanying regional specificity of evidence accumulation schemes in the brain, one underlying mechanism becomes apparent: the breach of sensory attenuation (i.e., attentional filtering) by exogenously or endogenously generated cues that underwrite belief updating about states of the world and our active engagement with that world. Clearly, in many instances, this intrusion is part of normal perceptual synthesis and subsequent planning. For example, a loud noise is salient because it offers a person the opportunity to "look over there" and resolve any uncertainty associated with the surprising sensory signal.

The pathology implicit in the examples above rests on aberrant salience that maintains irreducible uncertainty incurred through a failure to attenuate interoceptive signals (as in the case of overt compulsive behavior in OCD). It can also rest on the failure of sensory attenuation to be attributed to, and subsequent failure to relearn, the right kind of attentional response to triggers (as in the case of PTSD). As discussed above, the notion of a breach in sensory attenuation is a key aspect of higher-order models of intrusive experiences that consider the evaluation (i.e., the appraisal) of inferred states and subsequent metacognitive influences. At present, three conclusions follow from the formal analysis of this section that are remarkably consistent with the treatments offered in other chapters in this volume:

- Intrusive experiences are inherently *interruptive* in the sense that they induce a quantitative change in the selection of narratives or sequential policies, which underwrite overt or covert (mental) action.

- These intrusive episodes (events) are *experienced* in virtue of being manifest in terms of beliefs about (overt or covert) action. This follows because there is an egocentric aspect to action generated by these beliefs; that is, the only thing that can act is "me."
- Finally, intrusive experiences have, at some level, a *salience*, either an irreducible epistemic affordance that cannot be dispelled or in terms of aberrant precision; namely, the failure to suspend or attenuate attention to certain kinds of cues. In OCD, for example, the inability to attenuate arousal sensations manifests in the repetition of salient acts (such as checking), despite the fact that these acts do not produce a lasting reduction in uncertainty about the state of the world. (A computational model of the neuronal underpinnings of recurrent intrusions in OCD is provided in Appendix 13.1.)

The Insula and Functional Anatomy of Salience and Value

Now let us consider the above account from a systems neuroscience perspective. In this setting, salience can be thought of as an attribute of a cue (i.e., internal or external stimulus) deemed important to the individual *in a given context* (Uddin 2014; Kahnt and Tobler 2017; Miyata 2019)—it is salient because of the potential for information gain and thus belief updating. Salience is distinct from value in that the latter is a valenced or signed currency that varies monotonically from negative to positive, whereas the former is an unsigned currency (i.e., something is salient or not). This means that value and salience are dissociable in terms of what they mean for behavior: both negative and positive outcomes can be salient in the sense that experiences can change our beliefs, even if they are unpleasant (Kahnt and Tobler 2017). As such, intrusive experiences can be thought of as arising from an aberrant processing of internal and external stimuli with respect to the current (belief) state of the individual. This salience misattribution leads to an overemphasis of one thought or action over the current, ongoing cognitive process and subsequently influences attentional capture, motivation, and goal-directed cognition. Importantly, the unsigned nature of salience calculations necessitates that both appetitive and aversive stimuli can sway the calculations of salience that ultimately influence behavior.

There are many potential points at which biases can enter salience calculation. Ascribing salience to a given stimulus at a given time scale (Kennerley et al. 2011) and within a given context (Heilbronner and Hayden 2016) results from integration across a wide range of processes, including attentional (Menon and Uddin 2010), reward (Olney et al. 2018), affective (Etkin et al. 2011), and homeostatic regulation (Craig 2009).

Neurobiological instantiation of both ongoing and intrusive, highly salient events occurs at many levels of the neuraxis. One highly interconnected hub that seems to play a major role as an integrator or transmitter of the interoceptive and exteroceptive environment is the insula. More specifically, the anterior insula (and the von Economo neurons it contains) possesses the anatomical linkages to support awareness and (together with its connections to, e.g., the anterior cingulate; Craig 2009) the monitoring of the environment that is necessary (when combined with the calculation of value) to assign behavioral relevance to the event. In Tourette syndrome, for example, the insula may play a role in assigning salience and aversiveness to premonitory urges (Conceição et al. 2017).

There are important, mostly bidirectional connections between the anterior insula and key affective, cognitive, autonomic, and regulatory systems—components which place the anterior insula in a unique position in the calculation of salience (Critchley et al. 2005; Craig 2010; Nieuwenhuys 2012). In addition to the posterior regions of the insula that receive predominantly somatosensory inputs, homeostatic regulators enter via the hypothalamus and amygdala, hedonic inputs from the nucleus accumbens and orbitofrontal cortex, and motivational, social, and cognitive information from anterior cingulate, ventromedial, and dorsolateral prefrontal cortex (Craig 2010). While anatomically and functionally simplistic, this schema provides a framework uniquely placing the insula in the position to assess the relative weights of the environmental processes in the assessment of attentional capture. The insula also has important connections to motor regions that allow it to then drive behavior. In fact, in Tourette syndrome, it may play a role in driving tics (Conceição et al. 2017), whereas in addiction it may be driving craving (Naqvi and Bechara 2010; Naqvi et al. 2014).

Importantly, there appears to be a transdiagnostic component to the dysregulation of salience attribution (McTeague et al. 2016), as neuroimaging studies have demonstrated anterior insula involvement across a number of diagnostic assignments in various disorders characterized by intrusive events, including addiction (Naqvi et al. 2014), ADHD (Klein et al. 2013; Bubenzer-Busch et al. 2016; Norman et al. 2016), autism (Gu et al. 2018), OCD (Zhu et al. 2016), psychosis (Brosey and Woodward 2017), anxiety (Paulus and Stein 2006; Shiba et al. 2017), and depression (Ellard et al. 2018). This is potentially important because it suggests that interoception plays a key role in all forms of aberrant salience or precision attribution, as illustrated by the OCD example above, and will be discussed in more detail below.

Interoceptive Contributions to Intrusive Experiences

Understanding intrusive experience at the level of brain systems will be incomplete without a consideration of the systems that underlie the self. Selfhood is fundamental to the phenomenology of intrusive experiences (see Liu and Lau, this volume). If intrusive experiences are to be understood as involuntary mental phenomena that disrupt ongoing psychological narrative flow (see above), one needs to have a sense of oneself as both an observer, experiencing such intrusions, and as an agent perceiving the intrusions as unsolicited interruptions of one's sense of agency (e.g., I did not intend to have these thoughts, or perform these actions, even though I recognize them as my thoughts, or my actions). Importantly, the brain systems that regulate and represent bodily physiology, or interoception, are considered to be at the core of selfhood.

Increasingly, there is appreciation that the self as a continuous coherent, unitary representation is not the output of one specific single specialized system within the brain. Experiences of selfhood are embodied and require coordination between dissociable brain systems, as revealed by careful experimentation and in the symptomatic expression of particular psychiatric and neurological "disorders of self," such as depersonalization disorder (for a review, see Fletcher and Fotopoulou 2015). Within a broad taxonomy, selfhood can be parsed into minimal (embodied or biological) and extended (reflective and narrative) components (akin to the first- and higher-order mechanisms described by Liu and Lau, this volume). The sense of a core minimal self is proposed to emerge from the integrative processing of sensory and motor signals from the body. The frequent concomitant occurrence of sensory signals on the body eventually gives rise to mental, predictive models of "owned" first-person feelings of (bodily) sentience and presence (e.g., I exist and feel alive in this body), agency (e.g., I was the author of this action), and ownership (e.g., this bodily experience belongs to me) (Gallagher 2005; Seth et al. 2012). Extended concepts of the self are built on embodied self-representation to encompass the notion of the narrative or autobiographical self (Damasio 1999). Extended selfhood affords the ability to make one's self the object of explicit thoughts irrespective of any particular experience or perspective in the here and now (e.g., self-reference, I am a woman). More generally, this enables reflection on one's experiences across time, space, and person in counterfactual ways: I have always been a woman, I anticipate being a woman tomorrow, I imagine that I am a woman in the mind of others (Fotopoulou 2015). These notions rest on the idea that while perception can be understood as the unconscious process of hierarchical Bayesian inference on the (hidden) causes of sensory input, more higher-order abilities for self (metacognition) or other mentalization or reflection rest upon similar unconscious inferential processes of greater depth, whereby the generative models refer not only to current sensory predictions but also to predictions about the effects of actions, not yet executed, and bodily or external situations, not yet encountered (e.g., Palmer et al. 2015).

Interoception is at the core of this hierarchical view of the self and, by extension, of psychopathological disorders of self-representation (for an overview, see Khalsa et al. 2018). Indeed, there is growing theoretical acknowledgment that core aspects of selfhood (Seth 2013; Fotopoulou and Tsakiris 2017) and emotion (Gu et al. 2013; Seth 2013; Barrett et al. 2016) can be formalized as the inferential processing of interoceptive signals, implemented within a Bayesian/predictive coding framework (discussed above). Correspondingly, models of how interoceptive inference is instantiated or regulated within the brain can inform understanding of emotional and psychosomatic disorders, such as anxiety, depression, and fatigue (Paulus and Stein 2006; Barrett et al. 2016; Stephan et al. 2016). Here we offer an overview of the neurocognitive mechanisms by which interoceptive processes underpin self-representation, psychopathology, and, in particular, intrusive experiences. We conclude with an example of an eating disorder as an instance in which interoception influences the psychopathological expression of intrusive experiences within a hierarchical predictive framework that encompasses self-conceptualization.

What Is Interoception?

Interoception encompasses afferent signaling, integrative processing, and central representation of the internal physiological state of the body (Quadt et al. 2018; for a discussion of alternative definitions, see Ceunen et al. 2016). Interoception is the sensory component of homeostatic and allostatic control. Homeostasis refers to the regulation of internal physiology, through which life is sustained by maintenance of a more or less constant internal environment, through supporting the dynamic metabolic needs of bodily tissues while excluding potential toxic or other threats to the integrity of the body (homeostatic regulation; Cannon 1929). Cardiac output, blood oxygenation, hydration, temperature regulation, and blood glucose are among the many parameters regulated homeostatically. However, homeostatic interoceptive autonomic reflex arcs alone are inefficient: better control of internal state is achieved through allostasis, wherein the future state of the body is predicted and responses are made in anticipation of future physiological states to mitigate unpredicted dyshomeostatic states that threaten life (Sterling 2012).

Allostasis is informed by the integration of interoceptive information with exteroceptive (about the external world) information for the predictive selection of autonomic/physiological and behavioral action or "policies," which ultimately ensure longer-term survival. For example, the set point of the homeostatic baroreflex, which stabilizes blood perfusion of organs by regulating the heart's beat-to-beat output, is allostatically adjusted to meet actual and anticipated physical demands (e.g., if you see a snake or bear in the woods, baroreflex suppression allows your heart rate and blood pressure to rise together to enhance skeletomuscular perfusion, facilitating the capacity for fight and flight). From a more computational perspective, the most efficient way to regulate homeostatic risk is to build a model of the body as separate from its external environment, following the cybernetic idea that "every good regulator of a system must be a model of that system" (Conant and Ashby 1970). Ultimately, physiological control combines allostatic and homeostatic mechanisms, but both can be subsumed under homeostasis (Ramsay and Woods 2014). Allostatic anticipatory control requires an inferential model (hypotheses about the causes of interoceptive inputs) of our own current and future (counterfactual) bodily states in relation to states of the external world (including other agents). The complexity is reduced by holding a set of prior "beliefs" or more broadly generative models. Deviations from homeostatic ranges are avoided by choosing in advance an appropriate sequence of actions ("policies"). These can be autonomic as well as behavioral and can cross different systems. For example, you need to eat before you faint and you need to store fat for future metabolic needs when resources may need to be allocated to other tasks. These ideas are coherent with formal frameworks of brain function, such as the Bayesian brain and active inference (discussed above). Details and discussion of the neural organization supporting interoceptive processing can be found in Appendix 13.1.

Interoception and Intrusive Experiences

There are at least three ways in which interoception can impact upon intrusive experiences:

1. It can provide *context* which can (a) have an impact on the permissive threshold for the occurrence of intrusions (discussed above) and (b) influence or constrain the content of what intrudes.
2. It can affect appraisal and control processes engaged by the intrusive experience.
3. It can also act as content itself.

Moreover, these can interact to produce a self-sustained cycle of intrusive experiences. In conceptualizing the impact of interoception on intrusive experiences, it is helpful to conceptualize it within a hierarchical or dimensional framework (see Table 13.1). Lowest in the hierarchy are the levels of physiological arousal (indexed by heart rate, blood pressure, or electrodermal activity) and the bodily changes governed by homeostatic reflex arcs. These signal the integrity and arousal state of the body through visceral afferent pathways. Fluctuations in central signaling of bodily physiology (including both engagement of ascending neuromodulatory systems and representation within primary "viscerosensory" insula, a cortical level) can thus provide the context (Pt. 1 from the above list).

As a context, psychophysiological states (e.g., sickness, arousal, and alertness) gate what enters the sensorium (Pt. 1a). For example, a heightened state of cardiovascular arousal enhances the detection and appraisal of threat (Garfinkel et al. 2014; Pezzulo et al. 2018) associated with symptoms of anxiety; increased sympathetic electrodermal tone enhances occurrence of tics in Tourette syndrome (Nagai et al. 2009). In addition, however, a particular homeostatic context, such as hunger, can motivate relevant intrusions about food (Pt. 1b; a specific example is given in the next section). Affective state represents a more elaborated interoceptive context that can again change the permissible threshold of intrusion.

Table 13.1 Dimensions of interoceptive measurement, adapted after Garfinkel et al. (2015). Psychological dimensions of interoception are given in boldface; self-referential dimensions are capitalized.

Dimensional level	Nature	Index measures
EXECUTIVE	Behavioral	Shifting from interoceptive to exteroceptive attention (e.g., within dual tasks or between tasks)
METACOGNITIVE	Correspondence between subjective self-report and objective performance accuracy	Receiver operating characteristic curves between task performance and rated confidence
		Correlational measures of task and confidence scores
		Trait measures (e.g., correspondence between task performance and body perception questionnaire score)
SENSIBILITY	Subjective self-report	Confidence measures on interoceptive tasks
		Questionnaires probing interoceptive sensitivity
Accuracy	Objective behavioral performance score	Heartbeat detection tasks
		Respiratory resistance load detection
		Water load task
		Balloon dilation of stomach/colon
Preconscious impact on other processes	Behavioral, neural	Cardiac modulation of eyeblink startle
		Cardiac modulation of fear
		Respiratory modulation of memory
Afferent signal	Neural	Visceral afferent nerve recording
		Intracranial recording
		Heartbeat evoked potential
		Respiratory evoked potential
		Neuroimaging
Bodily response	Organ-level response	Heart rate, heart rate variability, tachygastria, blood pressure, glucose, O_2 and CO_2 levels, etc.
		Autonomic psychophysiology

Second, appraisal and control processes engaged by the intrusive experience are impacted by higher-order cognitive levels of interoceptive representation, likely supported within insula and cingulate cortices (Pt. 2). Higher-order cognitive or psychological levels of interoception (highlighted in bold in Table 13.1) refer to attention and appraisal directed at bodily processes themselves. These encompass measures of interoceptive accuracy of objective (behavioral) sensitivity to bodily responses, subjective (i.e., self-reported) interoceptive sensibility to bodily signals, and metacognitive interoceptive insight (Garfinkel et al. 2015). The latter two align with notions of expectation and interoceptive prediction error ("surprise") and the precision weighting of interoceptive inputs, beliefs, and policies. Interoceptive self-efficacy (Stephan et al. 2016) is a metacognitive representation of self-efficacy.

Third, such a mental representation of bodily sensation may act as the content of the intrusion (Pt. 3). Salient bodily signals (e.g., breathlessness, heart arrhythmia, urge to void, or visceral pain) necessarily attract attention and appraisal. Upon appraisal, prior experience will determine if the intrusion per se represents a major concern or acts as a driver for subsequent general perseverative intrusions associated with overall health (e.g., health anxiety). Related to this are the so-called quasi-interoceptive signals, such as rib pain (a somatic sensation), which can be misinterpreted as a prelude to a heart attack, with anxiety again becoming amplified by the accompanying interoceptive sensations of cardiorespiratory arousal as a consequence of the appraisal process (Clark et al. 1997). Moreover, ephemeral interoceptive sensations can (through prior associations) trigger emotional (e.g., panic or fear response PTSD) or drug-related intrusive experiences such as craving (Goldstein et al. 2009; Garavan 2010). Similarly, the interoceptive feelings of premonitory urge, linked again to representation within insular cortex, will trigger tics in Tourette syndrome (Rae et al. 2018, 2019b).

Finally, an executive dimension of interoception contributes to intrusions mostly through appraisal control processes (Pt. 2) which in turn can affect the stickiness of the context (Pt. 1) and the interoceptive content (Pt. 3). The executive dimension encompasses the capacity to shift between interoceptive representations or away from interoceptive representations, aligned with both precision weighting and policy selection. Such a capacity may be evident in measures of lower levels of interoceptive signaling: for instance, heart rate variability (a product of baroreflex regulation) is linked to more general psychophysiological flexibility and is positively associated with success in suppressing unwanted intrusive thoughts and memories, like in PTSD (Gillie and Thayer 2014; Gillie et al. 2015). Conversely, intrusive perseverative cognition (worries and ruminations) and the capacity for thought control are coupled to the inflexibility associated with blunted heart rate variability, both in wakefulness and during sleep (Brosschot et al. 2010; Meeten et al. 2016; Ottaviani et al. 2016; Ottaviani et al. 2017). It should also be noted that in addition to all of these direct and indirect effects of interoception on intrusive experiences, interoceptive signals may be evoked under some circumstances as countermeasures to control intrusion, often through physiological relaxation but sometimes using physiological arousal (Nagai 2015).

Given this intimate relationship between interoception and intrusive experiences, it is perhaps not surprising that disordered interoceptive processing is reported across conditions associated with intrusive thinking. In anxiety disorders, increasing evidence indicates an association between anxiety symptoms and a mismatch between subjective (sensibility) and objective (accuracy) measures of cardiac interoception—a metacognitive interoceptive deficit (trait interoceptive prediction error) (Garfinkel et al. 2015) that is also relevant to symptoms in Tourette syndrome (Rae et al. 2019b), autism (Garfinkel et al. 2016), and, if extended to measures of choice, addiction (Moeller et al. 2014). Moreover, intrusive dissociative experiences, consistent with a fundamental self-disturbance in self-representation, are associated with lower-level interoceptive abnormalities (Schulz et al. 2016). Below we present an example of how abnormalities in interoception can act as the content and character of an intrusive experience.

Intrusive Experiences in an Ego-Syntonic Disorder Exemplified by Anorexia Nervosa

Patients with anorexia nervosa report thoughts, bodily experiences, and mental images that they consider as involuntary and intrusive to other goals, even though these may not always be unpleasant in themselves and may, in fact, constitute most people's everyday experiences. For example, a patient described the feeling of a full stomach as intruding on her mental concentration (Skårderud 2007:127):

> Some days ago, I should have had a meeting with my boss. I was anxious about this. Then I decided to vomit. I couldn't stand having the lunch in my stomach. I cannot have anything in my stomach, because then I cannot concentrate. I need to be empty to feel alert.

Similar experiences of hunger or satiation and other interoceptive sensations are frequently experienced as intrusive by individuals with anorexia, while their attempts to control their eating and body weight and to "silence" any relevant bodily needs are seen as compatible with the goal of building a coherent and stable self. This treatment-resistant concordance in eating disorders between symptoms and a sense of self is referred to as ego-syntonicity (Gregertsen et al. 2017). Unlike in (ego-dystonic) disorders like OCD, where symptoms are seen as intruding into one's other everyday goals, anorexia exemplifies a psychiatric disorder where symptoms are not viewed as intrusions into one's life; instead, necessary bodily functions, and particularly interoceptive experiences, are experienced as intrusive.

Drawing on the framework outlined earlier in the chapter as well as from knowledge of the brain systems that support homeostatic and allostatic control (outlined above), intrusive experiences in eating disorders, particularly in anorexia nervosa, can be understood as a failure to model interoceptive states at a homeostatic level, due to a deeper failure in the regulation of metabolism (notably adiposity or fat storage) at an allostatic level (i.e., a failure to optimize flexibly the precision weighting of allostatic control policies). In other words, patients with anorexia nervosa may not be able to correctly predict and regulate adiposity (and metabolism more generally), leading to a chronic dyshomeostatic state that evokes aberrant metacognitive beliefs about the low efficacy of their autobiographical self: "I cannot eat now because I will then lose control over my eating and store excessive fat" (see Figure 13.6). Recent converging evidence highlights wide dysregulation across neuromodulatory systems in eating disorders, including hormones and neuropeptides involved in the regulation of metabolic states (see Figure 13.6; Gorwood et al. 2016), and a large-scale genetic study implicating metabolic (alongside psychiatric) factors in pathoetiology of anorexia nervosa (Watson et al. 2019). Neurocomputational formulations of allostasis, that is, predictive, counterfactual interoceptive control (Stephan et al. 2016), suggest that allostasis requires a temporary change or suspension of homeostatic set points, effectively altering the priors (beliefs) of the relevant homeostatic reflex arc (e.g., the expectation of a meal will drop

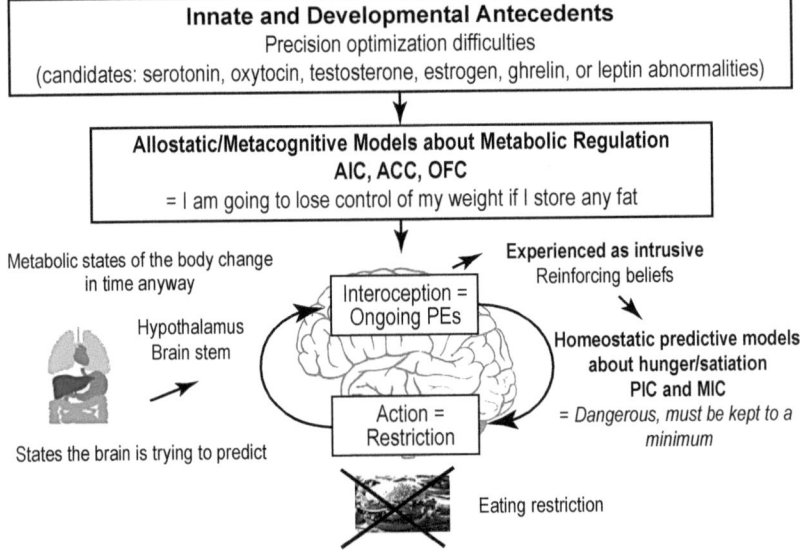

Figure 13.6 Schematic depiction of a predictive coding account of intrusive interoceptive experiences in anorexia nervosa: AIC (anterior insular cortex), ACC (anterior cingulate cortex), OFC (orbitofrontal cortex), PIC (posterior insulate cortex), MIC (medial insular cortex), PE (prediction error).

blood glucose levels to mitigate the hyperglycemia that follows eating). In the brain, allostatic coupling of behavioral policy with internal physiology is supported by regions including the anterior insula and dorsal and subgenual anterior cingulate cortices. These manifest three properties (Stephan et al. 2016): (a) access to estimates of bodily state (interoception), (b) the capacity to generate predictions over longer time scales, and (c) anatomical connections (descending visceromotor outputs) that can convey sustained changes in homeostatic beliefs instantiated by more reactive humoral/autonomic reflex response arcs within hypothalamus, brain stem, and periphery. Thus, functional abnormalities within these regions may lead to inappropriate adjustments to specific physiological parameters (e.g., glucose levels before or after meals), leading to persistent prediction errors driving abnormal eating habits. For anorexia nervosa, eating control will always be suboptimal for regulating metabolic and interoceptive states since these necessarily fluctuate in time. Persistent exacerbated interoceptive feelings of hunger and satiation are experienced as ongoing intrusive experiences that interfere with the ego-syntonic goal of a rigid control of body fat, achieved by eating restraint, exercise, and/ or vomiting. These acts in themselves and their interoceptive consequences reinforce the homeostatic beliefs of patients regarding the unpredictable and intrusive nature of hunger and satiation signals.

Several studies have indeed shown abnormalities in correctly predicting and experiencing interoceptive states in anorexia nervosa, including, for example, cardiac signals, satiation and affective touch, and the related brain function abnormalities best tracked by the anterior insular cortex and related limbic and prefrontal areas (Crucianelli et al. 2016; Bischoff-Grethe et al. 2018; Khalsa et al. 2018). Such abnormalities have been linked to persistent prediction errors about interoception and a dysregulated ability to adequately sense what is happening in the body resulting in a turbulent reference state; that is, a "noisy baseline" (Paulus and Stein 2010). This may explain why patients experience all those states as *intrusive experiences of the body* that need to be controlled by eating restriction, exercise, or vomiting (see Figure 13.6). These attempts to actively restrict and control hunger and satiation in turn lead to starvation and further maintenance mechanisms (starvation dampens hunger and slows down cognitive processing along with further complications). According to the above speculations, a fundamental difficulty in reducing interoceptive uncertainty via allostatic control would be at the heart of why otherwise normal feelings of hunger or satiation are experienced as intrusive and as "out of control."

Relevance of Stability and Flexibility to Intrusion Experiences

Balancing stability and flexibility in the brain is critical for individuals to maximize exploitation and exploration of their environment. Working memory and associative learning models provide a psychological and neural framework in which the concepts of flexibility and stability can be understood (Hochreiter and Schmidhuber 1997; Frank et al. 2001; Oberauer 2013). Biophysically detailed computational models have also investigated how dynamical interactions between different neuronal populations in cortex may promote stability versus flexibility (Durstewitz et al. 2000; Wang 2001). In addition, a large body of empirical evidence has implicated specific neural structures and neuromodulators in behavioral flexibility (Robbins 2005; Cools and D'Esposito 2011). In particular, a central role is played by the prefrontal cortex and its interactions with the rest of the brain, especially the specialized processing modules of the posterior cortex, including parietal (spatial attention) and inferotemporal (feature attention) areas; the declarative memory systems in the temporal lobes, including the rhinal cortex (recognition memory) and hippocampus (scene/episodic memory); and the language processing modules such as Wernicke's area, specialized in the comprehension of speech, and Broca's speech and production area. In addition, the prefrontal cortex interacts with subcortical structures such as the limbic structures involved in the processing of motivational and emotional cues as well as the orchestration of behavioral, autonomic, and endocrine responses, including the amygdala, hypothalamus, and brain stem centers; the basal ganglia, which are involved in the higher-order control of thought (see below) and action; and the neuromodulatory systems of the reticular core of the brain, including monoamine and cholinergic cell groups in the midbrain and hindbrain.

Working Memory Models

As an example of how these interactions could support a balance between stability and flexibility, let us consider the case of working memory. In models of working memory, stability (i.e., stable goal-oriented performance) can be maintained by holding temporally stable representations of our goals. Goals for action can reside at different levels of task abstraction and unfold over different timescales. Importantly, however, a goal held in working memory can include the goal of meeting the requirement of specific tasks. As such, our ability to hold this goal in memory, available for use as a control signal, allows for stable task performance. Likewise, our ability to update working memory (i.e., to shift goals as context demands) is important for flexibility.

The control of working memory is often conceptualized as a gate that is distinct from the memory store itself (Hochreiter and Schmidhuber 1997). Closing an input gate against distracting information prevents its access to working memory, keeping the current contents available as control signals; this gating function promotes stability. In contrast, opening the gate enables the updating of working memory and allows new contextual information to modify behavior; this gating function promotes flexibility.

Though different mechanisms for working memory gating have been proposed (e.g., Wang et al. 2004; Zhu et al. 2018), one influential model has focused on frontostriatal interactions (Frank et al. 2001). This circuit is schematized in Figure 13.7 (O'Reilly 2006). The corticostriatal model of working memory gating proposes that the prefrontal cortex supports information maintenance, whereas the striatum-pallidal-thalamic pathway implements gating by regulating what information is allowed in and out of working memory.

Based on this model, spontaneous, unwanted events could be experienced as intrusions when the gate to working memory is breached and the intrusion supplants ongoing working memory processes. Once this occurs, intrusive events serve as signals to drive other cognitive processes and actions. Thus, the integrity of the working memory gating is paramount for mitigating against intrusive experiences. For example, by preventing an unwanted experience from updating to working memory or by inhibiting their influence on output control signals, one could stop the negative cycle of behaviors that can result from intrusive experiences. These gating mechanisms could be global (like the fast, inhibitory mechanisms supported by the hyperdirect pathway that can affect multiple processes simultaneously) or selective, supported by both the direct and indirect pathways, schematized in Figure 13.7 as the Go and No-Go pathways. Coordination among multiple corticostriatal loops can also be a mechanism for working memory operations in separate prefrontal areas to carry out complex, sequential, and hierarchically structured tasks (for a review, see Badre and Nee 2018).

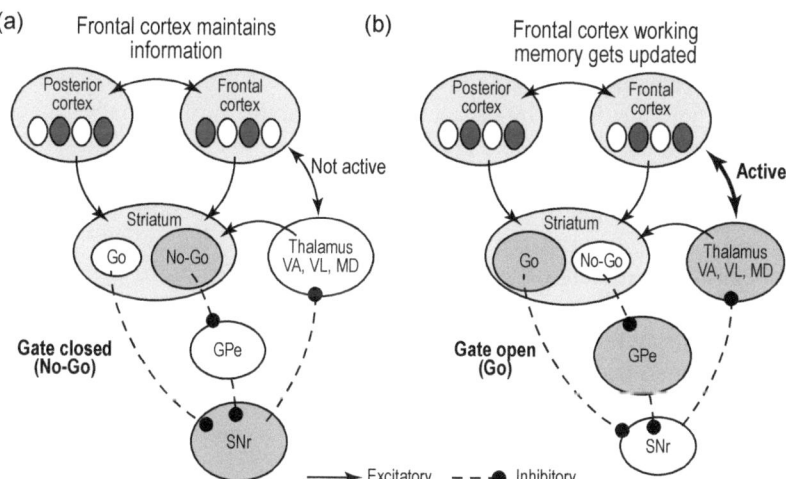

Figure 13.7   Schematic depicting a mechanism of working memory gating through corticostriatal interactions. Inhibition (a) or disinhibition (b) of thalamocortical dynamics through the striatum can regulate gate closing and opening, respectively: VA (ventral area), VL (ventrolateral), MD (medial dorsal), GPe (globus pallidus external), SNr (substantia nigra pars reticulata). Reprinted with permission from O'Reilly (2006).

The type of gating one selects can be thought of as a *gating policy*, in the same sense as defined above. As such, the match of the right gating policy to the particular dynamics of the situation is a key determinant of successful control (Bhandari and Badre 2018). For example, when confronted with an unwanted memory, one could deploy a global suppression to prevent it from entering working memory or instead attempt to selectively input another thought into working memory in its place. The consequences of these policies on memory or the ongoing impacts of the triggering event (both in this instance or in the future) might differ depending on the gating strategy that is selected. Thus, pathologies could arise as a result of any of the following:

- Items seeking to enter working memory are sufficiently salient or valued and will therefore breach the gating mechanism to update working memory (breach intrusion).
- Gating itself is weak and thus items access working memory, even if they are not adaptive or helpful to the individual (permissive intrusion).
- Mechanisms involved in maintaining stability (other than gating) are too strong and thus do not allow working memory to be updated once an intrusive experience has occurred.
- The wrong gating policy is selected given the nature or dynamics of the intrusion.

Associative Learning Models

Corticostriatal circuits are also central to *associative learning models* (Balleine and Dickinson 1998). Two control processes have been identified that are engaged in the control of goal-directed and habitual actions and which are mediated by distinct parallel circuits through the basal ganglia; in some circumstances, they compete with one another (Balleine et al. 2009). The goal-directed network is engaged rapidly with changes in the environment, incorporates the cortical working memory process described above, and utilizes this network to encode the action–outcome associations that mediate goal-directed action in a region of dorsomedial striatum. Generally speaking, this network relies on this prefrontal-dorsomedial-striatal (or caudate) pathway and feedback to the cortex via the substantia nigra pars reticulata and mediodorsal thalamus (Balleine and O'Doherty 2010) to encode and utilize novel solutions to problems presented by a changing environment. It also functions to inhibit older, more routine and outdated solutions, particularly the performance of habitual actions centered on the sensorimotor cortices and putamen or dorsolateral striatum (Graybiel 2008), when these have or are likely to have aversive consequences. If the goal-directed circuit is altered (e.g., through damage, disease, or drugs), inhibition can be reduced or mistimed, resulting in dysregulation of habits (even in the presence of aversive consequences), and producing intrusive experiences. An increased reliance on habits may not only apply to the compulsive acts seen in OCD (Saxena et al. 1998; Robbins et al. 2019), but also the persistent motor habits (tics) associated with Tourette syndrome (Maia and Conceição 2017, 2018) as well as craving and compulsive drug use in addiction (Everitt and Robbins 2016; Furlong et al. 2017).

There are, however, other important features that are controlled by the goal-directed circuit, particularly by the dorsomedial striatal component of that circuit. As mentioned, considerable evidence suggests that the prefrontal working memory systems provide inputs to the striatum that mediate the plasticity necessary to encode goal-directed actions in the posterior segment of the dorsomedial striatum (reviewed in Balleine and O'Doherty 2010). However, to allow this large structure to encode more than one action–outcome association, plasticity associated with new action–outcome learning needs to be segregated from prior learning. It appears that this segregation is achieved via state-related information provided by inputs to the striatum from the parafascicular thalamus (Bradfield et al. 2013). This input onto the tonically active striatal cholinergic interneurons causes them to pause, allowing the principal neurons (the spiny projection neurons) relief from inhibition induced by tonic acetylcholine release. During this pause, cortical and midbrain dopaminergic inputs to the dorsomedial striatum can combine to induce plasticity in the spiny projection neurons. Accordingly, changes in action–outcome contingency provoke changes in the patterned input from the parafascicular thalamus, leading to plasticity changes in the targeted dorsomedial-striatal region.

Importantly, evidence suggests that the retrieval of specific action–outcome ensembles for performance is mediated by state-related information, based largely on outcome-related information (Bradfield et al. 2015) conveyed to the striatum, not by the parafascicular thalamus but via inputs from the orbitofrontal cortices (Gremel and Costa 2013; Bradfield et al. 2015; Stalnaker et al. 2016). Thus, accurate retrieval of specific action–outcome associations will be determined by the fidelity of this orbitofrontal cortical input: as a consequence, changes in orbitofrontal cortex activity (e.g., in OCD) could result in faulty retrieval, causing changes in flexibility (described above) and leading to the intrusion of unwanted information. Retrieval can become "frozen" if the orbitofrontal cortex gets "stuck" in a given state (see Appendix 13.1); alternatively, it could become highly, temporally disparate if activity in the orbitofrontal cortex fluctuates rapidly and unpredictably.

This type of state information features heavily in computational accounts, particularly model-based reinforcement learning accounts of goal-directed action. Such accounts provide information about state transitions for retrieval and could be seen as the computational implementation of these ideas (Wilson et al. 2014). See Appendix 13.1 for further computational modeling of recurrent intrusions in OCD focusing on neuromodulation within orbitofrontal cortex.

Summary

Translating findings across different levels of analysis, including computational, psychological, neurobiological, and physiological, is challenging. However, to understand the nature of intrusive experiences and to develop effective treatments, such translation is essential. A first step in this translation is to use a common language. To this end, we have attempted to define all terms and concepts, especially where multiple, related, but somewhat distinct meanings exist. Salience is one such example of a term that is often used broadly to refer to the quality of being particularly noticeable, but in a Bayesian framework is specifically used to refer to the value afforded to uncertainty resolution. The models we discuss provide explanations for a range of intrusive experiences: from the obsessions and compulsions of OCD and drug and emotion-related intrusions in addiction and PTSD to intrusions of thoughts, bodily experiences, and mental images in anorexia nervosa. We focused on two major networks, frontostriatal and insula-cingulate, to illustrate how imbalances in these networks can lead to intrusive experiences. Whenever possible, overlap between different models and levels of analysis have been highlighted to provide a systems overview of how intrusive experiences across a range of distinct psychiatric, neurodevelopmental, and neurological disorders may emerge as a consequence of dysfunction at different levels of the nervous system.

Appendix 13.1

Active Inference

This appendix provides a technical description of belief updating under active inference. One useful aspect of treating "trains of thought" as "planning" under a generative model is that one can always express a generative model as a *graphical model* (Figure 13.A1). This is important because a graphical model can be used to understand the computational architecture of neuronal message passing in the brain. For every graphical model that specifies the states and outcomes in play and their conditional dependencies, there is an associated *factor graph* that provides, and must be supported by, unambiguous specifications of the architecture (e.g., neuronal connectivity) and message passing (e.g., neurophysiology); for details, see Figure 13.A2 and Friston et al. (2017).

In brief, the sorts of generative models commonly used to explain planning as inference are usually based on partially observed Markov decision process models. Crucially, in these generative models, discrete states of the world evolve over time in a way that *depends upon action*.

Expected Free Energy, Salience, and Value

For technical readers, expected free energy can be decomposed into an epistemic, information-seeking, uncertainty-reducing part (*intrinsic value*) and a pragmatic, goal-seeking, instrumental part (*extrinsic value*). Formally, the expected free energy for a particular policy π at time τ in the future can be expressed as in terms of beliefs $Q\left(s_\tau, o_\tau \mid \pi\right)$ about future states s_τ and outcomes o_τ:

$$G(\pi,\tau) = -\underbrace{E\left[\ln Q\left(s_\tau \mid o_\tau, \pi\right) - \ln Q\left(s_\tau \mid \pi\right)\right]}_{intrinsic\ value} - \underbrace{E\left[\ln P(o_\tau)\right]}_{extrinsic\ value}. \quad (13.\text{A}1)$$

Extrinsic (instrumental) value is simply the expected value of a policy defined in terms of outcomes that are preferred a priori. The more interesting part is the uncertainty-resolving or intrinsic (epistemic) value, variously referred to as relative entropy, mutual information, information gain, Bayesian surprise, intrinsic motivation, or value of information expected under a particular policy (Barlow 1961; Howard 1966; Optican and Richmond 1987; Linsker 1990; Itti and Baldi 2009).

Intrinsic (epistemic) value can be regarded as *salience*. Formally, this means that salience is the Kullback-Leibler (KL) divergence between posterior beliefs about hidden states with and without observations solicited by a particular act (or policy). The reason this divergence is associated with salience stems from the visual neurosciences, where the salience of a potential location for a saccadic fixation is known as *Bayesian surprise* (Itti and Baldi 2009; Sun et al. 2011; Barto et al. 2013). In robotics and machine learning, the information gain or Bayesian surprise is known as intrinsic motivation or value (Ryan and Deci 1985; Eccles and Wigfield 2002; Oudeyer and Kaplan 2007; Schmidhuber 2010; Barto et al. 2013). It is also referred to as *epistemic value* or *epistemic affordance* (Parr and Friston 2017). Epistemic affordance appeals to Gibsonian notions of affordance: it is the resolution of uncertainty afforded by a particular act: "What would I learn by looking over there?" On a psychological interpretation, intrinsic value can also be associated with *incentive salience* (Berridge and Robinson 1998; McClure et al. 2003). Exactly the same kind of mathematical arguments can be applied not just to beliefs about states in the world but also the parameters of the generative model. These parameters encode contingencies and laws governing the evolution of states or their mapping to observations. In this setting, salience becomes *novelty*; namely, the information gain afforded by knowing "what would happen if I did that?"

The factor graph in Figure 13.A2 is used to pass messages among the nodes (e.g., neuronal populations) to minimize free energy per se; in other words, to maximize the evidence for any given generative model of how outcomes were generated. This leads to biologically plausible message-passing schemes of the sort studied in terms of evidence accumulation and predictive coding

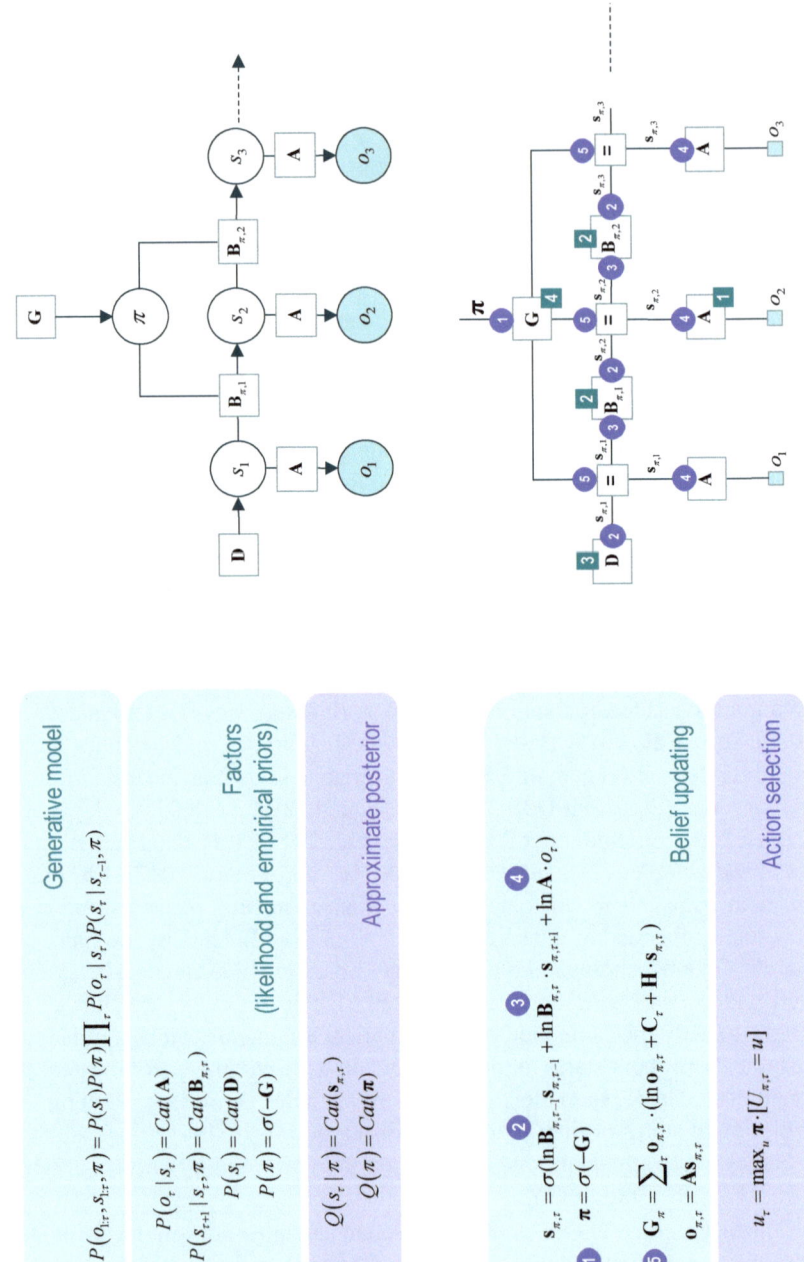

Figure 13.A1 A generative model for discrete states and outcomes. Upper left panel: these equations specify a generative model. A generative model is the joint probability, P, of outcomes or consequences and their (latent or hidden) causes, see first equation. Usually, the model is expressed in terms of a likelihood (the probability of consequences given causes) and priors over causes. When a prior depends upon a random variable it is called an empirical prior. Here, the likelihood is specified by a matrix $\mathbf{A}$, whose elements are the probability of an outcome under every combination of hidden states. Cat denotes a categorical probability distribution. The empirical priors pertain to probabilistic transitions (in the $\mathbf{B}$ matrix) among hidden states that can depend upon actions, which are determined by policies (sequences of actions encoded by π). The key aspect of this generative model is that policies are more probable a priori if they minimize the (time integral of) expected free energy $\mathbf{G}$, which depends on prior preferences about outcomes or costs encoded in $\mathbf{C}$ and the uncertainty or ambiguity about outcomes under each state, encoded by $\mathbf{H}$. Finally, the vector $\mathbf{D}$ specifies the initial state. This completes the specification of the model in terms of model parameters that constitute $\mathbf{A}$, $\mathbf{B}$, $\mathbf{C}$, and $\mathbf{D}$. Bayesian model inversion refers to the inverse mapping from consequences to causes (i.e., estimating the hidden states and other variables that cause outcomes). In approximate Bayesian inference, one specifies the form of an approximate posterior distribution Q. This particular form in this figure uses a mean field approximation, in which posterior beliefs are approximated by the product of marginal distributions over time points. Subscripts index time (or policy), italic variables represent hidden states, and bold variables indicate expectations about those states. Upper right panel: this Bayesian network or graphical model represents the conditional dependencies among hidden states and how they cause outcomes. Open circles are random variables (hidden states and policies), filled circles denote observable outcomes, and squares indicate fixed or known variables, such as the model parameters. Lower left panel: these equalities are the belief updates mediating approximate Bayesian inference and action selection. Lower right panel: this is an equivalent representation of the Bayesian network in terms of a Forney or normal style factor graph. Here the nodes (square boxes) correspond to factors and the edges are associated with unknown variables. Filled squares denote observable outcomes. The edges are labeled in terms of the sufficient statistics of their marginal posteriors (see approximate posterior). Factors have been labeled intuitively in terms of the parameters encoding the associated probability distributions (on the upper left). The circled numbers correspond to the messages that are passed from nodes to edges (the labels are placed on the edge that carries the message from each node). These correspond to the messages implicit in the belief updates (lower left). This figure is based on Friston et al. (2017).

278

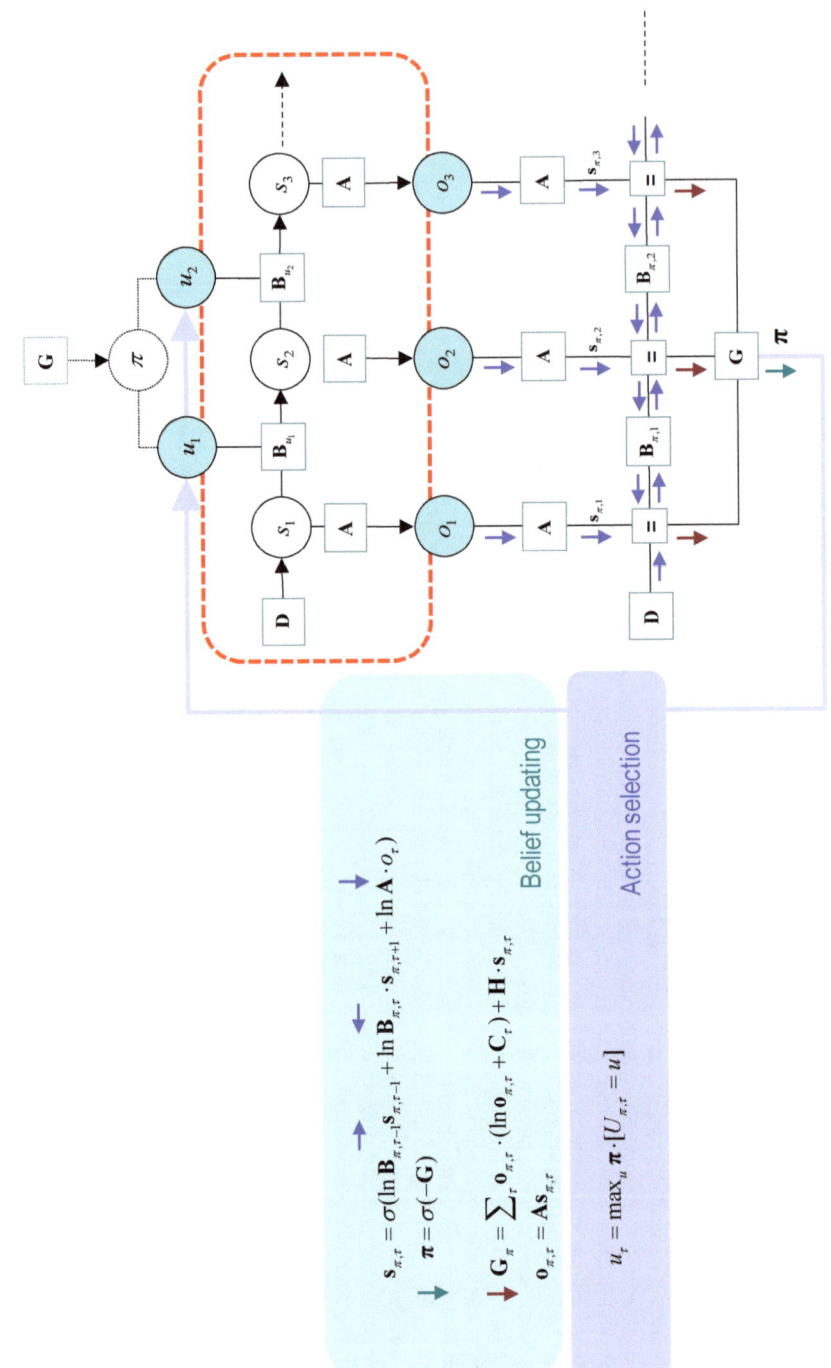

Figure 13.A2 The generative process and model. This figure reproduces the Bayesian network and Forney factor graph of Figure 13.A1; however, here the Bayesian network describes the process that generates data, as opposed to the generative model of data. This means that we can link the two graphs to show how the policy half-edge of Figure 13.A1 couples back to the generative process (by generating an action that determines state transitions). The selected action corresponds to the most probable action under posterior beliefs about action sequences or policies. Here, the message labels have been replaced with little arrows to emphasize the circular causality implicit in active inference: the real world (red box) generates a sequence of outcomes that induce message passing and belief propagation to inform (approximate) posterior beliefs about policies (that also depend upon prior preferences and epistemic value). These policies then determine action, which generates new outcomes as time progresses, thereby closing the action perception cycle. This figure is based on Friston et al. (2017).

(Srinivasan et al. 1982; Rao and Ballard 1999; Huk and Shadlen 2005; Beck et al. 2008; Bastos et al. 2012; Egner and Summerfield 2013; de Lafuente et al. 2015; Kira et al. 2015; Shipp 2016). In terms of the parameters of the generative model, associative plasticity is the corresponding belief update for neuronal connections (Friston et al. 2016).

Precision and Parameters

Of particular interest here are the parameters that link states to outcomes and states at one point in time to states at the next point in time. In Figure 13.A1, these are simply matrices of probabilities encoding the likelihood mapping from states to outcomes **A** and the transition probabilities to one state to the next **B** (which depend upon a particular policy).

Neurobiologically, these matrices play the role of *connectivity* matrices, which play an important role in sensory data assimilation and subsequent planning based on beliefs about the consequences of any action. Furthermore, each column of these matrices has a *precision*. Precision, in this instance, reflects the fidelity or confidence about the outcome (or subsequent state) given the current state of the world. A very precise mapping means that we can be almost 100% confident that this will happen given that state, while a very imprecise mapping means that all outcomes (all subsequent states) are equally likely. For discrete space models, one can express the likelihood and priors in terms of inverse temperature or softmax parameters with the following form, where $\sigma(\cdot)$ is a softmax function or normalized exponential:

$$P\left(o_\tau \mid s_\tau\right) = \sigma\left(\gamma_o \cdot \ln \mathbf{A}\right)$$
$$P\left(s_\tau \mid s_{\tau-1}, \pi\right) = \sigma\left(\gamma_s \cdot \ln \mathbf{B}\right) \qquad (13.A2)$$
$$P\left(\pi\right) = \sigma\left(-\gamma_\pi \cdot \mathbf{G}\right).$$

Neural Organization of Interoceptive Processing

The control and representation of internal bodily physiology is instantiated throughout the neuraxis (for reviews, see Craig 2003; Critchley and Harrison 2013). While ganglionic and spinal reflexes support proximate physiological regulation, the brain orchestrates homeostatic control and allostatic responses across bodily organs, integrating control with behavioral demand. The brain receives interoceptive information about the internal state of the body via neural afferent and humoral interoceptive routes (for details, see Figure 13.A3). Somatosensory pathways also contribute to quasi-interoceptive sensation of bodily physiology (e.g., via heart beating against the chest wall, rib motion, pharyngeal airflow) and to referred pain (e.g., angina felt in shoulder). Chemosensory signaling (including O_2/CO_2, hormones, cytokines, blood pH, glucose, and hydration) occurs through central blood sampling at

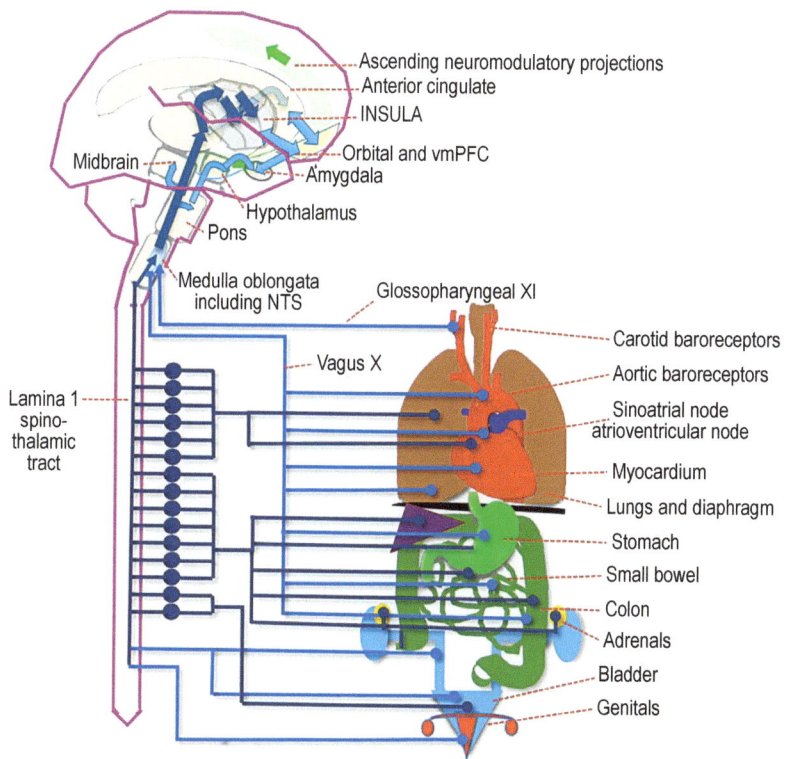

Figure 13.A3 Schematic illustration of feed-forward neural interoceptive pathways. Peripheral afferents in cranial nerves X (vagus) and X1 (glossopharyngeal) and those following sympathetic nerves to spine (ascending laminar1) converge in the medullary nucleus of the solitary tract (NTS). Here, local connections support homeostatic reflexes (e.g., baroreflex) modulating autonomic outflow. Interoceptive information passes via a primary thalamocortical route to insula (shown in gray; viscerosensory cortex) where integrative processing builds a representation within anterior insular that is consciously accessible and can give rise to potentially intrusive interoceptive and affective feelings. Secondary interoceptive channels include (1) a subcortical route to hypothalamic and basal ganglia (including amygdala), modulating ascending widespread monoamine projections from midbrain (shown in light green) and (2) a thalamocortical route to visceromotor anterior cingulate cortex. Connections to orbitofrontal and ventromedial prefrontal cortices (vmPFC) offer another putative source of intrusive motivated feelings related to selection of action policies.

paraventricular organs and hypothalamus and may engender powerful motivational and arousal states (e.g., air hunger) with correspondingly intense feelings. The interoceptive representation within insular cortex shows a partial viscerotopy and connections follow a posterior-anterior and dorsal-ventral progression with increasing opportunity for cross-modal integration (Craig 2003, 2009; Evrard 2019). Anterior insula is most implicated in supporting conscious access to interoceptive sensations and associated emotional and motivational feelings (Critchley et al. 2004), including the urge-to-tic in Tourette syndrome (Conceição et al. 2017) and drug cravings (Goldstein et al. 2009; Garavan 2010). Reciprocal connections between anterior insula and "visceromotor" rostral cingulate regions (both "allostatic" dorsal anterior cingulate and "homeostatic" subgenual cingulate) represent a putative functional architecture for higher-order predictive regulation of bodily states (Critchley et al. 2004; Critchley et al. 2005; Medford and Critchley 2010). As described above, anterior insula and dorsal anterior cingulate are key hubs within the so-called salience network (highlighting the motivational primacy of interoception, where salience is the epistemic value afforded by uncertainty resolution of Bayesian surprise; see above and Fedota and Stein, this volume).

Example of a Computational Model to Explain Recurrent Intrusions in Obsessive-Compulsive Disorder

In addition to being intrusive, obsessions in OCD are both recurrent and "sticky" in the sense that they are difficult to shake from mind. From a dynamical systems perspective, these characteristics seem to suggest that obsessions correspond to attractors (Rolls et al. 2008; Maia and McClelland 2012; Rolls 2012; Maia and Cano-Colino 2015); that is, states toward which a system tends and from which it may have difficulty escaping.

OCD prominently involves disturbances in the orbitofrontal cortex and connected regions (Maia et al. 2008). Neurochemically, OCD may be associated with low serotonin and/or high glutamate. A biophysically detailed computational model of serotonin and glutamate modulation of the orbitofrontal cortex showed that both low serotonin and high glutamate tend to create excessively strong attractors in the orbitofrontal cortex (Figure 13.A4). The network tends to fall into these attractors and then has difficulty escaping from them. This is consistent with the perseverative responding to a previously rewarded visual stimulus displayed by marmoset monkeys following depletions of serotonin in the orbitofrontal cortex, either following reversal (Clarke et al. 2006) or extinction (Walker et al. 2009) of the association between the stimulus and reward.

In these simulations, neuronal activity was elicited by "manually" activating subsets of neurons. A more complete design would also have to incorporate the endogenous gating of information into (and out of) this local network, as was described above in the context of working memory. In addition, there are complex interactions between neuromodulatory levels and their effects across interacting brain structures. For example, the extent to which a monkey displays perseverative responding depends not only on low levels of serotonin in the orbitofrontal cortex but also on high levels of dopamine in the striatum (Groman et al. 2013). Moreover, alterations in these neuromodulators at the level of the orbitofrontal cortex can have profound opposing influences on the levels of the same or different neuromodulators in other structures, including the striatum (Roberts et al. 1994; Clarke et al. 2014) and the amygdala

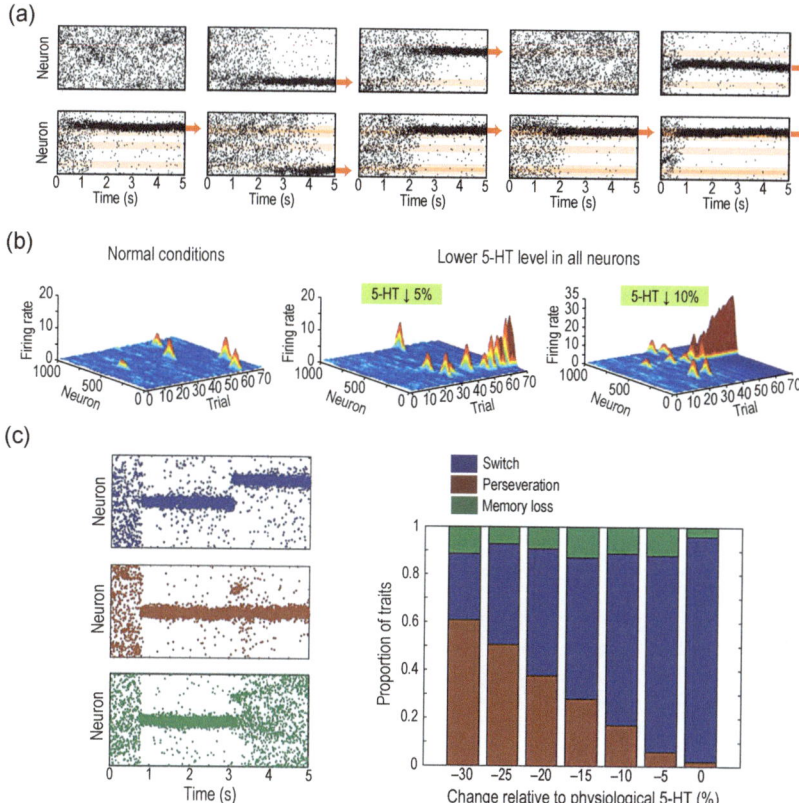

Figure 13.A4 A computational model of the role of serotonin (5-HT) in the orbito-frontal cortex in OCD, adapted after Maia and Cano-Colino (2015). (a) Illustration of the process of entrenchment of patterns of neuronal activity, taken to correspond to obsessions or, in less pathological cases, "habits of thought." Each plot represents a population of neurons; the dots along each line represent the action potentials for one neuron. The network stochastically develops patterns of activity ("bumps"). Each bump elicits strengthening of the synapses between the neurons that were active in that bump through Hebbian learning, thereby developing attractors (orange bands). The more frequently a bump occurs, the more likely it is that it will reoccur (see last three plots). (b) Effects of reducing serotonin on the tendency to develop and fall into excessively strong attractors. Under normal circumstances, the network develops bumps at varying places over time (left panel). Under low levels of serotonin, however, the network tends to develop excessively strong attractors into which it repeatedly falls (middle panel). Moreover, there is a dose-response effect, such that reducing serotonin further causes even stronger attractors to develop (right panel). Increasing glutamate has the same effect as decreasing serotonin (not shown). (c) Low levels of serotonin cause the attractors to become excessively stable. Simulated activation of a set of neurons elicited a bump, followed by activation of a different set of neurons. Under normal conditions, the network's pattern of activity flexibly shifts to the state represented by the new bump (blue). Under low levels of serotonin, the network fails to shift to the new bump, resulting in perseverative activation of the prior bump (brown). (continued on next page)

Figure 13.A4 (continued) Importantly, low levels of serotonin increase such perseverative errors (brown) without affecting a different type of error in which the network simply loses the memory of what it was initially representing (green). The latter error, which is more reminiscent of disorders in which there is difficulty in keeping items in working memory (e.g., ADHD), is not affected by the serotonin manipulations.

(Roberts and colleagues, unpublished), which may exacerbate the possibility of intrusions occurring or becoming sticky. Understanding these effects and their implications for obsessions, if any, will require more complex models that incorporate the interactions between various regions and neuromodulators.

Acknowledgments

Figures 13.A1 and 13.A2, from Friston et al. (2017), are used under the terms of the Creative Commons Attribution 4.0 International License (CC-BY).

Interventions and Treatments

14

Psychological Interventions as They Relate to Intrusive Thinking

Intrusive, Emotional Mental Imagery after Traumatic and Negative Events

Emily A. Holmes, Lisa Espinosa, Renée M. Visser,
Michael B. Bonsall, and Laura Singh

Abstract

Common across psychological disorders, intrusive, emotional mental images are sensory-perceptual representations that intrude involuntarily into the mind. Mental health treatments typically focus on entire disorders with multiple symptoms. This chapter suggests focusing on core clinical symptoms (i.e., intrusive imagery). Existing psychological therapy techniques (e.g., imagery rescripting) are promising, but underlying treatment mechanisms need to be better understood.

Precise treatments and preventions are required. Using the example of psychological trauma, this chapter argues that psychological interventions can be developed in the laboratory: effective experimental analogues of trauma can generate intrusions so that putative interventions that modulate intrusions can be explored at various mechanistic levels (e.g., molecular, cognitive, social). Examples of targeting "new" (i.e., Day 1 of the traumatic event) memories include a simple cognitive interference intervention that holds promise for preventing intrusive images after trauma (a behavioral protocol including *Tetris* game play). This intervention specifically targets intrusive involuntary memories while leaving voluntary memory intact. Work on targeting "old" (as of Day 2) memories is at an earlier stage. Research on reconsolidation update mechanisms appears valuable in reducing older trauma memories via interference interventions, again with a behavioral task interference technique. To understand mechanisms across different levels (e.g., molecular, cognitive, or social), mathematical models can aid the identification of causal mechanisms involved in memory formation. Questions are posed to instigate discussion of future science-driven psychological interventions for intrusive images.

Introduction

What Is Intrusive Thinking?

One of the rare clinical psychology textbooks dedicated to intrusive thinking (Clark 2005:4) states:

> we define unwanted clinically relevant intrusive thoughts, images or impulses as any distinct, identifiable cognitive event that is unwanted, unintended, and recurrent. It interrupts the flow of thought, interferes in task performance, is associated with negative affect, and is difficult to control.

In this chapter, we focus on intrusive thoughts in the form of mental images. This is because intrusive images occur across mental disorders. Further, it has been shown that, compared to verbal thought, imagery has a more powerful impact on emotion (Holmes and Mathews 2010) and thus carries the most weight in psychopathology. We suggest that a focus on mental imagery opens up novel angles for treatment innovation.

Mental images are sensory-perceptual representations; that is, like perception in the absence of percept (Pearson et al. 2015), as if "seeing in the mind's eye." They can occur in any sensory modality, not just visual. Imagery can be of past memories or simulations of future events. When images intrude involuntarily into the mind, they can carry strong emotion and influence behavior.

Intrusive image-based memories of a traumatic event are a core clinical feature of both acute stress disorder and posttraumatic stress disorder (PTSD) (American Psychiatric Association 2013). For instance, a person who experienced a traumatic road traffic accident may develop a distressing intrusive image of a red truck coming toward them right before the accident happened (Iyadurai et al. 2018). Mothers who have experienced traumatic childbirth may see an intrusive image of the hospital lights above them when they went into surgery (Horsch et al. 2017).

Intrusive images are not only present after trauma, such as in PTSD, they can appear in numerous other psychological disorders (Holmes and Mathews 2010). For instance, people with social phobia may repeatedly see themselves performing badly in a social situation, and such negative self-images play a causal role in maintaining symptoms in social phobia. People with depression may experience intrusive images of themselves being rejected or socially isolated, whereas they usually experience impoverished positive future imagery (Hales et al. 2011; Newby and Moulds 2012; Holmes et al. 2016a). Intrusive mental imagery may act as an emotional amplifier of all mood states in people with bipolar disorder (Holmes et al. 2016b): Vivid negative mental imagery, such as seeing oneself committing suicide in the future, may drive despair. Vivid overly positive imagery, such as seeing others responding extremely well to one in social situations, may drive mania. These are only some examples of intrusive images across mental disorders; the list could even be further extended to account for intrusive imagery in obsessive-compulsive disorder (OCD) (Coughtrey et al. 2015), body dysmorphic disorder (Osman et al. 2004), and agoraphobia (Day et al. 2004).

Here, we focus on intrusive mental imagery associated with psychological trauma and depression. Thus, the mechanisms, treatments, and mathematical framework discussed may or may not apply to other psychological disorders characterized by intrusive images, such as OCD, which we do not address in detail.

What Are Psychological Interventions?

Psychological therapy is described as an interpersonal intervention, usually provided by a mental health professional, such as a clinical psychologist who employs any of a range of psychological techniques. Various schools of therapy include cognitive behavioral therapy (CBT), psychoanalysis, systemic therapy, and so forth. The CBT model was derived from a combination of principles from behavioral and cognitive psychology alongside clinical experience. In brief, CBT focuses on challenging and changing unhelpful cognitive distortions (e.g., thoughts, beliefs, and attitudes) and behaviors and improving emotional regulation. The individual learns personal coping strategies to solve current problems. The therapist assists the individual in identifying strategies to address goals and improve symptoms. CBT asserts that maladaptive thoughts and behaviors influence the development and maintenance of psychological disorders, and thus symptoms can be reduced through new information-processing skills and coping strategies.

Is there evidence that these methods actually work? Critically, how do we choose which psychological or pharmacological intervention to use? Evidence-based guidelines, such as put out by the U.K. National Institute for Health and Care Excellence (NICE), critically review the full clinical literature and make recommendations on the basis of how effectively treatments work. There have now been hundreds of trials of psychological treatments. These can be classified by mental disorder; in the case of intrusive thinking, PTSD provides an example.

From such a review of the clinical research evidence, the main recommendation for PTSD, in terms of a first-line treatment, is individual *trauma-focused cognitive therapy* (National Institute for Health Care Excellence 2018). This is a tailored form of cognitive behavioral psychological therapy which follows a clear, validated manual; typically 8–12 sessions, each an hour long, are held with a mental health professional who is highly trained in its specific delivery. It includes elaboration and processing of the trauma memories, processing trauma-related emotions, and restructuring trauma-related meanings for the individual. The evidence-based guidelines for PTSD also recommend another psychological treatment: *eye movement desensitization and reprocessing.* CBT interventions can also be targeted at specific symptoms, such as sleep disturbance or anger. Drug treatments are only recommended as a secondary approach if patients prefer drugs over therapy (National Institute for Heath and

Care Excellence 2018), as the evidence is less strong for clinical effectiveness. Drug treatments after trauma are not recommended as a preventive strategy; in particular, benzodiazepines can worsen symptoms.

Since the inception of CBT in the 1960s, great advances have been made. For instance, we now have highly effective evidence-based CBT treatments for full-blown PTSD. This is one of the areas in which we have the best treatments: some of the best PTSD trials have recovery rates of 75%, whereas the norm in CBT is 50% (Holmes et al. 2014). However, there are still many ways in which we need to improve psychological treatments (Holmes et al. 2018). How can we adapt or simplify them to reach more people? How do we understand the critical ingredients in a psychological treatment in so doing? Can we also "prevent" rather than "cure"? That is, can we prevent intrusive memories after trauma rather than only have treatments once the full-blown disorder has been established?

When we think, we can think in the form of words (verbal thought) or mental images (sensory representations in any modality such as visual, olfactory, or auditory images). The dominant focus in psychological treatment research and therapy, such as CBT, has been on verbal thoughts. A focus on *mental imagery* is one alternative and may open up opportunities for both research and treatment innovation.

What Can We Do at the Moment?

CBT Techniques That Target Intrusive Mental Imagery

In a handbook for clinicians and patients, we have described four face-to-face mental imagery-focused techniques in CBT: imagery rescripting, metacognitive techniques, imagery-competing tasks, and enhancement of positive imagery (Holmes et al. 2019). In contrast to the CBT techniques that typically focus on a whole disorder (rather than on one symptom), these techniques specifically focus only on distressing intrusive mental images. During imagery rescripting, a person is asked to bring an intrusive negative image to mind and describe it in detail (e.g., an image of oneself in a car crash). An alternative image is then introduced to transform and update the negative image (e.g., an image of oneself being well today despite the car crash). Metacognitive techniques aim at lessening an image's impact on a person by emphasizing that it is not real (e.g., switching attention from the internal image to the outside world or reinforcing the image's unreality by making it look comical). Imagery-competing techniques aim at disrupting or interfering with intrusive images with a competing task (e.g., a visuospatial computer game) while the image is active in a person's mind. Positive imagery enhancement aims at encouraging people to generate positive future images or repeatedly train them to interpret ambiguous scenarios in a more positive way (Holmes et al. 2019).

Need to Understand Psychological Interventions in the Lab and Be Able to Disseminate Them Globally

To understand how existing treatments for intrusive images work, we need to clarify the key mechanisms driving these treatments. An experimental psychopathology approach allows a direct test of whether an experimental manipulation of a specific mechanism leads to a change in intrusions where key processes that maintain or change aspects of psychopathology can be identified. Thus, we need to take current treatments back to the laboratory and examine specific treatment mechanisms modulating intrusions in isolation and under controlled conditions. We can then remove irrelevant strategies from current treatments and develop novel approaches that target the essential causal mechanisms modulating intrusive images and are more effective and precise (Holmes et al. 2018). Such novel approaches need to be easily scalable to meet the global need for mental health treatments. Thus, we need to develop simple, brief, and flexible interventions that can be adapted to people's needs across cultures. Such interventions should ideally not require highly trained mental health professionals but should be able to be delivered by trained lay mental health care workers via innovative and remotely accessible online platforms (Holmes et al. 2018).

What Do We Need to Do Next?

Research Paradigms to Study and Generate Intrusions

To improve treatments that target intrusive thinking, we need to be able to generate intrusions and study their crucial underlying mechanisms in controlled laboratory settings. Going back to the lab allows us to change focus from complex real-life situations with clinical populations to simpler experimental procedures with nonclinical human (or nonhuman) populations (Visser et al. 2018). That is, we need experimental models that incorporate the dynamic nature of intrusive memories (Visser et al. 2018). Different paradigms can be used to generate intrusions in the laboratory. Here, we focus on two commonly used analogs of stressful events/trauma in anxiety and PTSD research: fear conditioning (Pittig et al. 2018) and the trauma film paradigm (James et al. 2016a).

Fear Conditioning

One well-known experimental method to investigate involuntary expressions of aversive memory in both human and nonhuman animals is Pavlovian fear conditioning. This paradigm has been used to investigate aversive associative learning, considered to be an important mechanism in the etiology of anxiety disorders (Pittig et al. 2018). Indeed, in real-life settings, learning what is threatening or safe is important for survival. However, the association of neutral cues with threat can become maladaptive when such conditioned fear fails to extinguish long after the danger has passed or overgeneralizes to a wider range of contexts.

Fear-conditioning paradigms allow for an investigation of the emergence, persistence, and resurgence of maladaptive fear responses (Pittig et al. 2018). It has been suggested that intrusion, hyperarousal, and hypervigilance symptoms, which characterize PTSD, may arise as a result of conditioned fear responses (for a review, see Norrholm and Jovanovic 2018). Note, however, that fear-conditioning experiments have mainly investigated arousal or hypervigilance (e.g., skin conductance, startle reflex–fear responses) that are mostly relevant to anxiety disorders.

Intrusions, however, are the hallmark feature of PTSD, which is no longer classified as an anxiety disorder but as a trauma- and stressor-related disorder in DSM-5 (American Psychiatric Association 2013). As fear-conditioning paradigms do not specifically account for the image-based episodic nature of intrusive thoughts (Visser et al. 2018), it remains an interesting open question whether this paradigm could also be used to generate intrusive memories of a stressor in the laboratory. A more ecologically valid and clinically relevant experimental model may be needed to generate and study intrusions per se in the laboratory.

Trauma Film Paradigm

In the trauma film paradigm, participants are asked to view a composition of short, distressing film clips with traumatic content (James et al. 2016a). This paradigm has been shown to induce intrusive memories to clips of the film (i.e., with image-based episodic nature).

Trauma is defined as exposure to death, threatened death, actual or threatened serious injury, or actual or threatened sexual violence (American Psychiatric Association 2013). Notably, in addition to *direct* exposure (e.g., as a victim or witness), repeated or extreme *indirect* exposure to aversive details of trauma, usually over the course of work (e.g., when a police officer has to *view pictures* of murder), is now included as part of the diagnostic criterion for what comprises a traumatic event in the DSM-5 (American Psychiatric Association 2013). This recent inclusion of indirect exposure to trauma underscores the ecological validity of the trauma film paradigm (James et al. 2016a).

Of note, intrusive memories can also be induced by overly positive film stimuli (Davies et al. 2012) or depression-linked film material (Lang et al. 2009). Thus, the trauma film paradigm is not only useful for studying intrusive images related to PTSD but also for studying intrusive thinking in depression or bipolar disorder.

In studies using the trauma film paradigm, intrusive memories of the film are usually monitored in a paper-and-pencil diary directly when they occur over the course of daily life (James et al. 2016a). This method records intrusion frequency data over longer time frames and carefully matches intrusions to the trauma film (participants usually record whether they had an intrusion or not for several time periods per day over the course of one week and briefly describe each intrusion's content). Mixture models can be a useful tool to analyze such diary data because they model intrusion and non-intrusion data differently (see discussion on mechanisms and mathematics below for further details).

Additional Methods

Watching visual stills of distressing content (e.g., injured people) has also been shown to generate intrusions two days later (Battaglini et al. 2016). In addition, listening to negative arousing stories while watching a slide show of pictures can generate negative emotional memories (Galarza Vallejo et al. 2019).

Levels of Mechanism to Modulate Intrusive Emotional Images

Using controlled and standardized experimental procedures and the possibility to focus on specific clinical targets is an essential step toward understanding complex clinical disorders (Visser et al. 2018). Once one has successfully generated intrusions in the laboratory, it is possible to study specific mechanisms that could modulate them (e.g., reduce intrusion frequency or distress/ vividness of intrusions). At this point it is important to note that such intrusions can be modulated at any level of mechanism (e.g., molecular, cognitive, or social). Here we discuss examples of paradigms modulating intrusions at various levels: pharmacological approaches operating at the molecular level, visuospatial interference interventions operating at the cognitive level, social support operating at the social level, and other examples such as sleep and wakeful rest.

Molecular Level: Pharmacological Approaches

Pharmacological approaches may offer a way to modulate intrusive memories. For instance, inhibiting N-methyl D-aspartate receptor (NMDAR)-dependent memory consolidation through antagonistic drugs may reduce the frequency of intrusive memories after trauma. In line with this idea, inhaling the NMDAR antagonist gas nitrous oxide (N_2O) shortly after a laboratory analog of trauma *fastened* the reduction of intrusive trauma-related memories compared to inhaling medical air over the course of the following week. Of note, N_2O led to an increase in intrusion frequency in those individuals who were highly dissociated at baseline, urging caution regarding the use of N_2O in dissociated individuals (Das et al. 2016).

Cognitive Level: Visuospatial Task Interference

Research in cognitive psychology and experimental psychopathology has shown that cognitive interference interventions (memory orientation/reminder cue and visuospatial task) may be a promising technique to reduce both the frequency of intrusive images as well as the level of distress and vividness associated with them (Iyadurai et al. 2019). The working mechanisms behind this intervention are based on three assumptions:

1. Intrusive memories can be altered shortly after an event or at retrieval (Visser et al. 2018).
2. The capacity of people's working memory is limited (Baddeley 2003).
3. Visuospatial tasks occupy working memory resources that would be needed to (re)consolidate intrusive mental images (James et al. 2015).

Thus, engaging in a visuospatial task such as a visuospatial computer game like *Tetris* (James et al. 2015; Iyadurai et al. 2018), or a complex finger tapping exercise, at a time when mental images of the event are active, may disrupt these distressing images. It is hypothesized that the intervention works because the two processes compete for visual processing resources and the brain cannot attend equally to the distressing image and the visuospatial task. Importantly, such task interference has to take place at a time when the memory is labile and vulnerable to alteration (McGaugh 2000; Nader 2003).

Even though visuospatial interference interventions have mostly been investigated in relation to distressing trauma memories, they also work with overly positive material (Davies et al. 2012). This suggests that the mechanisms apply to intrusive emotional memories in a more general sense rather than only to trauma-related intrusive images.

Social Level: Social Support

There has been an increased interest in the impact of social factors on emotion regulation. Both human and nonhuman experiments have shown that the presence of another during an aversive experience may work as a buffer by reducing fear responses (Thorsteinsson and James 1999; Mikami et al. 2016). Experiences of social support could increase the process of learning what is *safe* in the environment (social safety learning) through social support interactions, which in turn decrease stress reactivity to stressful experiences (Ditzen and Heinrichs 2014).

After a psychologically traumatic event, social support (i.e., supportive interactions with family and friends) is believed to be associated with having fewer posttraumatic cognitions (e.g., trauma-related thoughts and beliefs), which in turn is associated with PTSD symptoms (Woodward et al. 2015). These results signal a need to investigate social interactions and social support after a negative or stressful experience as a potential causal mechanism for the development or maintenance of psychological disorders and to study this in the laboratory. For example, if the absence of social support after trauma leads to people having more intrusions, whereas perceived social support causes people to have less intrusions, we could specifically target this mechanism in future preventive treatments for people who experienced a traumatic event.

Studying the mechanisms at a social level may have relevance for public mental health. Brief and low-intensity social support interventions may not need to be delivered by highly trained professionals but could instead be implemented by members of the public. *Social prescription* refers to the idea of linking patients in primary care with sources of support within the community, for instance, enabling health care professionals to refer patients to a service provided by the voluntary and community sector alongside existing treatments to improve health and well-being (Bickerdike et al. 2017). In line with the idea of social prescription, social support interventions including emotional, instrumental, and informational support could be delivered by volunteers (e.g., hospital volunteers) who are already present in many medical facilities, thus allowing us to scale up preventive interventions.

Other Levels: Sleep or Wakeful Rest

An example of how memory could be boosted rather than blocked is wakeful rest (Dewar et al. 2014). A brief wakeful rest period after learning may actually *enhance* memory in the short (after 15 minutes) and long term (after seven days) compared to performing a nonverbal task (note that these studies tested participants' declarative memory). The wakeful-resting period is thought to boost recently acquired memories by isolating the memory trace of the story (or nonwords) from competing memories, making the memory easier to retrieve at a later stage (Dewar et al. 2014). These results confirm the crucial role of the memory consolidation period in the strengthening of new memories, here through spontaneous reactivation during wakeful resting.

What remains to be further explored is the possible involvement of similar processes in the maintenance of intrusive thoughts during the consolidation period of emotional material. Wakeful rest might actually be what trauma patients usually do when waiting for medical care in the emergency department after a traumatic experience. Thus, investigating the effects of wakeful rest on intrusive memories after a traumatic event could have clinical implications and guide the development of future interventions. In line with this idea, a few studies have already investigated the role of sleep and sleep deprivation after trauma on intrusive memories (e.g., Porcheret et al. 2015, 2019, 2020). As these initial investigations revealed mixed results, further research on the role of sleep and wakeful rest as candidates to modulate intrusive memories is clearly warranted.

Modulating the Frequency of Intrusive Memories: From Lab to Clinic

New Memories of a Traumatic Event

By "new" we refer to Day 1 of the (experimental or real) traumatic event. Intrusive memories of trauma have a clear onset (i.e., the time of the traumatic event), making them amenable to study. This allows us to investigate ways to intervene with the initial memory consolidation of a problematic image before it causes further distress at any of the above described levels of mechanism, such as the molecular level (Das et al. 2016).

Several studies using the trauma film paradigm (James et al. 2016a) indicate that after the experimental trauma (30 min or 4 hr), performing a brief cognitive interference intervention (comprised of a memory orientation/reminder cue, mental rotation instructions, and playing the visuospatial computer game *Tetris*) reduces intrusive images compared to not performing any task (e.g., Lau-Zhu et al. 2019). Two proof of principle randomized control studies have recently extended this effect to a clinical setting that involves (a) road traffic accident survivors who are waiting in the emergency department (Iyadurai et al. 2018) and (b) mothers who experienced traumatic childbirth (Horsch et al. 2017), both within the first six hours after the traumatic event.

Psychological interventions for traumatic memories should ideally interfere with the involuntary, intrusive aspect of a memory but should not impair voluntary memory expression (Lau-Zhu et al. 2019). A person who has experienced sexual abuse by a piano teacher would, for instance, not want images of the abuse to intrude on their mind involuntarily, whereas they may want to be able to recall episodes and facts about the event when required for legal reports (see Figure 14.1). Experimental studies in the laboratory make it possible to investigate such a distinction. Findings suggest that a visuospatial interference task intervenes with the involuntary (intrusive) memory, whereas the voluntary memory remains intact when controlling for potential other task characteristics (Lau-Zhu et al. 2019).

This data raises the intriguing possibility that intrusive image-based memories are in fact "special" and can be selectively targeted by visuospatial interference interventions, whereas voluntary memory remains unaltered (Lau-Zhu et al. 2019). In contrast to traditional single trace theories of memory, which argue that involuntary and voluntary memories are derived from the same memory system, this data conforms to *separate trace theories*, stating that different memory traces underlie involuntary and voluntary memories. Thus, intrusive reexperiences may be supported by a specialized perceptual memory system that is functionally dissociable from the episodic memory system supporting voluntary recall of the same event, in line with dual representation theory (Brewin 2014). Visuospatial interference intervention (e.g., reminder cue and *Tetris*) may then preferentially disrupt this sensory-perceptual memory system, whereas the episodic memory system remains unaffected (Lau-Zhu et al.

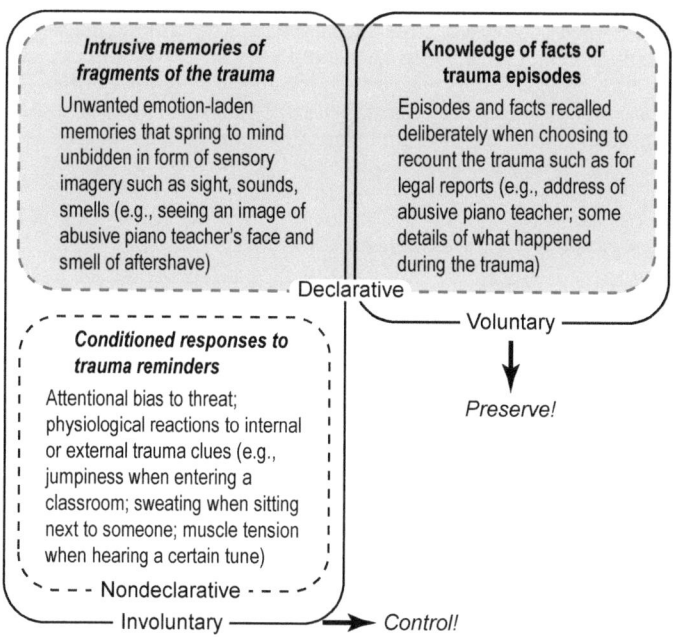

Figure 14.1 Diagram depicting how different memory systems may represent various aspects of a traumatic event (e.g., sexual abuse by a piano teacher). In general, clinically beneficial interventions should aim to target the maladaptive involuntary expression of trauma memories (e.g., intrusive memories) while preserving its voluntary recall (e.g., ability to testify in court). Adapted after Visser et al. (2018).

2019). In contrast to widely used fear-conditioning paradigms, the trauma film paradigm may be particularly useful to assess these different aspects (readouts) of trauma memory in the laboratory (Table 14.1). Future work is warranted.

Old Memories of a Traumatic Event

By "old" we refer to Day 2 onward of the (experimental or real) traumatic event, and in one study many years later. Most work discussed above has focused on the time window that is thought to overlap mainly with (synaptic) consolidation. Consolidation refers to a strengthening of local neural circuits via a cascade of molecular processes involving protein synthesis and the formation of new synaptic connections necessary for a memory to persist in the long term. As illustrated above, interventions delivered during this time period (i.e., minutes to hours after an event) are able to interfere with the newly formed memory and reduce its intrusiveness (McGaugh 1966, 2000). Promisingly, recent research suggests that, under the right circumstances, even established memories can be modified. Rather than there being a one-off opportunity, memories can, upon their retrieval, enter a transient labile state; that is, they

Table 14.1 Overview of different aspects of trauma memory that can be targeted and associated research approaches, from animal models (bottom) to clinical populations (top). Left: different levels at which trauma can be modeled. Middle: potential targets for intervention. Right: memory readouts: (1) occurrence of intrusive images (e.g., diary, provocation task), (2) event details (e.g., interview), (3) learning episode details (e.g., recognition test), (4) self-report of symptoms, (5) rating of subjective distress, (6) unconditioned stimulus expectancy, (7) attentional bias, (8) approach/avoidance behavior, (9) noninvasive physiology, (10) invasive physiology. Note: voluntary memory recall (e.g., trauma details) can be measured in humans but is not the key clinical target of a treatment. Adapted after Visser et al. (2018).

Level	Targets	Memory readouts
Complex real-life emotional memory, PTSD patients: • Heterogeneous, possible multiple or sustained trauma exposure • Clinical >1 month posttrauma	• Intrusive image-based memories or other experiences (DSM5 cluster B) • Physiological reactivity to internal/external reminders (DSM5 cluster B) • Unwanted avoidance of internal/external reminders (DSM5 cluster C) • Negative mood/cognition (DSM5 cluster D) • Hyperarousal (DSM5 cluster E) • Functional impairment (DSM5 cluster G)	1, 2, 4, 9
Simpler real-life emotional memory, humans • Single trauma exposure soon after event • Simple phobia • Possibly subclinical	• Intrusive images • Subjective distress • Unwanted avoidance • Physiological reactivity Associated with (reminders of) specific real-life situations	1, 2, 4, 6–9
Complex experimental emotional memory, humans • Trauma analog from viewing of aversive film clips	• Intrusive images • Subjective distress • Physiological reactivity Associated with (reminders of) aversive lab stimuli	1, 3, 5, 9
Simpler experimental emotional memory, humans • Aversive conditioning, still pictures paired to electric shocks	• Subjective distress • Avoidance • Physiological reactivity Associated with conditioned cues and contexts	3, 5–9
Simpler experimental emotional memory, nonhuman animals: • Fear conditioning, tones paired to electric shocks	• Avoidance • Physiological reactivity Associated with conditioned cues and contexts	8–10

can become malleable again (e.g., Sara 2000; Alberini 2005; Nader and Hardt 2009). The restabilization of a memory is a putative process termed memory reconsolidation (Nader et al. 2000). This process is dependent on *de novo* protein synthesis; interventions that directly or indirectly target this process thus have the potential to change maladaptive emotional memories (Milton and Everitt 2012), including those giving rise to intrusive images of trauma. Figure 14.2 depicts different time windows of memory malleability.

Different interventions can interfere with the reconsolidation of a memory on different levels. On a molecular level, fear-conditioning studies in rodents have shown the potential of pharmacologically disrupting one-day-old (Nader et al. 2000; Lee et al. 2006; Ortiz et al. 2015) and even one-month-old (Gräff et al. 2014) memories, resulting in a persistent attenuation of conditioned fear responses. Even though less consistent (Lonergan et al. 2013), these types of findings have been translated to studies in humans, where the beta-adrenergic antagonist propranolol was used to disrupt pharmacologically one-day-old fear memories (Kindt et al. 2009) as well as much older memories; that is, fear memories that underlie simple phobias for spiders (Soeter and Kindt 2015). On a cognitive level, behavioral interventions, including extinction after memory retrieval procedure, have been shown to attenuate one-day-old fear memory in rodents (Monfils et al. 2009). This finding has been translated to one-day-old (Schiller et al. 2010) and one-week-old (Steinfurth et al. 2014) memories in humans, and more recently also to older memories such as those underlying simple phobia for spiders (Björkstrand et al. 2016) and snakes (Telch et al. 2017). For further details, we refer the reader to recent overviews on memory reconsolidation literature (Lee et al. 2017; Elsey et al. 2018; Monfils and Holmes 2018).

With regard to intrusive memories, a visuospatial interference intervention administered after a reminder cue was effective in reducing intrusive memories for established (24-hour-old) memory of experimental trauma (James et al. 2015). In this study, individuals who underwent a memory reactivation procedure and performed an intervention, including *Tetris* game play, had fewer intrusive memories than a no-reactivation/no-*Tetris* group. More recently, two studies used a similar reactivation and cognitive task interference procedure, administered three days (Kessler et al. 2020) or four days (Hagenaars et al. 2017) after trauma film viewing; again, a reduction in subsequent intrusive memories was demonstrated. While both studies showed that an active control condition (verbal task) also reduced intrusions compared to a no-task control, in one study the effect was significantly larger for the visuospatial interference intervention compared to a verbal control task (Kessler et al. 2020). Interestingly, and again in line with *separate trace theories* (Lau-Zhu et al. 2019), both Kessler et al. (2020) and James et al. (2015) showed that the intervention left voluntary memory (i.e., performance on a recognition task) intact. Still, more work is warranted.

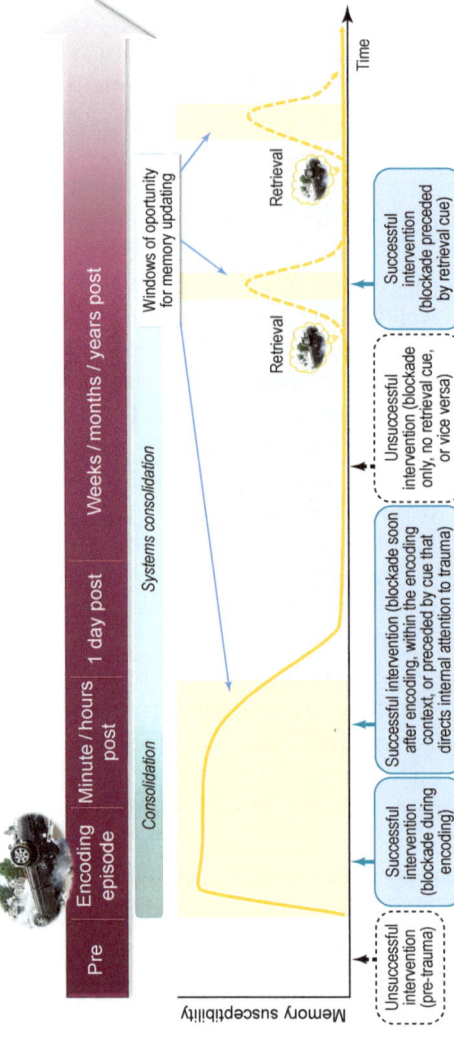

Figure 14.2 In the hours after an experience, memories are believed to go through an initial labile phase before being stored into stable long-term memory (i.e., consolidation). The purple arrow depicts different time intervals with respect to the encoding of an aversive episode. The gradients below indicate the putative processes of memory encoding and consolidation that occur during these different intervals. Recent insights suggest that certain aspects of memories, including the intrusiveness, are not necessarily permanent. Instead, they may become transiently malleable upon reactivation, rendering them susceptible to interference or updating before returning to a fixed state, a process referred to as "memory reconsolidation." This offers a second window of opportunity to interfere with consolidated memories (shown as yellow background shades), making them less intrusive. Successful interventions (blue arrows) need to be timed such that the blockade interferes with memory when it is in an active, susceptible state (indicated by the dotted yellow line), either in the first hours after trauma or at later time intervals after a retrieval procedure (e.g., reactivation through reminder cues). In the first hours after an experience, blockade procedures may also need to be preceded by cues that orient attentional resources to the event in order for procedures to successfully interfere with it (e.g., when the intervention is delivered in a context other than the one in which the trauma occurred). Unsuccessful interventions, timed when memories do not yet exist or are in a fixed state (i.e., not recently retrieved), are shown. Adapted after Visser et al. (2018).

Real trauma memories are typically stronger and broader than aversive memories formed in the laboratory. Finding the optimal conditions and reminder cues to reactivate and render a memory labile (a first step for successful interference) is assumed to be much more challenging (Monfils and Holmes 2018) for real memories of trauma. Yet, a recent study on inpatients with complex trauma (Kessler et al. 2018) has shown promise in attenuating the intrusiveness of memories for old trauma some years previously. Twenty patients monitored the occurrence of intrusive trauma memories over the course of their admission (5–10 weeks). Weekly interventions involved a memory reminder for a selected (particularly distressing) memory, followed by 25 minutes of playing *Tetris*. A within-subjects multiple baseline design was used, in which the pre-intervention period was varied. Further, some intrusions were never targeted by the intervention. The frequency of targeted intrusive memories reduced, on average, by 64% from baseline to the postintervention phase, whereas never-targeted intrusions reduced in frequency, on average, by 11% over a comparable time period. This shows that even persistent, older memories of real-life trauma can be changed using memory interference techniques.

Despite its clear promise for clinical translation, it should be noted that a number of potential limitations and boundary conditions of reconsolidation-based clinical applications have been raised (Treanor et al. 2017; Monfils and Holmes 2018). Moreover, at present, it is not possible to attribute conclusively therapeutic gains to reconsolidation mechanisms (Elsey et al. 2018). Nevertheless, the notion of memory plasticity has proved useful in inspiring new avenues for intervention for older memories of trauma (Figure 14.2). Of particular interest to our current discussion is the potential to modify *intrusive features* of memory during time windows of memory plasticity.

To be able to interfere with older trauma memories, the memory trace has to be activated in working memory via a reminder cue. According to reconsolidation theory, there is an optimal duration for a reminder cue. When memory is retrieved via a brief learning experience (e.g., one unreinforced conditioned stimulus) it enters a labile state. However, if retrieval is prolonged (e.g., four unreinforced conditioned stimuli), the memory might enter a "limbo state," and if it is prolonged further, finally extinction. In short, if the reminder cue "dosage" (duration, instances) is too little, nothing happens (no labilization); if it is too big, the memory may enter a "limbo state" (nothing happens) or extinction learning—a new inhibitory trace is formed, fear/distress diminishes, but this effect may be temporary as it does not alter the original emotional memory trace (Lee et al. 2006; Merlo et al. 2014; Sevenster et al. 2014). All three possibilities (no labilization, limbo state, extinction learning) are different than the reconsolidation state, so the optimization of the retrieval procedure follows an inverted U-shape.

From a clinical perspective, experiencing an intrusion may even offer an opportunity to interfere with reconsolidation of this memory by engaging in a competing cognitive task (e.g., playing *Tetris*) within a specific time frame after the intrusion occurred (minutes). However, the question is whether spontaneous retrieval by means of experiencing an intrusion induces the required reconsolidation state, or instead any of the other states, in which case one cannot really interfere with it. This is an important empirical question which has yet to be tackled.

Recently, an experimental paradigm has been developed to capture intrusive memories as they occur in the fMRI scanner (Clark et al. 2015). After viewing scenes of traumatic events (trauma film paradigm; James et al. 2016a), particular scenes then intrude for an individual. A specific intrusive memory is triggered in the scanner by a reminder cue. The first results of experiencing an intrusive memory are shown in Figure 14.3. Understanding the neural mechanisms of experiencing an intrusive memory may yield insights for treatment (e.g., for neuromodulation strategies that could be combined with behavioral interference techniques, such as our *Tetris* procedure). Colleen Hanlon (this volume) discusses transcranial magnetic stimulation (TMS). It is possible that understanding the neural mechanisms of an intrusive event (e.g., Clark et al. 2015) alongside associated multivoxel pattern analysis (e.g., Clark et al. 2014b) will inform how best to apply TMS during an interference procedure (e.g., Kessler et al. 2018) to reduce the occurrence of intrusive memories. However, to date this has not been attempted.

Mechanisms and Mathematics

Mechanisms of cognition operate across the scales of brain organization (Bonsall et al. 2015). If multiple processes operate at different scales of organization in psychopathology (e.g., posttraumatic stress reactions or mood instability), aggregating the collective molecular and neuron interactions to higher levels of organization (such as the network level or cognitive level) might provide novel, emergent insights into the patterns associated with brain function within and among individuals.

Using mathematical approaches to scale (appropriately) across a hierarchy from cognitive and emotional processes through neuron firing patterns to candidate, molecular processes allows development of a mechanistic approach to cognition. This *mechacognitive* approach (i.e., using mathematical approaches to move down a hierarchy from symptoms to candidate, molecular processes) may allow insights through a mechanistic approach to cognition (Holmes et al. 2016b). By developing descriptions of psychopathology, it can also lead to novel approaches to understand the underlying neuroscience of brain function.

Memory Formation, Consolidation, and Reconsolidation Is a Probabilistic Process

As discussed above, memory consolidation and memory trace are not fixed (Müller and Pilzecker 1900; McGaugh 2000). Neural changes and reorganization operate through the time period from the perception of an event to the later memory retrieval. Given that memory trace is probabilistic (changes between states) and dynamic in time (and across spatial organization in the brain), we develop mechanistic frameworks to link across scales of organization to cognition. While the molecular basis of memory involves epinephrine and cortisol and protein synthesis to affect changes in synaptic consolidation (Dudai 2004), scaling this up to focus on the neural mechanism of systems-level consolidation requires appropriate tools. We argue that this is best achieved within the frameworks and architectures of mathematics.

To bridge the gap between (intrusive) memory consolidation and treatment through a *mechacognitive* lens, we use a mathematical framework coupled to data analysis. Here, our framework focuses on intrusive memory consolidation and cognitive interference interventions including a visuospatial task (Figure 14.4). In this way we consider how perception (bringing to mind) of a traumatic event (zM) following an orientation cue (also called reminder cue for consistency with the reconsolidation literature) might lead to memory consolidating into an intrusive memory (iM) or a more neutral task memory (tM) following a task (*T*).

To formalize this, we consider a discrete-time Markov chain (Kemeny and Snell 1960) where memory moves through a series of states, and the probability of moving from one state to the next is only dependent on present state. Discrete-time Markov chain models underwrite many common state space models of data (e.g., latent variable models, hidden Markov models, and Markov decision processes). In our current (illustrative) application, we assume that the states are directly observable and focus on the probability transitions from one state to another and their implications for understanding therapeutic interventions. Markov chains can be described by a *directed graph* where edges are the probabilities of moving between states and vertices represent the states (Figure 14.4). The directed graph illustrates the steps needed to move from trauma memory state to consolidated task memory state. For instance, this approach allows clear assumptions to be formulated; one assumption is that trauma memory needs to be in a labile state before tasks can be undertaken and affect consolidation of the task memory. We emphasize that this is all a probabilistic process as we learn how to scale up to aggregate memory processes, driven from the molecular, short- or long-term scale to the system level (Albo and Gräff 2018) to a level of organization at the cognitive scale.

As noted, in this exemplar we consider four states: initial trauma memory state, labile memory, intrusive memory, and consolidated neutral task memory. This can be represented by the following transition matrix (N):

304

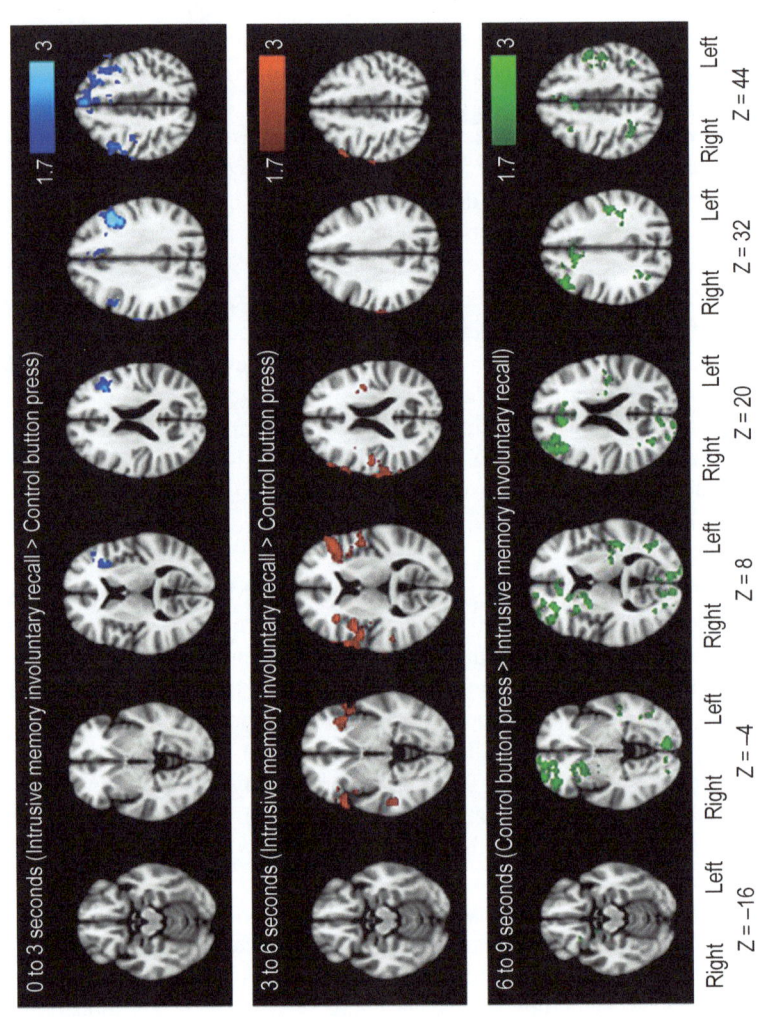

(a)

0 to 3 seconds (Intrusive memory involuntary recall > Control button press)

3 to 6 seconds (Intrusive memory involuntary recall > Control button press)

6 to 9 seconds (Control button press > Intrusive memory involuntary recall)

Right Z = −16 Left Right Z = −4 Left Right Z = 8 Left Right Z = 20 Left Right Z = 32 Left Right Z = 44 Left

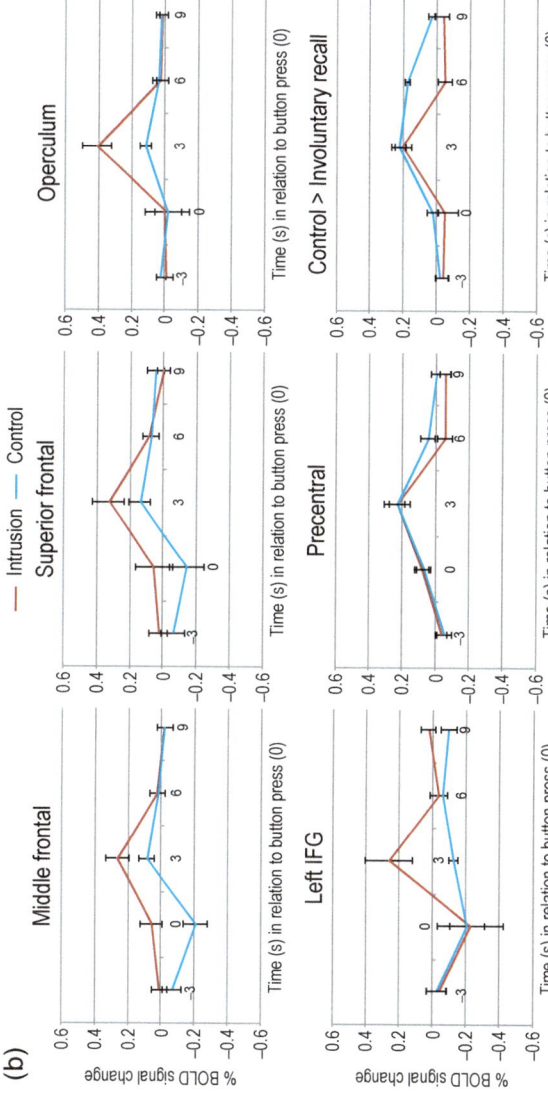

Figure 14.3 Intrusive memory involuntary recall. Top: Whole-brain analysis showing the increased blood oxygen level-dependent (BOLD) response for intrusive memory involuntary recall versus control button press group at the two time bins (0–3 s and 3–6 s in relation to the button press): note the significant differences in activation and the one time bin (6–9 s) of increased BOLD response for the control button press group versus intrusive memory involuntary recall. Bottom: Region-of-interest profile plots of the signal change observed across each time bin from –3 to +12 s in relation to the button press. Intrusive memory involuntary recall signal change activation is shown in red; control button press signal change activation in blue. Values are means; standard deviations represented by vertical bars. IFG: inferior frontal gyrus. Adapted after Clark et al. (2015).

$$N = \begin{pmatrix} 0 & p_1 & 1-p_1 & 0 \\ 0 & p_3/(1+p_3) & p_2/(1+p_3) & (1-p_2)/(1+p_3) \\ 0 & 0 & 1 & 0 \\ 0 & 0 & 0 & 1 \end{pmatrix}, \quad (14.1)$$

where p_{ij} is a transition probability (moving from a state in row i to a state in column j) such that rows of the matrix sum to one (by normalizing across the transition probabilities within a row). p_1 is the transition probability from trauma memory state to labile memory state and, in our context, is the conditional probability that memory is a labile state (lM) given a reminder cue (rC), (Pr(lM|rC)). The probability that the reminder cue fails and intrusive memory forms for initial trauma state is $1-$Pr(lM|rC). Here, we assume that this is a logistic function, $1/(1+\exp(-\alpha))$, where α represents the strength of the reminder cue). Similarly, p_2 is the transition probability from the labile memory state to consolidated iM. In our context this is represented as a conditional probability that a task intervention affects the formation of intrusive memories (Pr(iM|T)), and $1-$Pr(iM|T) is the probability that task intervention is effective and leads to a consolidated neutral task memory. Here we assume that

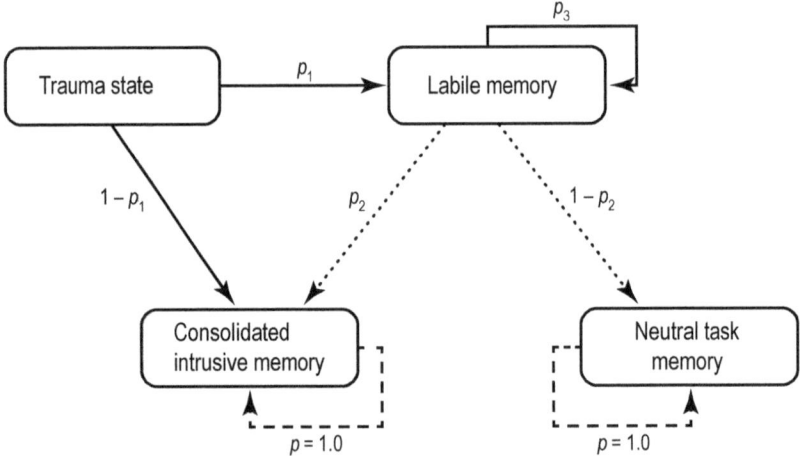

Figure 14.4 Markov chain model: Directed graph representing a Markov chain framework for exploring intrusive and more neutral task memory consolidation. Arrows (edges) represent transition probabilities between states, and boxes (nodes) represent different memory states. p_1 represents the probability that the reminder cue is successful. We describe this probability as $1/(1+\exp(-\alpha))$, where α is the strength of the reminder cue. p_2 represents the probability that task intervention is unsuccessful and parameterized here as $1/(1+T)$, where T is the strength of the task intervention. p_3 is the probability maintained in a labile state. Once memories enter an intrusive memory or neutral task memory they are fixed in these states ($p=1.0$).

Pr(iM | T) = $1/(1 + T)$, where T is a measure of task strength. p_3 is the probability that a memory is maintained in a labile state. With these assumptions and the stochastic process framework, to investigate the role of task interventions once a memory is either in an intrusive or task consolidated state, the memory is fixed (absorbed) into this state and not amenable to further alteration. While memory consolidation and reconsolidation are dynamic processes, here we focus on the role of task interventions in affecting memory states.

Analysis of the Markov chain allows a number of different metrics to be assessed including (a) expected time in a state, (b) variance of time in a state, (c) dynamics over finite time steps, and (d) probability of absorption into a final state as a function of covariates (e.g., Kemeny and Snell 1960).

Sensitivity analysis highlights the importance of particular probability transitions (or more specifically drivers of probability transitions) and the factors influencing memory consolidation at different levels of organization. Here, analysis of the Markov chain reveals that the *labile memory state* is transient. Expected time in this labile memory state is a function of reminder cue probability (p_1) and probability of staying in the labile memory state (p_3) (Figure 14.5a). Increases in the probability of keeping the memory labile (p_3), and increases in reminder cue strength (α) lead to longer expected times of memory in a labile state. However, the uncertainty (variance) in the length of time a memory is labile is most likely influenced by the probability of keeping the memory labile (p_3) rather than reminder cue strength, and most uncertainty in the length of time a memory is labile is greatest for low reminder cue strengths and high probabilities of keeping the memory labile (Figure 14.5b).

Potentially more important is the *probability that intrusive or task memories consolidate*; that is, since this is an absorbing Markov chain with two end states, whether memories persist in the iM or tM state. Analysis of the Markov chain reveals that the probability that an iM consolidates following trauma is:

$$\Pr(\text{iM} \mid \text{trauma}) = (1 - p) + p_1 p_2. \qquad (14.2)$$

Equation 14.2 expresses the probability that an iM consolidates given that the reminder cue fails ($1 - p_1$) or the probability that the reminder cue is successful and that the task intervention is not successful ($p_1 p_2$). For our parameterization, where $p_1 = 1/(1 + \exp(-\alpha))$ and $p_2 = 1/(1 + T)$:

$$\Pr(\text{iM} \mid \text{trauma}) = \left(1 - \left(\frac{1}{1 + \exp(-\alpha)}\right)\right) + \frac{1}{(1 + \exp(-\alpha))(1 + T)}$$
$$= \frac{1 + \exp(-\alpha) + T}{1 + \exp(-\alpha) + T + \exp(-\alpha)T}. \qquad (14.3)$$

The terms that contribute to this conditional probability can also be determined intuitively by looking at Figure 14.4 and tracing the paths from the trauma state to the intrusive memory state, accumulating the probabilities along all

(a)   (b)

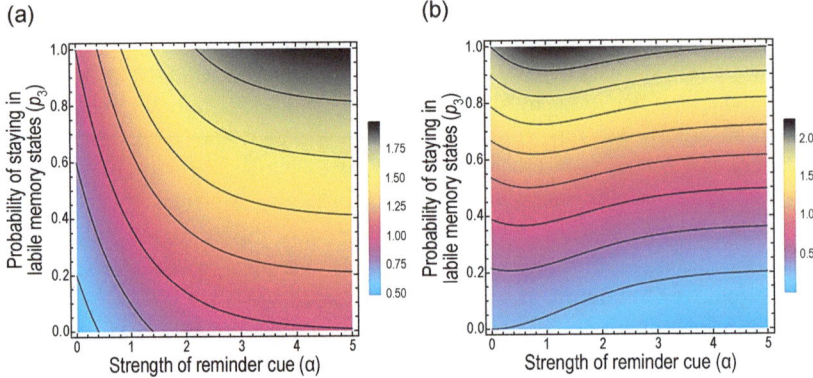

Figure 14.5 Markov chain analysis of transient memory states. (a) Expected times and (b) variances in expected times in labile memory states in terms of strength of reminder cue (α) and the probability of staying in the labile memory state (p_3). Strong reminder cues and high probability of maintaining a memory in a labile state favor long retention times in this transient (labile) memory state. However, most uncertainty in expected retention times is observed for low reminder cue strengths and high probability of maintaining a memory in a labile state.

direct and indirect paths. The marginal change in the probability of iMs is more sensitive to changes in the probability of task success (p_2) than the reminder cue success probability (p_1), as the following inequalities for the change in $d(\Pr(\text{iM} \mid trauma)/dp_i$ can be shown to hold:

$$\frac{d\left(\Pr\left(\text{iM} \mid \text{trauma}\right)\right)}{dp_2} = p_1 > \left(p_2 - 1\right) = \frac{d\left(\Pr\left(\text{iM} \mid \text{trauma}\right)\right)}{dp_1}. \quad (14.4)$$

The above inequality holds from our parameterization of p_1 and p_2:

$$1 + \frac{1}{1 + \exp\left(-\alpha\right)} > \frac{1}{1 + T}, \quad (14.5)$$

which, again, are the probabilities along the path from trauma to task memory consolidation (Figure 14.4). The probability that a task memory consolidates is:

$$\Pr\left(\text{tM} \mid \text{trauma}\right) = p_1\left(1 - p_2\right), \quad (14.6)$$

which is the probability that the reminder cue is successful (p_2), and the task intervention is successful ($1 - p_2$). Again, the terms that contribute to this conditional probability can also be determined intuitively from Figure 14.4, tracing paths from the trauma state to the consolidation state, and accumulating the probabilities along all direct and indirect paths. Two key predictions from this analysis are as follows:

1. The probability of maintaining a memory in a labile state (p_3) has no effect on the probability of iMs or tMs consolidating.

2. Most importantly, the probability of a successful reminder cue is critical: if this cue is unsuccessful (p_1 tends to zero), then the probability of an intrusive memory consolidating tends to 1 and the probability of consolidated task memory tends to 0.

Formulating memory consolidation as a stochastic process, aggregating across scales of organization, allows systems-level mechanisms associated with cognition to be investigated. Next, we briefly show how the model can be validated against empirical observations and/or experiments.

Model Validation: Statistical Approaches

Determining the accuracy and applicability of a mathematical framework centers around model validation. Validating a model involves appropriate parameterization, goodness of fit to data, uncertainty quantification, and prediction. Linking a mechanistic model, such as our Markov chain, to data involves statistics and statistical modeling.

While full model validation is beyond the scope of what we present here, we show ways in which Markov chains can be parameterized from trauma and iM studies and the likely predictions that arise from this parameterization, deriving the conditional probabilities that (a) a reminder cue places a memory in a labile state and (b) task intervention affects the probability of intrusive memories has been approached empirically (e.g., James et al. 2015; Iyadurai et al. 2018, 2019; Lau-Zhu et al. 2019; Visser et al. 2018).

To determine the efficacy of a reminder cue, a binary regression is needed with probability of a successful reminder cue as a response with a set of explanatory covariates. To determine the conditional probability that task affects probability of memory consolidation, we have advocated appropriately addressing statistical issues, such as correlation structures (James et al. 2015; Iyadurai et al. 2018) and/or heterogeneity (Iyadurai et al. 2019). One way to derive appropriate probability estimates on the efficacy of a task is through the use of mixture models (Cameron and Trivedi 2013), where statistical modeling of zero and nonzero intrusive memories (from diary data) is considered differently.

Mixture models represent the nonzero and the zero observations separately with two statistical models. First is to ask: Are the zero counts generated because of the iM/trauma process or something else (perhaps to do with data collection)? This is a Bernoulli process with a probability (say p, where this probability is to be determined by the set of explanatory variables) that the zeros are generated by alternative processes than those under observation. So $(1 - p)$ is the probability that the zeros are generated by the iM/trauma process. As iMs are count data (e.g., the number of intrusive memories per day), the nonzeros are modeled assuming Poisson errors. The probability of j intrusive memories can then be represented by the following mixture model:

$$\Pr(\mathrm{iM} = j) = \begin{cases} p + (1-p)\exp(-u) & \text{if } j = 0 \\ (1-p)\dfrac{u^{iM}\exp(u)}{\mathrm{iM}!} & \text{if } j > 0 \end{cases}. \qquad (14.7)$$

The mixture model applies a binary/binomial regression to determine p (top line, right-hand side of Equation 14.7) and a Poisson regression to model the nonzero counts (bottom line, right-hand side of Equation 14.7).

The binomial and Poisson components of the regression can consider different covariates (e.g., task/no-task, task strength, task quality) to determine probability of intrusive memories and appropriately address heterogeneity generated by an overinflated number of zeros.

To illustrate this approach for data analysis and application to the Markov chain model, a zero-inflated Poisson model analysis was undertaken on a group of patients involved in traumatic road traffic; these patients took part in a cognitive interference intervention that included a reminder cue and a visuospatial task (for details, see Iyadurai et al. 2018). This analysis reveals that the overall probability (across this group of patients) of a successful intervention $(1 - p_2)$ and no iMs is 0.542 (Figure 14.6). Together with the Markov chain analysis, this empirical estimate of no iMs (i.e., zeros), given a cognitive task, predicts that the probability of a consolidated task memory across this group of patients would range from 0 to 0.542 (depending on the probability of a successful reminder cue).

The opportunities for using mathematical approaches for linking across mechanisms of cognition, different illnesses and traumata, modalities of perception, and individual patients is a nascent approach. However, we believe this *mechacognitive* approach has value along the continuum from the basic through to clinical aspects of neuroscience and will provide a fuller understanding of memory consolidation.

Conclusions

Recent findings on intrusive mental images, reviewed in the first part of this chapter, as well as the mathematical model on intrusive memory consolidation and visuospatial interference interventions, introduced in the second part, give reason to take a step back and think about how current psychological interventions might be improved. We have proposed that to progress in this regard, we need to know more about specific processes involved in intrusive thinking and adopt a targeted treatment approach (Iyadurai et al. 2019). We conclude by raising the following questions to invigorate discussion on how we can make future psychological interventions more precise and effective.

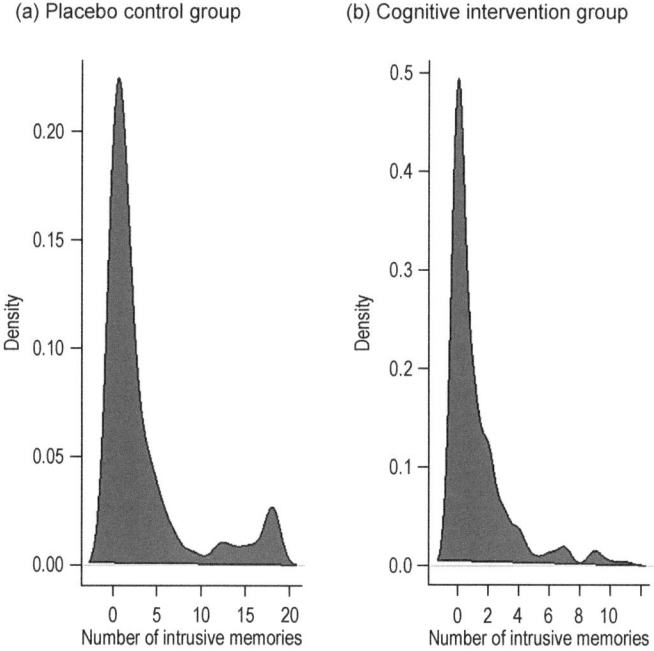

Figure 14.6 Density distribution of intrusive memories (adapted after Iyadurai et al. 2018) for (a) the attention placebo control group and (b) the active cognitive intervention (reminder cue + visuospatial interference task) group. Statistical analysis on (b), using a zero-inflated Poisson intercept-only model (Equation 14.7), reveals that the nonzero counts have an intercept that is significantly different from zero (intercept = 0.908 ± 0.144 (SE), z-value = 13.72, $p < 0.001$), while the zero counts do not (intercept = 0.0006 ± 0.144, z-value = 0.0004, $p = 0.997$). The expected probability of no intrusive memories in the active intervention group, determined from Equation 14.7 and across this group of patients, is 0.542.

What if we specifically target intrusive thinking as a primary outcome rather than a multitude of symptoms of a given disorder? Intrusive imagery is common across psychological disorders and appears to be an important transdiagnostic factor regardless of specific diagnosis (Iyadurai et al. 2019). By looking specifically at intrusive imagery rather than broad and fuzzy assemblies of symptoms clusters, we may be able to radically change current treatment approaches in mental health. For instance, while we may not be able to treat PTSD or depression reliably as a whole (especially at scale), we may be able to target a specific issue, such as intrusive memories, that is common to both disorders. This could also cross-fertilize treatment approaches across different disorders. Interventions targeting intrusive memories of trauma might inform us about potential methods to target intrusive thoughts in depression, as has been the case with imagery rescripting. Developing a simple intervention that is precise and can be delivered exactly to the intrusion trace may be a galvanizing aim for experimental medicine across disorders.

What if we can target involuntary intrusive memories yet leave voluntary memory intact?   One of the most heated debates in the literature on intrusive memories involves whether involuntary and voluntary memories of an event are best represented by single or separate memory trace accounts (Lau-Zhu et al. 2019). Based on a recent series of carefully controlled experiments investigating this question, we argue that memory is in fact dissociable, and involuntary intrusive memories stand out from voluntary memories. This dissociation has important implications in clinical settings as well as for society. We need to develop psychological interventions that can prevent involuntary distressing images from intruding on one's mind while still enabling people to voluntarily recall information about the event (e.g., to be able to testify regarding a traumatic event in a court of law). Visuospatial interference interventions are promising because they appear to target selectively and precisely the intrusive memory trace while leaving the voluntary memory intact (Lau-Zhu et al. 2019). A successful recovery posttrauma from a clinical perspective is being able to talk about the traumatic event(s) when one decides to or needs to, but not to have them continually intrude in one's mind against one's will.

What if we are able to prevent new intrusive memories as well as tackle older ones?   In addition to preventing the consolidation of "new" intrusive memories with visuospatial interference interventions directly after a traumatic event (the same day), we argue that a similar type of intervention could be adapted to target "old" intrusive memories. Importantly, when targeting older trauma memories (24 hr to several years after the traumatic event), studies have indicated that an approximately 10-minute gap has to be added between reminder cue and intervention, supposedly to make the memory trace malleable (Agren et al. 2012; Schiller et al. 2013; James et al. 2015; Kessler et al. 2018, 2020), although more research on this gap is needed. Methodologically, this opens experimental designs to study visuospatial interference interventions. For example, where there are older intrusive memories of several different events, one could target single intrusive images one after the other (i.e., one at a time) and compare frequency and distress of targeted and nontargeted intrusions over time (Kessler et al. 2018). Targeting old intrusive memories is clearly important for vulnerable patient groups (e.g., refugees). Creating rational approaches that look more like computer game play may be useful for those who do not seek traditional psychological help because of perceived stigma. Developing a brief, easily accessible, and nonstigmatized cognitive intervention that could potentially be self-administered would fill an important gap to reach such vulnerable patient groups (Holmes et al. 2018).

What if we could go global and target large sections of populations suffering trauma?   Trauma is a global health issue, and to address this we need innovative interventions that can be scaled up to overcome current barriers in psychological treatment. For instance, it is impossible for traditional psychological treatments to be delivered to the large number of people who need them globally due to the lack of trained psychologists. Thus, we need interventions that are of low intensity, require few resources to deliver, are culturally adaptive, and accessible to many (Iyadurai et al. 2019). Approaches such as the cognitive interference intervention (if shown effective in large-scale clinical trials) could potentially be readily delivered by nonspecialists or even be self-administered. Thus, an important aspect of bringing an intervention to a global level is how to train people successfully with different background knowledge in delivering the intervention (Holmes et al. 2018). Relatedly, at the global level, prevention may ultimately be as important as a cure. Rather than treating psychological disorders after they have developed and caused burden on the individual and society (in terms of suffering, health care costs, and loss of work force), selective prevention for high-risk groups (e.g., firefighters, paramedics, emergency department staff, war survivors) or universal prevention (i.e., everyone experiencing trauma could be treated no matter if they would actually develop intrusive images or not) would be a useful way to combat maladaptive intrusive thinking.

What if we could prevent intrusive memories as well as boost positivity at the same time?   We need to find a way to make interventions as effective as possible while keeping them simple. Rather than simply aiming for a reduction of intrusive memories, we might also want to boost positivity, if we could find a simple way to combine this within the same task procedure (e.g., increase optimistic mental images of the future and direct attention to adaptive information; see Kress and Aue 2017). Positive imagery and optimism can be two main targets, potentially at once. Notably, the mathematical model on memory (re)consolidation and task interference introduced here could help find the best way to test this (and related) questions. For instance, the model raises the intriguing possibility that, at least in some situations, two weak tasks can lead to a stronger outcome than one strong task, and this may help clarify at which point in time specific interventions are most effective.

Acknowledgments

EAH receives grant support from the Oak Foundation, The Lupina Foundation, and the Swedish Research Council (2017-00957). RMV is supported by a Netherlands Organization for Scientific Research (NOW Veni) grant (016.195.246). Figures 14.1 and 14.2, from Visser et al. (2018), are used under the terms of the Creative Commons Attribution 4.0 International License (CC-BY).

15

Pharmacological Interventions as They Relate to Intrusive Thinking

Harriet de Wit and Anya K. Bershad

Abstract

Intrusive thoughts are features of numerous psychiatric disorders. They vary widely in form, duration, frequency, and severity. They are associated with disorders with widely differing pathophysiology, and they are likely to respond to different pharmacological treatments. It is possible that intrusive thoughts represent a cross-diagnostic symptom that can be a pharmacological target in their own right, separate from the associated disorder. This chapter considers the challenges in studying intrusive thoughts as a separate entity. It examines intrusive thoughts that are symptoms of several different psychiatric disorders and reviews the medications that have been used to treat them. It holds that relatively little is known about the effects of psychiatric medications on intrusive thoughts, either within disorders (separate from other symptoms) or across disorders. A wide range of medications is used to treat intrusive thoughts that target different neurotransmitter systems. In addition to the psychopharmacological armamentarium, new, single dose treatments (e.g., ketamine, psilocybin, and MDMA) have emerged that may specifically address intrusive thoughts across the psychiatric spectrum. In conclusion, possible directions are discussed for identifying subcategories of intrusive thoughts that could advance research and treatment in this area.

Introduction

While most of the time we have a sense of control over our own thoughts, sometimes it seems as if our thoughts control us. Under these circumstances, thoughts appear to assert themselves into our consciousness, influencing mood and sometimes behavior. These unwanted images and urges, or "intrusive thoughts," can interfere with everyday functioning (Clark 2005). In an early definition, Rachman (1981) identified three criteria for intrusive thoughts: (a) the thought interrupts ongoing activity, (b) the thought is recognized to be of internal origin, and (c) the thought is difficult to control. When intrusive thoughts become severe enough to interfere with normal function, they become clinically significant psychiatric symptoms and thus potential targets for treatment. In this chapter we review several psychiatric disorders in which intrusive thoughts are a prominent symptom and summarize the medications that have been used to treat these disorders. We also comment on future directions of pharmacological treatments for this symptom.

In the context of psychiatric disorders, intrusive thoughts vary widely in their nature, form, frequency, and controllability, and this variance can make them difficult to study as well as to treat. Intrusive thoughts may occur as part of normal human experience, such as in the case of grief or intense romantic love, or they may indicate a serious psychiatric disorder such as psychosis (Table 15.1). They vary in the degree to which they interrupt normal thinking and activity, whether they are perceived to originate from within the individual or from outside, and the extent to which the individual feels they can be controlled. Some intrusive thoughts may be experienced as pervasive emotional experiences, such as worry or rumination in anxiety and depression, whereas others may be experienced as either spontaneous intrusive experiences, such as compulsions in obsessive-compulsive disorder (OCD). Others are elicited by discrete environmental stimuli, such as triggers in posttraumatic stress disorder (PTSD) or craving elicited by cues in drug users. Intrusive thoughts also differ in content along several modalities, including verbal, visual imagery, and sensorimotor phenomena (e.g., urges and compulsions) and the degree to which they can be suppressed or controlled by various behavioral procedures. Although studying these dimensions may shed light on the psychological and neural processes underlying intrusive thoughts, there have been few attempts to categorize intrusive thoughts along any of these dimensions.

An obvious follow-up to recognizing this variability is to ask: To what extent do pharmacological or behavioral treatments depend on the nature of the intrusive thoughts? Certain drugs, for example, may be effective for intrusive thoughts related to depressive rumination and suicidality, whereas others may more effectively target stimulus-elicited urges. Careful examination of intrusive thoughts as separate entities may reveal shared mechanisms across diagnoses, which can be revealed by focusing on medications that are efficacious across disorders. Naltrexone, for example, is used for opioid use as well as for OCD, and aripiprazole is used for suicidal thoughts and for gambling. Such analysis is consistent with the RDoC approach, designed to identify the neurobiological and cellular mechanisms underlying psychopathological states. Increasingly, researchers have focused on the neural features of intrusive thoughts. For example, Popa et al. (2016) report that stimulation in the dorsolateral prefrontal cortex in patients with epilepsy produces the persistent thoughts that often presage frontal seizures. Other studies (e.g., Hellerstedt et al. 2016) have used a Think/No-Think procedure to track neural activity related to unwanted memories using electrophysiological measures. Kühn et

Table 15.1 Examples of intrusive events that occur in different psychiatric disorders.

Disorder	Symptoms of Intrusive Thinking
OCD	Recurrent and persistent thoughts, urges, or images that are experienced as intrusive or unwanted; cause marked anxiety and distress; individual attempts to reduce or ignore thoughts and urges, or neutralize them with an action
	Compulsions: repetitive behaviors that the individual feels driven to perform
Body dysmorphic disorder	Preoccupation with one or more perceived defects or flaws in physical appearance
Eating disorders (bulimia nervosa, anorexia nervosa)	Intense fear of gaining weight; undue influence of body weight on self-evaluation
Gambling	Preoccupation with gambling, persistent thoughts of reliving past gambling experiences, thinking of ways to get money to gamble; often gambles when distressed
Pyromania	Fascination with, interest in, curiosity about, or attraction to fire; pleasure, gratification, or relief when setting fires
Kleptomania	Recurrent failure to resist impulses to steal objects
Intermittent explosive disorder	Failure to control aggressive impulses (verbal or damage to property)
Major depressive disorder	Rumination on feelings of worthlessness or guilt
Generalized anxiety disorder	Recurrent worries about various dimensions of life, including work, family, romance, or others
PTSD	Recurrent, involuntary, and intrusive distressing memories of a traumatic event; hyperreactivity to cues and reminders of the event; avoidance of reminders of the event
Psychosis	Preoccupation with delusions or false beliefs*
Substance use disorder	Craving, strong desire to use a drug

*Psychosis may be part of many different disorders

al. (2013) used fMRI to show that there was greater activity in brain regions involved in language production in healthy individuals who reported high, compared to low, habitual tendencies for intrusive thought. This focus on intrusive thoughts as entities in their own right provides an opportunity in the future to investigate pharmacological interventions that target such thoughts.

Here we review the main medications that are used for psychiatric disorders for which intrusive thoughts are a major feature (Table 15.2). We separate the medications into categories based on first-line treatment, second-line treatment, and experimental approaches. The medications listed under the first two categories are considered standard care and were drawn from the online clinical decision support resource site UpToDate. The medications listed under experimental approaches were drawn from a review of the literature (PubMed).

318

Table 15.2 Drugs used to treat psychiatric disorders with intrusive thoughts in first-line treatment, second-line treatment (where several randomized control trials have shown strong evidence), and experimental approaches (where some small studies may show effects). MAOI (monoamine oxidase inhibitor), MDMA (3,4-methylenedioxymethamphetamine), SNRI (serotonin and norepinephrine reuptake inhibitors), SSRI (selective serotonin reuptake inhibitors), TCA (tricyclic antidepressants).

Disorder	First-Line Treatment	Second-Line Treatment	Experimental Approaches
OCD	SSRIs (all have been FDA approved except citalopram and escitalopram), clomipramine (TCA)	SNRIs (e.g., venlafaxine), augmentation with antipsychotic (e.g., risperidone)	Psilocybin, eszopiclone, substance P, neuropeptide Y, vasopressin, riluzole, ketamine, D-cycloserine
Body dysmorphic disorder	SSRIs (escitalopram or fluoxetine have the most evidence)	Clomipramine	Augmentation with atypical antipsychotics, oxytocin
Eating disorders			
Bulimia nervosa	SSRIs (e.g., fluoxetine)	TCA (e.g., desipramine), trazodone, MAOI (e.g., phenelzine), topiramate	
Anorexia nervosa	Psychotherapy	TCA (e.g., desipramine), trazodone, MAOI (e.g., phenelzine), topiramate	D-cycloserine plus exposure-based therapy
Gambling	SSRIs	Opioid antagonists	Glutamatergic agents, lithium (for comorbid bipolar disorder), topiramate
Pyromania	None	Mood stabilizers (e.g., lithium, carbamazepine, valproate, SSRIs), opioid antagonists	
Kleptomania	None	SSRIs, opioid antagonists	

Table 15.2 (continued)

Disorder	First-Line Treatment	Second-Line Treatment	Experimental Approaches
Intermittent explosive disorder	SSRIs	Mood stabilizer; phenytoin, oxcarbazepine, carbamazepine, lamotrigine, valproate, lithium	Atypical antipsychotic (e.g., ziprasidone, clozapine, risperidone, and olanzapine)
Major depressive disorder	SSRIs	SNRIs, serotonin modulator, TCA, MAOI, bupropion, mirtazapine	Ketamine, psilocybin, LSD, opioid medications
Generalized anxiety disorder	SSRIs, SNRIs	Buspirone, pregabalin, benzodiazepines	Eszopiclone, substance P, neuropeptide Y, vasopressin, riluzole, ketamine
PTSD	Psychotherapy, then SSRIs, SNRIs	Second-generation antipsychotics (e.g., quetiapine), prazosin	Mood stabilizers (e.g., tigabine, topiramate, divalproex), beta-adrenergic blockers, D-cycloserine, ketamine, cannabis, MDMA
Substance use disorder	*Opioids:* Buprenorphine, methadone *Alcohol:* Naltrexone, acamprosate *Tobacco:* Varenicline, bupropion	*Opioids:* Naltrexone *Cocaine:* Amphetamines, modafinil, disulfiram, topiramate, galantamine *Alcohol:* Disulfiram *Tobacco:* Nortryptiline, cysticine (not available in the U.S.)	*Multiple disorders:* LSD, psilocybin, ketamine, MDMA, oxytocin, n-acetylcysteine
*Psychosis**	Second-generation antipsychotics (e.g., quetiapine), first-generation antipsychotic (e.g., haloperidol)	Lumateperone (novel antipsychotic)	Cannabidiol, glutamate modulators, nicotine receptor agonists, D-cycloserine

*Psychosis may be part of many different disorders, some of which are treated differently than the above.

In the sections below, we group the intrusive thoughts into four broad categories based on their phenomenology, known biological or environmental origins, and involvement of emotional states:

1. Behavioral: intrusive thoughts that take the form of urges and compulsions to engage in repetitive or unwanted actions.
2. Affective: intrusive thoughts that appear to emerge from valenced emotional states, such as anxiety or depression.
3. Substance-induced: intrusive thoughts that are related to psychoactive drugs.
4. Cognitive: intrusive thoughts that involve delusional thoughts and hallucinations.

Intrusive thoughts are rarely a primary target of treatment, and when they are, they are usually treated with psychosocial interventions (e.g., Clark 2005; Ainsworth et al. 2017; van Schie and Anderson 2017; Rebetez et al. 2018; Iyadurai et al. 2019). However, as we note below, a small handful of studies have used pharmacological techniques to reduce intrusive thoughts.

Behavioral

For several disorders, intrusive thoughts take the form of strong urges to engage in harmful, apparently unnecessary, or socially unacceptable actions. This form is a key diagnostic symptom of OCD (Clark and O'Connor 2005) and may include thoughts about engaging in inappropriate sex acts, violence or harm to others, or thoughts related to family members, children, or death. They may also take the form of fear of germs or contamination as well as discomfort at having things out of order, accompanied by urges to act (e.g., cleaning or washing, repeated checking or repeated counting). In the case of body dysmorphic disorder, patients may be obsessively preoccupied with a particular aspect of their appearance and make repeated efforts to alleviate this discomfort by altering their bodies. Patients with OCD may worry about engaging in socially inappropriate behaviors, such as touching or kissing someone inappropriately, hurting someone, or engaging in actions that go against the individual's value system. OCD patients report escalating anxiety as they resist urges to act and a temporary sense of relief after acting on the urge. To varying degrees, intrusive thoughts that relate to urges to engage in socially unacceptable behaviors are also features of other impulsive control disorders, such as eating disorders, gambling, pyromania, kleptomania, and intermittent explosive disorder. In each of these cases, patients become preoccupied by thoughts of engaging in an inappropriate behavior; these thoughts are difficult to control and can interfere with their normal function.

For each of these disorders, selective serotonin reuptake inhibitors (SSRIs) are the first- or second-line of treatment. The behavioral processes through which SSRIs relieve OCD and associated disorders have been studied in some depth, but are still not understood. Drugs may reduce anxiety and dampen responses to stimuli that trigger the obsessions, or they may modulate the obsessive thoughts themselves. Tricyclic antidepressants are also used for OCD, and second-line treatments include other serotonin and norepinephrine reuptake inhibitors (SNRIs) with less serotonergic activity (Hirschtritt et al. 2017). Cognitive behavioral therapy is an accepted nonpharmacological treatment for OCD, and there is some support for adjuvant use of neuroleptics, which may be especially helpful in OCD patients with tics, though may not act to reduce obsessive thoughts (Bloch et al. 2006).

Affective

Intrusive thoughts are key features of several affective disorders, which are characterized by strong emotional states such as depression, anxiety, or mania. Patients with major depressive disorder report ruminative intrusive thoughts that are congruent with their negative mood states, such as self-deprecating thoughts of worthlessness or guilt, delusions of guilt, or paranoia in psychotic depression. Patients with generalized anxiety disorder report intrusive thoughts that are congruent with their anxious mood states, such as worries about life, including finances, family, work performance, or accomplishing everyday tasks. In another disorder in which anxiety is a prominent feature, PTSD, intrusive thoughts relate directly to the experienced trauma, typically in the form of vivid memories related to the traumatic event, including the people, context, and emotions experienced. Individuals with PTSD exhibit a heightened sensitivity to cues related to the trauma, which may generalize to other stress-related cues. In PTSD, intrusive thoughts are typically associated with strong negative affect, including both anxiety and depression. The intrusive thoughts themselves are highly distressing and may be accompanied by an urge to act, in the form of fighting or fleeing from the situation. They are also a feature of intermittent explosive disorder in which patients experience explosive outbursts of anger and violence. These may be spontaneous or elicited by inconsequential events perceived as provocation. The final category of mood disorders, those including symptoms of mania, may be accompanied by strong urges or recurrent thoughts that lead to impulsive behaviors, such as financial spending or risky sexual behavior, but it is not clear to what extent these urges are distressing to the individual experiencing them.

In major depressive disorder, intrusive thoughts which occur as part of ruminations are typically treated with SSRIs. While ruminative thoughts are not part of the diagnostic criteria for a major depressive episode, they frequently occur as a part of the disorder. Commonly used instruments, like the Hamilton Depression Rating Scale and the Beck Depression Inventory, do not directly assess ruminative thinking, so little is known about the direct effect of antidepressant medications on these types of thoughts. Future research may utilize instruments that specifically assess ruminations, such as the response scale Penn State Worry Questionnaire (Meyer et al. 1990), or specific questions in standardized clinical scales, such as GAD-7 (two worry questions) and PHQ-9 (thoughts of being dead or self-harm). Beyond ruminative thoughts of worthlessness or guilt that may present during a major depressive episode, depression may also have psychotic features, characterized by delusions of guilt and paranoia, and these require different pharmacological treatments. A recent meta-analysis suggested that the most effective treatment for depression with psychotic features consists of combination therapy with both an antidepressant and antipsychotic medication (Wijkstra et al. 2013). While initially there was a concern that novel treatments for depression, such as ketamine, may not be appropriate for patients with psychotic features, due to potentially psychotomimetic effects, case reports have suggested otherwise (Ribeiro et al. 2016).

To address the persistent worries involved in generalized anxiety disorder, SSRIs and SNRIs are first-line treatments. Over the past several years, there has been a shift away from treating such disorders with benzodiazepines due to the higher risk of tolerance and higher rate of withdrawal symptoms seen with this class of medications (Offidani et al. 2013). Other second-line treatments include the azapirone buspirone and pregabalin, both of which may act to reduce persistent worries. Buspirone may also act at 5HT-1A receptors (Howland 2015). Pregabalin, though similar in structure to GABA, may act by binding voltage-gated calcium channels and reducing downstream glutamatergic signaling (Baldwin et al. 2013). Antipsychotics have also been studied in the treatment of generalized anxiety disorder (Gao et al. 2006). Neuropeptides, such as vasopressin and neuropeptide Y, have shown some promise in the treatment of the intrusive thoughts associated with generalized anxiety disorder, based on their efficacy in facilitating fear extinction (Tasan et al. 2016). For many of these treatments, it is not clear whether the medication targets the affective state of fear or the anxious thoughts associated with the disorder, and it can be hard to isolate these components in a research setting. In a promising novel line of research, cancer patients with anxious thoughts related to their terminal illness have been treated with single doses of psychedelic drugs, such as psilocybin or LSD (Griffiths et al. 2016b; Ross et al. 2016). Reportedly, 80% of patients showed a reduction in anxious thoughts after only one to two experiences with the medication. There is an urgent need for further research to determine how single administration of psychoactive substances can induce lasting changes in mood and psychiatric symptomatology.

In PTSD, treatments have focused on blocking different stages of the processing of traumatic memories. Graebener et al. (2017) examined the effect of cortisol on intrusive memories in patients with PTSD but found no effect. Taylor and Torregrossa (2015) discuss the highly promising approach of pharmacologically blocking the reconsolidation of maladaptive and intrusive memories, but this intriguing idea has not yet led to effective treatments. Hill et al. (2017) discuss the possibility of targeting PTSD-related intrusive thoughts with drugs that act on the endocannabinoid system, but such medications have not yet reached clinical application.

Drug Induced

In substance use disorders (SUDs), intrusive thoughts are a key feature. They typically take the form of preoccupation with drug use, reactivity to drug-related stimuli, and craving or urges to use drugs. Indeed, "craving" was recently added to DSM-5 (American Psychiatric Association 2013) as a symptom in the diagnostic criteria for SUDs, and it is among the most frequent symptom reported by drug users attempting to quit. Cravings are problematic not only because they predict or presage the occurrence of actual drug use, but also because they can be sufficiently distressing to drug users to be a target of treatment in their own right, independently of actual drug use (Green and Ray 2018). The origin of drug-related intrusive thoughts in SUDs is not fully understood and may be multidimensional, reflecting both underlying physiological conditions and responses to Pavlovian conditioned stimuli. For example, craving is a central feature of acute withdrawal from a drug after regular use, and it is also elicited by drug-related cues in the environment or memories of drug use. Cravings may be elicited in abstinent users, even long after withdrawal, by drug-related stimuli, by positive or negative emotional events, or by ingestion of a small amount of the drug itself. In other cases, cravings or urges occur without any definable precipitant. Whether all these forms of drug-related intrusive thoughts and cravings reflect a single underlying neural process remains unknown. Notably, however, there are intrusive thoughts specific to each drug, and withdrawal from one drug does not induce craving for other drugs.

Although intrusive thoughts are a key feature of SUDs, they are not usually selected as a separate target symptom for pharmacological treatment. Thus, as with other psychiatric disorders, pharmacological treatments for SUDs target the full constellation of the disorder rather than individual symptoms. Further, pharmacological treatments of SUDs are typically specific for the class of drug that is abused (e.g., opioids, nicotine). Nevertheless, several studies have examined the effects of medications specifically on ratings of craving for a particular drug. For example, Courtney et al. (2016) reported that naltrexone (opioid antagonist) blocked conditioned craving to heroin cues, and Green and Ray (2018) reported that varenicline (nicotinic partial agonist) dampened cravings for tobacco cigarettes. In other studies, oxytocin blocked cravings elicited by cues related to both cannabis (McRae-Clark et al. 2013) and tobacco cigarettes (Miller et al. 2016). One promising new pharmacological intervention for SUDs, as well as other psychiatric disorders characterized by intrusive thoughts, is n-acetylcysteine (Dean et al. 2011; McClure et al. 2014; Minarini et al. 2017). N acetylcysteine is a precursor to the antioxidant glutathione, and it acts as a modulator of glutamatergic, dopaminergic, neurotropic, and inflammatory pathways. Although this medication has not been approved for treatment, Dean et al. (2011) review promising results with this drug in the treatment of cannabis use, cigarette smoking, cocaine addiction, and several other psychiatric conditions in which intrusive thoughts play a role. However, larger studies on the effects of n-acetylcysteine for cocaine addiction or cannabis addiction have been disappointing (Mardikian et al. 2007; Gray et al. 2017).

Cognitive

Intrusive thoughts may be a part of several different psychotic disorders in the form of delusions or cognitive distortions. Psychotic disorders, including schizophrenia and schizoaffective disorder, may present with patients experiencing uncontrolled and sometimes disturbing thoughts, images, or perceptions. Delusions, or beliefs that are inconsistent with reality, are the most obvious form of intrusive thoughts manifesting in psychotic disorders. Perceptual hallucinations (auditory, visual, olfactory, tactile) may or may not fall into the category of "thoughts," depending on whether patients recognize them as originating within their own mind. Both hallucinations and delusions are considered "positive symptoms" of these disorders, suggesting that they are additional thoughts and perceptions that patients struggle with, as opposed to "negative symptoms," which reflect a relative deficiency of volition, speech, or other constructs.

The typical treatment of positive symptoms of psychotic disorders consist of second-generation antipsychotic medications. Commonly used second-generation antipsychotic medications, such as olanzapine, ziprasidone, and quetiapine, are remarkably effective at reducing hallucinations in about 92% of patients who take them for a year, though they have many side effects (Sommer et al. 2012). First-generation antipsychotic medications, such as haloperidol, are thought to target and block D2 receptors, while second-generation medications are believed to have complex actions involving D2 receptor blockade and activity at 5HT2A receptors as well (Abi-Dargham and Laruelle 2005). Delusions in schizophrenia and depression appear to respond to antipsychotics, although delusions in delusional disorder and bipolar disorder are less responsive to antipsychotic treatment. One study found that only 22% of psychotic patients with schizophrenia spectrum disorders reported improvement in "cognitive preoccupations" after two weeks of antipsychotic treatment (Mizrahi et al. 2006).

New treatments for intrusive thoughts in psychotic disorders are being investigated. Cannabidiol (CBD), a component of cannabis, has been tested in the treatment of schizophrenia. One recent study showed reductions in positive symptoms of schizophrenia and improved cognition after six weeks of 1000 mg CBD/day adjunctive treatment (McGuire et al. 2018). Glutamate modulators have also shown promise in the treatment of intrusive thoughts in schizophrenia (Hashimoto et al. 2013), presumably as a result of the importance of learning and the integration of new information in reducing delusional thought patterns. Glutamate-modulator treatments are being used to facilitate cognitive behavioral therapy, which can be used to target intrusive thoughts. For example, the partial agonist at the glycine site of the NMDA receptor, d-cyclosterine (50 mg), has shown promise in reducing delusional severity in patients with schizophrenia and schizoaffective disorder (Gottlieb et al. 2011). Sodium nitroprusside, another potential NMDA modulator, was not effective in the treatment of schizophrenia (Brown et al. 2019a). More research is needed to determine exactly which pharmacological treatments are effective, and in which contexts.

Neuropharmacological Considerations

It is worth considering the possibility that different neurotransmitter systems play distinct roles in the generation of intrusive thoughts or in the ability to control or treat them. Not surprisingly, given the heterogeneity of the intrusive thoughts described above, almost every neurotransmitter system is implicated in either the generation or the treatment of intrusive thoughts, including dopamine, norepinephrine, serotonin, acetylcholine, glutamate, endogenous opioid, and endocannabinoid systems, as well as hormonal systems such as oxytocin or stress hormones. With respect to the behavioral intrusive thoughts described above, no single neurotransmitter system has been implicated in "impulse control" disorders or OCD (Williams and Potenza 2008). Impulse control disorders, such as pathological gambling, have been linked to serotonin, dopamine (by modulating reward pathways), and norepinephrine dysfunction (arousal and excitement). The pathophysiology of OCD is poorly understood, but serotonergic drugs are commonly used to treat this disorder, and there is evidence that glutamatergic drugs may be effective as well (Goodman et al. 2014). With respect to affective intrusive thoughts, serotonin circuits are most strongly implicated in both the etiology and treatment. Yet other transmitter systems, such as GABA, norepinephrine, and most recently glutamatergic systems, are also implicated. As with the other disorders, little is known about the receptor mechanisms underlying intrusive thoughts, independent of the broader range of symptoms of the disorders. In the case of drug-induced or substance-abuse disorders, the focus of most treatments is on the neurotransmitter system involved with each specific drug (e.g., cholinergic for cigarette smoking, dopaminergic for stimulants). Yet, there is also some evidence that certain pharmacological treatments may target other systems, such as oxytocin receptors or stress hormone receptors. Other possible treatments target the reconsolidation of drug-related memories (cues), independent of the specific drug type. With respect to cognitive intrusive thoughts described above, the dopamine system is most strongly implicated in the pathophysiology of psychotic disorders, and drugs that block dopamine function are also the primary basis for pharmacological treatment. As noted above, however, the role of dopamine specifically in intrusive thoughts, separate from other symptoms of psychotic disorders, is not known.

Another consideration in the pharmacological treatment of intrusive thoughts is the temporal characteristics of dosing. Many psychiatric medications, such as antidepressant and antipsychotic drugs, are prescribed for daily use over extended periods of time. Recently there has been a growing interest in medications that can be used with a single or small number of administrations. Examples of these are the use of ketamine for depression and, more controversially, the use of psychedelic drugs in the treatment of anxiety, depression, or substance abuse. How exactly single, high doses of psychedelic or other psychoactive drugs improve persistent psychiatric symptoms is an exciting and challenging new direction of research. Do their effects depend on the psychological experience, or are they related to neuropharmacological changes in brain function? Finally, there is also an important role for drugs used in combination with a behavioral intervention, such as the use of MDMA during psychotherapy sessions or the use of beta-adrenergic blockers during presentation of drug-related memory cues. These different modes of administration of psychiatric medications offer promising new approaches to combining pharmacological with behavioral interventions.

Conclusions

Intrusive thoughts are symptoms of a wide range of psychiatric disorders with markedly different pathophysiology, and they take many forms: from the urge to act, ruminations, and cravings to delusional thoughts. Perhaps because they are associated with so many different disorders, each with a different pathophysiology, a wide range of medications have been used to treat intrusive thoughts. Given the heterogeneity of their symptoms and associated disorders, it is unlikely that any single neural circuit mediates intrusive thoughts, or that any single medication can be used to target them across diagnostic categories. Nevertheless, there may be commonalities associated with different disorders (e.g., cravings of food in eating disorders or drugs in SUDs). Some intrusive thoughts may be mainly a disorder of memory (e.g., symptoms in PTSD or addiction), whereas others originate from disordered cognition or strong emotional states. By examining intrusive thoughts both within and across disorders, it may be possible to identify underlying processes that could be targets for new pharmacological treatments.

A handful of issues warrant consideration in studying pharmacological interventions for this elusive symptom. Although certain medications are classically associated with certain psychiatric disorders, in practice a wide range of medications are used, even for the same disorder. For example, dopamine antagonists are typically the first line of treatment for schizophrenia or serotonin reuptake blockers for depression, yet physicians often sample from a range of medications on an individualized basis to find effective treatments. The range of pharmacological treatments is partly because patients vary in their response to medications, as well as because the symptoms within a disorder vary across individuals and over time. Further, diagnoses and disorders are in reality the composite of individual symptoms, and thus the individual symptoms are rarely the target of medication development. It is thus difficult to determine which medications are effective specifically for an individual symptom, such as intrusive thoughts. Further, most psychiatric medications target a constellation of symptoms involved in the disorder rather than any specific symptom. For example, SSRIs are prescribed for depression to improve all the components of the disorder, which may include mood, sleep, appetite, and energy levels. This makes it difficult to examine published studies into the efficacy of various medications in the treatment of intrusive thoughts. Prospective, controlled studies with specific intrusive thought-related outcome measures are necessary as a next step in investigating these questions.

We conclude with a comment about the recent introduction of a novel form of pharmacological treatment for psychiatric disorders: the use of single, high doses of "psychedelic" drugs. Drugs such as ketamine, MDMA, and psilocybin, which were formerly only considered in the context of nonmedical use, are now being tested in therapeutic settings. Remarkably, these drugs appear to be effective after just one or two single administrations, in contrast to other medications that are used on a daily basis. It remains to be determined how single doses of these drugs can produce lasting beneficial effects on, for example, major depression, end-of-life anxiety, or PTSD. It also remains to be determined whether, and how, these novel treatments affect intrusive thoughts in particular. This is an exciting and promising new direction for psychiatric medications.

Acknowledgments

HdW was supported in part by NIDA DA02812 and AKB was supported by a training grant from the National Institute of General Medical Sciences (2T32GM007281). We thank Larry Price for helpful comments on the manuscript.

16

Developing a Neuromodulation Tool to Suppress Intrusive Thinking

Things We (Think We) Know and Things We Need to Figure Out

Colleen A. Hanlon and Lisa M. McTeague

Abstract

Intrusive thinking is a core feature in multiple psychiatric diseases, including obsessive-compulsive disorder (OCD), posttraumatic stress disorder (PTSD), substance use disorder (SUD), and Tourette syndrome. These diseases are not only bound by intrusive thinking, they also share similar disruptions in the functional architecture of the brain, including frontal-striatal-thalamic circuitry which is involved in salience attribution and shifting attention. As more is learned about the neural circuit dysfunctions involved in the initiation, maintenance, and attention to intrusive thoughts, it may become possible to develop noninvasive neuromodulation approaches to attenuate the presence of these thoughts or the morbidity associated with their existence in individuals. This chapter focuses on transcranial magnetic stimulation (TMS) as a tool to induce causal change in behavior, cortical excitability, and frontal-striatal activity. An overview is provided of the cortical and subcortical areas that are often implicated in intrusive thinking, using examples from Tourette syndrome, OCD, PTSD, and SUD. The hypotheses presented can be generalized past TMS to other invasive and noninvasive forms of neuromodulation. In conclusion, key questions are posed to move the field forward.

Introduction

As discussed in the other chapters of this book, there are many operational definitions for intrusive thoughts. While it is difficult to unify these definitions into one common framework, a familiar theme is present in all of them:

a persistent, stereotyped mental pattern that aversively interrupts the flow of competing mental processes despite attempts to inhibit and/or counter these thoughts. One particularly compelling example is that intrusive thoughts are like a mental "tic" wherein, as in the case of Tourette syndrome, an individual may be able to suppress intrusive thoughts for some period of time, but eventually that cognitive buffer is broken down and the intrusive thought pattern floods the neural systems that kept it in check.

The clinical disorder that is most easily characterized as an impairment of intrusive thinking is obsessive-compulsive disorder (OCD), wherein recurrent thoughts or urges lead to debilitating levels of anxiety, distress, and resultant compulsive actions that patients typically fail to willfully suppress. The anxiety and distress associated with intrusive, unwanted thoughts are also a hallmark of posttraumatic stress disorder (PTSD). Similarly, intrusive thoughts about avoiding opiate or alcohol withdrawal or having time to "take the edge off" with a smoking break, fuel the growth of substance use disorder (SUD) and impair the ability for treatment-seekers to remain abstinent.

All four of these psychiatric conditions (Tourette syndrome, OCD, PTSD, and SUD) have intrusive thoughts at their core, yet existing behavioral and pharmacological treatment strategies for these diseases are very different. Modern psychiatry has only recently begun to approach disease treatment in a manner that focuses on core transdiagnostic symptoms of psychiatric disease rather than discrete disease labels. Inasmuch as intrusive thinking is a core symptom common to these disorders, it is certainly a research domain worthy of focus for treatment development.

In addition, these four diseases have something else in common: they all entail disruptions in frontal-striatal circuitry involved in limbic drive and cognitive control. Through recent advances in dosing and coil design, it appears that a noninvasive brain stimulation technique known as transcranial magnetic stimulation (TMS), approved by the U.S. Food and Drug Administration (FDA) to treat depression and OCD, may be a promising tool to target the functional neurocircuit substrates of intrusive thinking in these patients. TMS is one of several noninvasive techniques that can be used to modulate neural circuitry (see Figure 16.1). Here we introduce TMS as a tool to induce causal change in behavior, cortical excitability, and frontal-striatal activity. We provide an overview of the cortical and subcortical areas that are often implicated in intrusive thinking (using examples from Tourette, OCD, PTSD, and SUD) and outline several key questions that should be addressed to move the field forward.

The Application of Transcranial Magnetic Stimulation to Diseases of Intrusive Thinking

TMS is a noninvasive brain stimulation technique that can induce changes in neural activity in the cortex and in monosynaptic afferent projections. When

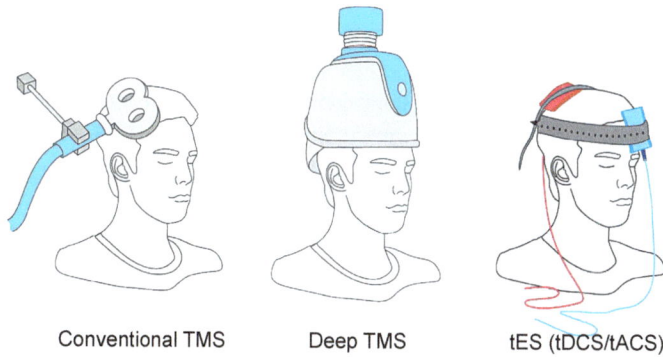

Conventional TMS   Deep TMS   tES (tDCS/tACS)

Figure 16.1 Noninvasive neuromodulation techniques used in individuals with psychiatric disorders that involve intrusive thoughts. The most common technique is the conventional transcranial magnetic stimulation (TMS), which is done by placing a figure-of-eight coil over a specific cortical location. This technique has been used to modulate craving in substance use disorder, impulse control in Tourette syndrome, obsessions in obsessive-compulsive disorder (OCD), and general symptoms of post-traumatic stress disorder. A unique form of TMS, known as "deep TMS," uses similar technology to modulate a wider, deeper area of cortex. This technique was approved for treatment of OCD in 2018. Transcranial electrical stimulation (tES) includes transcranial direct current (tDCS) and alternating current (tACS) approaches and has been used in these disorders as well, although there is still not clear evidence of its clinical efficacy. Reprinted with permission from Ekhtiari et al. (2019).

delivered repetitively (e.g., 600–3000 pulses every 40 seconds to 20 minutes), TMS can change cortical excitability and various behavioral phenomena for 30 minutes to several hours. When these repetitive sessions are given sequentially over a series of days (e.g., 10–30 sessions over 2–6 weeks), there may be lasting changes in functional connectivity in the brain as well as behavioral symptom resolution for several months to a year.

TMS was approved by the FDA as a treatment for major depressive disorders in 2008, and there are now TMS clinics in all 50 states in the United States, throughout Europe, Asia, Australia, and South America as well as a few new clinics in Africa. While the majority of the research in TMS has focused on optimizing treatment protocols for depression, there has been an exponential growth in the application of TMS to investigate and modulate these networks in populations with Tourette syndrome, OCD, PTSD, and SUD. The data has been growing fast, such that in 2018 the FDA approved a unique form of TMS to treat OCD. There is already approval for its use as a tool in SUD and OCD in Europe.

In developing a noninvasive neuromodulation solution for intrusive thinking, however, many questions remain:

- Is there a common neural circuitry that drives intrusive thinking across disease states? If so, can we use the same stimulation protocol for everyone or will there be biotypes that should be considered?

- Assuming that we focus on TMS (as it is the only FDA-approved, non-invasive neurostimulation technique), what is the best cortical location and frequency?
- What is the best way to combine neurostimulation with pharmaco-therapy and behavioral therapy to attenuate intrusive thoughts? Should these techniques be given simultaneously or in serial?
- Should we be pursuing closed-loop neuromodulation strategies for intrusive thinking (rather than current open-loop approaches)? Can we do that noninvasively?
- What stage of intrusive thinking is the optimal target for remediating intrusive thinking? Should the focus be on *prevention* (i.e., prevent the initiation or exacerbation of intrusive thoughts), *inhibition* (i.e., suppress intrusive thoughts), *reframing* (i.e., change the valence of a positive/negative thought), or *distraction* (i.e., enable a patient to shift attention away from the thoughts)?

Below, we will attempt to provide some insight into these questions. As a basis for this discussion, we begin with a review of several key principles that are important to understand, in terms of the capabilities and restrictions of current-generation TMS devices.

What Is Transcranial Magnetic Stimulation?

TMS can modulate neural excitability. It is a noninvasive form of brain stimulation that induces a depolarization of neurons through electromagnetic induction. Although a comprehensive review of studies that have demonstrated the principles of TMS is beyond the scope of this chapter, prior behavioral, electrophysiological, and neuroimaging work in this area is well described and summarized in several review articles (Fitzgerald and Daskalakis 2008; Hoogendam et al. 2010). The majority of our knowledge regarding the basic electrophysiological effects of TMS on the brain are from studies in the motor system. When applied over the hand knob of the primary motor cortex, a single, transient pulse of current through the TMS coil induces a reliable contraction of the contralateral hand, proportional to the amplitude of the induced electrical field (Barker et al. 1986). The amplitude of this motor-evoked potential (MEP) in the contralateral hand can be manipulated by pharmaceutical agents that effect voltage-gated sodium channels and glutamate (Ziemann and Rothwell 2000; Di Lazzaro et al. 2008). There is a dose-response relationship between the amplitude of the TMS pulse and the amplitude of the MEP. This dose-response relationship is referred to as the "recruitment curve" in brain stimulation literature and can be used as a measure of cortical excitability.

TMS can modulate neural pharmacology. Although hundreds of studies have evaluated the effects of various repetitive TMS protocols on behavior and cortical excitability (via EEG and functional MRI), very little is known about the effects of rTMS on neuropharmacology. The most cited studies in this domain have been done using positron emission tomography (PET) imaging, wherein the radioligand is given to the participant and then the TMS stimulation is delivered before the participant goes into the PET scanner. Using PET imaging, Strafella et al. (2001) have demonstrated that dorsolateral prefrontal cortex (dlPFC) stimulation leads to an increase in dopamine binding in the caudate. They also showed that when 10 rTMS is delivered to the left primary motor cortex, increases in dopamine are seen in the ipsilateral, left putamen. Using magnetic resonance spectroscopy, several studies have demonstrated the effects of TMS on cortical γ-aminobutyric acid (GABA) and glutamate (Stagg et al. 2009; Vidal-Pineiro et al. 2015; Iwabuchi et al. 2017). One of the most cited studies, by Stagg et al (2009), demonstrated that the attenuating effects of inhibitory, continuous theta burst stimulation (cTBS) on cortical excitability are related to an increase in GABA at the area of stimulation rather than a change in glutamate. Recently, Iwabuchi et al. (2017) showed that a single session of excitatory, intermittent theta burst stimulation (iTBS) to the dlPFC leads to a decrease in the GABA/glutamate ratio in both the dlPFC and in the insula, suggesting that it is possible to modulate paralimbic cortex through superficial PFC stimulation.

The following principles need to be considered as TMS therapeutic strategies are developed to address intrusive thinking.

Principle 1: Stimulation Depth

With a growing number of TMS coil designs, the depth at which stimulation should occur has become increasingly complex. The focality of TMS is related to the shape of the coil. There is a substantial body of literature devoted to computational modeling of electric field distributions associated with different coil shapes. In one of the most comprehensive papers, Deng et al. (2013) investigated the focality and penetration depth of 50 existing TMS coils. Their computational models revealed that typical figure-of-eight coil designs affected approximately 10 cm^2 of cortical surface, circular coils affected approximately 50 cm^2, and H-coil designs affected approximately 100 cm^2. Most flat figure-of-eight and circular coil designs had penetration depths of 1–2 cm^2, whereas the H-coil designs had consistently higher depths of 2–3 cm^2. A single TMS pulse from a standard figure-of-eight coil stimulates a 12.5 cm^2 area, which is approximately 1/125 (0.8%) of the cortical surface area. By comparison, deep brain stimulation can be at least an order of magnitude more precise than the most focal TMS coils available, with stimulation volumes ranging from 1–20 cm, depending on the electrode configuration (Wei and Grill 2005). Electroconvulsive therapy, on the other hand, appears to effect 94% of the brain and magnetic seizure therapy effects 21% of the brain (Lee et al. 2016). To put the focality of TMS in context with something that is meaningful to the average curious member of the public, 1/125th of the cortical surface is roughly analogous to the surface area of India (or half of Australia) relative to Earth (Hanlon 2017).

Principle 2: Polysynaptic Transmission

Beyond the direct cortical effects of TMS, it is possible to modulate monosynaptic (and possibly polysynaptic) targets of these cortical areas (Figure 16.2). The indirect effects of cortical TMS on monosynaptic afferent targets can be demonstrated through a behavioral assessment of the recruitment curve. When TMS is applied to the hand area of the primary motor cortex there is a dose-dependent change in the MEP of the hand contralateral to the TMS coil. This pathway from the motor cortex to the hand requires at least two neurons: the upper motor neuron, which originates in the motor cortex and terminates in the spinal cord, and the lower motor neuron, which originates in the spinal cord and terminates in the muscles that will contract to produce the MEP. The majority of upper motor neurons, however, terminate on interneurons, which then facilitate lower motor neuron activity. This suggests that TMS can have polysynaptic effects.

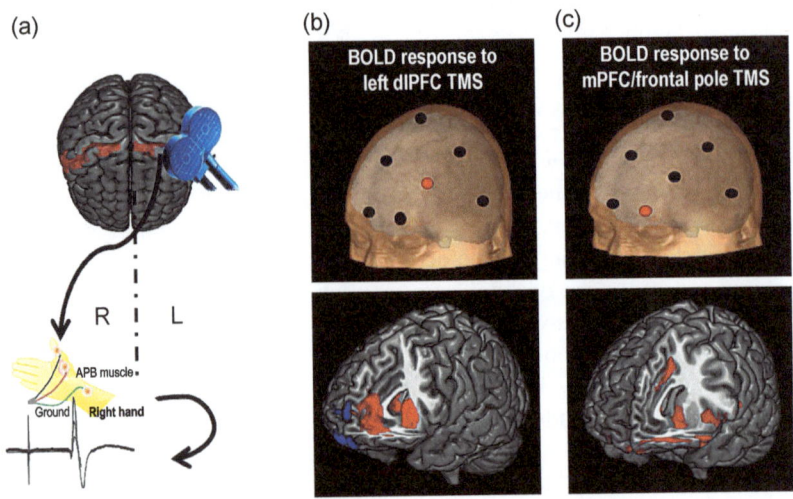

Figure 16.2 Polysynaptic effects: (a) Single pulses of TMS delivered to the hand knob of the primary motor cortex are able to transmit information down the cortical spinal tract, which crosses the synapse in the ventral horn, leading to contraction of the efferent target muscle in the hand, measured with motor-evoked potentials. (b) This polysynaptic engagement can be demonstrated in the cortex as well, wherein single pulses of TMS delivered to the left dorsolateral prefrontal cortex (dlPFC) lead to elevated BOLD signal in the dorsal striatum and ventral cingulate, whereas (c) TMS to the left frontal pole leads to BOLD signal in the ventral striatum, dorsal cingulate, and anterior insula. Adapted after Hanlon (2017).

Principle 3: Frequency-Dependent Modulation

As stated above, when single pulses of TMS are delivered in rapid succession (rTMS), it is possible to change cortical excitability and various behavioral phenomena for a relatively brief period of time (e.g., 30 minutes to several hours; see Figure 16.3). These effects appear to be frequency dependent: low-frequency, continuous stimulation decreases cortical excitability whereas high-frequency, intermittent stimulation leads to an increase in cortical excitability (reviewed in Fitzgerald et al. 2006; Thickbroom 2007).

One of the first studies in this field was conducted by Pascual-Leone et al. (1994), who discovered that 20 pulses at 10 Hz and 20 Hz stimulation over the motor cortex produced an increase in the amplitude of the MEP, suggesting this frequency increases cortical excitability. Chen et al. (1997) then demonstrated that 15 minutes of 0.9 Hz TMS stimulation (810 pulses) to the motor cortex would decrease motor cortex excitability. In a sample of 14 individuals, 1 Hz TMS to the motor cortex for 15 minutes decreased the MEP by 20% for at least 15 minutes after stimulation. These data are compatible with preclinical electrophysiology studies which have demonstrated that 1 Hz stimulation induces long-term depression of neural activity in slice preparations of the motor cortex, visual cortex, and hippocampus.

While 10 Hz and 1 Hz TMS are still widely used, a unique bursting frequency known as *human theta burst stimulation* (TBS) has now gained significant traction in the field. Human TBS was first evaluated by Huang et al. (2005). Leveraging data from preclinical literature, which demonstrated that

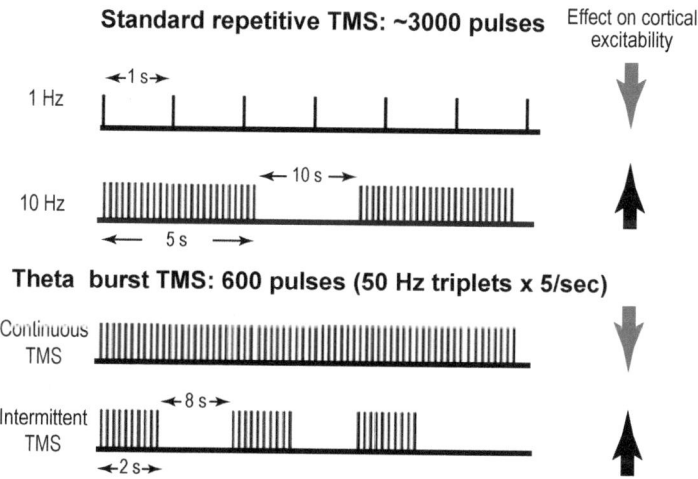

Figure 16.3 Frequency-dependent effects: When delivered in a repetitive manner, a single session can effect cortical excitability for 30–60 minutes.

electrical stimulation of cortical slices in 100 Hz bursts five times per second (known as theta burst) can induce long-term plasticity (Bear and Malenka 1994; Malenka and Bear 2004), Huang et al. performed a clinical TMS study wherein TMS pulses were delivered to the motor cortex in 50 Hz bursts five times per second (human TBS). When TBS was delivered continuously for 600 total pulses, it decreased motor cortex excitability. When TBS was delivered in an intermittent pattern (2 sec on, 8 sec off) for 600 pulses, excitability increased. The effect sizes of these brief continuous TBS (40 sec) and intermittent TBS (190 sec) paradigms are comparable to studies of 1 Hz and 10 Hz. However, many publications have recently shown that there is high interindividual variability in TBS response, which has led to some caution in the reliance on this stimulation protocol (Vernet et al. 2014; Jannati et al. 2017).

Principle 4: Priming and State-Dependent Effects

A large body of literature demonstrates that the effects of TMS on behavior are brain-state dependent and may be amplified by priming the brain with either a behavioral task or brain stimulation (Opie and Cirillo 2017). One of the earliest studies in this field was by Iyer et al. (2003), who demonstrated that the attenuation of cortical activity with 1 Hz TMS can be amplified by priming the motor cortex with 6 Hz TMS. This was expanded to studies in the motor system and visual system, which demonstrated that there were brain-state dependent effects of TMS on cortical excitability (Silvanto et al. 2007, 2008a, b). Additionally, priming the brain with continuous TBS may enhance efficacy of intermittent TBS (Opie et al. 2017).

Although this body of research existed in sensory and motor control literature, it has only recently been harnessed by the clinical TMS research field. Whereas the recent FDA approval of TMS for OCD requires a behavioral prime, for example, neither the brain state nor the behavioral state of the individual was accounted for during the initial multicenter clinical trials of TMS for depression. This represents a latent opportunity for us to improve outcomes and minimize some of the interindividual variability that is observed in patients receiving clinical TMS treatment.

Within the addiction literature, a large clinical trial demonstrated that exposing a smoker to smoking cues (behavioral prime) before TMS amplified the effects of TMS on smoking cessation (Dinur-Klein et al. 2014). In this prospective, double-blind, sham-controlled study, 115 regular cigarette smokers were randomized to receive ten daily treatments of TMS. Immediately before each session, half of the participants were presented with visual smoking cues: cigarette consumption and nicotine dependence were reduced, and the effects were greatest in individuals that were exposed to smoking cues. In PTSD treatment, priming a trauma memory at the outset of each rTMS session has also been shown to enhance TMS effect sizes (Isserles et al. 2013). In this study, thirty PTSD patients were randomized to one of three groups: sham rTMS, real rTMS following exposure to a 30-second patient-tailored trauma script, or real rTMS following exposure to a 30-second patient-tailored neutral script. Participants received 12 sessions of real or sham rTMS (three sessions per week for four weeks, deep TMS H-coil). The only group with a significant improvement in the Clinician-Administered PTSD Scale was the group that received exposure to trauma scripts before each rTMS treatment.

The aforementioned studies all demonstrate that a priming stimulus amplifies the effects of a form of TMS intended to increase excitability in targeted networks. Although the mechanism through which cue exposure enhances the behavioral effects of rTMS are not clear, one possibility is that cue exposure reactivates latent memory traces, frequently referred to as an engram (Vernet et al. 2014), enabling them to be manipulated and reconsolidated (Opie et al. 2017). If this were true, priming may also be effective for TMS paradigms designed to decrease cortical excitability (e.g., 1Hz, cTBS). A recent study by our group demonstrated that, on average, individuals with cocaine use disorder, who were exposed to cocaine cues before and after continuous TBS, had a decrease in cocaine-cue reactivity following 3600 pulses of real but not sham TBS (Malenka and Bear 2004). Secondary analyses of the data, however, demonstrated that real cTBS decreased cue reactivity in individuals with a high baseline brain response to cues, whereas it increased cue reactivity in individuals with a low baseline brain response to cues. This bidirectional shift was not present following sham cTBS. While all individuals in this study received a behavioral prime (drug cue exposure), it seems that a behavioral prime was not alone sufficient. The directionality of TBS-induced effects was dependent on the baseline level of brain activity (neural state) in the TBS target; in this study (Malenka and Bear 2004), the medial PFC.

Potential Neural Circuit Targets for Neuromodulation of Intrusive Thinking

One of the key advances in the neuroimaging literature over the last twenty years is that brain regions organize their activity into coherent functional networks (Figure 16.4). Through functional magnetic resonance imaging (fMRI), these networks appear as correlations of the low-frequency fluctuations in BOLD signal between brain regions. Many networks were originally identified via data-driven methods from brain activity at rest, and are called *resting-state networks*. However, these networks reliably appear in ongoing brain activity during tasks, and meta-analyses of task-based activation also reveal consistent *functional networks* similar to those identified at rest.

Several functional networks have been studied extensively that have relevance to intrusive thinking: the default mode network (DMN), containing the medial PFC and posterior cingulate; the salience network (SN), containing the anterior cingulate and anterior insula; and the executive control network

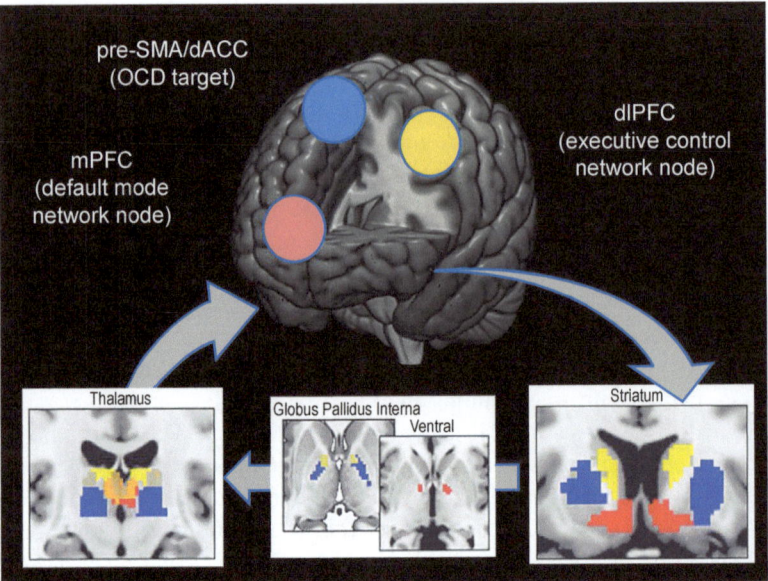

Figure 16.4 Candidate neural circuits amenable to modulation for intrusive thinking. Typically, therapeutic neuromodulation approaches require that a specific neural system be identified. The therapeutic strategy can then target the nodes of this neural system. This can be done invasively (e.g., through deep brain stimulation) or noninvasively (e.g., using TMS, transcranial electrical stimulation, pulsed ultrasound). Three cortical nodes may be putative targets for neuromodulation based on their role in intrusive thinking and their striatal-thalamic connectivity: the medial prefrontal cortex (mPFC, red), the left and right dorsolateral prefrontal cortex (dlPFC, yellow), and the presupplementary motor area (pre-SMA, blue). The striatal, pallidal, and thalamic nodes of these circuits are shown in the lower panels. Adapted after Morris et al. (2016).

(ECN), containing the dlPFC and posterior parietal cortex. DMN is the best known and most studied of these functional networks. It serves various introspective functions related to intrusive thinking, including mind wandering, recollection and prospection, rumination, and self-reflection. Other "task positive" networks act in opposition to the DMN. These networks activate during behaviorally regulated task performance and externally focused cognition. For example, the SN activates for transitions from introspection to task performance as well as during task initiation and switching. The ECN is involved in cognitive control, working memory, and in tasks governed by external stimuli, whereas functional connectivity in the DMN is typically high during tasks of internal monitoring. In this manner, the ECN and DMN are considered anticorrelated networks.

Etkin and colleagues demonstrated the central importance of the SN as a common neural substrate across psychiatric illness categories (Goodkind et al.

2015). They performed a meta-analysis of structural abnormalities across six psychiatric disorder categories, including OCD, PTSD, and SUD, and found that all of them showed gray matter volume reductions in the dorsal anterior cingulate cortex and anterior insula. In a parallel meta-analysis of functional neurocircuit anomalies during cognitive processing tasks across psychiatric disorders, Etkin and colleagues demonstrated that the SN, in conjunction with the broader frontoparietal ECN, is hyporeactive among these patients during cognitive demands. Importantly, the ECN is recruited for cognitive regulation as well as emotional regulation, and the dlPFC target most frequently utilized in therapeutic rTMS is seated within this network. As such, the transdiagnostic effects of rTMS may, in part, be attributable to ECN upregulation and its influence on attenuating intrusive thinking and associated negative affect. In addition to putatively increasing activity of the ECN with rTMS, we might predict that intrusive thinking could alternatively be attenuated by either decreasing the activity of the DMN or enabling the SN to switch more effectively from the DMN to the ECN. These hypotheses have not been directly tested but are amenable to systematic evaluation through TMS.

Questions and Hypotheses

1. Using TMS (because it is the only FDA-approved noninvasive technique), what is the best cortical location and frequency?

Strategy 1 Decrease the amplitude of intrusive thoughts at rest by dampening the medial orbitofrontal cortex (OFC) connectivity (a node in the DMN). Several studies have demonstrated that individuals with OCD have elevated activity in medial aspects of the OFC. Three studies that targeted the OFC with TMS showed improvements in OCD symptoms. These were all relatively small studies that applied 1–3 weeks of treatment and showed changes that lasted up to one month.

Strategy 2 Increase control over intrusive thoughts by increasing dlPFC connectivity (a node of the ECN). The dlPFC is the FDA-approved target for the treatment of depression and has been investigated as a treatment target for PTSD, SUD, and OCD. The first study to explore the use of TMS for OCD demonstrated that a single session of 10 Hz TMS decreased compulsions but not obsessions, yet the effects lasted for eight hours (n = 12 individuals). Although these results were promising, they have been difficult to replicate. Two recent studies have demonstrated that several weeks of TMS may improve OCD, but again, the obsessive component does not seem to respond very well. This may be because the obsessive component relies more heavily on subcortical structures such as the basal ganglia and amygdala. While 10 Hz TMS to the left dlPFC and 1 Hz TMS to the right dlPFC have been evaluated as a target for PTSD, with mixed success, intrusive thinking is not often reported as a primary outcome measure. Consequently, with mixed results from the dlPFC, it is still unclear if targeting this area is likely to improve intrusive thoughts.

Strategy 3   Target the supplementary motor area/pre-SMA. The first studies that targeted the pre-SMA examined the use of TMS for patients with OCD and Tourette syndrome. At the end of treatment, patients showed general reduction in OCD symptoms, an improvement in functioning, and reductions in depression and anxiety. Importantly, the improvements held for at least three months. This study was followed by a second study with 21 OCD patients and a more careful study design. After four weeks of TMS treatment, patients showed notable decreases in OCD symptoms as well as a reduction in depression and anxiety; benefits were still present for most patients three months later.

Although TMS targeting the pre-SMA has been shown to be the most effective, it is not clear whether this is indeed the only or best area of the brain to target, as pre-SMA studies are the only ones thus far that use doses and treatment protocols similar to the standard of care for depression. These positive results are, however, very encouraging and are helping us move forward.

In 2018, a unique form of deep TMS was approved by the FDA to treat OCD. This type of TMS has a wider cortical field and likely modulates a large portion of the medial PFC and cingulate cortex, including the SMA/pre-SMA. Although it is not clear exactly which of these brain regions is responsible for the clinical effect (or if all are necessary), these data suggest that the dorsal medial wall of the PFC may be a good target for modulation. In an interim analysis of a larger study, the research team evaluated the effects of 10 Hz, 1 Hz, and sham TMS on OCD symptoms in 23 individuals (25 sessions over five weeks). They demonstrated that although there was no significant interaction between group and time with this sample size, the effect size was higher with 10 Hz TMS compared to 1 Hz TMS. Hence, the remainder of the participants were randomized to 10 Hz TMS or sham TMS for a total sample of 30, wherein 10 Hz led to a significant reduction in OCD symptoms up to one month after the five weeks of TMS (Carmi et al. 2018). These data led to an 11-site clinical trial of 42 individuals who received six sessions of daily TMS to this medial PFC target.

2.   What is the best way to combine behavioral therapy (including mindfulness and neurofeedback) with TMS to maximally attenuate intrusive thoughts? Should this be given simultaneously or in serial?

As described earlier, there is growing evidence that the effects of TMS can be amplified by priming the individual (and perhaps pushing the brain into a specific state) before TMS is administered. One of the key components of the 2018 FDA approval of TMS for OCD was that TMS had to be given in the presence of a personalized visual cue that caused stress and anxiety for the patient (e.g., placing a purse on the dirty floor in front of someone with obsessional thoughts about dirt). It was assumed that this external cue places the brain in a primed state, which would then be modulated by the TMS.

Based on these empirical results from rTMS experiments and from a strong preclinical foundation regarding manipulation and reconsolidation of memories, it is likely that any neuromodulation approach for intrusive thinking should involve putting the individual in a state where the intrusive thoughts are present and perhaps bothersome.

3. Should we pursue closed-loop neuromodulation strategies for intrusive thinking (rather than current open-loop technologies)? Can we do this noninvasively?

Given that intrusive thoughts are frequently transient and share some of the same temporal properties of a seizure (e.g., largely unpredictable onset time but with some known triggers, an episodic disease feature rather than a stable state as is seen in mood disorders or chronic pain), a closed-loop system may be more effective and appropriate for intrusive thinking than an open-loop neuromodulation approach (Figure 16.5).

Open-loop neuromodulation typically refers to a device that provides a fixed stimulation protocol over a fixed period of time. Currently, most invasive and noninvasive neuromodulation approaches are open loop (e.g., TMS, DBS). It is easy to see the value of a closed-loop system, which could include "sensing technology" to detect changes in the brain state, and then dynamic stimulation settings, which could adjust to the individual patient's neural needs. Closed-loop stimulation technology has shown promise as a treatment for various diseases (Widge et al. 2017). The most successful closed-loop system in clinical trials thus far has been developed, FDA approved, and is now deployed for use in intractable epilepsy. The leads of this device (NeuroPace) are implanted into an epileptic focus in the brain. It monitors activity in the areas, can detect the prodrome of seizure activity, and when a seizure begins it will deliver stimulation to block the growth of that activity. This is referred to as responsive neurostimulation and received approval from the FDA in November, 2013. In the years since its approval, it has been successfully employed in hospitals throughout the United States: several trials demonstrated a 53% seizure reduction after two years and a 70% median seizure reduction after five years (Geller et al. 2017). Closed-loop stimulation has also been used in the spinal cord for pain management (the RestoreSensor™ System, Medtronic, Minneapolis, MN). Although no noninvasive, closed-loop brain stimulation devices have been approved for clinical use yet, there is extensive growth in this field (Bergmann et al. 2016).

There are also non-device-based closed-loop neuromodulation strategies. Simpler versions include a "human in the loop": clinicians observe the recorded brain signals and provide manual adjustment of the stimulation rather than the device automatically self-adjusting. Other techniques such as real-time

342

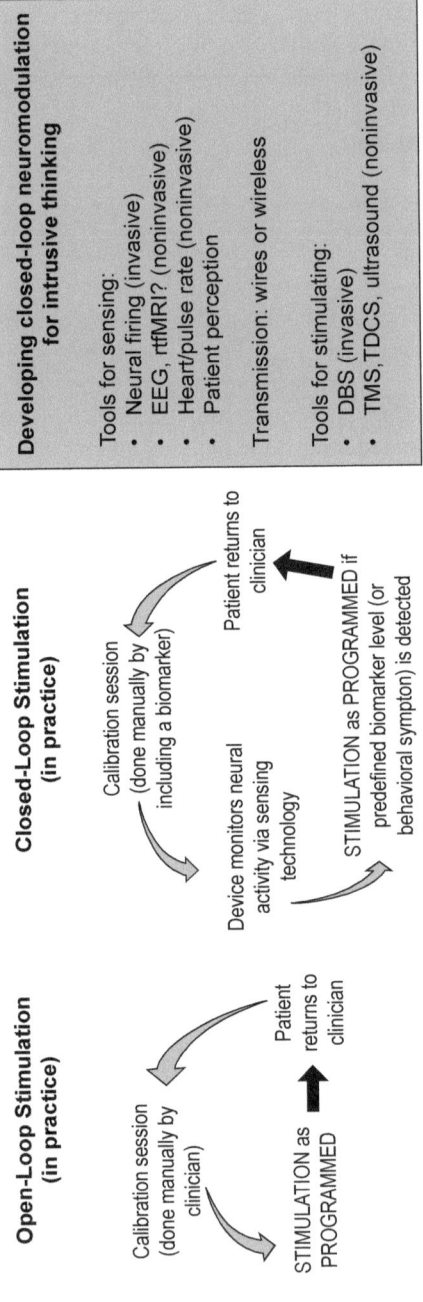

Figure 16.5 Open- and closed-loop neuromodulation options for intrusive thinking. Given the temporal profile of intrusive thinking, a closed-loop neuromodulation approach may be more beneficial—one that includes the ability to detect aberrant brain states (or behavioral states) and adapt its stimulation properties accordingly (e.g., turning itself on/off, modifying the frequency, changing the amplitude). A critical element involved in closed-loop technologies, however, is the reliability of the underlying biomarker it is trying to sense. In practice, standard "open-loop" techniques such as TMS and DBS are actually closed-loop systems, because the patient always returns to the clinician to have the parameters adjusted. New generation closed-loop techniques, however, may enable these adjustments to be made in real time, responding to endogenous neural changes and increasing the "agency" of patients by enabling them to modify the settings based on their own perceptions. This may be particularly useful for intrusive thinking.

fMRI feedback, EEG feedback, and mindfulness strategies may also be useful approaches for intrusive thinking.

4. What is the best treatment strategy: *preventing the initiation* of intrusive thoughts or increasing the ability of the patient to *shift their attention away* from the thoughts?

When designing a neuromodulation approach for intrusive thoughts, the neural regions targeted and the dosing parameters used will likely be different if the goal is to help individuals shift their attention away from intrusive thoughts than if the goal is to stop them from happening. If the goal is to stop the initiation of intrusive thoughts, it is possible that targeting the DMN (perhaps using TMS directed at the medial PFC) might be a fruitful strategy. Alternately, if the goal is to enable individuals to shift their attention away from thoughts which target the salience network, a deep form of TMS directed at the cingulate or insula might be the best strategy, given its role in set-shifting and attributing value.

Conclusion

Intrusive thoughts are a common, transdiagnostically relevant feature of many psychiatric conditions including Tourette syndrome, SUD, PTSD, and OCD. With the approval of TMS as a tool to treat OCD in 2018, we are in the early stages of an era of rapid discovery regarding the use of neuromodulation to alter intrusive thoughts that plague these patient populations. Although several concepts of rTMS treatment are robust and replicable (e.g., regional specificity, depth of the magnetic field, dose-dependent amplification of behavior, polysynaptic engagement, frequency-dependent effects), many key components of TMS treatment development have not yet been widely explored, especially for intrusive thinking (e.g., optimal number of sessions per day or in total, the use of behavioral primes to amplify TMS treatment effects, the effects of applying TMS before vs. after behavioral therapy, the use of TMS to amplify pharmacotherapy treatment). As study of the neural circuitry that underlies the initiation, maintenance, and distraction from intrusive thinking matures, we will be better prepared to design biologically informed and rigorous neuromodulation clinical trials in this domain.

Here, we have attempted to introduce TMS as an innovative new tool which can modulate brain activity in a circuit-specific, frequency-dependent manner as well as to review current knowledge regarding the pharmacologic effects of TMS. While development of TMS as a new treatment tool is still in its infancy, we hope to have sparked interest in the need to develop a neural circuit-based treatment tool—one that is available to our patients—and to increase our knowledge of the synergy between pharmacotherapeutics and brain stimulation interventions. A large body of knowledge suggests that frontos-triatal circuit activity is a significant biomarker involved in intrusive think-ing. Through closed-loop brain stimulation techniques, it may be possible to develop an adaptive, personalized neural circuit-based treatment for patients.

It is important to note that lasting behavioral change may require more than just brain stimulation. Just as the plasticity potential of a primed neuron is higher than an unprimed neuron, TMS may have higher efficacy when an in-dividual is engaged in the cognitive/emotional process they wish to amplify or attenuate. Hence, TMS is likely to be most effective when combined with a pharmacotherapeutic agent that lowers the threshold for cortical excitability or with behavioral interventions (e.g., exposure therapy or contingency manage-ment). Nonetheless, while these statements are based on preclinical literature and human learning theory, they await rigorous evaluation.

As the field continues to grow, we hope to see more interactions between clinical and preclinical neuroscience researchers from electrophysiological and pharmacological backgrounds. With any luck, through the continued re-finement of open- and closed-loop brain stimulation tools, we may soon be rigorously evaluating noninvasive brain stimulation solutions for intrusive thinking. The quest for a sustainable treatment solution will undoubtedly re-quire a complementary approach to modifying the pharmacology, neural cir-cuitry, and ultimately the behavioral manifestations of intrusive thinking in these complementary cohorts of patients.

Acknowledgments

Figure 16.4, from Morris et al. (2016), is used under the terms of the Creative Commons Attribution 4.0 International License (CC-BY).

17

Interventions and Implications

Judson Brewer, Harriet de Wit, Aurelio Cortese,
Damiaan Denys, Colleen A. Hanlon, Emily A. Holmes,
Martin P. Paulus, Jens Schwarzbach, and Peter Tse

Abstract

This chapter provides a framework for developing interventions that specifically target intrusive events. It describes the challenges in defining intrusive thoughts and the difficulty in distinguishing normal processes of cognition and emotion from indicators of dysfunction, defined from practical, neurobiological, or cultural points of view. Throughout, the term *intrusive events* is used to encompass both thoughts and images that become intrusive. Examples are explored as they occur in different psychiatric disorders to demonstrate their variance in form, frequency, and controllability. Treatment modalities that have been used to alleviate intrusive events in different psychiatric disorders are reviewed, including behavioral, pharmacological, and emerging electromagnetic brain interventions. Two clinical vignettes illustrate the nature and severity of intrusive events in patient populations as well as the complex, multidimensional nature of the clinical reality. Ways of measuring intrusive events are examined and deconstructed into components (e.g., sensory, motor, and cognitive features). By examining intrusive events across diagnostic categories, common basic biobehavioral processes may be revealed which, in turn, could facilitate the study of neural processes underlying the behaviors. A model of cognitive and emotional decision making is presented to provide a basis for understanding and studying intrusive events. Examples of how the model might account for the "failure modes" in intrusive events are used to formulate testable hypotheses, and future interventions that combine multiple treatment modalities are considered. The chapter concludes with a discussion of the broader cultural context of intrusive events.

Group photos (top left to bottom right) Judson Brewer, Harriet de Wit, Aurelio Cortese, Martin Paulus, Emily Holmes, Judson Brewer, Damiaan Denys, Colleen Hanlon, Peter Tse, Jens Schwarzbach, Martin Paulus, Colleen Hanlon, Aurelio Cortese, Harriet de Wit, Judson Brewer, Peter Tse, Damiaan Denys, Martin Paulus, Harriet de Wit, Jens Schwarzbach, Emily Holmes

Introduction

Defining the Phenomenon

What is an intrusion? To which entity does the concept of an intrusion correspond? Is it a single concept that refers to one and the same mental phenomenon, or are there multiple types that arise under different circumstances and across multiple mental states? Are intrusions different in animals than in humans, and do they vary across cultural contexts or even time?

The definition by Clark (2005:4) serves as a starting point for our analysis:

> ...unwanted, clinically relevant intrusive thoughts, images, or impulses [are] any distinct, identifiable cognitive event that is unwanted, unintended, and recurrent. It interrupts the flow of thought, interferes in task performance, is associated with negative affect, and is difficult to control.

Accordingly, an *intrusion* can be understood both as a clinical symptom and as a regular mental phenomenon. In as many as 85% of healthy individuals, for instance, intrusions have been observed in the form of thoughts, images, or impulses (Rachman and de Silva 1978), experienced as an individual event that is undesirable, unintentional, and recurrent. In comparison, intrusions as a clinical phenomenon may be fundamentally different. Their severity may increase over time, despite or even because of the interventions undertaken by the patient or caregiver, gradually consuming increased amounts of a patient's time and energy. In addition, pathological intrusions rarely diminish on their own. Unravelling the principle of "reinforcement" that leads to a snowball effect poses a challenge for both clinical and neurobiological researchers.

As suggested by Clark (2005) and confirmed in our discussions, future empirical studies need to clarify the boundaries of what constitutes an intrusion for psychiatric clinical and psychological purposes. While a *minimum definition* runs the risk of being too restrictive, and may not describe all types of intrusive phenomena, it permits us to distinguish between normal and abnormal intrusions. This, in turn, is needed by diagnosticians to distinguish between sick and healthy mental events and aid in treatment decision making. As a minimum definition, we propose the following:

> Intrusive events are unwanted, clinically relevant, intrusive thoughts, images, or impulses that an individual may attempt to resist, but which are out of their control.

A *maximum definition*, by contrast, needs to incorporate all intrusive phenomena, as exhaustively as possible, into a general descriptive approach. In this way, an intrusion will be able to be described as a specific mental event or experience and, regardless of its nature, be studied by different scientific disciplines. We propose the following as a maximum definition:

> An intrusive event is any interruption in the flow of mental events by an external (e.g., a ringing telephone) or internal (e.g., a thought) stimulus.

Many types of mental events can intrude into consciousness. Verbal and non-verbal thoughts, mental images, impulses, memories, emotions, desires, and dreams can all reset the contents of consciousness and be experienced as unwanted or intrusive. Do all such events have the potential to be clinically relevant? How does the minimum definition, which primarily encompasses psychiatric symptoms, relate to the maximum definition, which relates to everyday experienced phenomena? To what extent do they overlap or deviate from each other? To what extent are they qualitatively, or perhaps only quantitatively, different from each other? Importantly, if clinically relevant intrusive thinking takes different forms or domains (e.g., verbal vs. imagery), what are the implications for treatment? Different treatment approaches may be needed or optimized for different domains.

In our discussions, we juxtaposed these two definitions next to each other—one with a psychiatric clinical purpose, the other with a psychological fundamental purpose—but wish to emphasize that intermediate viewpoints are possible. To address our group's topic, however, we found these artificially contrasting definitions helpful. A discussion of the philosophical and social implications of defining the phenomenon is included later in the chapter.

Intrusive Events and Psychiatric Disorders

Across many common psychiatric disorders, intrusive events present as key symptoms (see Schlagenhauf et al., this volume). Recurrent unwanted thoughts and images occur in almost every psychiatric disorder and are explicitly described among the criteria that must be met for a formal diagnosis in several disorders. Notably, intrusive events may be problematic symptoms in themselves. As such, they are appropriate targets for interventions but have rarely been identified as a transdiagnostic clinical feature constituting a target for treatment (cf. Iyadurai et al. 2018). Given their importance across mental disorders, intrusive events may offer insight into the neural mechanisms involved in the pathophysiology of psychiatric conditions.

In our discussions we de-emphasized intrusive events that are manifest in perceptual or thought disorders (e.g., hallucinations or intrusive delusions associated with schizophrenia or psychotic depression) or tics (as in Tourette syndrome). These may fall into separate categories of events and may notably lack the "negative affect" component specified by Clark (2005). As a result, our focus here is on the specific symptoms of intrusive events, not on full psychiatric diagnoses.

Intrusive events vary in their prominence as defining symptoms for different diagnoses. Importantly, the pathophysiology of intrusive events may differ across diagnostic categories and may differ from other symptoms within a disorder. For some disorders, such as posttraumatic stress disorder (PTSD) and obsessive-compulsive disorder (OCD), intrusive events appear to be central defining features and may causally drive other symptoms (see Holmes et al. as well as Hanlon, this volume). In others, intrusive events may not be the defining feature of a disorder but rather one of many criteria used to reach a diagnosis. For example, craving is a symptom of substance use disorder (SUD), but SUDs can and do appear without cravings. Similarly, suicidal ideation appears in major depressive disorder, but the disorder can occur without it. By focusing on a single symptom, it may be possible to relate the clinical manifestation of an intrusive event to a dysfunction of neural circuits controlling normal brain function.

Given the heterogeneous nature of intrusive events associated with different psychiatric disorders, we recommend that a research program be developed to examine whether subcategories of intrusive events exist; this information is needed to establish an association with the underlying neurobiology. More importantly, intrusive events with a different neurobiological signature may require distinct treatment approaches. Given the different dimensions of an intrusive event (e.g., prior experience and expectancies, precipitating events, temporal sequence of events, emotional breadth, contextual features, psychological consequences), we offer two clinical vignettes to illustrate the clinical manifestations of intrusive events in actual patients and demonstrate the inherent complexities.

Posttraumatic Stress Disorder

From an index traumatic event, patients with PTSD typically experience two or three different, highly vivid intrusive memories both in visual and other sensory modalities (Grey and Holmes 2008). For example, after a traumatic road accident, a person might experience vivid intrusive visual mental images of an oncoming red truck, which originated from the moment just before the accident. This intrusive image is highly distressing and often associated with the strong emotions that occurred at the time of the trauma (fear, helplessness). Additional multimodal sensory images may originate from the same event. For example, the visual memory is also multimodal, comprising sight of the person's hand on the driving wheel accompanied by the sound of glass breaking and the smell of burning. The memory may also be associated with secondary emotions such as horror and guilt. Well after the event has occurred, an otherwise innocuous visual stimulus, such as a red front door, may remind the person of the red truck, thus triggering an emotional response. Emotional states may also serve as triggers for the traumatic event: when a person experiences a feeling of helplessness about an unrelated event, this may elicit the feelings of helplessness associated with the traumatic memory. These experiences may be especially disturbing because the patient has not associated these external or internal cues to the occurrence of the intrusive memory, so that they occur as if without warning and as both unpredictable and uncontrollable. Thus, even though these intrusive memory experiences (colloquially often referred to as "flashbacks") may be brief (i.e., seconds, followed by up to 30 min of emotional response), they can have a powerful impact on ongoing behaviors and disrupt attention and performance (Holmes et al. 2017). Intrusive memories not only interfere with normal activities, they also produce significant emotional distress and physiological arousal which can be highly disruptive. The experiences typically occur repeatedly, at unpredictable intervals and in unexpected settings (i.e., elicited by various contextual stimuli), and may occur at varying intervals (e.g., one per week to several per day). Efforts to "push the memory" from one's mind often fail. The unpredictable nature of the events can set off a cascade of other symptoms, including efforts to avoid reminders of the trauma which may lead to social withdrawal. A primary goal for treatment is to reduce the frequency and emotional valence of the intrusive image-based memories.

There are several treatments with a good evidence base for PTSD. Behavioral treatments provide an interesting example from the perspective of intrusive memory. Trauma-focused CBT (cognitive behavioral therapy) and EMDR (eye movement desensitization and reprocessing therapy) both involve repeated exposure to the trauma memory. For example, trauma-focused CBT, a particular form of CBT, focuses on the trauma memory but not on verbal thoughts "about" the trauma (as CBT for depression would). It does this by using imaginal or *in vivo* exposure to the trauma memory, which requires the patient to bring to mind the sensory image-based memory in rich detail. For instance, over a series of 12 sessions, the patient is encouraged to talk about the trauma in detail, to "relive" the memory in their mind's eye, and when possible to bring in adaptive information for memory updating (e.g., feeling of safety, that they did not die in the car crash). Initially, this process is typically highly emotional (and the patient may become upset and cry in reliving sessions), but it becomes less so over repeated sessions. Patients are taught not to "avoid" reminders or to push the intrusive memory from their mind. In summary, the emphasis is on deliberately retrieving the emotional memory in vivid sensory detail. Over time the memory becomes less vivid, the emotions and meaning updated, and the number of intrusive memories declines.

EMDR is a related behavioral treatment that requires somewhat less detailed recall of the trauma: when the memory is recalled in a therapy session, it is done so in the presence of a concurrent task, such as side-to-side eye movements or bilateral beeps. The patient deliberately brings the trauma to mind and simultaneously performs the task guided by the therapist. Research indicates that the success of EMDR is related to impact of the concurrent task in working memory. This is not unlike the brief procedure being developed to reduce trauma intrusions by using a memory reminder plus *Tetris* computer game play (Holmes et al. 2009, 2010; James et al. 2015; Horsch et al. 2017; Iyadurai et al. 2018; Kessler et al. 2020; see also section below on Future Interventions).

At the end of successful behavioral treatment, the patient should be able to recall the traumatic event at will, if they wish to, without becoming overwhelmed. Critically, they will no longer be experiencing frequent involuntary, unwanted intrusive imagery-based memories that impair their daily life. If you have not experienced intrusive events yourself, it may be hard to imagine how powerful brief intrusive events can be for an individual, and thus how beneficial it can be to ameliorate them. After trauma, intrusive thoughts can carry damaging and toxic meanings for the patient. For example, a rape victim may have an image of the rapist telling them they are worthless. Rationally, a victim may know this is not true but still suffer under extreme distress and shame, brought about by the intrusive events, which cause them to relive this toxic message vividly. Verbal discussion does not change the meaning carried by the emotional image (presumably because it differs neurally), but strategies to change the intrusive event (the image itself) can, as illustrated here.

It is clinically compelling to see how successful PTSD treatments are centered on ameliorating the "hub" symptoms of intrusive sensory memories of the traumatic event. Reducing the emotional efficacy of intrusive events and the meanings they carry, as well as the frequency of their occurrence, can lead to substantial improvements in a patient's quality of life.

Many questions remain: How can models explain the impact of existing treatments on the reduction in frequency of intrusive memories? How can treatments be improved to make them briefer and even more effective? Ideally, we should develop simpler, focused treatments that can help more people globally (Holmes et al. 2014, 2018).

Obsessive-Compulsive Disorder

In OCD, intrusiveness coincides with the feeling of "being out of control" and is experienced in different phenomenological domains, as OCD develops over time. OCD is a process with different clinical stages rather than one single stage. These stages follow a dialectic interaction in which an intrusive event elicits a response and the response amplifies the intrusive event. Through different neurobiological adaptations, the course of OCD eventually worsens. Thus, OCD should be regarded as a disease *process* that develops through the amplifying *interaction* between (the reflection and resistance of) the person (mind) and the disorder (brain).

Take, for instance, a young mother who recently gave birth to her first child. She carries an enormous burden of being solely responsible for a helpless and vulnerable life. Her husband leaves daily for work, leaving her alone at home with the child. Seeing her young baby in the crib, a thought appears in her mind: she imagines that she could strangle her baby in the crib and that because she is alone, no one could prevent her from acting on that thought. In summary:

1. The mere *presence* of this thought is intrusive because it occurs against her will. She feels out of control and is unable to volitionally control her thinking.

2. She is worried because the idea of strangling her baby does not match the ideal of motherhood she maintains and strives to achieve. The *content* of the thought is intrusive because it is not in line with her identity or expectancy. It is *ego-dystonic* in that it does not match her self-image, thereby giving rise to mental discomfort and distress.

3. She wonders whether she really could strangle her baby. How can she be certain, given the fact that humans are notoriously unpredictable, that she won't in fact destroy that which she most treasures? It seems that the freedom to commit such a terrible act is in itself so disturbing that it causes the thought repeatedly to reoccur. In addition, *moral implications* of the initial thought intrude as well.

4. She feels anxious because the thought confronts her with feelings of being out of control. The *emotional value* (anxiety) of the thought is intrusive. *The presence, content, implication, and emotional value of the thought all have an intrusive quality.* Although she actively resists the thought because it annoys her and feels intrusive, the very process of reflecting and/or resisting the thought *reinforces the frequency and strength* of the intrusive thought. Her efforts to eliminate the thought may, in fact, enhance its occurrence.

5. The thought becomes obsessional, and her attention is completely drawn to that one single thought. Obsessionality is a dysfunction of *intentionality*: the incapacity to shift focus or attention to another topic, due to a stronger and longer intentional relation with the mental act. The thought is intrusive because of its *obsessive* nature.

6. She cannot suppress the thought; moreover, she is compelled to think about her obsession. Compulsivity is a dysfunction of *sense of agency*: she is forced to think about the intrusion, contrary to her willpower. The thought is intrusive because of its *compulsive* nature.

7. Gradually the thought becomes more present and repetitive; it loses its original meaning, but remains an intrusion because of its duration and repetition. The thought is intrusive because of its new form or *appearance*; it is now a full-blown *obsession*.

8. Obsessions are answered with compulsions. (Note: both obsessions and compulsions are intrusive, with both having obsessional and compulsive qualities.) Though initially successful in reducing anxiety, these compulsions gradually become intrusive since the acts have to be performed *compulsively*.

9. Eventually, the anticipatory power of the intrusion becomes so overwhelming that reality testing is disturbed. She does not know anymore whether she has or has not strangled her baby. Thoughts may become *delusion-like*, with psychotic features.

Assessment and Domains of Intrusive Events

Because intrusive thoughts occur in healthy individuals as well as pathological states, it is important to develop sensitive tools to quantify these experiences. Healthy individuals experience unwanted intrusive thoughts (e.g., of dirt, contamination, doubt, harm, injury, sex, religion, order, symmetry, superstition) that are structurally and content-wise similar to clinical obsessions in OCD patients (Freeston et al. 1991; Langlois et al. 2000a, b) but less severe and disruptive. Intrusions vary in both the structure and content across individuals (Clark and Inozu 2014), as well as in terms of frequency, intensity (or distress), the degree to which the event is being perceived as an intrusion, unexpectedness, persistence (duration), controllability, vividness, valence (positive vs. negative), adhesiveness (durability), and modality (verbal vs. imagery based). These features provide dimensions that can be used to quantify intrusive events and to relate them to the underlying neurobiology. Thus, a refined multidimensional quantitative assessment of an intrusive event is critical for developing quantitative models and to assess the efficacy of interventions.

As reviewed by Clark and Purdon (1995), a number of investigators have developed questionnaires to assess intrusive events. These questionnaires distinguish between intrusive events that appear to be triggered by external stimuli and those that occur spontaneously. In one analysis, it was estimated that approximately 80% of intrusive events are provoked by an external trigger (e.g., Edwards and Dickerson 1987a). Clark and Inozu (2014) note that whereas intrusive events in nonclinical samples are context dependent (i.e., have external precipitants), clinical obsessions in patient populations appear to be more spontaneous. They also claim that avoidance of triggers in patients is associated with more adverse impact of both intrusive events and obsessions. Clark and Radomsky (2014) identify several aspects that need to be considered in assessing intrusive events:

- Content and process characteristics (the degree to which the event is unwanted)
- Discriminant validity
- Relationship to measures of worry and/or rumination
- Degree of self-relevance
- Appraisal variables (e.g., controllability, unacceptability, discomfort, guilt, dismissibility, unpleasantness)
- Degree of personal responsibility

Investigations of intrusions have taken several approaches, including questionnaires, diaries, and procedures that assess the impact of intentional mental control on unwanted intrusive thoughts. Each approach has limitations. The most common approach has been self-report questionnaires:

- The *Experience of Intrusions Scale* is a five-item measure that assesses the frequency, unpredictability, and unwantedness of intrusive thoughts, as well as the interference and distress caused by the intrusions, each on a five-point Likert-type scale (Salters-Pedneault et al. 2009).

- The *Interpretation of Intrusions Inventory* consists of 31 items that refer to interpretations of intrusions that have occurred recently. Three of the above domains are represented: importance of thoughts, control of thoughts, and responsibility (Obsessive Compulsive Cognitions Working Group 2001).

- The *Obsessive Beliefs Questionnaire* focuses primarily on obsessive intrusion (Obsessive Compulsive Cognitions Working Group 2003).

- The *Obsessive Intrusions Inventory* consists of a 52-item self-report instrument designed to assess intrusive thoughts, images, and impulses that are similar to the aggressive, sexual, and disease-related thinking characteristic of clinical obsessions (Purdon and Clark 1993).

- The *Cognitive Intrusion Questionnaire* examines the following domains: frequency, duration, percentage of verbal and image content, interference, ego-dystonic nature, stimuli awareness, and associated emotions (Langlois et al. 2000a).

- The *Cognitive Intrusions Questionnaire–Transdiagnostic Version* items are grouped based on theoretical criteria into categories labeled intrusiveness, appraisals, emotions, and strategies, which were selected as components of a model that encompasses the different ways in which intrusive thoughts are processed (Romero-Sanchiz et al. 2017).

Whereas each of these questionnaires has a place in a specific context, they are limited as measures of intrusive events across diagnostic categories. Further, they were not designed with the goal of investigating the neurobiological basis of intrusive events. Some of the questionnaires include a broad range of negative thought content (e.g., anxiety and depressive thoughts) which are not, strictly speaking, intrusive events. Questionnaires are also limited because intrusions are often idiosyncratic and triggered by external cues that are difficult to describe: their dependence on retrospective self-report of unwanted intrusions may not be fully reliable. The questionnaires have not been used in the context of transdiagnostic, dimensional psychopathology, which are necessary to determine the heterogeneity of these events. In addition, they have not been designed to connect to the underlying neuroscience (i.e., the use of domains that can be mapped to specific brain systems), and the item response characteristics have not been rigorously assessed. Thus, future item bank-based questionnaires might substantially reduce subject burden by using an adaptive measurement framework similar to the PROMIS system (Cella et al. 2010).

An alternative to questionnaire-based approaches is the use of patient diaries. Clinicians using behavioral therapies for intrusive events typically ask patients to monitor their intrusions in a diary (Grey and Holmes 2008). In experimental studies with healthy volunteers, a range of diary measures for intrusive events have been developed so that participants can report on their intrusive experience in daily life (James et al. 2016c; Lau-Zhu et al. 2019). Recording modes can range from pen and paper, to an SMS or an online interface. Diaries have the merits of being able to be precisely tailored to the research question in mind; they are sampled in real time and are thus less prone to memory biases inherent in retrospective self-report. One could imagine that apps will become useful in this regard.

Although intrusive events may be a component of worry, simultaneous administration of worry and intrusive event questionnaires yielded different factor structures. Nevertheless, the factor structure for the strategies used to counter the thoughts were highly similar for both types of thought. Furthermore, regression analysis identified interesting relationships between the strategies, the thought characteristics, and appraisal of intrusive thoughts and worry (Langlois et al. 2000a, b).

It appears that concern about the personal meaning of the thought is a unique dimension for obsessive intrusive thoughts (Clark and Claybourn 1997). There is, however, fundamental disagreement as to whether clinical and nonclinical intrusive events can be conceptualized along a continuum. Some researchers focus on thought content: Belloch et al. (2004) found that the ten most frequently occurring thoughts were related to accident, harm, sex, and aggression. Others focus on the process characteristics, thus emphasizing the intrusive aspect of the thought (Rachman and de Silva 1978). Interestingly, Lee et al. (2005) found that the most upsetting intrusive thought is often autogenous; that is, such intrusions come abruptly into consciousness without identifiable evoking stimuli and are perceived as ego-dystonic, aversive enough to be repelled, and include sexual, aggressive, and immoral thoughts or impulses.

Intrusive thoughts are thought to be closely related to dysfunctional beliefs (Obsessive Compulsive Cognitions Working Group 2003). Thus, there is an urgent need to assess underlying belief systems associated with intrusive events. These include:

- Over-importance of thought: beliefs that the mere occurrence of an intrusive thought marks its significance.
- Need to control thoughts: beliefs that one can and should exercise complete control over unwanted intrusive thoughts, images, and impulses.
- Perfectionism: beliefs that a perfect response or solution to every problem is necessary and that even a minor mistake can lead to serious consequences.
- Inflated responsibility: beliefs that one is liable for causing and/or preventing significant negative outcomes for self or others.
- Overestimated threat: beliefs involving exaggerated estimates of the probability and/or severity of harm to self or others.
- Intolerance of uncertainty: beliefs that it is necessary to be certain and that unpredictability and ambiguity should be minimized as much as possible.

Another key aspect of intrusive thoughts relates to an appraisal, or how the intrusive events relate to important goals, values, and concerns. Appraisals follow intrusive thoughts and have been linked to future increased frequency. Appraisal can affect the ability to dismiss the thought, guilt, uncontrollability, and belief that the thought could come true. It could also affect responsibility, perceived consequences of the thought, beliefs about the importance of the thought, and worry that the thought may reflect something about one's personality (Berry and Laskey 2012). Moreover, in response to intrusive thoughts, individuals frequently engage in the following strategies:

- Reasoning that focuses on the thought being irrational or unimportant
- Thought replacement geared to distract or stop the worrying
- Social support, talking through, reassurance seeking, physical action, or doing nothing

While the themes of intrusive thinking are similar across clinical and nonclinical populations, the appraisal strategies in clinical populations seem to focus on responsibility and subsequent avoidance strategies. Results indicate that the more distressing a thought was perceived to be, the more likely participants were to recommend unhelpful strategies (Bomyea and Lang 2016). Conversely, the less distressing an intrusive thought was, the more likely participants were to recommend helpful strategies (Levine and Warman 2016).

Consistent with the appraisal model, Purdon and Clark (1994) found that the belief that one could act on the intrusive thought and a perceived uncontrollability of the thought were important predictors of the frequency or persistence of the distressing intrusion. Freeston et al. (1991) note three distinctive, dominant response styles: (a) no-effort response (26%), (b) attentive thinking (34%), and (c) escape or avoidance (40%). Moreover, they found that intrusions eliciting escape-avoidance strategies were evaluated more disapprovingly than thoughts eliciting attentive thinking. Rachman (2014) pointed out, however, that the following issues still need to be addressed:

- Additional information is needed on prevalence in clinical/nonclinical samples.
- The variable content of intrusions needs to be examined as a function of culture and environment.
- The nature and effect of repugnant versus nonrepugnant intrusions require examination.
- Further research is necessary on intrusive images and other percepts.
- Experimental investigations of intrusions are needed.
- Randomized clinical trials are needed to examine the effect of different treatments on intrusions.

Taken together, it will be important to assess appraisal domains and to quantify response styles associated with intrusive events.

Following work and suggestions by others (summarized above), one approach to measure and quantify intrusive events is to decompose the experience into different domains or dimensions. For example, an intrusive event brought on by a ruminative thought that refers to a past experience associated with a strong negative emotion could be quantified along the following dimensions:

1. The degree to which the intrusive event (e.g., a short utterance or an elaborate verbal instruction) is characterized by a verbal thought
2. The degree to which the intrusive event (e.g., a vivid image, smell, sound, or other internal perceptual experience) is characterized by an internal sensory representation in the absence of a percept
3. The degree to which the intrusive event is associated with a positively or negatively valenced affect (e.g., severe anxiety, guilt, or shame), and the quality of that affect
4. The degree to which the intrusive event refers to an experience in the past, present, or is focused on possible future events

This approach would allow us to characterize the degree to which an intrusive event recalls a distant or recent past event or refers to an immediate or remote future event. For instance, the intrusive image experienced by an individual with PTSD might be characterized as relatively low intensity on Pt. 1, high intensity on the sensory representation (Pt. 2), with an association of guilt or shame (Pt. 3), and related to the past (Pt. 4). In comparison, an intrusive thought (worry) in an individual with generalized anxiety disorder might rate high intensity on Pt. 1, low intensity on Pt. 2, with a focus on anxiety (Pt. 3), and a primary focus on the future (Pt. 4). Such a decomposition could be used in large-scale surveys to begin to delineate the frequency of the phenomenology of intrusive events and the association of these events with a particular disorder. Subsequent statistical analyses (e.g., latent variable analyses and cluster analyses) could then be used to develop an empirically derived taxonomy of intrusive events. This approach would permit the severity of the intrusive event to be quantified and could be an outcome measure for the success of intervention studies.

Finally, this type of decomposition lends itself as a covariate for neuroimaging studies to delineate the neural circuitry associated with intrusive events. For instance, it would be extremely interesting to determine whether individuals who have experienced predominantly visual sensory intrusive events show changes in activation patterns in the visual processing stream, such as the occipital cortex, compared to individuals with similarly intense intrusive verbal thoughts hypothesized to show primary changes in the left ventrolateral prefrontal cortex.

How could these domains be neurologically implemented? In Appendix 17.1, we present a model that decomposes intrusive events into domains and then considers how these domains might be instantiated in the brain.

Treatment Approaches

To date, most existing interventions consist of treatments that address each psychiatric disorder as a whole, rather than intrusive events. One notable exception is an innovative behavioral treatment for bipolar disorder, which specifically targets mental imagery-based intrusions through imagery-based cognitive therapy (Holmes et al. 2019). This same treatment has shown promise in reducing overall bipolar mood instability (Di Simplicio et al. 2016; Holmes et al. 2016a). In addition, a preventive intervention has been developed for PTSD that directly targets intrusive image-based memories of an experienced trauma (Iyadurai et al. 2018). The development of treatments follows three primary modalities:

First, *pharmacological treatments* exist for most psychiatric disorders that involve intrusive thoughts, although their efficacy varies widely across individuals and stages of the disease process. Selective serotonin reuptake inhibitors (SSRIs) are the first line of treatment for a broad range of disorders, from OCD to eating disorders, major depressive disorder, and PTSD. Patients who do not respond to SSRIs may be treated with serotonin and norepinephrine reuptake inhibitors (SNRIs), mood stabilizers, or anticonvulsants. Beyond these classes of drugs, there are numerous experimental treatments (e.g., hallucinogens, ketamine, atypical antipsychotics, and opioid drugs). We note that both SSRIs and hallucinogens, such as psilocybin, DMT and LSD, operate primarily on the serotinergic pathway: SSRIs increase the presence of serotonin in the synapse, whereas hallucinogens bind to the 5HT-2a receptor as a serotonin agonist. In both cases, a postsynaptic neuron will respond as if more serotonin is present. Some neural network modeling has suggested that an increase of serotonin in a network can facilitate the dynamic firing patterns of that network from getting stuck in local minima of processing behavior. There is a growing popular movement, albeit with limited empirical evidence, of using low doses (i.e., microdosing) of psilocybin and other hallucinogens to treat mood disorders, as well as high doses of psilocybin and ayahuasca to treat depression (Pollan 2018). For SUDs, a different class of pharmacological treatments exists: drugs used to treat SUDs are typically specific to the class of abused drug. For example, transdermal or oral nicotine may be used for smoking cessation, and opioid partial agonists are used for opioid disorder. Generally, these drugs take the form of agonists, partial agonists, or antagonists that target the receptor system where the drug acts. The idea is that the drug alters the receptor function so that the drug-related stimulus loses its incentive value. The intrusive event in SUDs often takes the form of strong cravings or urges to use a drug. Thus, drugs under development to treat substance abuse address these cravings as well as the actual drug use.

Second, *behavioral interventions* include a broad range of procedures: CBT, whose techniques include *exposure* to salient stimuli and reminders, cognitive restructuring, as well as mindfulness and contingency management for SUDs, among others. Typical treatment targets are (a) to lessen reactivity to triggers that initiate the intrusive event or to update the underlying memory (e.g., after trauma) and (b) to develop strategies to control the user's response to the intrusive event. Treatments used to lessen the initial impact and/or update the memory include exposure therapies or repeated experiences with the thoughts or images in a safe environment. This may be seen as akin to extinction procedures in operant conditioning. A new form of treatment technique is being developed by Holmes and colleagues that targets the memory aspect of an intrusive event by interfering with consolidation or reconsolidation of the memory of a traumatic event (Holmes et al. 2009, 2010; James et al. 2015; Horsch et al. 2017; Iyadurai et al. 2018; Kessler et al. 2020). Treatments used to control the responses include cognitively reframing the meaning of the eliciting event, developing competing behavioral strategies that are incompatible with the immediate response, utilizing social support to minimize the emotional response, or learning to decrease emotional reactivity. For example, mindfulness training has been shown specifically to decrease habitual reactivity to intrusive events (e.g., craving for cigarettes and food) and has led in some cases to significant reductions in unwanted behaviors: five times the quit rates in smoking cessation and 40% reduction in craving-related eating (Elwafi et al. 2013; Brewer and Pbert 2015; Brewer et al. 2018; Garrison et al. 2018; Mason et al. 2018).

Third, *nonpharmacological brain interventions* include neuromodulation such as transcranial magnetic stimulation (TMS), deep brain stimulation (DBS), lesions to specific brain areas (e.g., anterior cingulate cortex), and electroconvulsive shock. These techniques have been used under limited circumstances and in highly selected patient populations. DBS, for example, is used only in severe cases of OCD (Tyagi et al. 2019). DBS targeting the ventral limb of the internal capsule and the nucleus accumbens is an effective treatment strategy for treatment-refractory OCD. TMS is approved to treat depression (Mutz et al. 2018) and is under study for the treatment of OCD (Rapinesi et al. 2019), Tourette syndrome, PTSD (Kozel et al. 2019), and SUDs (Zhang et al. 2019).

In many cases, pharmacological, behavioral, and neuromodulatory modalities are used in combination in various forms. Drugs are used to facilitate the psychotherapeutic process, such as MDMA and psychotherapy for PTSD (Mithoefer et al. 2019), whereas behavioral treatments are used in combination with nicotine replacement therapy for smoking. These different treatments may occur simultaneously or sequentially. In one case, the FDA has approved a treatment for depression using all three modalities: TMS, SSRI, and behavioral treatment. As noted, we have no information about the effects of these treatments specifically on the frequency or severity of intrusive events, separate from other symptoms of the disorder, and future research is needed here.

Future Interventions for Psychiatric Disorders

As discussed, very few treatments exist that explicitly target intrusive events as an aspect of a disease, and for those that do exist, it is not clear how they actually modify intrusive events. Although intrusive events are included in the DSM-5 criteria for many disorders, they have not yet been targeted as a distinct treatment domain.

Treatment innovation is essential and may benefit from being mechanistically driven and by combining treatment modalities, such as combining pharmacology with a psychological/behavioral approach (Holmes et al. 2018). Before going into detail about specific treatment modalities, we discuss how interventions that target intrusive events might be used to modify core aspects of an intrusive event: gating, error correction, salience, and evaluation.

Gating Errors

Within psychiatric diseases, intrusive events may result from too much information being allowed into awareness, either in a global sense or by a selective memory gaining access to awareness through a selective gate. This broad construct could be applied to many psychiatric diseases. Intrusive verbal thoughts associated with general anxiety disorder (see earlier discussion) may be due to a "weak" gate that allows a lot of information in and ultimately leads to stress and anxiety. Intrusive image-based thoughts associated with PTSD may be due to a faulty assignment of salience and/or evaluation that emerges for the traumatic event, which allows them to enter into or persist in awareness more easily than nontrauma-related thoughts. Intrusive thoughts of drug cues in SUD may be construed as a combination of faulty gating and salience, compounded by reinforcement learning.

Error Signal/Detection

Intrusive events may also result from a heightened error signal during otherwise normal thought. To achieve ordinary activities of daily living, mental processes must stay on course and not be derailed by irrelevant streams of information that constantly come into our brain. A lot of information is processed on a subconscious level, wherein only the stimuli that are most different from our expectations are processed consciously. For example, when walking down the street, an individual is able to maintain posture and balance, typically via subconscious processing that has become highly automatized. Only when an event occurs that violates expectations (e.g., when we stumble or suddenly see an unexpected object in our path) is the walking process brought to the level of consciousness via bottom-up mechanisms. Alternatively, our own cascade of ongoing neural processing related to other aspects of life, which are not directly associated with the goal, can also interrupt automatic behaviors, as when we suddenly remember that we were supposed to pick up the kids at school. Thus, bottom-up and top-down internal interrupts exist as well as external interrupts. The difference between our expectations of internal and external stimuli and the actual perception of these stimuli can be considered a prediction error.

It is possible, for example, that verbal or visual intrusions in PTSD may be related to a momentary yet large prediction error. That is, at times, the perceived difference between the thoughts associated with the trauma and the typical subconscious narrative that is ongoing in an individual's mind is large (high prediction error) and triggers the individual to switch their attention to those momentary thoughts. This is a particularly attractive hypothesis given the sparse temporal nature of intrusive events in these patients. In daily living, for instance, many individuals with PTSD are able to function relatively well in between intrusive events of their trauma. They are often able to work and care for families, and to conduct their lives for some days without having an intrusive event. The frequency of occurrence of an intrusive memory of trauma for some patients may be infrequent (e.g., once a fortnight) whereas for others with more severe levels of PTSD, it may be up to every hour. Typically, measurement tools of PTSD capture this rate of occurrence, although more research is needed.

Here we address emerging ideas for future treatment innovation based on the model proposed in the Appendix. Assuming that these reward prediction errors are coded in a specific neural network, a technology capable of sensing the magnitude of the violation between expectations and actual input may be able to change activity in this network in a dynamic manner, pushing it back into the intended state. This type of closed-loop neuromodulation is currently used in the treatment of epilepsy. Briefly, sensing and stimulating electrodes are placed in the brain of individuals with intractable epilepsy in the vicinity of the seizure focus. The device has a certain "tolerance" for background variability in the neural activity in the vicinity of the electrode. Once a critical level of variance is detected (high variance), the device is able to stimulate the brain and push it back into a healthy state (low variance), thus avoiding the cascade of a seizure (for further discussion, see Hanlon and McTeague, this volume). It is easy to see how a similar approach could be used to abolish intrusive events in a dynamic manner in individuals with PTSD. In veterans with PTSD, for example, a device could be trained to identify the "background" levels of activity present in normal life as well as activity associated with intrusive war-related memories, and permit the device to push the system back into the healthy range of error. One problem with this approach, however, is that large prediction errors are a crucial element for flexible human behavior. So, an autonomous closed-loop system would have to be able to detect differences specific to the trauma memory. While this seems like a tall order, because these intrusive events are so debilitating to the patients, it is reasonable to imagine that the amplitude of their prediction error is so large that the device would be able to have a very high tolerance threshold, and thus only produce stimulation during the most extreme examples of prediction errors. Alternatively, individuals could be trained to self-administer the stimulation when they start to become aware of an intrusive event. More basic neuroscience research with a fast temporal sampling profile (e.g., EEG, MEG, *in vivo* recording) is necessary to evaluate if prediction error is a viable treatment.

Evaluation

One of the most widely implicated behavioral domains thought to be responsible for the maintenance of intrusive events is an aberrant evaluation system. Evaluation brings together goal hierarchy (including the current task, homeostatic goals, etc.), salience, affective, and reinforcement properties of previously learned behaviors to determine whether to stay on a certain task or switch. In terms of intrusive events, which can be seen as an alternate task, evaluation would help to determine whether to switch, and for how long, to the new task. Clinically relevant downstream manifestations of the evaluation system include frequency and duration of intrusive events in consciousness. In light of emerging understanding with regard to intrusive events, these manifestations can now be linked to specific neural systems, informed by what is known with current treatment paradigms. For example, treatments such as CBT target the cognitive elements of intrusive events, theoretically changing belief systems related to intrusive events (e.g., "Is this true?"), whereas mindfulness training targets the relational affective component of the intrusive event (e.g., "How caught up am I?"). These examples provide concrete treatment modalities that can be more critically studied with regard to efficacy. For example, does mindfulness training (as a modality to improve evaluative accuracy) change the "stickiness" of an intrusive event, and does this change the frequency, duration, and salience of the intrusive event in the future (see Brewer et al. 2013, 2018, 2019)? Further, the neural systems that are affected by these interventions can be more specifically studied. For instance, recent work with mindfulness training has linked reduction in default mode network activity with a reduction in cigarette smoking (Janes et al. 2019), yet the frequency and duration of intrusive events related to smoking have not yet been evaluated. Future studies can use experience sampling or other modalities to determine if intrusive events reduce in either of these domains.

A Neuroscience-Derived Neuromodulatory Approach to Treatment

Pharmacological treatments may change the sensitivity of the gate that allows items to enter into conscious awareness. For example, benzodiazepines may blunt emotional response to negative thoughts in a nonselective manner. Alternatively, there may be potential for combination treatments that couple a behavioral approach (e.g., one that evokes intrusive events and thoughts, such as cue exposure), to open the gate, with a pharmacologic agent, to blunt the response.

Imagine if one could simply "zap" the brain and magically remove the intrusive event causing the distress. Such an idyllic scenario is (unfortunately) still far away, but promising research is moving us closer. A whole new field of treatment research has emerged around specific brain stimulation techniques due to their noninvasive nature and simple theoretical motivation. The central idea behind brain stimulation and modulation techniques takes a systems perspective, which considers the brain as the source of any behavior (the outcome of brain processes). Behavior is therefore the dependent variable whereas brain circuits, activity, or neurons are the independent variables. If we are to change a pathological behavior, this systems-level interpretation implies that we have to find, target, and modulate the brain process or instantiation that is generating this specific behavior.

Two main approaches have produced important results in recent years: TMS and neuroimaging-based neurofeedback. TMS is a technique that generates weak electrical currents within the brain through direct application of electromagnetic flux over the scalp, noninvasively into brain tissue (see also Hanlon and McTeague, this volume). At the electrophysiological level, TMS induces changes in neuronal excitability, which can cause changes in behavior. TMS has been recently approved by the FDA for use in the treatment of several psychiatric disorders, such as addiction and OCD. Future studies need to evaluate how TMS in combination with other therapeutic forms might be used specifically to resolve intrusive events.

In neuroimaging-based neurofeedback, "neurofeedback" is defined as a closed-loop procedure whereby online feedback of ongoing neural activity is given to the participant for the purpose of self-regulation. Noninvasive neurofeedback can be implemented with several neuroimaging modalities. Some of the most promising proof-of-concept cases derive from functional magnetic resonance imaging (fMRI) and electroencephalography (EEG)-based neurofeedback (Sitaram et al. 2016; Taschereau-Dumouchel et al. 2018a; Keynan et al. 2019). We will first consider fMRI, due to its potential as an acute intervention (used once or a few times).

With fMRI, one simple option is to use the overall signal strength in one predefined brain region. This may work well when the target is general (e.g., motor action initiation vs. no motor activity); in other cases, however, one needs to access a specific representation, which by definition would not be retrieved through the overall activation level. Therefore, rather than focusing on the overall activity level of a region (*unspecific*), machine-learning algorithms now allow us to infer the precise activity *pattern* that corresponds to a unique stimulus, object, or category (*specific*). This approach, borrowed from machine learning, is termed multivoxel pattern analysis (MVPA) (Kamitani and Tong 2005; Norman et al. 2006). Going even further, methods have now reached the point where we can infer the pattern of brain activity in a target participant from brain activity patterns in surrogate participants (Haxby et al. 2011).

Although these advances have often remained outside the realm of clinical applications, recent innovative efforts have been directed toward utilizing MVPA in the field of neuropsychiatric disorders. While mainly centered around the development of markers for certain disorders or their subtypes, with some remarkable results (e.g., Yahata et al. 2016; Etkin et al. 2019), some studies have explicitly targeted the prediction of intrusive events from fMRI activity patterns (see Holmes et al., this volume; Clark et al. 2014b, 2016).

Bringing MVPA to a real-time setting leads us to the concept of neurofeedback, which essentially monitors brain activity patterns over time while using the machine-learning prediction as neuromodulatory input (e.g., to provide rewards). If an algorithm is able to detect intrusive events, the system may then provide a reward to remodel the association between brain state and appraisal or some other form of modulation to disrupt the cascade of neural events following the emergence of the intrusive event.

In terms of clinical development, due to the intrinsic nature of MVPA-based designs, neurofeedback interventions can be easily used in a double-blind, placebo-controlled way. Since we can find activation patterns that correspond to specific, distinct mental representations, we can also let the algorithm randomly choose which representation will be the target and which the control. Neither the experimenter nor the patient has to be aware of the category for the software procedure to be deployed effectively.

Research has shown that neurofeedback may work via reward processing, learning, and control networks (Sitaram et al. 2016), and that it depends on reinforcement learning processes (Shibata et al. 2018). During a typical experiment, the machine-learning algorithm monitors activity in a selected brain region and, at predefined time intervals, computes a (monetary) score reflecting the likelihood that the current activity pattern resembles a template, target mental representation. Over time, the brain learns to associate the occurrence of such representation with rewards.

In two proof-of-concept studies, this technique was used to reduce fear responses in healthy individuals conditioned toward simple visual stimuli (Koizumi et al. 2016) as well as in participants with subclinical phobia toward their feared object (Taschereau-Dumouchel et al. 2018a). The physiological fear responses (amygdala reactivity and skin conductance) were diminished only for the targeted representation, while a control stimulus elicited unchanged fear responses. More recently, the same group reported on the feasibility of using decoded neurofeedback as a target treatment in PTSD (Chiba et al. 2019). Importantly, throughout these studies, participants were not told about the link between their brain activity and the amount of reward they received on a trial-by-trial basis. When asked to make a forced choice about the target and control categories, participants answered randomly (Shibata et al. 2018).

The aspects introduced so far raise one crucial point: the entire neurofeedback intervention can be applied without ever mentioning or displaying the upsetting content or event to the patient. The enormous advantage that this could bring is evident: reducing implicit, physiological fear responses could create the foundation for subsequent, and more effective, behavioral or cognitive therapies. Neurofeedback could then, as a second step, be applied for maximal effects and learning.

Alternatively, neurofeedback modulation may provide the basis for dampening a physiological, autonomic reaction, which may then facilitate subsequent cognitive or behavioral therapies. Current work on fMRI, MVPA-based neurofeedback can be expanded to include the notion of dynamic brain state. Indeed, utilizing a very simple correlation of activity fluctuation between two brain areas, or a pattern of activity representing an object or category, may be too simplistic, in particular if we are to target states that are more global (e.g., attention, emotions, arousal, interoception). Spatiotemporal oscillatory patterns describe relatively well these global states, distinguishing between rest and task-based efforts (Vidaurre et al. 2017; Bolton et al. 2018).

Rather than acting on the intrusive events themselves, dynamics-based neurofeedback interventions could target and modulate an overall affective or cognitive state. Here, the rationale is different: prepare the brain to receive the treatments targeting specific aspects of the pathology in the individual.

fMRI neurofeedback carries high costs, immobility, and the requirement for specialized operating personnel. This greatly hampers the scalability of this approach, particularly if we think about going beyond acute treatments.

As such, EEG-based neurofeedback holds potential from a different perspective. New generation EEG headsets have relatively low costs, are small in size, and are almost "plug-and-play" ready for use. We can envisage EEG products in the near future that could be used autonomously by patients on a daily basis and essentially without many constraints in terms of location or function, allowing for real scalability. Proof-of-concept EEG neurofeedback has been demonstrated with specific mental states while targeting deep brain structures, such as the posterior cingulate cortex which is part of the default mode network (van Lutterveld et al. 2017), training stress resilience through electrical fingerprint (Keynan et al. 2019), as well as reduction, consolidation, and personalization of lead placement (Pal et al. 2019).

A different form of feedback might therefore involve generating an external interrupt when an undesirable thought pattern has emerged. For example, MVPA of EEG signals might allow ruminative thoughts associated with depression to be decoded. If detected, an EEG headset could then be designed that might vibrate or emit a tone whenever rumination had exceeded, say, ten seconds, bringing the patient back to the present.

Combination Treatments

A significant future direction in the treatment of intrusive events is the possibility of combining treatment modalities. The interventions used in the treatment of intrusive events fall broadly into three categories: behavioral, pharmacological, and neuromodulatory. Some of these treatments have already been combined. For example, MDMA is currently under investigation for the treatment of PTSD, but a central feature of the treatment is that the drug is administered in the context of psychotherapy. Other examples include the use of behavioral treatments combined with nicotine replacement therapy for smoking cessation or other pharmacological-behavioral combinations for treatment of other addictions.

An Innovative Approach to Targeting Intrusive Events in Trauma-Related Disorders

In PTSD, trauma is induced through vivid, emotionally laden intrusive memories that take the form of sensory multimodal mental images of specific moments experienced during the traumatic event. Current treatments are available to reduce these memories or their consequences (e.g., trauma-focused CBT), but they can be hard to implement on a large scale (to reach more people), and evidence-based preventive interventions after trauma are lacking. Recently, Holmes and colleagues developed a brief, noninvasive behavioral treatment that reduces both the establishment and the maintenance of intrusive memories after a traumatic event (Iyadurai et al. 2018). It should be noted that the treatment focuses on specific intrusive memories, rather than the whole disorder of PTSD. The type of intervention developed may be applied either shortly after the trauma, or later, well after the memories have been established (Holmes et al. 2009, 2010; James et al. 2015). The procedure consists of the following steps:

1. Subject is instructed to recollect the event briefly via images and thoughts.
2. Subject participates in a 15-minute *Tetris* task that involves mental rotation.

This simple procedure can be implemented shortly after the actual event or after a retrieval of the event later in the process. It may be administered in a hospital emergency room or acute ward within six hours of the traumatic event (Horsch et al. 2017; Iyadurai et al. 2018). In these studies, the intervention reduced the number of intrusive memories in the following week by approximately two-thirds. It can also be used to treat intrusive memories of traumatic events that occurred many years ago. In one study (Kessler et al. 2018), inpatients with complex PTSD received the treatment with intrusive memory, resulting in a reduction in the frequency of those intrusions (compared to nontargeted intrusions). Further research is required.

This novel treatment approach is derived from cognitive psychology and experimental psychopathology related to memory processes. Performing a visually demanding task shortly after encoding or retrieving a remembered mental image, while holding the image in mind, competes for working memory resources in a way that interferes with consolidation/storage (during the initial experience) or reconsolidation (after the image is retrieved later). Remarkably, performing this simple *Tetris* task immediately after thinking about the image reduces the frequency of intrusive mental images (Iyadurai et al. 2018). The technique is based on various assumptions:

- Intrusive memories of trauma comprise sensory mental imagery (Grey and Holmes 2008).
- Intrusive memories can be altered shortly after an event or at retrieval: memory consolidation/reconsolidation (Visser et al. 2018).
- The capacity of people's working memory is limited (Baddeley 2003).
- Visuospatial tasks compete for resources in working memory with a mental image; that is, those that would be needed to (re)consolidate intrusive mental images (James et al. 2015).

Thus, engaging in a visuospatial task, such as a highly visually demanding computer game like *Tetris*, at a time when the mental image of the intrusive event is active may reduce the reoccurrence of distressing images later, as well as the level of distress and vividness associated with them. Further research is needed to develop this relatively simple and brief behavioral intervention approach. It lends itself to be studied as part of a combination treatment approach, as discussed above. The difference between this theoretical approach and many others is the focus on a single symptom (intrusive events) and a neuroscientific account about the underlying mechanisms bringing about effects.

Implications Beyond Neuroscience and Psychiatry

Philosophical and Social Implications of Definitions

Although the original definition by Clark (2005) and its minimum and maximum derivatives are intuitively very recognizable, the appearance of clarity is misleading. The problems that come to light reflect some important philosophical and social assumptions.

What exactly is meant by unwanted? Is it a mental event that is unwanted by the individuals themselves? Or does this represent what is unwanted by members of society, relative to what counts as unacceptable, abnormal, or even moral or immoral (e.g., expressions of sexuality or aggression)? Is what is unwanted not wanted relative to short-term individual goals, such as wanting to drink water, eat a marshmallow, or court a colleague? Or is what is unwanted defined relative to interference with the fulfillment of long-term goals, such as completing a PhD program or maintaining a marriage? Are intrusions perhaps unwanted in light of the conflicts they present to a person's ideal image, such as becoming a good scientist? Certainly the notion of being unwanted depends on a broader philosophical perspective: What kind of human being do we have in mind as our ideal? Are mental phenomena and interventions that foster this ideal more wanted than those that lead away from this ideal? In sum, there are multiple ways to be unwanted, whether by the individual or society. Should we privilege one type of being unwanted over another?

Inherent in the definition of intrusive thinking is the notion that a lack of control over the flow of mental events is unwanted. The implication is that intrusive mental events are unwanted at least in part because they are not subject to volitional control. Overcoming a loss of control, in turn, implies that mental events could be made subject to volitional control. But what makes a mental or neural process volitional versus nonvolitional? In the motoric domain of eye movements, for instance, changes in pupillary size (or nystagmus or microsaccades) seem to proceed automatically, regardless of intentions or plans held in working memory or reportable by a subject. In contrast, saccades and smooth pursuit eye movements are subject to flexibly updateable plans or intentions held in working memory, which in turn can be reported by a subject (e.g., "I was looking around for my child"). Implicit in this is the notion that different intentions would lead to different eye movements. Similarly, volitional control of mental events implies that different mental events could and would have arisen had intentions or plans been different. There seems to be an implicit assumption here that events could have turned out otherwise than they did, had volitional control of actions or thoughts been different.

This raises the age-old problem of free will. If, under a deterministic worldview, mental events could not have turned out otherwise, then whether a particular intrusion would happen at a given time is dictated by the laws of physics before one was even born. The occurrence of involuntary intrusions seems consistent with the possibility that they could not have turned out otherwise. It is the notion that they can be brought under voluntary control that may be undermined by determinism. But how would indeterminism save the possibility of voluntary control of mental events? It might seem that randomness is as little subject to agentic control as events determined to happen before one was even born. Is there a middle path between the apparent lack of voluntary control that arises either under determinism or utter indeterminism? If yes, what might that solution be?

Our definitions of intrusion demand that we specify what it means to be clinically relevant. No absolute answer is possible, only one relative to a given clinical approach, of which there are many. Moreover, relevance depends not only on the clinical picture, but also on the person being treated, as well as the diagnostician and the social context or culture in which the patient and diagnostician find themselves. This implies that it is never possible to determine with objective certainty what an intrusion means in the psychopathological sense. It is therefore impossible to find a single neurobiological substrate of an intrusion without first specifying with great precision what exactly one means by seemingly basic terms such as "unwanted," "control," "volition," or "clinically relevant."

This definition of an intrusion refers to the fact that an intrusive mental event should be distinct or temporally punctate. We may try to apply this definition to "mind wandering," which is sometimes considered an intrusive event. Mind wandering, however, is often difficult to distinguish from regular thought patterns or normal free associations. It is certainly not punctate, but rather durationally extended. Would we regard the transition to mind wandering as an intrusion?

Moreover, is an intrusion by definition negative? Not every intrusion is accompanied by a negative emotion. Some intrusions, such as love or manic thoughts, are experienced as pleasurable, or may be concurrently both wanted and unwanted. Further, some types of intrusion are wanted by the subject, but viewed as intrusive by others in society. For example, in a traditional Nepali village, where marriages are typically arranged, the statement "I have fallen in love" might be met with concern rather than happiness, whereas in the West this transition to a potentially obsessive mental state is commonly regarded as positive. Patients with bipolar disorder report experiencing highly positive mental imagery intrusions associated with their mania, but these may also be diagnostic for a major psychiatric disorder (Ivins et al. 2014). Interestingly, one study has demonstrated that positive intrusive thoughts can be induced even in healthy adults, suggesting that they are open to experimental manipulation and study (Davies et al. 2012).

Although our focus in this chapter has been on the neuroscientific basis of intrusive events, their clinical relevance, and interventions, consideration should also be given to the degree to which intrusive events affect our society or can be understood outside of a solely biological or medical perspective. Radomsky et al. (2014) found that intrusive events are experienced across a large variety of cultures. They concluded that there were far more similarities than differences across different cultural sites and that the contents centered worldwide on themes of contamination, aggression, doubt, blasphemy, immorality, sex, victimization, and miscellaneous intrusions. Culture seems to influence the content of intrusions to some extent but not its prevalence (Clark and Inozu 2014). Moreover, the relationship between appraisals, control strategies, and the frequency and distress of intrusive events appears invariant across countries. This has been confirmed by others, who conclude that there is a certain degree of universality regarding the prevalence of obsessive, dysmorphic, hypochondriacal, and eating-related intrusions across a variety of countries and cultural contexts (Pascual-Vera et al. 2019). Nonetheless, there are cultural differences in how intrusive events are interpreted. For example, according to Luhrmann et al. (2015), voices are experienced by schizophrenics in the United

States as primarily negative, in particular, as violations of thinking, whereas in India and Ghana they tend to be interpreted in a more positive light, involving relationships with the presumed speakers. Understanding intrusive events in different cultural contexts will constrain the degree of generalizability of any neuroscience-based model of intrusive events. Future research should examine individuals who report similarly frequent or intense types of intrusive events in different disorders, but who are members of different cultures and have varied responses to intrusive events. Such an approach is not unlike extracting different computational models from psychotic and nonpsychotic hallucinators (Powers et al. 2017). Taken together, intrusive events can be a fruitful and important topic of research that extends the biological sciences and can provide important information about how society should consider managing these phenomena in the future.

Of particular interest is the change in the nature of intrusions experienced by people in our modern societies. Those who came of age in the 1980s or earlier effectively grew up in an analog world. Since the 1990s, however, with the advent of personal computers, the Internet, and smartphones, our society has become digital. Because the last generation to be raised without digital devices is still alive, now may be an opportune time for this generation, which has experienced both the analog and the digital world, to reflect on the pros and cons of this societal and personal transformation.

Modern society is now deeply penetrated by mobile devices, which can be viewed as "intrusion machines." Screen media activity (SMA) is ubiquitous worldwide and among the most salient recreational activities of children and adolescents. Children and adolescents spend about 40–60% of their time after school engaged in SMA (Arundell et al. 2016), and nearly 97% of U.S. youth have at least one electronic item in their bedroom (Hale and Guan 2015). A heated debate has emerged on whether SMA is associated with psychological and social problems (Ferguson 2017; Twenge et al. 2017). However, media behavior is complex, encompassing a variety of activities, such as social and nonsocial Internet use, gaming, as well video or TV viewing. For example, whereas males are more likely to engage in video games with a higher potential for excessive use (Choi et al. 2015), females engage more with social media (Schou Andreassen et al. 2016) and exhibit more excessive cell phone use. Moreover, gaming has replaced sedentary screen time, such as TV viewing, Internet usage, and nonactive gaming (Simons et al. 2012). In the near future, with 5G technology, individuals may be able to interact with numerous devices on an almost constant basis. For instance, whereas currently a cell phone might signal the arrival of a text, email, or message (the intrusive event), we may in the future be alerted by our refrigerator, car, air-conditioning system, or other devices that we use in our daily life. Thus, technologically based intrusive events could have the potential to seriously affect healthy individuals and possibly to a greater extent any individual who is cognitively or affectively compromised (i.e., a person with a psychiatric disorder). Moreover, the constant engagement with SMA may affect both brain structure and function, and in turn make individuals more susceptible to the experience of both physiological and pathological intrusive events. The investigation of mobile technology on cognition is still in its infancy (Wilmer et al. 2017), and our knowledge about the impact of this technology on intrusive events is nonexistent. Thus, an important goal for future studies should be to determine whether such devices could contribute to the exacerbation of psychiatric disorders, characterized by frequent and/or severe intrusive events.

Another perspective on intrusive events and their pathology is the notion that an individual who experiences a low level of need for control might not experience intrusive events as problematic or in need of treatment. This raises the question whether the fact that we are focusing on neuroscience, clinical consequences, and interventions associated with intrusive events is a by-product of our society's focus on "over control." Interestingly, individuals with strong beliefs about controlling thoughts are more likely to experience distressing intrusions, both with and without meta-awareness, compared to people with weaker beliefs (Takarangi et al. 2017). Thus, it would be interesting to examine the frequency, severity, and clinical consequence of intrusive events in different societies that place different emphases on cognitive control. Along similar lines, an intrusive event might only be assessed as an unwanted perturbation if the individual has a concept of causes and effects, which is a historical consequence of the Enlightenment period of the seventeenth and eighteenth centuries. In prior societies there was a greater emphasis on a teleological framework within which experiences and events were interpreted through the lens of their potential function, end, purpose, or goal. In this context, an intrusive event may be experienced as something that is necessary to lead to a particular goal rather than an unwanted distraction. Thus, it may be interesting to conduct a historical literature analysis that focuses on the characterization of intrusive events before and after the Enlightenment period.

Let us return now to intrusive events and tie them in with the notion of frameworks of meaning. If efforts to suppress intrusive events tend to exacerbate intrusive events, then perhaps efforts should be made concerning how best to suppress such suppression, or, on the contrary, how best to facilitate the expression of intrusive events so that the salient issue so expressed can be processed in a healthy manner. Other cultures have created modes for the expression of "forbidden" emotions. Ancient Greek drama often centered on creating tensions that would then lead to emotional catharsis. Rather than suppress unwanted emotions, such as lust for forbidden objects of desire, as in the play Oedipus Rex, these emotions were vented in a manner that was safe for society. Even modern European cultures have aspects that are reminiscent of catharsis. For example, Carnival is a venue for the expression of carnal desires and behaviors that in other times would be regarded as deviant or dangerous. How might catharsis be exploited in the context of existing or new therapeutic methods? One possibility would be role-playing, where a traumatic event or idea is expressed in a manner such that emotions can be safely expressed. Indeed, by pretending to reenact a particular traumatic event or relationship, the possibility emerges of having it not only expressed, but of expressing it in a new way, perhaps having the event turn out differently than it in fact did. Another avenue might be to build on the tragedies and plays of Ancient Greece by creating "virtual worlds," perhaps exploiting movies or virtual reality technology, where pent-up and suppressed emotions and desires could be released virtually, rather than through real acts in the life of the patient.

According to this view, intrusive events are analogous to salience signals arising from exogenous attentional circuitry. Just as the sudden motion of a tiger demands an interruption of the current plan engaging conscious thought and planning, so that a new plan (in this case, to escape the tiger) can be generated, intrusive events are salience signals that enter consciousness because there is an unresolved issue that requires executive control circuitry to come up with a plan to resolve the unresolved issue. If this view has validity, then suppressing the salience signal might be about as effective as attempts to suppress hunger or thirst signals. The reason these signals barge into consciousness is so that goals can be generated by executive planning areas that will resolve them (e.g., coming up with a plan to get food or water). If unresolved emotional or cognitive issues barge into consciousness, a better approach than suppression may be to find ways to resolve the unresolved issue. Catharsis and role-playing have already been mentioned. Other possible methods may involve unorthodox techniques that are considered quite orthodox in non-Western traditions. For example, according to Kundalini yoga, emotions are stored in the body and can be activated with certain bodily actions, such as breathing patterns of physical exercises. By intentionally invoking the breathing pattern associated with fear or calmness, say, the mental state that normally accompanies such breathing patterns can also be invoked and perhaps processed in a manner subject to volitional control. If unresolved emotional issues are stored in part in the body, or in bodily patterns of action, perhaps the Western tradition can gain insights from other traditions concerning the "cleansing" of stored psychological tension and pain.

Conclusion

Intrusive events are emerging as an area of interest that can help further our collective understanding of basic brain function, psychiatric conditions, and their treatment. Characterizing domains in which intrusive events manifest in psychiatric disorders may help their characterization, diagnosis, and treatment. Basic heuristic models can inform how normal brain function can go awry due to faulty systems, including gating, salience, evaluation, and prediction error detection. From these models, current and evolving treatments (and potential combinations) can be tested for target engagement, specificity of effect and efficacy with regard to reduction of frequency, duration, and salience of intrusive events. Future research would benefit from mechanistically framed and neuroscience-based approaches with specified targets and tangible target engagement and clinical outcomes. Our proposed model (see Appendix 17.1) provides an example of one such emerging approach of a behavioral intervention that specifically targets intrusive events based on theories of memory (re) consolidation and cognitive task interference.

Appendix 17.1: Proposed Model

Our model begins with a sequence of goals that are represented in the brain (Figure 17.A1). A mental workspace keeps the present goal in mind, allows operations to take place over representations held in the workspace (area 4, working memory), and takes into account prior knowledge, context, and other system constraints. Salience helps the organism determine how important a potential interrupt is and interacts with an evaluative system to determine whether to stay on task or to interrupt it and, if so, which task to then prioritize as the task to do next. A goal maintenance system helps the mental workspace maintain the present goal using feedback loops that afford the minimization of prediction error signals. This may happen via the enhancement of gating of potential interrupt inputs, or the inhibition of the salience of potential interrupt signals. The goal maintenance system evaluates deviations from the trajectory leading toward fulfillment of the goal. Such prediction error signals are used dynamically and cybernetically to correct the present trajectory to minimize that error, similar to when a heat-seeking missile alters its path to hit its target.

External interrupts (not internally generated) may orient the organism to unexpected inputs from the external world (e.g., when we hear a loud sound or see a sudden motion): if the magnitude of the prediction error (i.e., between what was expected and what in fact occurs) is large enough, the organism orients to the external stimulus.

Internal interrupts can be both bottom up and top down. Bottom-up systems that can generate interrupts include systems that maintain physiological (e.g., hydration) and nonphysiological (e.g., happiness) goals that are separate from the current goal. When salient enough, these interrupts provide inputs to the mental workspace to force a reprioritization of what to do next (stay on task or switch tasks). Subjectively these signals are experienced as, for example, thirst, hunger, lust, or a need for oxygen, salt, or sleep. Other bottom-up systems include reward/punishment and other evaluative systems but may not have intrinsic homeostatic functions. These may be experienced as, for example, fear or other emotions, such as anger. In addition, the memory systems may automatically retrieve memories which can then appear to "pop" into consciousness.

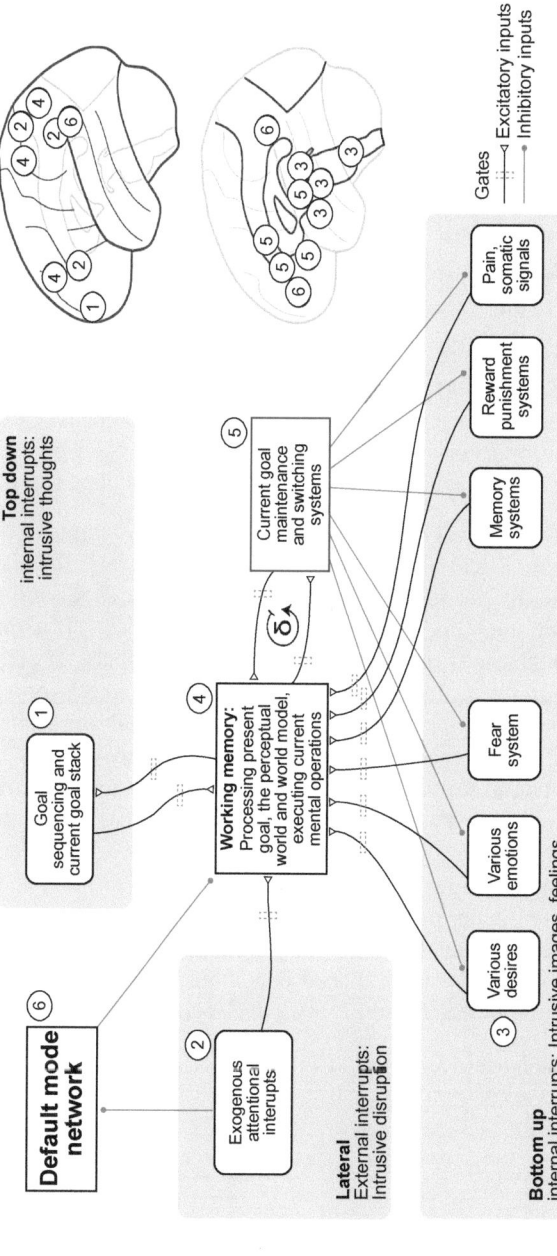

Figure 17.A1 (1) Neural substrates: frontopolar cortex (e.g., Brodmann area 10). (2) Ventral attentional network: ventrolateral prefrontal cortex, temporoparietal junction (3) Numerous subcortical systems including amygdala, hippocampus, nucleus accumbens and ventral tegmental area, brain stem, and thalamus. (4) Dorsal attentional network: dorsolateral prefrontal cortex, posterior parietal cortex, superior parietal lobule, and intraparietal sulcus. (5) Cingulo-opercular control network: dorsal anterior cingulate, frontal operculum, basal ganglia; for emotional regulation ventromedial prefrontal cortex (BA 12) and orbitofrontal cortex (BA 11). (6) Default mode network: anterior medial prefrontal cortex, posterior cingulate, and temporoparietal junction.

Thus, under normal conditions, a goal is maintained using gating, evaluation, and prediction error monitoring and minimization. Under normal conditions, gating is appropriate, evaluation is accurate, and prediction error is accurate and dynamically corrected. Under pathological conditions, gating can become too weak or too strong, and prediction error and evaluation can become inaccurate. This model suggests that any one of these processes, or a combination of gating, evaluation, and prediction error, can go awry and thereby cause a pathological condition.

Failure Modes

Pathological conditions may arise when an individual fails to stay on target in the pursuit of the current goal because

- the prediction error (Figure 17.A1, between areas 4 and 5) does not arise (e.g., one should feel guilt for a misdeed, but does not) or it is inaccurate,
- evaluation (i.e., signals generated from the different components of Figure 17.A1, area 3) becomes inaccurate, or
- prediction errors do not get eliminated after a course correction takes place.

A system can have a number of modes of failure when one attempts to fulfill one's goals (see Table 17.A1). For instance, in a normal state, if the current goal is to go to a movie, the following thought might arise as a bottom-up interrupt: "You didn't study enough for your exam next week!" If gating is appropriate, that thought need not get into the mental workspace. If the gate is too strong, appropriate evaluation does not enter the mental workspace when it should, and one does not study. Once in the mental workspace, if the thought is deemed to be accurate ("Yes, you did not study enough"), the system changes its plan; namely, to study instead of going to see the movie.

Consider a case of anxiety, where the same thought arises: "You didn't study enough!" If gating is too weak, the thought arises too easily or comes in more frequently, moving it toward the spectrum of intrusive events. If there is an excessively strong gate, the thought does not enter consciousness and one does not feel a need to reevaluate the present plan, leading to the unhappy result that one does not study when indeed one should. Once in the mental workspace, if evaluation of the thought is inaccurate (you have indeed studied

Table 17.A1 Modes of failure.

	Normal	Pathological
Gating	Appropriate	Too weak, too strong
Salience	Appropriate	Too little, too much, etc.
Evaluation	Accurate	(In)accurate
Prediction error	Accurate	(In)accurate, failure to reset

enough when you think you didn't), the thought can become pervasive, lead to excessive worry, or inappropriately change the task to studying.

Salience signals can be faulty when studying seems much more appropriate than necessary, and the task is changed from going to a movie to studying. Another example of inappropriate salience occurs when one is studying, and studying appears so much more important than sleep that one continues studying until the moment of the exam, paradoxically hampering performance.

An inaccurate prediction error signal arises when one is on task, but nonetheless gets a signal that one is off task. For example, when someone is studying for a test, one feels anxiety despite engaging in studying, which can paradoxically undermine test preparation through perseverative worry (instead of studying).

An additional pathway that can compound mode failures emerges from reinforcement learning to combine with the faulty elements described above. If the threshold of the gate is too low, the interrupt changes the ongoing plan. The change itself (because it is new) can be reinforcing. This reinforcement leads to increased salience of the event that produced the change in plan. As salience iteratively increases, the likelihood increases that the interrupt that is linked to the new plan is going to disrupt other plans in the future (positive feedback loop). If the new plan is now in place, the stronger salience of the interrupt supports its maintenance. For example, with perseverative worry, the worry led to a change in plan (didn't go to the movie), which then led the person to study more and indeed feel better, which then functions as a reward signal that leads to a reinforcement of perseverative worry in the future. Such a simple mechanism could account for the common finding that OCD worsens with time. An important question for future research into effective interventions will be how to rein in this positive feedback loop afforded by reinforcement learning.

Bibliography

Note: Numbers in square brackets denote the chapter in which an entry is cited.

Abi-Dargham, A., and M. Laruelle. 2005. Mechanisms of Action of Second Generation Antipsychotic Drugs in Schizophrenia: Insights from Brain Imaging Studies. *Eur. Psychiatry* **20**:15–27. [15]

Abramovitch, A., and A. Schweiger. 2009. Unwanted Intrusive and Worrisome Thoughts in Adults with Attention Deficit\Hyperactivity Disorder. *Psychiatry Res.* **168**:230–233. [9]

Abramowitz, J. S., D. F. Tolin, and G. P. Street. 2001. Paradoxical Effects of Thought Suppression: A Meta-Analysis of Controlled Studies. *Clin. Psychol. Rev.* **21**:683–703. [6, 9]

Adam, D. 2014. The Man Who Couldn't Stop: OCD, and the True Story of a Life Lost in Thought. London: Picador. [9]

Adam, Y., J. J. Kim, S. Lou, et al. 2019. Voltage Imaging and Optogenetics Reveal Behaviour-Dependent Changes in Hippocampal Dynamics. *Nature* **569**:413–417. [4]

Adams, R. A., K. E. Stephan, H. R. Brown, C. D. Frith, and K. J. Friston. 2013. The Computational Anatomy of Psychosis. *Front. Psychiatry* **4**:47. [8]

Adler, C. M., P. McDonough-Ryan, K. W. Sax, et al. 2000. fMRI of Neuronal Activation with Symptom Provocation in Unmedicated Patients with Obsessive Compulsive Disorder. *J. Psychiatr. Res.* **34**:317–324. [8]

Agren, T., J. Engman, A. Frick, et al. 2012. Disruption of Reconsolidation Erases a Fear Memory Trace in the Human Amygdala. *Science* **337**:1550–1552. [14]

Ahmari, S. E., T. Spellman, N. L. Douglass, et al. 2013. Repeated Cortico-Striatal Stimulation Generates Persistent OCD-Like Behavior. *Science* **340**:1234–1239. [3]

Ahrens, A. M., B. F. Singer, C. J. Fitzpatrick, J. D. Morrow, and T. E. Robinson. 2016. Rats That Sign-Track Are Resistant to Pavlovian but Not Instrumental Extinction. *Behav. Brain Res.* **296**:418–430. [5]

Ainley, V., M. A. Apps, A. Fotopoulou, and M. Tsakiris. 2016. Bodily Precision: A Predictive Coding Account of Individual Differences in Interoceptive Accuracy. *Philos. Trans. R. Soc. Lond. B. Biol. Sci.* **371**:20160003. [13]

Ainsworth, B., H. Bolderston, and M. Garner. 2017. Testing the Differential Effects of Acceptance and Attention-Based Psychological Interventions on Intrusive Thoughts and Worry. *Behav. Res. Ther.* **91**:72–77. [15]

Airan, R. D., K. R. Thompson, L. E. Fenno, H. Bernstein, and K. Deisseroth. 2009. Temporally Precise *in Vivo* Control of Intracellular Signalling. *Nature* **458**:1025–1029. [4]

Aitchison, L., and M. Lengyel. 2017. With or without You: Predictive Coding and Bayesian Inference in the Brain. *Curr. Opin. Neurobiol.* **46**:219–227. [10]

Akerboom, J., N. Carreras Calderón, L. Tian, et al. 2013. Genetically Encoded Calcium Indicators for Multi-Color Neural Activity Imaging and Combination with Optogenetics. *Front. Mol. Neurosci.* **6**:2. [4]

Alberini, C. M. 2005. Mechanisms of Memory Stabilization: Are Consolidation and Reconsolidation Similar or Distinct Processes? *Trends Neurosci.* **28**:51–56. [14]

Alberini, C. M., and J. E. Ledoux. 2013. Memory Reconsolidation. *Curr. Biol.* **23**:R746–750. [14]

Albertella, L., M. E. Le Pelley, S. R. Chamberlain, et al. 2019. Reward-Related Attentional Capture Is Associated with Severity of Addictive and Obsessive-Compulsive Behaviors. *Psychol. Addict. Behav.* **33**:495–502. [5]

Albo, Z., and J. Gräff. 2018. The Mysteries of Remote Memory. *Philos. Trans. R. Soc. Lond. B. Biol. Sci.* **373**:20170029. [14]

Alcaraz, F., A. R. Marchand, G. Courtand, E. Coutureau, and M. Wolff. 2016. Parallel Inputs from the Mediodorsal Thalamus to the Prefrontal Cortex in the Rat. *Eur. J. Neurosci.* **44**:1972–1986. [3]

Alexander, B. K., R. B. Coambs, and P. F. Hadaway. 1978. The Effect of Housing and Gender on Morphine Self-Administration in Rats. *Psychopharmacology* **58**:175–179. [12]

Alexander, G. E., M. R. DeLong, and P. L. Strick. 1986. Parallel Organization of Functionally Segregated Circuits Linking Basal Ganglia and Cortex. *Annu. Rev. Neurosci.* **9**:357–381. [10, 11]

Alexander, W. H., and J. W. Brown. 2011. Medial Prefrontal Cortex as an Action-Outcome Predictor. *Nat. Neurosci.* **14**:1338–1344. [11]

———. 2015. Hierarchical Error Representation: A Computational Model of Anterior Cingulate and Dorsolateral Prefrontal Cortex. *Neural Comput.* **27**:2354–2410. [11]

———. 2017. The Role of the Anterior Cingulate Cortex in Prediction Error and Signaling Surprise. *Top. Cogn. Sci.* **11**:119–135. [10]

Al-Hasani, R., J.-M. T. Wong, O. S. Mabrouk, et al. 2018. In Vivo Detection of Optically-Evoked Opioid Peptide Release. eLife 7:e36520. [4]

Al-Shawaf, L., D. Conroy-Beam, K. Asao, and D. M. Buss. 2016. Human Emotions: An Evolutionary Psychological Perspective. Emotion Rev. 8:173–186. [7]

American Psychiatric Association. 2013. Diagnostic and Statistical Manual of Mental Disorders: Fifth edition. Arlington, VA: American Psychiatric Publishing. [6, 8, 9, 13–15]

Anderson, M. C. 2001. Active Forgetting: Evidence for Functional Inhibition as a Source of Memory Failure. *J. Aggress. Maltreat. Trauma* 4:185–210. [9]

———. 2003. Rethinking Interference Theory: Executive Control and the Mechanisms of Forgetting. *J. Mem. Lang.* 49:415–445. [9]

Anderson, M. C., E. L. Bjork, and R. A. Bjork. 2000. Retrieval-Induced Forgetting: Evidence for a Recall-Specific Mechanism. *Psychon. Bull. Rev.* 7:522–530. [9]

Anderson, M. C., R. A. Bjork, and E. L. Bjork. 1994. Remembering Can Cause Forgetting: Retrieval Dynamics in Long-Term Memory. *J. Exp. Psychol. Learn. Mem. Cogn.* **20**:1063–1087. [9]

Anderson, M. C., J. G. Bunce, and H. Barbas. 2016. Prefrontal–Hippocampal Pathways Underlying Inhibitory Control over Memory. *Neurobiol. Learn. Mem.* **134**:145–161. [9]

Anderson, M. C., and C. Green. 2001. Suppressing Unwanted Memories by Executive Control. *Nature* **410**:366–369. [6, 9]

Anderson, M. C., and S. Hanslmayr. 2014. Neural Mechanisms of Motivated Forgetting. *Trends Cogn. Sci.* **18**:279–292. [9]

Anderson, M. C., and E. Huddleston. 2012. Towards a Cognitive and Neurobiological Model of Motivated Forgetting. In: True and False Recovered Memories, pp. 53–120. New York: Springer. [9]

Anderson, M. C., and J. H. Neely. 1996. Interference and Inhibition in Memory Retrieval. In: Memory, pp. 237–313. Elsevier. [9]

Anderson, M. C., K. N. Ochsner, B. Kuhl, et al. 2004. Neural Systems Underlying the Suppression of Unwanted Memories. *Science* **303**:232–235. [9]

Anderson, M. C., and B. A. Spellman. 1995. On the Status of Inhibitory Mechanisms in Cognition: Memory Retrieval as a Model Case. *Psychol. Rev.* **102**:68–100. [9]

Andrews-Hanna, J. R., R. H. Kaiser, A. E. J. Turner, et al. 2013. A Penny for Your Thoughts: Dimensions of Self-Generated Thought Content and Relationships with Individual Differences in Emotional Wellbeing. *Front. Psychol.* **4**:900. [6]

Andrews-Hanna, J. R., J. S. Reidler, C. Huang, and R. L. Buckner. 2010. Evidence for the Default Network's Role in Spontaneous Cognition. *J. Neurophysiol.* **104**:322–335. [13]

Andrews-Hanna, J. R., R. Saxe, and T. Yarkoni. 2014. Contributions of Episodic Retrieval and Mentalizing to Autobiographical Thought: Evidence from Functional Neuroimaging, Resting-State Connectivity, and fMRI Meta-Analyses. *NeuroImage* **91**:324–335. [6]

Aramakis, V. B., C. Y. Hsieh, F. M. Leslie, and R. Metherate. 2000. A Critical Period for Nicotine-Induced Disruption of Synaptic Development in Rat Auditory Cortex. *J. Neurosci.* **20**:6106–6116. [5]

Archer, J. 2003. The Nature of Grief: The Evolution and Psychology of Reactions to Loss. New York: Routledge. [7]

Aron, A. R. 2011. From Reactive to Proactive and Selective Control: Developing a Richer Model for Stopping Inappropriate Responses. *Biol. Psychiatry* **69**:e55–e68. [9]

Aron, A. R., T. E. Behrens, S. Smith, M. J. Frank, and R. A. Poldrack. 2007. Triangulating a Cognitive Control Network Using Diffusion-Weighted Magnetic Resonance Imaging (MRI) and Functional MRI. *J. Neurosci.* **27**:3743–3752. [11]

Aron, A. R., W. Cai, D. Badre, and T. W. Robbins. 2015. Evidence Supports Specific Braking Function for Inferior Pfc. *Trends Cogn. Sci.* **19**:711–712. [11]

Aron, A. R., D. M. Herz, P. Brown, B. U. Forstmann, and K. Zaghloul. 2016. Frontosubthalamic Circuits for Control of Action and Cognition. *J. Neurosci.* **36**:11489–11495. [11]

Aron, A. R., and R. A. Poldrack. 2006. Cortical and Subcortical Contributions to Stop Signal Response Inhibition: Role of the Subthalamic Nucleus. *J. Neurosci.* **26**:2424–2433. [11]

Aron, A. R., T. W. Robbins, and R. A. Poldrack. 2004. Inhibition and the Right Inferior Frontal Cortex. *Trends Cogn. Sci.* **8**:170–177. [9, 11]

———. 2014. Inhibition and the Right Inferior Frontal Cortex: One Decade On. *Trends Cogn. Sci.* **18**:177–185. [9]

Arundell, L., E. Fletcher, J. Salmon, J. Veitch, and T. Hinkley. 2016. A Systematic Review of the Prevalence of Sedentary Behavior during the after-School Period among Children Aged 5–18 Years. *Int. J. Behav. Nutr. Phys. Act.* **13**:93. [17]

Aso, T., K. Nishimura, T. Kiyonaka, et al. 2016. Dynamic Interactions of the Cortical Networks During Thought Suppression. *Brain Behav.* **6**:e0503. [6]

Assem, M., M. F. Glasser, D. C. Van Essen, and J. Duncan. 2020. A Domain-General Cognitive Core Defined in Multimodally Parcellated Human Cortex. *Cereb. Cortex* doi: 10.1093/cercor/bhaa023. [11]

Attias, H. 2003. Planning by Probabilistic Inference. In: Proc. of the 9th Intl. Workshop on Artificial Intelligence and Statistics. Key West: AISTATS [13]

Averbeck, B. B., J. Lehman, M. Jacobson, and S. N. Haber. 2014. Estimates of Projection Overlap and Zones of Convergence within Frontal-Striatal Circuits. *J. Neurosci.* **34**:9497–9505. [5]

Baddeley, A. D. 2003. Working Memory: Looking Back and Looking Forward. *Nat. Rev. Neurosci.* **4**:829–839. [14, 17]

Badre, D., and M. D'Esposito. 2007. Functional Magnetic Resonance Imaging Evidence for a Hierarchical Organization of the Prefrontal Cortex. *J. Cogn. Neurosci.* **19**:2082–2099. [11]

———. 2009. Is the Rostro-Caudal Axis of the Frontal Lobe Hierarchical? *Nat. Rev. Neurosci.* **10**:659–669. [11]

———. 2020. On Task: How Our Brain Gets Things Done. Princeton: Princeton Univ. Press. [11]

Badre, D., and M. J. Frank. 2012. Mechanisms of Hierarchical Reinforcement Learning in Cortico-Striatal Circuits 2: Evidence from fMRI. *Cereb. Cortex* **22**:527–536. [11]

Badre, D., J. Hoffman, J. W. Cooney, and M. D'Esposito. 2009. Hierarchical Cognitive Control Deficits Following Damage to the Human Frontal Lobe. *Nat. Neurosci.* **12**:515–522. [11]

Badre, D., and D. E. Nee. 2018. Frontal Cortex and the Hierarchical Control of Behavior. *Trends Cogn. Sci.* **22**:170–188. [11, 13]

Badry, R., T. Mima, T. Aso, et al. 2009. Suppression of Human Cortico-Motoneuronal Excitability during the Stop-Signal Task. *Clin. Neurophysiol.* **120**:1717–1723. [9]

Baier, B., H. O. Karnath, M. Dieterich, et al. 2010. Keeping Memory Clear and Stable: The Contribution of Human Basal Ganglia and Prefrontal Cortex to Working Memory. *J. Neurosci.* **30**:9788–9792. [11]

Baird, B., S. A. Mota-Rolim, and M. Dresler. 2019. The Cognitive Neuroscience of Lucid Dreaming. *Neurosci. Biobehav. Rev.* **100**:305–323. [12]

Baird, B., M. D. Mrazek, D. T. Phillips, and J. W. Schooler. 2014. Domain-Specific Enhancement of Metacognitive Ability Following Meditation Training. *J. Exp. Psychol. Gen.* **143**:1972–1979. [9]

Baird, B., J. Smallwood, D. J. F. Fishman, M. D. Mrazek, and J. W. Schooler. 2013a. Unnoticed Intrusions: Dissociations of Meta-Consciousness in Thought Suppression. *Conscious. Cogn.* **22**:1003–1012. [9]

Baird, B., J. Smallwood, K. J. Gorgolewski, and D. S. Margulies. 2013b. Medial and Lateral Networks in Anterior Prefrontal Cortex Support Metacognitive Ability for Memory and Perception. *J. Neurosci.* **33**:16657–16665. [12]

Baird, B., J. Smallwood, M. D. Mrazek, et al. 2012. Inspired by Distraction: Mind Wandering Facilitates Creative Incubation. *Psychol. Sci.* **23**:1117–1122. [9]

Baird, B., J. Smallwood, and J. W. Schooler. 2011. Back to the Future: Autobiographical Planning and the Functionality of Mind-Wandering. *Conscious. Cogn.* **20**:1604–1611. [9]

Baker, C. L., R. Saxe, and J. B. Tenenbaum. 2009. Action Understanding as Inverse Planning. *Cognition* **113**:329–349. [13]

Baker, C. L., and J. B. Tenenbaum. 2014. Modeling Human Plan Recognition Using Bayesian Theory of Mind. In: Plan, Activity, and Intent Recognition, ed. G. Sukthankar et al., pp. 177–204. Boston: Morgan Kaufmann. [13]

Baldwin, D. S., K. Ajel, V. G. Masdrakis, M. Nowak, and R. Rafiq. 2013. Pregabalin for the Treatment of Generalized Anxiety Disorder: An Update. *Neuropsychiat. Dis. Treat.* **9**:883–892. [15]

Bale, T. L., T. Abel, H. Akil, et al. 2019. The Critical Importance of Basic Animal Research for Neuropsychiatric Disorders. *Neuropsychopharmacology* **44**:1349–1353. [5]

Baler, R. D., and N. D. Volkow. 2006. Drug Addiction: The Neurobiology of Disrupted Self-Control. *Trends Mol. Med.* **12**:559–566. [12]

Balleine, B. W. 2001. Incentive Processes in Instrumental Conditioning. In: Handbook of Contemporary Learning Theories, ed. R. Klein and S. Mowrer, pp. 307–366. Hillsdale, NJ: LEA. [3]

———. 2005. Neural Bases of Food-Seeking: Affect, Arousal and Reward in Corticostriatolimbic Circuits. *Physiol. Behav.* **86**:717–730. [3]

Balleine, B. W., and A. Dickinson. 1998. Goal-Directed Instrumental Action: Contingency and Incentive Learning and Their Cortical Substrates. *Neuropharmacology* **37**:407–419. [3, 13]

Balleine, B. W., M. Liljeholm, and S. B. Ostlund. 2009. The Integrative Function of the Basal Ganglia in Instrumental Conditioning. *Behav. Brain Res.* **199**:43–52. [3, 13]

Balleine, B. W., and J. P. O'Doherty. 2010. Human and Rodent Homologies in Action Control: Corticostriatal Determinants of Goal-Directed and Habitual Action. *Neuropsychopharmacology* **35**:48–69. [3, 8, 13]

Balleine, B. W., and S. B. Ostlund. 2007. Still at the Choice-Point: Action Selection and Initiation in Instrumental Conditioning. *Ann. N.Y. Acad. Sci* **1104**:147–171. [3]

Banich, M. T. 2009. Executive Function: The Search for an Integrated Account. *Curr. Dir. Psychol. Sci.* **18**:89–94. [9]

Banich, M. T., K. L. Mackiewicz, B. E. Depue, et al. 2009. Cognitive Control Mechanisms, Emotion and Memory: A Neural Perspective with Implications for Psychopathology. *Neurosci. Biobehav. Rev.* **33**:613–630. [9]

Banich, M. T., K. L. Mackiewicz Seghete, B. E. Depue, and G. C. Burgess. 2015. Multiple Modes of Clearing One's Mind of Current Thoughts: Overlapping and Distinct Neural Systems. *Neuropsychologia* **69**:105–117. [6, 9]

Banks, W. P., and E. A. Isham. 2009. We Infer Rather Than Perceive the Moment We Decided to Act. *Psychol. Sci.* **20**:17–21. [12]

Barbas, H., and N. Rempel-Clower. 1997. Cortical Structure Predicts the Pattern of Corticocortical Connections. *Cereb. Cortex* **7**:635–646. [11]

Barfield, E. T., K. J. Gerber, K. S. Zimmermann, et al. 2017. Regulation of Actions and Habits by Ventral Hippocampal Trkb and Adolescent Corticosteroid Exposure. *PLoS Biol.* **15**:e2003000. [5]

Barfield, E. T., and S. L. Gourley. 2019. Glucocorticoid-Sensitive Ventral Hippocampal-Orbitofrontal Cortical Connections Support Goal-Directed Action - Curt Richter Award Paper 2019. *Psychoneuroendocrinology* **110**:104436. [5]

Bargh, J. A., K. L. Schwader, S. E. Hailey, R. L. Dyer, and E. J. Boothby. 2012. Automaticity in Social-Cognitive Processes. *Trends Cogn. Sci.* **16**:593–605. [12]

Bari, A., and T. W. Robbins. 2013. Inhibition and Impulsivity: Behavioral and Neural Basis of Response Control. *Prog. Neurobiol.* **108**:44–79. [9, 13]

Barker, A. T., I. L. Freeston, R. Jabinous, and J. A. Jarratt. 1986. Clinical Evaluation of Conduction Time Measurements in Central Motor Pathways Using Magnetic Stimulation of Human Brain. *Lancet* **1**:1325–1326. [16]

Barlow, H. 1961. Possible Principles Underlying the Transformations of Sensory Messages. In: Sensory Communication, ed. W. Rosenblith, pp. 217–234. Cambridge, MA: MIT Press. [13]

Barrett, L. F., K. S. Quigley, and P. Hamilton. 2016. An Active Inference Theory of Allostasis and Interoception in Depression. *Philos. Trans. R. Soc. Lond. B. Biol. Sci.* **371**:20160011. [13]

Barretto, R. P. J., and M. J. Schnitzer. 2012. *In Vivo* Optical Microendoscopy for Imaging Cells Lying Deep within Live Tissue. *Cold Spring Harb. Protoc.* **2012**:1029–1034. [4]

Barto, A. G. 1995. Adaptive Critics and the Basal Ganglia. In: Computational Neuroscience: Models of Information Processing in the Basal Ganglia, ed. J. C. Houk et al., pp. 215–232. Cambridge, MA: MIT Press. [2]

Barto, A. G., M. Mirolli, and G. Baldassarre. 2013. Novelty or Surprise? *Front. Psychol.* **4**:907. [10, 13]

Bartra, O., J. T. McGuire, and J. W. Kable. 2013. The Valuation System: A Coordinate-Based Meta-Analysis of BOLD fMRI Experiments Examining Neural Correlates of Subjective Value. *NeuroImage* **76**:412–427. [10]

Bastos, A. M., W. M. Usrey, R. A. Adams, et al. 2012. Canonical Microcircuits for Predictive Coding. *Neuron* **76**:695–711. [13]

Battaglini, E., B. Liddell, P. Das, et al. 2016. Intrusive Memories of Distressing Information: An fMRI Study. *PLoS One* **11**:e0140871. [9, 14]

Baumeister, R. F., E. J. Masicampo, and K. D. Vohs. 2011. Do Conscious Thoughts Cause Behavior? *Ann. Rev. Psychol.* **62**:331–361. [12]

Bäuml, K.-H., B. Pastötter, and S. Hanslmayr. 2010. Binding and Inhibition in Episodic Memory: Cognitive, Emotional, and Neural Processes. *Neurosci. Biobehav. Rev.* **34**:1047–1054. [9]

Baxter, L. R., Jr., M. E. Phelps, J. C. Mazziotta, et al. 1987. Local Cerebral Glucose Metabolic Rates in Obsessive-Compulsive Disorder. A Comparison with Rates in Unipolar Depression and in Normal Controls. *Arch. Gen. Psychiatry* **44**:211–218. [8]

Beadel, J. R., J. S. Green, S. Hosseinbor, and B. A. Teachman. 2013. Influence of Age, Thought Content, and Anxiety on Suppression of Intrusive Thoughts. *J. Anxiety Disord.* **27**:598–607. [6]

Bear, M. F., and R. C. Malenka. 1994. Synaptic Plasticity: LTP and LTD. *Curr. Opin. Neurobiol.* **4**:389–399. [16]

Beck, A. T. 1967. Depression: Clinical, Experimental, and Theoretical Aspects. New York: Harper & Row. [7]

———. 1976. Cognitive Therapy and the Emotional Disorders. New York: International Universities Press. [9]

Beck, D. M., and S. Kastner. 2009. Top-Down and Bottom-up Mechanisms in Biasing Competition in the Human Brain. *Vision Res.* **49**:1154–1165. [13]

Beck, J. M., W. J. Ma, R. Kiani, et al. 2008. Probabilistic Population Codes for Bayesian Decision Making. *Neuron* **60**:1142–1152. [13]

Behrens, T. E. J., P. Fox, A. Laird, and S. M. Smith. 2013. What Is the Most Interesting Part of the Brain? *Trends Cogn. Sci.* **17**:2–4. [10]

Behrens, T. E. J., H. Johansen-Berg, M. W. Woolrich, et al. 2003. Non-Invasive Mapping of Connections between Human Ahalamus and Cortex Using Diffusion Imaging. *Nat. Neurosci.* **6**:750–757. [10]

Beilock, S. L., and S. Gonso. 2008. Putting in the Mind versus Putting on the Green: Expertise, Performance Time, and the Linking of Imagery and Action. *Q. J. Exp. Psychol.* **61**:920–932. [9]

Bekinschtein, P., N. V. Weisstaub, F. Gallo, M. Renner, and M. C. Anderson. 2018. A Retrieval-Specific Mechanism of Adaptive Forgetting in the Mammalian Brain. *Nat. Commun.* **9**:4660. [9]

Belin, D., A. C. Mar, J. W. Dalley, T. W. Robbins, and B. J. Everitt. 2008. High Impulsivity Predicts the Switch to Compulsive Cocaine-Taking. *Science* **320**:1352–1355. [5]

Belin-Rauscent, A., M. Fouyssac, A. Bonci, and D. Belin. 2016. How Preclinical Models Evolved to Resemble the Diagnostic Criteria of Drug Addiction. *Biol. Psychiatry* **79**:39–46. [5]

Belloch, A., C. Morillo, and A. Gimenez. 2004. Effects of Suppressing Neutral and Obsession-Like Thoughts in Normal Subjects: Beyond Frequency. *Behav. Res. Ther.* **42**:841–857. [17]

Benoit, R. G., and M. C. Anderson. 2012. Opposing Mechanisms Support the Voluntary Forgetting of Unwanted Memories. *Neuron* **76**:450–460. [6, 9]

Benoit, R. G., D. J. Davies, and M. C. Anderson. 2016. Reducing Future Fears by Suppressing the Brain Mechanisms Underlying Episodic Simulation. *PNAS* **113**:E8492. [9]

Benoit, R. G., J. C. Hulbert, E. Huddleston, and M. C. Anderson. 2015. Adaptive Top-Down Suppression of Hippocampal Activity and the Purging of Intrusive Memories from Consciousness. *J. Cogn. Neurosci.* **27**:96–111. [6, 9]

Bergmann, T. O., A. Karabanov, G. Hartwigsen, A. Thielscher, and H. R. Siebner. 2016. Combining Non-Invasive Transcranial Brain Stimulation with Neuroimaging and Electrophysiology: Current Approaches and Future Perspectives. *NeuroImage* **140**:4–19. [16]

Bergstrom, J., G. Andersson, A. Karlsson, et al. 2009. An Open Study of the Effectiveness of Internet Treatment for Panic Disorder Delivered in a Psychiatric Setting. *Nor. J. Psychiatry* **63**:44–50. [9]

Bergström, Z. M., J. W. de Fockert, and A. Richardson-Klavehn. 2009. ERP and Behavioural Evidence for Direct Suppression of Unwanted Memories. *NeuroImage* **48**:726–737. [9]

Berntsen, D. 1996. Involuntary Autobiographical Memories. *Appl. Cogn. Psychol.* **10**:435–454. [9]

Berntsen, D., and D. C. Rubin. 2008. The Reappearance Hypothesis Revisited: Recurrent Involuntary Memories after Traumatic Events and in Everyday Life. *Mem. Cogn.* **36**:449–460. [9]

———. 2013. Involuntary Memories and Dissociative Amnesia: Assessing Key Assumptions in Posttraumatic Stress Disorder Research. *Clin. Psychol. Sci.* **2**:174–186. [9]

Berridge, K. C., and T. E. Robinson. 1998. What Is the Role of Dopamine in Reward: Hedonic Impact, Reward Learning, or Incentive Salience? *Brain Res. Brain Res. Rev.* **28**:309–369. [13]

———. 2003. Parsing Reward. *Trends Neurosci.* **26**:507–513. [5]

Berry, L.-M., and B. Laskey. 2012. A Review of Obsessive Intrusive Thoughts in the General Population. *J. Obsessive Compuls. Relat. Disord.* **1**:125–132. [17]

Bertran-Gonzalez, J., B. C. Chieng, V. Laurent, E. Valjent, and B. W. Balleine. 2012. Striatal Cholinergic Interneurons Display Activity-Related Phosphorylation of Ribosomal Protein S6. *PLoS One* **7**:e53195. [3]

Betzel, R. F., S. Gu, J. D. Medaglia, F. Pasqualetti, and D. S. Bassett. 2016. Optimally Controlling the Human Connectome: The Role of Network Topology. *Sci. Rep.* **6**:30770. [11]

Bhandari, A., and D. Badre. 2018. Learning and Transfer of Working Memory Gating Policies. *Cognition* **172**:89–100. [13]

Bickerdike, L., A. Booth, P. M. Wilson, K. Farley, and K. Wright. 2017. Social Prescribing: Less Rhetoric and More Reality: A Systematic Review of the Evidence. *BMJ Open* **7**:e013384. [14]

Birrer, E., T. Michael, and S. Munsch. 2007. Intrusive Images in PTSD and in Traumatised and Non-Traumatised Depressed Patients: A Cross-Sectional Clinical Study. *Behavioral Research and Therapy* **45**:2053–2065. [7]

Bisby, J. A., and N. Burgess. 2017. Differential Effects of Negative Emotion on Memory for Items and Associations, and Their Relationship to Intrusive Imagery. *Curr. Opin. Behav. Sci.* **17**:124–132. [9]

Bischoff-Grethe, A., C. E. Wierenga, L. A. Berner, et al. 2018. Neural Hypersensitivity to Pleasant Touch in Women Remitted from Anorexia Nervosa. *Transl. Psychiatry* **8**:161. [13]

Björkstrand, J., T. Agren, F. Ahs, et al. 2016. Disrupting Reconsolidation Attenuates Long-Term Fear Memory in the Human Amygdala and Facilitates Approach Behavior. *Curr. Biol.* **26**:2690–2695. [9, 14]

Bland, A. R., J. P. Roiser, M. A. Mehta, et al. 2016. Emoticom: A Neuropsychological Test Battery to Evaluate Emotion, Motivation, Impulsivity, and Social Cognition. *Front. Behav. Neurosci.* **10**:25. [9]

Blanshard, B., and B. F. Skinner. 1967. The Problem of Consciousness: A Debate. *Phil. Phenom. Res.* **27**:317–337. [12]

Bloch, M. H., A. Landeros-Weisenberger, B. Kelmendi, et al. 2006. A Systematic Review: Antipsychotic Augmentation with Treatment Refractory Obsessive-Compulsive Disorder. *Mol. Psychiatry* **11**:622–632. [15]

Block, N. 1995. On a Confusion About a Function of Consciousness. *Behav. Brain Sci.* **18**:227–247. [12]

Block, N. J., and J. A. Fodor. 1972. What Psychological States Are Not. *Philos. Rev.* **81**:159–181. [11]

Blumberg, S. J. 2000. The White Bear Suppression Inventory: Revisiting Its Factor Structure. *Pers. Individ. Dif.* **29**:943–950. [9]

Bobadilla, A. C., J. A. Heinsbroek, C. D. Gipson, et al. 2017. Corticostriatal Plasticity, Neuronal Ensembles, and Regulation of Drug-Seeking Behavior. *Prog. Brain. Res.* **235**:93–112. [5]

Boelen, P. A., and R. J. Huntjens. 2008. Intrusive Images in Grief: an Exploratory Study. *Clin. Psychol. Psychother.* **15**:217–226. [7]

Boes, A. D., S. Prasad, H. Liu, et al. 2015. Network Localization of Neurological Symptoms from Focal Brain Lesions. *Brain* **138**:3061–3075. [10]

Bolger, N., A. Davis, and E. Rafaeli. 2003. Diary Methods: Capturing Life as It Is Lived. *Ann. Rev. Psychol.* **54**:579–616. [9]

Bolton, T. A. W., A. Tarun, V. Sterpenich, S. Schwartz, and D. Van De Ville. 2018. Interactions between Large-Scale Functional Brain Networks Are Captured by Sparse Coupled Hmms. *IEEE Trans. Med. Imaging* **37**:230–240. [17]

Bomyea, J., and A. J. Lang. 2016. Accounting for Intrusive Thoughts in PTSD: Contributions of Cognitive Control and Deliberate Regulation Strategies. *J. Affect. Disord.* **192**:184–190. [6, 17]

Bond, R. M., C. J. Fariss, J. J. Jones, et al. 2012. A 61-Million-Person Experiment in Social Influence and Political Mobilization. *Nature* **489**:295–298. [12]

Bonsall, M. B., J. R. Geddes, G. M. Goodwin, and E. A. Holmes. 2015. Bipolar Disorder Dynamics: Affective Instabilities, Relaxation Oscillations and Noise. *J. R. Soc. Interface* **12**: 20150670. [14]

Bonvicini, C., S. V. Faraone, and C. Scassellati. 2016. Attention-Deficit Hyperactivity Disorder in Adults: A Systematic Review and Meta-Analysis of Genetic, Pharmacogenetic and Biochemical Studies. *Mol. Psychiatry* **21**:872–884. [13]

Borkovec, T. D., W. J. Ray, and J. Stober. 1998. Worry: A Cognitive Phenomenon Intimately Linked to Affective, Physiological, and Interpersonal Behavioral Processes. *Cogn. Ther. Res.* **22**:561–576. [7]

Borkovec, T. D., E. Robinson, T. Prudinsky, and J. A. DePree. 1983. Preliminary Investigation of Worry: Some Characteristics and Processes. *Behav. Res. Ther.* **21**:9–16. [7]

Borkovec, T. D., and L. Roemer. 1995. Perceived Functions of Worry among Generalized Anxiety Disorder Subjects: Distraction from More Emotionally Distressing Topics. *J. Behav. Ther. Exp. Psychiatry* **26**:25–30. [9]

Botvinick, M. M. 2007. Conflict Monitoring and Decision Making: Reconciling Two Perspectives on Anterior Cingulate Function. *Cogn. Affect. Behav. Neurosci.* **7**:356–366. [10]

Botvinick, M. M., J. D. Cohen, and C. S. Carter. 2004. Conflict Monitoring and Anterior Cingulate Cortex: an Update. *Trends Cogn. Sci.* **8**:539–546. [11]

Botvinick, M. M., Y. Niv, and A. G. Barto. 2009. Hierarchically Organized Behavior and Its Neural Foundations: A Reinforcement Learning Perspective. *Cognition* **113**:262–280. [13]

Botvinick, M. M., and M. Toussaint. 2012. Planning as Inference. *Trends Cogn. Sci.* **16**:485–488. [13]

Bourne, C., F. Frasquilho, A. D. Roth, and E. A. Holmes. 2010. Is It Mere Distraction? Peri-Traumatic Verbal Tasks Can Increase Analogue Flashbacks but Reduce Voluntary Memory Performance. *J. Behav. Ther. Exp. Psychiatry* **41**:316–324. [9]

Bourne, C., C. E. Mackay, and E. A. Holmes. 2013. The Neural Basis of Flashback Formation: The Impact of Viewing Trauma. *Psychol. Med.* **43**:1521–1532. [9]

Bouvard, M., N. Fournet, A. Denis, A. Sixdenier, and D. Clark. 2017. Intrusive Thoughts in Patients with Obsessive Compulsive Disorder and Non-Clinical Participants: A Comparison Using the International Intrusive Thought Interview Schedule *Cogn. Behav. Ther.* **46**:287–299. [6, 13]

Bozhilova, N. S., G. Michelini, J. Kuntsi, and P. Asherson. 2018. Mind Wandering Perspective on Attention-Deficit/Hyperactivity Disorder. *Neurosci. Biobehav. Rev.* **92**:464–476. [13]

Bradfield, L. A., J. Bertran-Gonzalez, B. Chieng, and B. W. Balleine. 2013. The Thalamostriatal Pathway and Cholinergic Control of Goal-Directed Action: Interlacing New with Existing Learning in the Striatum. *Neuron* **79**:153–166. [13]

Bradfield, L. A., A. Dezfouli, M. van Holstein, B. Chieng, and B. W. Balleine. 2015. Medial Orbitofrontal Cortex Mediates Outcome Retrieval in Partially Observable Task Situations. *Neuron* **88**:1268–1280. [3, 13]

Braun, N., S. Debener, N. Spychala, et al. 2018. The Senses of Agency and Ownership: A Review. *Front. Psychol.* **9**:535. [13]

Braver, T. S. 2012. The Variable Nature of Cognitive Control: A Dual Mechanisms Framework. *Trends Cogn. Sci.* **16**:106–113. [13]

Bravo-Rivera, C., C. Roman-Ortiz, M. Montesinos-Cartagena, and G. J. Quirk. 2015. Persistent Active Avoidance Correlates with Activity in Prelimbic Cortex and Ventral Striatum. *Front. Behav. Neurosci.* **9**:184. [5]

Breiter, H. C., and S. L. Rauch. 1996. Functional MRI and the Study of OCD: From Symptom Provocation to Cognitive-Behavioral Probes of Cortico-Striatal Systems and the Amygdala. *NeuroImage* **4**:S127–138. [8]

Brewer, J. A. 2019. Mindfulness Training for Addictions: Has Neuroscience Revealed a Brain Hack by Which Awareness Subverts the Addictive Process? *Curr. Opin. Psychol.* **28**:198–203. [17]

Brewer, J. A., J. H. Davis, and J. Goldstein. 2013. Why Is It So Hard to Pay Attention, or Is It? Mindfulness, the Factors of Awakening and Reward-Based Learning. *Mindfulness* **4**:75–80. [17]

Brewer, J. A., and L. Pbert. 2015. Mindfulness: An Emerging Treatment for Smoking and Other Addictions? *J. Fam. Med.* **2**:1035. [17]

Brewer, J. A., A. Ruf, A. L. Beccia, et al. 2018. Can Mindfulness Address Maladaptive Eating Behaviors? Why Traditional Diet Plans Fail and How New Mechanistic Insights May Lead to Novel Interventions. *Front. Psychol.* **9**:1418. [17]

Brewin, C. R. 2011. The Nature and Significance of Memory Disturbance in Posttraumatic Stress Disorder. *Ann. Rev. Clin. Psychol.* **7**:203–227. [6]

———. 2014. Episodic Memory, Perceptual Memory and Their Interaction: Foundations for a Theory of Posttraumatic Stress Disorder. *Psychol. Bull.* **140**:69–97. [9, 14]

———. 2016. Coherence, Disorganization, and Fragmentation in Traumatic Memory Reconsidered: A Response to Rubin et al. (2016). *J. Abnorm. Psychol.* **125**:1011–1017. [9]

Brewin, C. R., M. Reynolds, and P. Tata. 1999. Autobiographical Memory Processes and the Course of Depression. *J. Abnorm. Psychol.* **108**:511–517. [7]

Brewin, C. R., and J. Saunders. 2001. The Effect of Dissociation at Encoding on Intrusive Memories for a Stressful Film. *Br. J. Med. Psychol.* **74**:467–472. [6]

Brewin, C. R., and L. Smart. 2005. Working Memory Capacity and Suppression of Intrusive Thoughts. *J. Behav. Ther. Exp. Psychiatry* **36**:61–68. [6]

Brewin, C. R., M. Watson, S. McCarthy, P. Hyman, and D. Dayson. 1998. Intrusive Memories and Depression in Cancer Patients. *Behav. Res. Ther.* **36**:1131–1142. [7]

Brosey, E. A., and N. D. Woodward. 2017. Neuroanatomical Correlates of Perceptual Aberrations in Psychosis. *Schizophr. Res.* **179**:125–131. [13]

Brosschot, J. F., B. Verkuil, and J. F. Thayer. 2010. Conscious and Unconscious Perseverative Cognition: Is a Large Part of Prolonged Physiological Activity Due to Unconscious Stress? *J. Psychosom. Res.* **69**:407–416. [13]

Brown, H. E., O. Freudenreich, X. Fan, et al. 2019a. Efficacy and Tolerability of Adjunctive Intravenous Sodium Nitroprusside Treatment for Outpatients with Schizophrenia. *JAMA Psychiatry* **76**:691–699. [15]

Brown, R., H. C. Lau, and J. E. LeDoux. 2019b. The Misunderstood Higher-Order Approach to Consciousness. *PsyArXiv Preprints*, https://doi.org/10.31234/osf.io/xpy8h (accessed Feb. 6, 2020). [12]

Bubenzer-Busch, S., B. Herpertz-Dahlmann, B. Kuzmanovic, et al. 2016. Neural Correlates of Reactive Aggression in Children with Attention-Deficit/Hyperactivity Disorder and Comorbid Disruptive Behaviour Disorders. *Acta Psychiat. Scand.* **133**:310–323. [13]

Buckner, R. L., and L. M. DiNicola. 2019. The Brain's Default Network: Updated Anatomy, Physiology and Evolving Insights. *Nat. Rev. Neurosci.* **20**:593–608. [10]

Buckner, R. L., F. M. Krienen, and B. T. Yeo. 2013. Opportunities and Limitations of Intrinsic Functional Connectivity MRI. *Nat. Neurosci.* **16**:832–837. [11]

Buckner, R. L., J. Sepulcre, T. Talukdar, et al. 2009. Cortical Hubs Revealed by Intrinsic Functional Connectivity: Mapping, Assessment of Stability, and Relation to Alzheimer's Disease. *J. Neurosci.* **29**:1860–1873. [5]

Bugg, J. M., and E. Streeper. 2019. Fate of Suspended and Completed Memory Intentions. In: Prospective Memory, ed. J. Rummel and M. A. McDaniel, pp. 44–59. London: Routledge. [9]

Burguiere, E., P. Monteiro, G. Feng, and A. M. Graybiel. 2013. Optogenetic Stimulation of Lateral Orbitofronto-Striatal Pathway Suppresses Compulsive Behaviors. *Science* **340**:1243–1246. [3]

Burnham, W. H. 1903. Retroactive Amnesia: Illustrative Cases and a Tentative Explanation. *Am. J. Psychol.* **14**:118–118. [9]

Burt, K. B., R. Whelan, P. J. Conrod, et al. 2016. Structural Brain Correlates of Adolescent Resilience. *J. Child Psychol. Psychiatry* **57**:1287–1296. [3]

Buss, D. M. 2000. The Dangerous Passion: Why Jealousy Is a Necessary as Love and Sex. New York: The Free Press. [7]

Buss, D. M., and M. G. Haselton. 2005. The Evolution of Jealousy. *Trends Cogn. Sci.* **9**:506–507. [7]

Butts, K. A., J. Weinberg, A. H. Young, and A. G. Phillips. 2011. Glucocorticoid Receptors in the Prefrontal Cortex Regulate Stress-Evoked Dopamine Efflux and Aspects of Executive Function. *PNAS* **108**:18459–18464. [2]

Cabeza, R., E. Ciaramelli, I. R. Olson, and M. Moscovitch. 2008. The Parietal Cortex and Episodic Memory: an Attentional Account. *Nat. Rev. Neurosci.* **9**:613–625. [9]

Cahill, E. N., and A. L. Milton. 2019. Neurochemical and Molecular Mechanisms Underlying the Retrieval-Extinction Effect. *Psychopharmacology* **236**:111–132. [2, 9]

Cahill, L., and M. T. Alkire. 2003. Epinephrine Enhancement of Human Memory Consolidation: Interaction with Arousal at Encoding. *Neurobiol. Learn. Mem.* **79**:194–198. [9]

Cahill, L., B. Prins, M. Weber, and J. L. McGaugh. 1994. Beta-Adrenergic Activation and Memory for Emotional Events. *Nature* **371**:702–704. [9]

Cai, W., T. Chen, S. Ryali, et al. 2015. Causal Interactions Within a Frontal-Cingulate-Parietal Network During Cognitive Control: Convergent Evidence from a Multisite–Multitask Investigation. *Cereb. Cortex* **26**:2140–2153. [10]

Cai, W., C. L. Oldenkamp, and A. R. Aron. 2011. A Proactive Mechanism for Selective Suppression of Response Tendencies. *J. Neurosci.* **31**:5965–5969. [9]

———. 2012. Stopping Speech Suppresses the Task-Irrelevant Hand. *Brain Lang.* **120**:412–415. [9]

Cameron, A. C., and P. R. Trivedi. 2013. Regression Analysis of Count Data. Cambridge: Cambridge Univ. Press. [14]

Campbell, J. 1999. Schizophrenia, the Space of Reasons, and Thinking as a Motor Process. *Monist* **82**:609–625. [8]

Campbell, J. N., E. Z. Macosko, H. Fenselau, et al. 2017. A Molecular Census of Arcuate Hypothalamus and Median Eminence Cell Types. *Nat. Neurosci.* **20**:484–496. [5]

Campese, V., M. McCue, G. Lázaro-Muñoz, J. E. LeDoux, and C. K. Cain. 2013. Development of an Aversive Pavlovian-to-Instrumental Transfer Task in Rat. *Front. Behav. Neurosci.* **7**:176. [9]

Cannon, W. B. 1929. Organisation for Physiologial Homeostasis. *Physiol. Rev.* **9**:399–431. [13]

Cardinal, R. N., J. A. Parkinson, J. Hall, and B. J. Everitt. 2002. Emotion and Motivation: The Role of the Amygdala, Ventral Striatum, and Prefrontal Cortex. *Neurosci. Biobehav. Rev.* **26**:321–352. [5]

Carew, C. L., A. M. Milne, F. L. Tatham, G. M. MacQueen, and G. B. C. Hall. 2013. Neural Systems Underlying Thought Suppression in Young Women with, and At-Risk, for Depression. *Behav. Brain Res.* **257**:13–24. [6]

Carmi, L., U. Alyagon, N. Barnea-Ygael, et al. 2018. Clinical and Electrophysiological Outcomes of Deep TMS over the Medial Prefrontal and Anterior Cingulate Cortices in OCD Patients. *Brain Stimul.* 11:158–165. [16]

Cartoni, E., B. Balleine, and G. Baldassarre. 2016. Appetitive Pavlovian-Instrumental Transfer: A Review. *Neurosci. Biobehav. Rev.* 71:829–848. [3, 9]

Caruso, G. D. 2012. Free Will and Consciousness : A Determinist Account of the Illusion of Free Will. Lanham, MD: Lexington Books. [12]

Caspi, A., R. M. Houts, D. W. Belsky, et al. 2013. The p Factor: One General Psychopathology Factor in the Structure of Psychiatric Disorders? *Clin. Psychol. Sci.* 2:119–137. [10]

Cassini, L. F., C. R. Flavell, O. B. Amaral, and J. L. C. Lee. 2017. On the Transition from Reconsolidation to Extinction of Contextual Fear Memories. *Learn. Mem.* 24:392–399. [9]

Castellanos, F. X., and E. Proal. 2012. Large-Scale Brain Systems in ADHD: Beyond the Prefrontal–Striatal Model. *Trends Cogn. Sci.* 16:17–26. [13]

Castiglione, A., J. Wagner, M. Anderson, and A. R. Aron. 2019. Preventing a Thought from Coming to Mind Elicits Increased Right Frontal Beta Just as Stopping Action Does. *Cereb. Cortex* 29:2160–2172. [6, 9, 11]

Castro, D. C., and M. R. Bruchas. 2019. A Motivational and Neuropeptidergic Hub: Anatomical and Functional Diversity within the Nucleus Accumbens Shell. *Neuron* 102:529–552. [5]

Catani, M., and D. H. Ffytche. 2005. The Rises and Falls of Disconnection Syndromes. *Brain* 128:2224–2239. [5]

Catarino, A., C. S. Küpper, A. Werner-Seidler, T. Dalgleish, and M. C. Anderson. 2015. Failing to Forget: Inhibitory-Control Deficits Compromise Memory Suppression in Posttraumatic Stress Disorder. *Psychol. Sci.* 26:604–616. [6, 9]

Cavada, C., and P. S. Goldman-Rakic. 1991. Topographic Segregation of Corticostriatal Projections from Posterior Parietal Subdivisions in the Macaque Monkey. *Neuroscience* 42:683–696. [5]

Cella, D., W. Riley, A. Stone, et al. 2010. The Patient-Reported Outcomes Measurement Information System (PROMIS) Developed and Tested Its First Wave of Adult Self-Reported Health Outcome Item Banks: 2005–2008. *J. Clin. Epidemiol.* 63:1179–1194. [1, 17]

Ceunen, E., J. W. S. Vlaeyen, and I. Van Diest. 2016. On the Origin of Interoception. *Front. Psychol.* 7:743. [13]

Chabris, C. F., P. R. Heck, J. Mandart, D. J. Benjamin, and D. J. Simons. 2019. No Evidence That Experiencing Physical Warmth Promotes Interpersonal Warmth. *Soc. Psychol.* 50:127–132. [12]

Chakrabarty, T., J. Ogrodniczuk, and G. Hadjipavlou. 2016. Predictive Neuroimaging Markers of Psychotherapy Response: A Systematic Review. *Harv. Rev. Psychiatry* 24:396–405. [5]

Chalmers, D. J. 1996. The Conscious Mind: In Search of a Fundamental Theory. Philosophy of Mind Series. New York: Oxford Univ. Press. [12]

Chamberlain, S. R., B. L. Odlaug, V. Boulougouris, N. A. Fineberg, and J. E. Grant. 2009. Trichotillomania: Neurobiology and Treatment. *Neurosci. Biobehav. Rev.* 33:831–842. [3]

Chambon, V., H. Thero, C. Findling, and E. Koechlin. 2018. Believing in One's Power: A Counterfactual Heuristic for Goal-Directed Control. *bioRxiv*: 498675. [12]

Chand, G. B., and M. Dhamala. 2016. The Salience Network Dynamics in Perceptual Decision-Making. *NeuroImage* 134:85–93. [10]

Chatham, C. H., and D. Badre. 2015. Multiple Gates on Working Memory. *Curr. Opin. Behav. Sci.* **1**:23–31. [11]

Chatham, C. H., M. J. Frank, and D. Badre. 2014. Corticostriatal Output Gating during Selection from Working Memory. *Neuron* **81**:930–942. [11]

Chen, B. T., H. J. Yau, C. Hatch, et al. 2013a. Rescuing Cocaine-Induced Prefrontal Cortex Hypoactivity Prevents Compulsive Cocaine Seeking. *Nature* **496**:359–362. [5]

Chen, L. W., D. Sun, S. L. Davis, et al. 2018a. Smaller Hippocampal CA1 Subfield Volume in Posttraumatic Stress Disorder. *Depr. Anxiety* **35**:1018–1029. [6]

Chen, R., J. Classen, C. Gerloff, et al. 1997. Depression of Motor Cortex Excitability by Low-Frequency Transcranial Magnetic Stimulation. *Neurology* **48**:1398–1403. [16]

Chen, T., B. Becker, J. Camilleri, et al. 2018b. A Domain-General Brain Network Underlying Emotional and Cognitive Interference Processing: Evidence from Coordinate-Based and Functional Connectivity Meta-Analyses. *Brain Struct. Funct.* **223**:3813–3840. [9, 13]

Chen, T., L. Michels, K. Supekar, et al. 2014. Role of the Anterior Insular Cortex in Integrative Causal Signaling During Multisensory Auditory-Visual Attention. *Eur. J. Neurosci.* **41**:264–274. [10]

Chen, T.-W., T. J. Wardill, Y. Sun, et al. 2013b. Ultrasensitive Fluorescent Proteins for Imaging Neuronal Activity. Nature 499:295–300. [4]

Chiba, T., T. Kanazawa, A. Koizumi, et al. 2019. Current Status of Neurofeedback for Post-Traumatic Stress Disorder: A Systematic Review and the Possibility of Decoded Neurofeedback. *Front. Hum. Neurosci.* **13**:233. [17]

Chikama, M., N. R. McFarland, D. G. Amaral, and S. N. Haber. 1997. Insular Cortical Projections to Functional Regions of the Striatum Correlate with Cortical Cytoarchitectonic Organization in the Primate. *J. Neurosci.* **17**:9686–9705. [10]

Choi, E. Y., G. K. Drayna, and D. Badre. 2018. Evidence for a Functional Hierarchy of Association Networks. *J. Cogn. Neurosci.* **30**:722–736. [11]

Choi, E. Y., Y. Tanimura, P. R. Vage, E. H. Yates, and S. N. Haber. 2017. Convergence of Prefrontal and Parietal Anatomical Projections in a Connectional Hub in the Striatum. *NeuroImage* **146**:821–832. [10]

Choi, S. W., D. J. Kim, J. S. Choi, et al. 2015. Comparison of Risk and Protective Factors Associated with Smartphone Addiction and Internet Addiction. *J. Behav. Addict.* **4**:308–314. [17]

Christakis, N. 2019. Blueprint: The Evolutionary Origins of a Good Society. New York: Little Brown. [7]

Christoff, K., A. M. Gordon, J. Smallwood, R. Smith, and J. W. Schooler. 2009. Experience Sampling during fMRI Reveals Default Network and Executive System Contributions to Mind Wandering. *PNAS* **106**:8719–8724. [9]

Christoff, K., Z. C. Irving, K. C. R. Fox, R. N. Spreng, and J. R. Andrews-Hanna. 2016. Mind-Wandering as Spontaneous Thought: A Dynamic Framework. *Nat. Rev. Neurosci.* **17**:718–731. [10]

Clark, D. A. 2005. Intrusive Thoughts in Clinical Disorders: Theory, Research, and Treatment. New York: Guilford Press. [5, 7–9, 14, 15, 17]

Clark, D. A., J. Abramowitz, G. M. Alcolado, et al. 2014a. Part 3: A Question of Perspective: The Association between Intrusive Thoughts and Obsessionality in 11 Countries. *J. Obsessive Compuls. Relat. Disord.* **3**.292–299. [6]

Clark, D. A., and M. Claybourn. 1997. Process Characteristics of Worry and Obsessive Intrusive Thoughts. *Behav. Res. Ther.* **35**:1139–1141. [17]

Clark, D. A., and M. Inozu. 2014. Unwanted Intrusive Thoughts: Cultural, Contextual, Covariational, and Characterological Determinants of Diversity. *J. Obsessive Compuls. Relat. Disord.* **3**:195–204. [17]

Clark, D. A., and K. O'Connor. 2005. Thinking Is Believing: Ego-Dystonic Intrusive Thoughts in Obsessive-Compulsive Disorder. In: Intrusive Thoughts in Clinical Disorders: Theory, Research, and Treatment, ed. D. A. Clark, pp. 145–174. New York: Guilford Press. [1, 15]

Clark, D. A., and C. L. Purdon. 1995. The Assessment of Unwanted Intrusive Thoughts: A Review and Critique of the Literature. *Behav. Res. Ther.* **33**:967–976. [7, 17]

Clark, D. A., and A. S. Radomsky. 2014. Introduction: A Global Perspective on Unwanted Intrusive Thoughts. *J. Obsessive Compuls. Relat. Disord.* **3**:265–268. [17]

Clark, D. M., P. M. Salkovskis, L. G. Ost, et al. 1997. Misinterpretation of Body Sensations in Panic Disorder. *J. Consult. Clin. Psychol.* **65**:203–213. [13]

Clark, I. A., E. A. Holmes, M. W. Woolrich, and C. E. Mackay. 2015. Intrusive Memories to Traumatic Footage: The Neural Basis of Their Encoding and Involuntary Recall. *Psychol. Med.* **46**:505–518. [14]

Clark, I. A., C. E. Mackay, M. W. Woolrich, and E. A. Holmes. 2016. Intrusive Memories to Traumatic Footage: The Neural Basis of Their Encoding and Involuntary Recall. *Psychol. Med.* **46**:505–518. [9, 17]

Clark, I. A., K. E. Niehaus, E. P. Duff, et al. 2014b. First Steps in Using Machine Learning on fMRI Data to Predict Intrusive Memories of Traumatic Film Footage. *Behav. Res. Ther.* **62**:37–46. [6, 14, 17]

Clark, J. J., N. G. Hollon, and P. E. M. Phillips. 2012. Pavlovian Valuation Systems in Learning and Decision Making. *Curr. Opin. Neurobiol.* **22**:1054–1061. [2]

Clarke, H. F., R. N. Cardinal, R. Rygula, et al. 2014. Orbitofrontal Dopamine Depletion Upregulates Caudate Dopamine and Alters Behavior via Changes in Reinforcement Sensitivity. *J. Neurosci.* **34**:7663–7676. [13]

Clarke, H. F., S. Walker, J. Dalley, T. Robbins, and A. Roberts. 2006. Cognitive Inflexibility after Prefrontal Serotonin Depletion Is Behaviorally and Neurochemically Specific. *Cereb. Cortex* **17**:18–27. [13]

Clem, R. L., and R. L. Huganir. 2010. Calcium-Permeable AMPA Receptor Dynamics Mediate Fear Memory Erasure. *Science* **330**:1108–1112. [5]

Clem, R. L., and D. Schiller. 2016. New Learning and Unlearning: Strangers or Accomplices in Threat Memory Attenuation? *Trends Neurosci.* **39**:340–351. [9]

Cohen, A. L., J. Kantner, R. A. Dixon, and D. S. Lindsay. 2011. The Intention Interference Effect: The Difficulty of Ignoring What You Intend to Do. *Exp. Psychol.* **58**:425–433. [9]

Cohen, J. Y., S. Haesler, L. Vong, B. B. Lowell, and N. Uchida. 2012. Neuron-Type-Specific Signals for Reward and Punishment in the Ventral Tegmental Area. *Nature* **482**:85–88. [2]

Cole, M. W., T. Ito, and T. S. Braver. 2015. Lateral Prefrontal Cortex Contributes to Fluid Intelligence through Multinetwork Connectivity. *Brain Connect.* **5**:497–504. [11]

Cole, M. W., J. R. Reynolds, J. D. Power, et al. 2013. Multi-Task Connectivity Reveals Flexible Hubs for Adaptive Task Control. *Nat. Neurosci.* **16**:1348–1355. [10, 11]

Collins, A. G., and M. J. Frank. 2013. Cognitive Control over Learning: Creating, Clustering, and Generalizing Task-Set Structure. *Psychol. Rev.* **120**:190–229. [11]

————. 2014. Opponent Actor Learning (OpAL): Modeling Interactive Effects of Striatal Dopamine on Reinforcement Learning and Choice Incentive. *Psychol. Rev.* **121**:337–366. [11]

Conant, R. C., and R. W. Ashby. 1970. Every Good Regulator of a System Must Be a Model of That System. *Int. J. Systems Sci.* **1**:89–97. [13]

Conceição, V. A., Á. Dias, A. C. Farinha, and T. V. Maia. 2017. Premonitory Urges and Tics in Tourette Syndrome: Computational Mechanisms and Neural Correlates. *Curr. Opin. Neurobiol.* **46**:187–199. [13]

Cools, R., and M. D'Esposito. 2011. Inverted-U–Shaped Dopamine Actions on Human Working Memory and Cognitive Control. *Biol. Psychiatry* **69**:e113–e125. [13]

Corbit, L. H., B. C. Chieng, and B. W. Balleine. 2014. Effects of Repeated Cocaine Exposure on Habit Learning and Reversal by N-Acetylcysteine. *Neuropsychopharmacology* **39**:1893–1901. [3]

Corcoran, K. M., and S. R. Woody. 2008. Appraisals of Obsessional Thoughts in Normal Samples. *Behav. Res. Ther.* **46**:71–83. [13]

Cornblath, E. J., D. M. Lydon-Staley, and D. S. Bassett. 2019. Harnessing Networks and Machine Learning in Neuropsychiatric Care. *Curr. Opin. Neurobiol.* **55**:32–39. [11]

Cortese, A., K. Amano, A. Koizumi, M. Kawato, and H. C. Lau. 2016. Multivoxel Neurofeedback Selectively Modulates Confidence without Changing Perceptual Performance. *Nat. Commun.* **7**:13669. [12]

Cortese, A., H. C. Lau, and M. Kawato. 2019. Metacognition Facilitates the Exploitation of Unconscious Brain States. *bioRxiv*: 548941. [12]

Coughtrey, A. E., R. Shafran, M. Lee, and S. J. Rachman. 2012. It's the Feeling inside My Head: A Qualitative Analysis of Mental Contamination in Obsessive-Compulsive Disorder. *Behav. Cogn. Psychother.* **40**:163–173. [8]

Coughtrey, A. E., R. Shafran, and S. J. Rachman. 2015. Imagery in Mental Contamination. *Behav. Cogn. Psychother.* **43**:257–269. [14]

Courtney, K. E., D. G. Ghahremani, and L. A. Ray. 2016. The Effects of Pharmacological Opioid Blockade on Neural Measures of Drug Cue-Reactivity in Humans. *Neuropsychopharmacology* **41**:2872–2881. [15]

Craig, A. D. 2003. Interoception: The Sense of the Physiological Condition of the Body. *Curr. Opin. Neurobiol.* **13**:500–505. [13]

————. 2009. How Do You Feel—Now? The Anterior Insula and Human Awareness. *Nat. Rev. Neurosci.* **10**:59–70. [10, 13]

————. 2010. The Sentient Self. *Brain Struct. Funct.* **214**:563–577. [13]

————. 2011. Significance of the Insula for the Evolution of Human Awareness of Feelings from the Body. *Ann. N.Y. Acad. Sci* **1225**:72–82. [9]

Craik, K. J. W. 1948. Theory of the Human Operator in Control Systems, I: The Operator as an Engineering System. *Br. J. Psychol.* **38**:56–61. [11]

Cremers, H. R., T. D. Wager, and T. Yarkoni. 2017. The Relation between Statistical Power and Inference in fMRI. *PLoS One* **12**:e0184923. [6]

Creswell, A., T. White, V. Dumoulin, et al. 2018. Generative Adversarial Networks: An Overview. *IEEE Signal Process. Mag.* **35**:53–65. [12]

Critchley, H. D., and N. A. Harrison. 2013. Visceral Influences on Brain and Behavior. *Neuron* **77**:624–638. [10, 13]

Critchley, H. D., J. Tang, D. Glaser, B. Butterworth, and R. J. Dolan. 2005. Anterior Cingulate Activity during Error and Autonomic Response. *NeuroImage* **27**:885–895. [13]

Critchley, H. D., S. Wiens, P. Rotshtein, A. Öhman, and R. J. Dolan. 2004. Neural Systems Supporting Interoceptive Awareness. *Nat. Neurosci.* **7**:189–195. [10, 13]

Crittenden, B. M., D. J. Mitchell, and J. Duncan. 2016. Task Encoding across the Multiple Demand Cortex Is Consistent with a Frontoparietal and Cingulo-Opercular Dual Networks Distinction. *J. Neurosci.* **36**:6147–6155. [11]

Crucianelli, L., V. Cardi, J. Treasure, P. M. Jenkinson, and A. Fotopoulou. 2016. The Perception of Affective Touch in Anorexia Nervosa. *Psychiatry Res.* **239**:72–78. [13]

Cuc, A., J. Koppel, and W. Hirst. 2007. Silence Is Not Golden: A Case for Socially Shared Retrieval-Induced Forgetting. *Psychol. Sci.* **18**:727–733. [9]

Cuthbert, B. N., and T. R. Insel. 2013. Toward the Future of Psychiatric Diagnosis: The Seven Pillars of RDoC. *BMC Med.* **11**:126. [9]

Dabney, W., M. Rowland, M. G. Bellemare, and R. Munos. 2017. Distributional Reinforcement Learning with Quantile Regression. *ArXiv* 1710.10044v10041. [9]

Dalley, J. W., and T. W. Robbins. 2017. Fractionating Impulsivity: Neuropsychiatric Implications. *Nat. Rev. Neurosci.* **18**:158–171. [11]

Damasio, A. R. 1999. The Feeling of What Happens: Body and Emotion in the Making of Consciousness. London: Vintage Books. [13]

d'Angelo, C., D. M. Eagle, C.-M. Coman, and T. W. Robbins. 2017. Role of the Medial Prefrontal Cortex and Nucleus Accumbens in an Operant Model of Checking Behaviour and Uncertainty. Brain Neurosci. Adv. 1:2398212817733403. [5]

d'Angelo, C., D. M. Eagle, J. E. Grant, et al. 2014. Animal Models of Obsessive-Compulsive Spectrum Disorders. *CNS Spectr.* **19**:28–49. [5]

Daniels, J. K., and E. Vermetten. 2016. Odor-Induced Recall of Emotional Memories in PTSD-Review and New Paradigm for Research. *Exp. Neurol.* **284**:168–180. [6]

Das, R. K., A. Tamman, V. Nikolova, et al. 2016. Nitrous Oxide Speeds the Reduction of Distressing Intrusive Memories in an Experimental Model of Psychological Trauma. *Psychol. Med.* **46**:1749–1759. [14]

Dauwan, M., M. M. J. Linszen, A. W. Lemstra, et al. 2018. EEG-Based Neurophysiological Indicators of Hallucinations in Alzheimer's Disease: Comparison with Dementia with Lewy Bodies. *Neurobiol. Aging* **67**:75–83. [5]

Davies, C., A. Malik, A. Pictet, S. E. Blackwell, and E. A. Holmes. 2012. Involuntary Memories after a Positive Film Are Dampened by a Visuospatial Task: Unhelpful in Depression but Helpful in Mania? *Clin. Psychol. Psychother.* **19**:341–351. [14, 17]

Davis, J., and M. E. Bitterman. 1971. Differential Reinforcement of Other Behavior (Dro): A Yoked-Control Comparison. *J. Exp. Anal. Behav.* **15**:237–241. [3]

Daw, N. D., S. J. Gershman, B. Seymour, P. Dayan, and R. J. Dolan. 2011. Model-Based Influences on Humans' Choices and Striatal Prediction Errors. *Neuron* **69**:1204–1215. [1]

Daw, N. D., S. Kakade, and P. Dayan. 2002. Opponent Interactions between Serotonin and Dopamine. *Neural Netw.* **15**:603–616. [2]

Daw, N. D., Y. Niv, and P. Dayan. 2005. Uncertainty-Based Competition between Prefrontal and Dorsolateral Striatal Systems for Behavioral Control. *Nat. Neurosci.* **8**:1704–1711. [9]

Day, S. J., E. A. Holmes, and A. Hackmann. 2004. Occurrence of Imagery and Its Link with Early Memories in Agoraphobia. *Memory* **12**:416–427. [14]

Dayan, P., and K. C. Berridge. 2014. Model-Based and Model-Free Pavlovian Reward Learning: Revaluation, Revision, and Revelation. *Cogn. Affect. Behav. Neurosci.* **14**:473–492. [12]

Dean, O., F. Giorlando, and M. Berk. 2011. N-Acetylcysteine in Psychiatry: Current Therapeutic Evidence and Potential Mechanisms of Action. *J. Psychiatry Neurosci.* **36**:78–86. [15]

Deco, G., and E. T. Rolls. 2005. Neurodynamics of Biased Competition and Cooperation for Attention: A Model with Spiking Neurons. *J. Neurophysiol.* **94**:295–313. [13]

Deecke, L., P. Scheid, and H. H. Kornhuber. 1969. Distribution of Readiness Potential, Pre-Motion Positivity, and Motor Potential of the Human Cerebral Cortex Preceding Voluntary Finger Movements. *Exp. Brain Res.* **7**:158–168. [12]

Dehaene, S. 2014. Consciousness and the Brain: Deciphering How the Brain Codes Our Thoughts. New York: Viking Adult. [12]

Dehaene, S., and J.-P. Changeux. 2004. Neural Mechanisms for Access to Consciousness. In: The Cognitive Neurosciences, 3rd edition, ed. M. S. Gazzaniga, pp. 1145–1158. [13]

Dehaene, S., M. Kerszberg, and J.-P. Changeux. 1998. A Neuronal Model of a Global Workspace in Effortful Cognitive Tasks. *PNAS* **95**:14529–14534. [13]

Dehaene, S., H. C. Lau, and S. Kouider. 2017. What Is Consciousness, and Could Machines Have It? *Science* **358**:486–492. [12]

Deisseroth, K., and M. J. Schnitzer. 2013. Engineering Approaches to Illuminating Brain Structure and Dynamics. *Neuron* **80**:568–577. [4]

de Lafuente, V., M. Jazayeri, and M. N. Shadlen. 2015. Representation of Accumulating Evidence for a Decision in Two Parietal Areas. J. Neurosci. 35:4306–4318. [13]

Delamater, A. R. 1995. Outcome-Selective Effects of Intertrial Reinforcement in a Pavlovian Appetitive Conditioning Paradigm with Rats. *Anim. Learn. Behav.* **23**:31–39. [3]

Deng, Z. D., S. H. Lisanby, and A. V. Peterchev. 2013. Electric Field Depth-Focality Tradeoff in Transcranial Magnetic Stimulation: Simulation Comparison of 50 Coil Designs. *Brain Stimul.* **6**:1–13. [16]

DePoy, L. M., L. P. Shapiro, H. W. Kietzman, K. M. Roman, and S. L. Gourley. 2019. β1-Integrins in the Developing Orbitofrontal Cortex Are Necessary for Expectancy Updating in Mice. *J. Neurosci.* **Jun 28**:3072–3018. [5]

DePoy, L. M., K. S. Zimmermann, P. J. Marvar, and S. L. Gourley. 2017. Induction and Blockade of Adolescent Cocaine-Induced Habits. *Biol. Psychiatry* **81**:595–605. [5]

Depue, B. E. 2012. A Neuroanatomical Model of Prefrontal Inhibitory Modulation of Memory Retrieval. *Neurosci. Biobehav. Rev.* **36**:1382–1399. [9]

Depue, B. E., M. T. Banich, and T. Curran. 2006. Suppression of Emotional and Nonemotional Content in Memory: Effects of Repetition on Cognitive Control. *Psychol. Sci.* **17**:441–447. [6]

Depue, B. E., T. Curran, and M. T. Banich. 2007. Prefrontal Regions Orchestrate Suppression of Emotional Memories via a Two-Phase Process. *Science* **317**:215–219. [6, 9]

Depue, B. E., N. Ketz, M. V. Mollison, et al. 2013. ERPs and Neural Oscillations During Volitional Suppression of Memory Retrieval. *J. Cogn. Neurosci.* **25**:1624–1633. [6]

Depue, B. E., J. M. Orr, H. R. Smolker, F. Naaz, and M. T. Banich. 2016. The Organization of Right Prefrontal Networks Reveals Common Mechanisms of Inhibitory Regulation across Cognitive Emotional and Motor Processes. *Cereb. Cortex* **26**:1634–1646. [6, 9]

Deroche-Gamonet, V., D. Belin, and P. V. Piazza. 2004. Evidence for Addiction-Like Behavior in the Rat. *Science* **305**:1014–1017. [5]

DeRubeis, R. J., G. J. Siegle, and S. D. Hollon. 2008. Cognitive Therapy versus Medication for Depression: Treatment Outcomes and Neural Mechanisms. *Nat. Rev. Neurosci.* 9:788–796. [12]

Desimone, R. 1998. Visual Attention Mediated by Biased Competition in Extrastriate Visual Cortex. *Philos. Trans. R. Soc. Lond. B. Biol. Sci.* 353:1245–1255. [13]

Desimone, R., M. Wessinger, L. Thomas, and W. Schneider. 1990. Attentional Control of Visual Perception: Cortical and Subcortical Mechanisms. *Cold Spring Harb. Symp. Quant. Biol.* 55:963–971. [13]

D'Esposito, M., and B. R. Postle. 2015. The Cognitive Neuroscience of Working Memory. Ann. Rev. Psychol. 66:115–142. [9]

Desrochers, T. M., C. H. Chatham, and D. Badre. 2015. The Necessity of Rostrolateral Prefrontal Cortex for Higher-Level Sequential Behavior. *Neuron* 87:1357–1368. [11]

DeVille, D. C., K. L. Kerr, J. A. Avery, et al. 2018. The Neural Bases of Interoceptive Encoding and Recall in Healthy Adults and Adults with Depression. *Biol. Psychiatry Cogn. Neurosci. Neuroimaging* 3:546–554. [10]

Dewar, M., J. Alber, N. Cowan, and S. Della Sala. 2014. Boosting Long-Term Memory via Wakeful Rest: Intentional Rehearsal Is Not Necessary, Consolidation Is Sufficient. *PLoS One* 9:e109542. [14]

de Wit, S., S. B. Ostlund, B. W. Balleine, and A. Dickinson. 2009. Resolution of Conflict between Goal-Directed Actions: Outcome Encoding and Neural Control Processes. J. Exp. Psychol. Anim. Behav. Process. 35:382–393. [3]

Dibbets, P. 2019. A Novel Virtual Reality Paradigm: Predictors for Stress-Related Intrusions and Avoidance Behavior. *J. Behav. Ther. Exp. Psychiatry* 2019:101449. [6]

Dickinson, A. 1994. Instrumental Conditioning. In: Animal Cognition and Learning, ed. N. J. Mackintosh, pp. 4–79. London: Academic Press. [3]

———. 2012. Associative Learning and Animal Cognition. *Philos. Trans. R. Soc. Lond. B. Biol. Sci.* 367:2733–2742. [3]

Dickinson, A., and B. Balleine. 1994. Motivational Control of Goal-Directed Action. *Anim. Learn. Behav.* 22:1–18. [3]

———. 2002. The Role of Learning in the Operation of Motivational Systems. In: Steven's Handbook of Experimental Psychology: Learning, Motivation, and Emotion, ed. H. Pashler and C. R. Gallistel, pp. 497–533, vol. 3. New York: John Wiley & Sons. [3]

Dickinson, A., B. Balleine, A. Watt, F. Gonzalez, and R. A. Boakes. 1995. Motivational Control after Extended Instrumental Training. *Anim. Learn. Behav.* 23:197–206. [3]

Dickinson, A., D. J. Nicholas, and C. D. Adams. 1983. The Effect of the Instrumental Training Contingency on Susceptibility to Reinforcer Devaluation. *Q. J. Exp. Psychol. Sect. B* 35:35–51. [3]

Dickinson, A., S. Squire, Z. Varga, and J. W. Smith. 1998. Omission Learning after Instrumental Pretraining. *Q. J. Exp. Psychol. Sect. B* 51:271–286. [3]

Di Lazzaro, V., U. Ziemann, and R. N. Lemon. 2008. State of the Art: Physiology of Transcranial Motor Cortex Stimulation. Brain Stimul. 1:345–362. [16]

Dinur-Klein, L., P. Dannon, A. Hadar, et al. 2014. Smoking Cessation Induced by Deep Repetitive Transcranial Magnetic Stimulation of the Prefrontal and Insular Cortices: A Prospective, Randomized Controlled Trial. *Biol. Psychiatry* 76:742–749. [16]

Di Simplicio, M., F. Renner, S. E. Blackwell, et al. 2016. An Investigation of Mental Imagery in Bipolar Disorder: Exploring "the Mind's Eye". Bipolar Disord. 18:669–683. [17]

Ditzen, B., and M. Heinrichs. 2014. Psychobiology of Social Support: The Social Dimension of Stress Buffering. *Restor. Neurol. Neurosci.* **32**:149–162. [14]

Dixon, M. L., A. De La Vega, C. Mills, et al. 2018. Heterogeneity within the Frontoparietal Control Network and Its Relationship to the Default and Dorsal Attention Networks. *PNAS* **115**:E1598–E1607. [11]

Dixon, M. L., K. C. R. Fox, and K. Christoff. 2014. A Framework for Understanding the Relationship between Externally and Internally Directed Cognition. *Neuropsychologia* **62**:321–330. [13]

Dolan, R. J., and P. Dayan. 2013. Goals and Habits in the Brain. *Neuron* **80**:312–325. [8]

Dosenbach, N. U., D. A. Fair, A. L. Cohen, B. L. Schlaggar, and S. E. Petersen. 2008. A Dual-Networks Architecture of Top-Down Control. *Trends Cogn. Sci.* **12**:99–105. [11]

Dosenbach, N. U., D. A. Fair, F. M. Miezin, et al. 2007. Distinct Brain Networks for Adaptive and Stable Task Control in Humans. *PNAS* **104**:11073–11078. [10, 11]

Dosenbach, N. U., K. M. Visscher, E. D. Palmer, et al. 2006. A Core System for the Implementation of Task Sets. *Neuron* **50**:799–812. [11]

Dougall, A. L., K. J. Craig, and A. Baum. 1999. Assessment of Characteristics of Intrusive Thoughts and Their Impact on Distress Among Victims of Traumatic Events. *Psychosom. Med.* **61**:38–48. [6]

Doya, K., S. Ishii, A. Pouget, and R. P. N. Rao, eds. 2007. Bayesian Brain: Probabilistic Approaches to Neural Coding. Cambridge, MA: MIT Press. [10]

Drevets, W. C., J. L. Price, and M. L. Furey. 2008. Brain Structural and Functional Abnormalities in Mood Disorders: Implications for Neurocircuitry Models of Depression. *Brain Struct. Funct.* **213**:93–118. [8]

Drevets, W. C., J. L. Price, J. R. Simpson, et al. 1997. Subgenual Prefrontal Cortex Abnormalities in Mood Disorders. *Nature* **386**:824–827. [9]

Drummond, S. P. A., M. P. Paulus, and S. F. Tapert. 2006. Effects of Two Nights Sleep Deprivation and Two Nights Recovery Sleep on Response Inhibition. *J. Sleep Res.* **15**:261–265. [9]

Dudai, Y. 2004. The Neurobiology of Consolidations, or, How Stable Is the Engram? *Ann. Rev. Psychol.* **55**:51–86. [9, 14]

———. 2012. The Restless Engram: Consolidations Never End. *Annu. Rev. Neurosci.* **35**:227–247. [9]

Dudley, R., C. Aynsworth, U. Mosimann, et al. 2019. A Comparison of Visual Hallucinations across Disorders. *Psychiatry Res.* **272**:86–92. [5]

Duncan, J. 2013. The Structure of Cognition: Attentional Episodes in Mind and Brain. *Neuron* **80**:35–50. [11]

Durkheim, E. 1951. Suicide, a Study in Sociology. Glencoe, IL: Free Press. [12]

Durstewitz, D., J. K. Seamans, and T. J. Sejnowski. 2000. Neurocomputational Models of Working Memory. *Nat. Neurosci.* **3**:1184–1191. [13]

Dweck, C. S. 1999. Self-Theories: Their Role in Motivation, Personality, and Development. Essays in Social Psychology. Philadelphia: Psychology Press. [12]

Eagle, D. M., C. Noschang, C. d'Angelo, et al. 2014. The Dopamine D2/D3 Receptor Agonist Quinpirole Increases Checking-Like Behaviour in an Operant Observing Response Task with Uncertain Reinforcement: A Novel Possible Model of OCD. *Behav. Brain Res.* **264**:207–229. [5]

Eccles, J. S., and A. Wigfield. 2002. Motivational Beliefs, Values, and Goals. *Ann. Rev. Psychol.* **53**:109–132. [13]

Economidou, D., Y. Pelloux, T. W. Robbins, J. W. Dalley, and B. J. Everitt. 2009. High Impulsivity Predicts Relapse to Cocaine-Seeking after Punishment-Induced Abstinence. *Biol. Psychiatry* **65**:851–856. [5]

Edwards, R. R., M. T. Smith, G. Stonerock, and J. A. Haythornthwaite. 2006. Pain-Related Catastrophizing in Healthy Women Is Associated with Greater Temporal Summation of and Reduced Habituation to Thermal Pain. *Clin. J. Pain* **22**:730–737. [9]

Edwards, S., and M. Dickerson. 1987a. Intrusive Unwanted Thoughts: A Two-Stage Model of Control. *Br. J. Med. Psychol.* **60**:317–328. [17]

———. 1987b. On the Similarity of Positive and Negative Intrusions. *Behav. Res. Ther.* **25**:207–211. [13]

Egner, T., and C. Summerfield. 2013. Grounding Predictive Coding Models in Empirical Neuroscience Research. *Behav. Brain Sci.* **36**:210–211. [13]

Ehlers, A., and D. M. Clark. 2000. A Cognitive Model of Posttraumatic Stress Disorder. *Behav. Res. Ther.* **38**:319–345. [9]

Ehlers, A., and R. Steil. 1995. Maintenance of Intrusive Memories in Posttraumatic Stress Disorder: A Cognitive Approach. *Behav. Cogn. Psychother.* **23**:217–249. [9]

Eisenreich, B. R., R. Akaishi, and B. Y. Hayden. 2017. Control without Controllers: Toward a Distributed Neuroscience of Executive Control. *J. Cogn. Neurosci.* **29**:1684–1698. [11]

Ekhtiari, H., H. Tavakoli, G. Addolorato, et al. 2019. Transcranial Electrical and Magnetic Stimulation (tES and TMS) for Addiction Medicine: A Consensus Paper on the Present State of the Science and the Road Ahead. *Neurosci. Biobehav. Rev.* **104**:118–140. [16]

Ellard, K. K., J. P. Zimmerman, N. Kaur, et al. 2018. Functional Connectivity between Anterior Insula and Key Nodes of Frontoparietal Executive Control and Salience Networks Distinguish Bipolar Depression from Unipolar Depression and Healthy Control Subjects. *Biol. Psychiatry Cogn. Neurosci. Neuroimaging* **3**:473–484. [13]

Elliott, M. L., A. Romer, A. R. Knodt, and A. R. Hariri. 2018. A Connectome-Wide Functional Signature of Transdiagnostic Risk for Mental Illness. *Biol. Psychiatry* **84**:452–459. [10]

Elsey, J. W. B., V. A. van Ast, and M. Kindt. 2018. Human Memory Reconsolidation: A Guiding Framework and Critical Review of the Evidence. *Psychol. Bull.* **144**:797–848. [9, 14]

Elua, I., K. R. Laws, and L. Kvavilashvili. 2012. From Mind-Pops to Hallucinations? A Study of Involuntary Semantic Memories in Schizophrenia. *Psychiatry Res.* **196**:165–170. [6]

Elwafi, H. M., K. Witkiewitz, S. Mallik, T. A. Thornhill, and J. A. Brewer. 2013. Mindfulness Training for Smoking Cessation: Moderation of the Relationship between Craving and Cigarette Use. *Drug Alcohol Depend.* **130**:222–229. [17]

Emerson, L.-M., C. Heapy, and G. Garcia-Soriano. 2017. Which Facets of Mindfulness Protect Individuals from the Negative Experiences of Obsessive Intrusive Thoughts? *Mindfulness* **9**:1170–1180. [6]

Engels, A. S., W. Heller, A. Mohanty, et al. 2007. Specificity of Regional Brain Activity in Anxiety Types during Emotion Processing. *Psychophysiology* **44**:352–363. [9]

Engen, H. G., and M. C. Anderson. 2018. Memory Control: A Fundamental Mechanism of Emotion Regulation. *Trends Cogn. Sci.* **22**:982–995. [9]

Esmaeeli, S., K. Murphy, G. M. Swords, et al. 2019. Visual Hallucinations, Thalamocortical Physiology and Lewy Body Disease: A Review. *Neurosci. Biobehav. Rev.* **103**:337–351. [5]

Esterlis, I., J. O. Hannestad, F. Bois, et al. 2013. Imaging Changes in Synaptic Acetylcholine Availability in Living Human Subjects. *J. Nucl. Med.* **54**:78–82. [5]

Etkin, A., T. Egner, and R. Kalisch. 2011. Emotional Processing in Anterior Cingulate and Medial Prefrontal Cortex. *Trends Cogn. Sci.* **15**:85–93. [13]

Etkin, A., A. Maron-Katz, W. Wu, et al. 2019. Using fMRI Connectivity to Define a Treatment-Resistant Form of Post-Traumatic Stress Disorder. *Sci. Transl. Med.* **11**:eaal3236. [17]

Everitt, B. J., and T. W. Robbins. 2016. Drug Addiction: Updating Actions to Habits to Compulsions Ten Years On. *Ann. Rev. Psychol.* **67**:23–50. [3, 8, 13]

Evrard, H. C. 2019. The Organization of the Primate Insular Cortex. *Front. Neuroanat.* **13**:43. [13]

Fanni, S., S. Scheggi, F. Rossi, et al. 2019. 5alpha-reductase Inhibitors Dampen L-DOPA-Induced Dyskinesia via Normalization of Dopamine D1-Receptor Signaling Pathway and D1-D3 Receptor Interaction. *Neurobiol. Dis.* **121**:120–130. [5]

Fawcett, J. M., R. G. Benoit, P. Gagnepain, et al. 2015. The Origins of Repetitive Thought in Rumination: Separating Cognitive Style from Deficits in Inhibitory Control over Memory. *J. Behav. Ther. Exp. Psychiatry* **47**:1–8. [9]

Fedorenko, E., J. Duncan, and N. Kanwisher. 2013. Broad Domain Generality in Focal Regions of Frontal and Parietal Cortex. *PNAS* **110**:16616–16621. [11]

Feinberg, I. 1978. Efference Copy and Corollary Discharge: Implications for Thinking and Its Disorders. *Schizophr. Bull.* **4**:636–640. [8, 13]

Feldmann-Wüstefeld, T., and E. K. Vogel. 2019. Neural Evidence for the Contribution of Active Suppression During Working Memory Filtering. *Cereb. Cortex* **29**:529–543. [9]

Feng, J., C. Zhang, J. R. Lischinsky, et al. 2019. A Genetically Encoded Fluorescent Sensor for Rapid and Specific *in Vivo* Detection of Norepinephrine. *Neuron* **102**:745–761.e748. [4]

Ferguson, C. J. 2017. Everything in Moderation: Moderate Use of Screens Unassociated with Child Behavior Problems. *Psychiatric Q.* **88**:797–805. [17]

Festinger, L. 1962. A Theory of Cognitive Dissonance. Stanford: Stanford Univ. Press. [9]

Fettes, P., L. Schulze, and J. Downar. 2017. Cortico-Striatal-Thalamic Loop Circuits of the Orbitofrontal Cortex: Promising Therapeutic Targets in Psychiatric Illness. *Front. Syst. Neurosci.* **11**:25. [3]

Fife, K. H., N. A. Gutierrez-Reed, V. Zell, et al. 2017. Causal Role for the Subthalamic Nucleus in Interrupting Behavior. *eLife* **6**: e27689. [11]

Fiorillo, C. D. 2013. Two Dimensions of Value: Dopamine Neurons Represent Reward but Not Aversiveness. *Science* **341**:546–549. [2]

Fischer, J. M. 2006. My Way: Essays on Moral Responsibility. New York: Oxford Univ. Press. [12]

Fisher, C. M. 1991. Visual Hallucinations on Eye Closure Associated with Atropine Toxicity. A Neurological Analysis and Comparison with Other Visual Hallucinations. *Can. J. Neurol. Sci.* **18**:18–27. [5]

Fisher, H. 2016. Anatomy of Love: A Natural History of Mating, Marriage, and Why We Stray. New York: WW Norton. [7]

Fitzgerald, P. B., and Z. J. Daskalakis. 2008. A Review of Repetitive Transcranial Magnetic Stimulation Use in the Treatment of Schizophrenia. *Can. J. Psychiatry* **53**:567–576. [16]

Fitzgerald, P. B., S. Fountain, and Z. J. Daskalakis. 2006. A Comprehensive Review of the Effects of rTMS on Motor Cortical Excitability and Inhibition. *Clin. Neurophysiol.* **117**:2584–2596. [16]

Flagel, S. B., H. Akil, and T. E. Robinson. 2009. Individual Differences in the Attribution of Incentive Salience to Reward-Related Cues: Implications for Addiction. *Neuropharmacology* **56 Suppl 1**:139–148. [5]

Flagel, S. B., C. M. Cameron, K. N. Pickup, et al. 2011a. A Food Predictive Cue Must Be Attributed with Incentive Salience for It to Induce C-Fos mRNA Expression in Cortico-Striatal-Thalamic Brain Regions. *Neuroscience* **196**:80–96. [5]

Flagel, S. B., J. J. Clark, T. E. Robinson, et al. 2011b. A Selective Role for Dopamine in Stimulus-Reward Learning. *Nature* **469**:53–U63. [2, 5]

Flavell, C. R., D. J. Barber, and J. L. C. Lee. 2011. Behavioural Memory Reconsolidation of Food and Fear Memories. *Nat. Commun.* **2**:504. [9]

Fleming, S. M., J. Ryu, J. G. Golfinos, and K. E. Blackmon. 2014. Domain-Specific Impairment in Metacognitive Accuracy Following Anterior Prefrontal Lesions. *Brain* **137**:2811–2822. [12]

Fletcher, P. C., and A. Fotopoulou. 2015. Sense of Agency and Its Disruption: Clinical and Computational Perspectives. In: The Sense of Agency, ed. P. Haggard and B. Eitam, pp. 347–370. Oxford: Oxford Univ. Press. [13]

Foland-Ross, L. C., J. P. Hamilton, J. Joorman, et al. 2013. The Neural Basis of Difficulties Disengaging from Negative Irrelevant Material in Major Depression. *Psychol. Sci.* **24**:334–344. [9]

Forstmann, B. U., M. C. Keuken, S. Jahfari, et al. 2012. Cortico-Subthalamic White Matter Tract Strength Predicts Interindividual Efficacy in Stopping a Motor Response. *NeuroImage* **60**:370–375. [11]

Fotopoulou, A. 2015. The Virtual Bodily Self: Mentalisation of the Body as Revealed in Anosognosia for Hemiplegia. *Conscious. Cogn.* **33**:500–510. [13]

Fotopoulou, A., and M. Tsakiris. 2017. Mentalizing Homeostasis: The Social Origins of Interoceptive Inference. *Neuropsychoanalysis* **19**:3–28. [13]

Fouragnan, E. F., B. K. H. Chau, D. Folloni, et al. 2019. The Macaque Anterior Cingulate Cortex Translates Counterfactual Choice Value into Actual Behavioral Change. *Nat. Neurosci.* **22**:797–808. [11]

Fox, E., K. Dutton, A. Yates, G. A. Georgiou, and E. Mouchlianitis. 2015. Attentional Control and Suppressing Negative Thought Intrusions in Pathological Worry. *Clin. Psychol. Sci.* **3**:593–606. [6]

Fox, M. D., A. Z. Snyder, J. L. Vincent, et al. 2005. From the Cover: The Human Brain Is Intrinsically Organized into Dynamic, Anticorrelated Functional Networks. *PNAS* **102**:9673–9678. [10, 13]

Fraley, R. C., and N. W. Hudson. 2014. Review of Intensive Longitudinal Methods: An Introduction to Diary and Experience Sampling Research. *J. Soc. Psychol.* **154**:89–91. [9]

Frank, M. J., and D. Badre. 2012. Mechanisms of Hierarchical Reinforcement Learning in Corticostriatal Circuits 1: Computational Analysis. *Cereb. Cortex* **22**:509–526. [11]

Frank, M. J., C. Gagne, E. Nyhus, et al. 2015. fMRI and EEG Predictors of Dynamic Decision Parameters during Human Reinforcement Learning. *J. Neurosci.* **35**:485–494. [11]

Frank, M. J., B. Loughry, and R. C. O'Reilly. 2001. Interactions between Frontal Cortex and Basal Ganglia in Working Memory: A Computational Model. *Cogn. Affect. Behav. Neurosci.* **1**:137–160. [13]

Frank, M. J., and R. C. O'Reilly. 2006. A Mechanistic Account of Striatal Dopamine Function in Human Cognition: Psychopharmacological Studies with Cabergoline and Haloperidol. *Behav. Neurosci.* **120**:497–517. [11]

Frankland, P. W., and S. A. Josselyn. 2014. Memory Allocation. *Neuropsychopharmacology* **40**:243. [5]

Franklin, M. S., M. D. Mrazek, C. L. Anderson, et al. 2017. Tracking Distraction: The Relationship between Mind-Wandering, Meta-Awareness, and ADHD Symptomatology. *J. Atten. Disord.* **21**:475–486. [9]

Franklin, M. S., M. D. Mrazek, C. L. Anderson, et al. 2013. The Silver Lining of a Mind in the Clouds: Interesting Musings Are Associated with Positive Mood While Mind-Wandering. *Front. Psychol.* **4**:583. [9]

Frau, R., L. J. Mosher, V. Bini, et al. 2016. The Neurosteroidogenic Enzyme 5α-Reductase Modulates the Role of D1 Dopamine Receptors in Rat Sensorimotor Gating. *Psychoneuroendocrinology* **63**:59–67. [5]

Freeman, S. M., I. Razhas, and A. R. Aron. 2014. Top-Down Response Suppression Mitigates Action Tendencies Triggered by a Motivating Stimulus. *Curr. Biol.* **24**:212–216. [9]

Freeston, M. H., R. Ladouceur, N. Thibodeau, and F. Gagnon. 1991. Cognitive Intrusions in a Non-Clinical Population, I: Response Style, Subjective Experience, and Appraisal. *Behav. Res. Ther.* **29**:585–597. [7, 13, 17]

Friedel, E., S. P. Koch, J. Wendt, et al. 2014. Devaluation and Sequential Decisions: Linking Goal-Directed and Model-Based Behavior. *Front. Hum. Neurosci.* **8**:587. [8]

Friston, K. J. 2017. Precision Psychiatry. *Biol. Psychiatry Cogn. Neurosci. Neuroimaging* **2**:640–643. [13]

Friston, K. J., T. FitzGerald, F. Rigoli, et al. 2016. Active Inference and Learning. *Neurosci. Biobehav. Rev.* **68**:862–879. [13]

Friston, K. J., M. Lin, C. D. Frith, et al. 2017. Active Inference, Curiosity and Insight. *Neural Comput.* **29**:2633–2683. [13]

Friston, K. J., T. Shiner, T. FitzGerald, et al. 2012. Dopamine, Affordance and Active Inference. *PLoS Comput. Biol.* **8**:e1002327. [10]

Frith, C. D. 2014. The Cognitive Neuropsychology of Schizophrenia. London: Psychology Press. [13]

Frith, C. D., S. J. Blakemore, and D. M. Wolpert. 2000. Abnormalities in the Awareness and Control of Action. *Philos. Trans. R. Soc. Lond. B. Biol. Sci.* **355**:1771–1788. [8]

Furlong, T. M., L. H. Corbit, R. A. Brown, and B. W. Balleine. 2018. Methamphetamine Promotes Habitual Action and Alters the Density of Striatal Glutamate Receptor and Vesicular Proteins in Dorsal Striatum. *Addict. Biol.* **23**:857–867. [3]

Furlong, T. M., J. R. Duncan, L. H. Corbit, et al. 2016. Toluene Inhalation in Adolescent Rats Reduces Flexible Behaviour in Adulthood and Alters Glutamatergic and GABAergic Signalling. *J. Neurochem.* **139**:806–822. [3]

Furlong, T. M., A. S. Supit, L. H. Corbit, S. Killcross, and B. W. Balleine. 2017. Pulling Habits out of Rats: Adenosine 2A Receptor Antagonism in Dorsomedial Striatum Rescues Meth-Amphetamine-Induced Deficits in Goal-Directed Action. *Addict. Biol.* **22**:172–183. [3, 13]

Fusi, S., E. K. Miller, and M. Rigotti. 2016. Why Neurons Mix: High Dimensionality for Higher Cognition. *Curr. Opin. Neurobiol.* **37**:66–74. [11]

Gable, S. L., E. A. Hopper, and J. W. Schooler. 2019. When the Muses Strike: Creative Ideas of Physicists and Writers Routinely Occur During Mind Wandering. *Psychol. Sci.* **30**:396–404. [9]

Gagnepain, P., R. N. Henson, and M. C. Anderson. 2014. Suppressing Unwanted Memories Reduces Their Unconscious Influence via Targeted Cortical Inhibition. *PNAS* **111**:E1310–E1319. [9]

Gagnepain, P., J. Hulbert, and M. C. Anderson. 2017. Parallel Regulation of Memory and Emotion Supports the Suppression of Intrusive Memories. *J. Neurosci.* **37**:6423–6441. [6, 9]

Galarza Vallejo, A., M. C. W. Kroes, E. Rey, et al. 2019. Propofol-Induced Deep Sedation Reduces Emotional Episodic Memory Reconsolidation in Humans. *Sci. Adv.* **5**:eaav3801. [14]

Gallagher, M., R. W. McMahan, and G. Schoenbaum. 1999. Orbitofrontal Cortex and Representation of Incentive Value in Associative Learning. *J. Neurosci.* **19**:6610–6614. [2]

Gallagher, S. 2005. How the Body Shapes the Mind. Oxford: Oxford Univ. Press. [13]

———. 2012. Multiple Aspects in the Sense of Agency. *New Ideas Psychol.* **30**:15–31. [13]

Gao, K., D. Muzina, P. Gajwani, and J. R. Calabrese. 2006. Efficacy of Typical and Atypical Antipsychotics for Primary and Comorbid Anxiety Symptoms or Disorders. *J. Clin. Psychiatry* **67**:1327–1340. [15]

Garavan, H. 2010. Insula and Drug Cravings. *Brain Struct. Funct.* **214**:593–601. [13]

Garbusow, M., D. J. Schad, M. Sebold, et al. 2016. Pavlovian-to-Instrumental Transfer Effects in the Nucleus Accumbens Relate to Relapse in Alcohol Dependence. *Addict. Biol.* **21**:719–731. [8]

Garcia-Keller, C., Y. M. Kupchik, C. D. Gipson, et al. 2016. Glutamatergic Mechanisms of Comorbidity between Acute Stress and Cocaine Self-Administration. *Mol. Psychiatry* **21**:1063–1069. [5]

Garcia-Soriano, G., A. Belloch, C. Morillo, and D. A. Clark. 2011. Symptom Dimensions in Obsessive-Compulsive Disorder: From Normal Cognitive Intrusions to Clinical Obsessions. *J. Anxiety Disord.* **25**:474–482. [8]

Garfinkel, S. N., L. Minati, M. A. Gray, et al. 2014. Fear from the Heart: Sensitivity to Fear Stimuli Depends on Individual Heartbeats. *J. Neurosci.* **34**:6573–6582. [13]

Garfinkel, S. N., A. K. Seth, A. B. Barrett, K. Suzuki, and H. D. Critchley. 2015. Knowing Your Own Heart: Distinguishing Interoceptive Accuracy from Interoceptive Awareness. *Biol. Psychol.* **104**:65–74. [13]

Garfinkel, S. N., C. Tiley, S. O'Keeffe, et al. 2016. Discrepancies between Dimensions of Interoception in Autism: Implications for Emotion and Anxiety. *Biol. Psychol.* **114**:117–126. [13]

Garrison, K. A., P. Pal, S. S. O'Malley, et al. 2018. Craving to Quit: A Randomized Controlled Trial of Smartphone App-Based Mindfulness Training for Smoking Cessation. *Nicotine Tob. Res.* **22**:324–331. [17]

Geisler, S., and R. A. Wise. 2008. Functional Implications of Glutamatergic Projections to the Ventral Tegmental Area. *Rev. Neurosci.* **19**:227–244. [2]

Geller, E. B., T. L. Skarpaas, R. E. Gross, et al. 2017. Brain-Responsive Neurostimulation in Patients with Medically Intractable Mesial Temporal Lobe Epilepsy. *Epilepsia* **58**:994–1004. [16]

Gentsch, A., S. Schütz-Bosbach, T. Endrass, and N. Kathmann. 2012. Dysfunctional Forward Model Mechanisms and Aberrant Sense of Agency in Obsessive-Compulsive Disorder. *Biol. Psychiatry* **71**:652–659. [13]

Gerfen, C. R., and D. J. Surmeier. 2011. Modulation of Striatal Projection Systems by Dopamine. *Annu. Rev. Neurosci.* **34**:441–466. [3]

Germeroth, L. J., M. J. Carpenter, N. L. Baker, et al. 2017. Effect of a Brief Memory Updating Intervention on Smoking Behavior a Randomized Clinical Trial. *JAMA Psychiatry* **74**:214–223. [9]

Gerraty, R. T., J. Y. Davidow, K. Foerde, et al. 2018. Dynamic Flexibility in Striatal-Cortical Circuits Supports Reinforcement Learning. *J. Neurosci.* **38**:2442–2453. [2]

Gershman, S. J. 2017. Predicting the Past, Remembering the Future. *Curr. Opin. Behav. Sci.* **17**:7–13. [13]

Gershman, S. J., and Y. Niv. 2010. Learning Latent Structure: Carving Nature at Its Joints. *Curr. Opin. Neurobiol.* **20**:251–256. [13]

Geschwind, N. 1965a. Disconnexion Syndromes in Animals and Man: I. *Brain* **88**:237–294. [5]

———. 1965b. Disconnexion Syndromes in Animals and Man: II. *Brain* **88**:585–644. [5]

Ghosh, K. K., L. D. Burns, E. D. Cocker, et al. 2011. Miniaturized Integration of a Fluorescence Microscope. *Nat. Methods* **8**:871–878. [4]

Gil-Jardiné, C., M. Née, E. Lagarde, et al. 2017. The Distracted Mind on the Wheel: Overall Propensity to Mind Wandering Is Associated with Road Crash Responsibility. *PLoS One* **12**:e0181327. [9]

Gillan, C. M., A. M. Apergis-Schoute, S. Morein-Zamir, et al. 2015. Functional Neuroimaging of Avoidance Habits in Obsessive-Compulsive Disorder. *Am. J. Psychiatry* **172**:284–293. [3, 9]

Gillan, C. M., M. Kosinski, R. Whelan, E. A. Phelps, and N. D. Daw. 2016. Characterizing a Psychiatric Symptom Dimension Related to Deficits in Goal-Directed Control. *eLife* **5**:e11305. [8]

Gillan, C. M., S. Morein-Zamir, G. P. Urcelay, et al. 2013. Enhanced Avoidance Habits in Obsessive-Compulsive Disorder. *Biol. Psychiatry* **75**:631–638. [9]

Gillan, C. M., and T. W. Robbins. 2014. Goal-Directed Learning and Obsessive-Compulsive Disorder. *Philos. Trans. R. Soc. Lond. B. Biol. Sci.* **369**:1655. [8]

Gillie, B. L., and J. F. Thayer. 2014. Individual Differences in Resting Heart Rate Variability and Cognitive Control in Posttraumatic Stress Disorder. *Front. Psychol.* **5**:758. [13]

Gillie, B. L., M. W. Vasey, and J. F. Thayer. 2014. Heart Rate Variability Predicts Control over Memory Retrieval. *Psychol. Sci.* **25**:458–465. [6]

———. 2015. Individual Differences in Resting Heart Rate Variability Moderate Thought Suppression Success. *Psychophysiology* **52**:1149–1160. [6, 13]

Glasner, S. V., J. B. Overmier, and B. W. Balleine. 2005. The Role of Pavlovian Cues in Alcohol Seeking in Dependent and Nondependent Rats. *J. Stud. Alcohol* **66**:53–61. [3]

Goetz, A. T., and K. Causey. 2009. Sex Differences in Perceptions of Infidelity: Men Often Assume the Worst. *Evol. Psychol.* **7**:253–263. [7]

Gola, M., M. Wordecha, G. Sescousse, et al. 2016. Can Pornography Be Addictive? An fMRI Study of Men Seeking Treatment for Problematic Pornography Use. *Neuropsychopharmacology* **42**:2021–2031. [10]

Goldstein, R. Z., A. D. Craig, A. Bechara, et al. 2009. The Neurocircuitry of Impaired Insight in Drug Addiction. *Trends Cogn. Sci.* **13**:372–380. [10, 13]

Goodkind, M., S. B. Eickhoff, D. J. Oathes, et al. 2015. Identification of a Common Neurobiological Substrate for Mental Illness. *JAMA Psychiatry* **72**:305–315. [10, 16]

Goodman, W. K., D. E. Grice, K. A. Lapidus, and B. J. Coffey. 2014. Obsessive-Compulsive Disorder. *Psychiatr. Clin. North Am.* **37**:257–267. [15]

Gordon, E. M., T. O. Laumann, A. W. Gilmore, et al. 2017. Precision Functional Mapping of Individual Human Brains. *Neuron* **95**:791–807. [11]

Gorwood, P., C. Blanchet-Collet, N. Chartrel, et al. 2016. New Insights in Anorexia Nervosa. *Front. Neurosci.* **10**:256. [13]

Goschke, T., and J. Kuhl. 1993. Representation of Intentions: Persisting Activation in Memory. *J. Exp. Psychol. Learn. Mem. Cogn.* **19**:1211–1226. [9]

Gottlieb, J. D., C. Cather, M. Shanahan, et al. 2011. D-cycloserine Facilitation of Cognitive Behavioral Therapy for Delusions in Schizophrenia. *Schizophr. Res.* **131**:69–74. [15]

Goulas, A., H. B. Uylings, and P. Stiers. 2014. Mapping the Hierarchical Layout of the Structural Network of the Macaque Prefrontal Cortex. *Cereb. Cortex* **24**:1178–1194. [11]

Gourley, S. L., A. Olevska, M. S. Warren, J. R. Taylor, and A. J. Koleske. 2012a. Arg Kinase Regulates Prefrontal Dendritic Spine Refinement and Cocaine-Induced Plasticity. *J. Neurosci.* **32**:2314–2323. [5]

Gourley, S. L., A. Olevska, K. S. Zimmermann, et al. 2013a. The Orbitofrontal Cortex Regulates Outcome-Based Decision-Making via the Lateral Striatum. *Eur. J. Neurosci.* **38**:2382–2388. [5]

Gourley, S. L., A. M. Swanson, A. M. Jacobs, et al. 2012b. Action Control Is Mediated by Prefrontal BDNF and Glucocorticoid Receptor Binding. *PNAS* **109**:20714–20719. [5]

Gourley, S. L., A. M. Swanson, and A. J. Koleske. 2013b. Corticosteroid-Induced Neural Remodeling Predicts Behavioral Vulnerability and Resilience. *J. Neurosci.* **33**:3107–3112. [5]

Gouwens, N. W., S. A. Sorensen, J. Berg, et al. 2019. Classification of Electrophysiological and Morphological Neuron Types in the Mouse Visual Cortex. *Nat. Neurosci.* **22**:1182–1195. [5]

Graebener, A. H., T. Michael, E. Holz, and J. Lass-Hennemann. 2017. Repeated Cortisol Administration Does Not Reduce Intrusive Memories: A Double Blind Placebo Controlled Experimental Study. *Eur. Neuropsychopharmacol.* **27**:1132–1143. [15]

Gräff, J., N. F. Joseph, M. E. Horn, et al. 2014. Epigenetic Priming of Memory Updating during Reconsolidation to Attenuate Remote Fear Memories. *Cell* **156**:261–276. [14]

Gramlich, M. A., S. M. Neer, D. C. Beidel, C. J. Bohil, and C. A. Bowers. 2017. A Functional near-Infrared Spectroscopy Study of Trauma-Related Auditory and Olfactory Cues: Posttraumatic Stress Disorder or Combat Experience? *J. Traum. Stress* **30**:656–665. [6]

Gratton, C., H. Sun, and S. E. Petersen. 2018. Control Networks and Hubs. *Psychophysiology* **55**(3):10.1111/psyp.13032. [10, 11]

Gray, K. M., S. C. Sonne, E. A. McClure, et al. 2017. A Randomized Placebo-Controlled Trial of N-Acetylcysteine for Cannabis Use Disorder in Adults. *Drug Alcohol Depend.* **177**:249–257. [15]

Graybiel, A. M. 2008. Habits, Rituals, and the Evaluative Brain. *Annu. Rev. Neurosci.* **31**:359–387. [13]

Graziano, M. S. A., and T. W. Webb. 2015. The Attention Schema Theory: A Mechanistic Account of Subjective Awareness. *Front. Psychol.* **06**:10.3389/fpsyg.2015.00500. [13]

Green, R., and L. A. Ray. 2018. Effects of Varenicline on Subjective Craving and Relative Reinforcing Value of Cigarettes. *Drug Alcohol Depend.* **188**:53–59. [15]

Greenberg, B. D., S. L. Rauch, and S. N. Haber. 2010. Invasive Circuitry-Based Neurotherapeutics: Stereotactic Ablation and Deep Brain Stimulation for OCD. *Neuropsychopharmacology* **35**:317–336. [3]

Gregertsen, E. C., W. Mandy, and L. Serpell. 2017. The Egosyntonic Nature of Anorexia: An Impediment to Recovery in Anorexia Nervosa Treatment. *Front. Psychol.* **8**:2273. [13]

Gregory, J. D., C. R. Brewin, W. Mansell, and C. Donaldson. 2010. Intrusive Memories and Images in Bipolar Disorder. *Behav. Res. Ther.* **48**:698–703. [7]

Gremel, C. M., and R. M. Costa. 2013. Orbitofrontal and Striatal Circuits Dynamically Encode the Shift between Goal-Directed and Habitual Actions. *Nat. Commun.* **4**:2264. [5, 13]

Grey, N., and E. A. Holmes. 2008. "Hotspots" in Trauma Memories in the Treatment of Post-Traumatic Stress Disorder: A Replication. *Memory* **16**:788–796. [9, 17]

Griffiths, K. R., J. Lagopoulos, D. F. Hermens, I. B. Hickie, and B. W. Balleine. 2015. Right External Globus Pallidus Changes Are Associated with Altered Causal Awareness in Youth with Depression. *Transl. Psychiatry* **5**:e653. [3]

Griffiths, K. R., J. Lagopoulos, D. F. Hermens, et al. 2016a. Impaired Causal Awareness and Associated Cortical-Basal Ganglia Structural Changes in Youth Psychiatric Disorders. *NeuroImage Clin.* **12**:285–292. [3]

Griffiths, K. R., R. W. Morris, and B. W. Balleine. 2014. Translational Studies of Goal-Directed Action as a Framework for Classifying Deficits across Psychiatric Disorders. *Front. Syst. Neurosci.* **8**:101. [3]

Griffiths, R. R., M. W. Johnson, M. A. Carducci, et al. 2016b. Psilocybin Produces Substantial and Sustained Decreases in Depression and Anxiety in Patients with Life-Threatening Cancer: A Randomized Double-Blind Trial. *J. Psychopharmacol.* **30**:1181–1197. [15]

Groman, S. M., A. S. James, E. Seu, et al. 2013. Monoamine Levels Within the Orbitofrontal Cortex and Putamen Interact to Predict Reversal Learning Performance. *Biol. Psychiatry* **73**:756–762. [13]

Gross, J. J., and O. P. John. 2003. Individual Differences in Two Emotion Regulation Processes: Implications for Affect, Relationships, and Well-Being. *J. Pers. Soc. Psychol.* **85**:348–362. [9]

Grupe, D. W., and J. B. Nitschke. 2013. Uncertainty and Anticipation in Anxiety: An Integrated Neurobiological and Psychological Perspective. *Nat. Rev. Neurosci.* **14**:488–501. [9]

Gu, S., R. F. Betzel, M. G. Mattar, et al. 2017. Optimal Trajectories of Brain State Transitions. *NeuroImage* **148**:305–317. [11]

Gu, S., F. Pasqualetti, M. Cieslak, et al. 2015. Controllability of Structural Brain Networks. *Nat. Commun.* **6**:8414. [11]

Gu, X., and F. Filbey. 2017. A Bayesian Observer Model of Drug Craving. *JAMA Psychiatry* **74**:419. [10]

Gu, X., P. R. Hof, K. J. Friston, and J. Fan. 2013. Anterior Insular Cortex and Emotional Awareness. *J. Comp. Neurol.* **521**:3371–3388. [13]

Gu, X., T. J. Zhou, E. Anagnostou, et al. 2018. Heightened Brain Response to Pain Anticipation in High-Functioning Adults with Autism Spectrum Disorder. *Eur. J. Neurosci.* **47**:592–601. [13]

Gunaydin, L. A., L. Grosenick, J. C. Finkelstein, et al. 2014. Natural Neural Projection Dynamics Underlying Social Behavior. *Cell* **157**:1535–1551. [4]

Guo, Y., T. W. Schmitz, M. Mur, C. S. Ferreira, and M. C. Anderson. 2018. A Supramodal Role of the Basal Ganglia in Memory and Motor Inhibition: Meta-Analytic Evidence. *Neuropsychologia* **108**:117–134. [9, 11]

Gürsel, D. A., M. Avram, C. Sorg, F. Brandl, and K. Koch. 2018. Frontoparietal Areas Link Impairments of Large-Scale Intrinsic Brain Networks with Aberrant Fronto-Striatal Interactions in OCD: A Meta-Analysis of Resting-State Functional Connectivity. *Neurosci. Biobehav. Rev.* **87**:151–160. [6]

Haber, S. N. 2003. The Primate Basal Ganglia: Parallel and Integrative Networks. *J. Chem. Neuroanat.* **26**:317–330. [11]

———. 2016. Corticostriatal Circuitry. In: Neuroscience in the 21st Century, ed. D. W. Pfaff and N. D. Volkow, pp. 1721–1741. New York: Springer. [10]

Haber, S. N., and T. E. J. Behrens. 2014. The Neural Network Underlying Incentive-Based Learning: Implications for Interpreting Circuit Disruptions in Psychiatric Disorders. *Neuron* **83**:1019–1039. [10]

Haber, S. N., K. S. Kim, P. Mailly, and R. Calzavara. 2006. Reward-Related Cortical Inputs Define a Large Striatal Region in Primates That Interface with Associative Cortical Connections, Providing a Substrate for Incentive-Based Learning. *J. Neurosci.* **26**:8368–8376. [5, 10]

Haber, S. N., and B. Knutson. 2009. The Reward Circuit: Linking Primate Anatomy and Human Imaging. *Neuropsychopharmacology* **35**:4–26. [10]

Hagenaars, M. A., E. A. Holmes, F. Klaassen, and B. Elzinga. 2017. Tetris and Word Games Lead to Fewer Intrusive Memories When Applied Several Days after Analogue Trauma. *Eur. J. Psychotraumatol.* **8**:1386959. [14]

Haight, J. L., and S. B. Flagel. 2014. A Potential Role for the Paraventricular Nucleus of the Thalamus in Mediating Individual Variation in Pavlovian Conditioned Responses. *Front. Behav. Neurosci.* **8**:79. [5]

Haight, J. L., K. M. Fraser, H. Akil, and S. B. Flagel. 2015. Lesions of the Paraventricular Nucleus of the Thalamus Differentially Affect Sign- and Goal-Tracking Conditioned Responses. *Eur. J. Neurosci.* **42**:2478–2488. [5]

Haight, J. L., Z. L. Fuller, K. M. Fraser, and S. B. Flagel. 2017. A Food-Predictive Cue Attributed with Incentive Salience Engages Subcortical Afferents and Efferents of the Paraventricular Nucleus of the Thalamus. *Neuroscience* **340**:135–152. [5]

Halassa, M. M., and S. Kastner. 2017. Thalamic Functions in Distributed Cognitive Control. *Nat. Neurosci.* **20**:1669–1679. [10]

Hale, L., and S. Guan. 2015. Screen Time and Sleep among School-Aged Children and Adolescents: A Systematic Literature Review. *Sleep Med. Rev.* **21**:50–58. [17]

Hales, S. A., C. Deeprose, G. M. Goodwin, and E. A. Holmes. 2011. Cognitions in Bipolar Affective Disorder and Unipolar Depression: Imagining Suicide. *Bipolar Disord.* **13**:651–661. [14]

Hamann, S. 2001. Cognitive and Neural Mechanisms in Emotional Memory. *Trends. Cogn. Neurosci.* **5**:394–400. [9]

Hamilton, J. P., M. Farmer, P. Fogelman, and I. H. Gotlib. 2015. Depressive Rumination, the Default-Mode Network, and the Dark Matter of Clinical Neuroscience. *Biol. Psychiatry* **78**:224–230. [8, 9]

Hammond, L. J. 1980. The Effect of Contingency Upon the Appetitive Conditioning of Free-Operant Behavior. *J. Exp. Anal. Behav.* **34**:297–304. [3]

Hanlon, C. 2017. Blunt or Precise? A Note About the Relative Precision of Figure-of-Eight rTMS Coils. *Brain Stimul.* **10**:338–339. [16]

Hannestad, J. O., K. P. Cosgrove, N. F. DellaGioia, et al. 2013. Changes in the Cholinergic System between Bipolar Depression and Euthymia as Measured with [^{123}I]5IA Single Photon Emission Computed Tomography. *Biol. Psychiatry* **74**:768–776. [5]

Hansen, J. 2018. Climate in a Nutshell: The Gathering Storm. New York: Columbia University. [9]

Hardt, O., E. Ö. Einarsson, and K. Nader. 2010. A Bridge over Troubled Water: Reconsolidation as a Link between Cognitive and Neuroscientific Memory Research Traditions. *Ann. Rev. Psychol.* **61**:141–167. [2]

Harrington, M. O., J. E. Ashton, S. Sankarasubramanian, M. C. Anderson, and S. A. Cairney. 2020. Losing Control: Sleep Deprivation Impairs the Suppression of Unwanted Thoughts. *bioRxiv*: 813121. [9]

Hart, C. L. 2013. High Price: A Neuroscientist's Journey of Self-Discovery That Challenges Everything You Know About Drugs and Society. New York: Harper. [12]

Hart, G., and B. W. Balleine. 2016. Consolidation of Goal-Directed Action Depends on Mapk/ERK Signaling in Rodent Prelimbic Cortex. *J. Neurosci.* **36**:11974–11986. [3]

Hart, G., L. A. Bradfield, and B. W. Balleine. 2018a. Prefrontal Corticostriatal Disconnection Blocks the Acquisition of Goal-Directed Action. *J. Neurosci.* **38**:1311–1322. [3]

Hart, G., L. A. Bradfield, S. Y. Fok, B. Chieng, and B. W. Balleine. 2018b. The Bilateral Prefronto-Striatal Pathway Is Necessary for Learning New Goal-Directed Actions. *Curr. Biol.* **28**:2218–2229.e2217. [3]

Hart, G., B. K. Leung, and B. W. Balleine. 2014. Dorsal and Ventral Streams: The Distinct Role of Striatal Subregions in the Acquisition and Performance of Goal-Directed Actions. *Neurobiol. Learn. Mem.* **108**:104–118. [3]

Hartley, C. A., and E. A. Phelps. 2010. Changing Fear: The Neurocircuitry of Emotion Regulation. *Neuropsychopharmacology* **35**:136–146. [9]

Haselton, M. G., and D. M. Buss. 2000. Error Management Theory: A New Perspective on Biases in Cross-Sex Mind Reading. *J. Pers. Soc. Psychol.* **78**:81–91. [7]

Haselton, M. G., and D. Nettle. 2006. The Paranoid Optimist: An Integrative Evolutionary Model of Cognitive Biases. *Pers. Soc. Psychol. Rev.* **10**:47–66. [7]

Hashimoto, K., B. Malchow, P. Falkai, and A. Schmitt. 2013. Glutamate Modulators as Potential Therapeutic Drugs in Schizophrenia and Affective Disorders. *Eur. Arch. Psychiatry Clin. Neurosci.* **263**:367–377. [15]

Haxby, J., J. Guntupalli, A. Connolly, et al. 2011. A Common, High-Dimensional Model of the Representational Space in Human Ventral Temporal Cortex. *Neuron* **72**:404–416. [17]

Heath, C. J., and M. R. Picciotto. 2009. Nicotine-Induced Plasticity during Development: Modulation of the Cholinergic System and Long-Term Consequences for Circuits Involved in Attention and Sensory Processing. *Neuropharmacology* **56 Suppl** 1:254–262. [5]

Heilbronner, S. R., and B. Y. Hayden. 2016. Dorsal Anterior Cingulate Cortex: A Bottom-up View. *Annu. Rev. Neurosci.* **39**:149–170. [10, 13]

Heinz, A. 1999. Neurobiological and Anthropological Aspects of Compulsions and Rituals. *Pharmacopsychiatry* **32**:223–229. [8]

———. 2002. Dopaminergic Dysfunction in Alcoholism and Schizophrenia: Psychopathological and Behavioral Correlates. *Eur. Psychiatry* **17**:9–16. [8]

Heinz, A. 2017. A New Understanding of Mental Disorders: Computational Models for Dimensional Psychiatry. Cambridge, MA: MIT Press. [8]

Heinz, A., G. K. Murray, F. Schlagenhauf, et al. 2019. Towards a Unifying Cognitive, Neurophysiological, and Computational Neuroscience Account of Schizophrenia. *Schizophr. Bull.* **45**:1092–1100. [8]

Heinz, A., H. Przuntek, G. Winterer, and A. Pietzcker. 1995. Clinical Aspects and Follow-up of Dopamine-Induced Psychoses in Continuous Dopaminergic Therapy and Their Implications for the Dopamine Hypothesis of Schizophrenic Symptoms. *Nervenarzt* **66**:662–669. [8]

Heinz, A., and F. Schlagenhauf. 2010. Dopaminergic Dysfunction in Schizophrenia: Salience Attribution Revisited. *Schizophr. Bull.* **36**:472–485. [8]

Heinz, A., M. Voss, S. M. Lawrie, et al. 2016. Shall We Really Say Goodbye to First Rank Symptoms? *Eur. Psychiatry* **37**:8–13. [8]

Hellerstedt, R., M. Johansson, and M. C. Anderson. 2016. Tracking the Intrusion of Unwanted Memories into Awareness with Event-Related Potentials. *Neuropsychologia* **89**:510–523. [6, 9, 15]

Henke, K. 2010. A Model for Memory Systems Based on Processing Modes Rather Than Consciousness. *Nat. Rev. Neurosci.* **11**:523–532. [8]

Hertel, P. T., and G. Calcaterra. 2005. Intentional Forgetting Benefits from Thought Substitution. *Psychon. Bull. Rev.* **12**:484–489. [9]

Hertel, P. T., D. Large, E. D. Stück, and A. Levy. 2012. Suppression-Induced Forgetting on a Free-Association Test. *Memory* **20**:100–109. [9]

Hertel, P. T., A. Maydon, A. D. Ogilvie, and N. Mor. 2018. Ruminators (Unlike Others) Fail to Show Suppression-Induced Forgetting on Indirect Measures of Memory. *Clin. Psychol. Sci.* **6**:872–881. [9]

Higgins, E. T. 1987. Self-Discrepancy: A Theory Relating Self and Affect. *Psychol. Rev.* **94**:319–340. [13]

———. 2011. Beyond Pleasure and Pain: How Motivation Works. New York: Oxford Univ. Press. [13]

Hill, M. N., P. Campolongo, R. Yehuda, and S. Patel. 2017. Integrating Endocannabinoid Signaling and Cannabinoids into the Biology and Treatment of Posttraumatic Stress Disorder. *Neuropsychopharmacology* **43**:80–102. [15]

Hillman, C. H., K. I. Erickson, and A. F. Kramer. 2008. Be Smart, Exercise Your Heart: Exercise Effects on Brain and Cognition. *Nat. Rev. Neurosci.* **9**:52–65. [9]

Hinton, E. A., D. C. Li, A. G. Allen, and S. L. Gourley. 2019. Social Isolation in Adolescence Disrupts Cortical Development and Goal-Dependent Decision-Making in Adulthood, Despite Social Reintegration. *eNeuro* **6**:0318–0319.2019. [5]

Hinton, G. E., P. Dayan, B. J. Frey, and R. M. Neal. 1995. The "Wake-Sleep" Algorithm for Unsupervised Neural Networks. *Science* **268**:1158–1161. [13]

Hirschtritt, M. E., M. H. Bloch, and C. A. Mathews. 2017. Obsessive-Compulsive Disorder. *JAMA* **317**:1358. [15]

Hochreiter, S., and J. Schmidhuber. 1997. Long Short-Term Memory. *Neural Comput.* **9**:1735–1780. [13]

Hoefer, C. 2016. Causal Determinism. In: The Stanford Encyclopedia of Philosophy, ed. E. N. Zalta, Stanford: The Metaphysics Research Lab, Center for the Study of Language and Information (CSLI), Stanford Univ. [12]

Holehonnur, R., A. J. Phensy, L. J. Kim, et al. 2016. Increasing the GluN2A/GluN2B Ratio in Neurons of the Mouse Basal and Lateral Amygdala Inhibits the Modification of an Existing Fear Memory Trace. *J. Neurosci.* **36**:9490–9504. [5]

Holland, P. C. 2004. Relations between Pavlovian-Instrumental Transfer and Reinforcer Devaluation. *J. Exp. Psychol. Anim. Behav. Process.* **30**:104–117. [3]

Holmes, E. A., S. E. Blackwell, S. Burnett Heyes, F. Renner, and F. Raes. 2016a. Mental Imagery in Depression: Phenomenology, Potential Mechanisms, and Treatment Implications. *Ann. Rev. Clin. Psychol.* **12**:249–280. [14, 17]

Holmes, E. A., M. B. Bonsall, S. A. Hales, et al. 2016b. Applications of Time-Series Analysis to Mood Fluctuations in Bipolar Disorder to Promote Treatment Innovation: A Case Series. *Transl. Psychiatry* **6**:e720. [14]

Holmes, E. A., and C. Bourne. 2008. Inducing and Modulating Intrusive Emotional Memories: A Review of the Trauma Film Paradigm. *Acta Psychol. (Amst.)* **127**:553–566. [6]

Holmes, E. A., C. R. Brewin, and R. G. Hennessy. 2004. Trauma Films, Information Processing, and Intrusive Memory Development. *J. Exp. Psychol. Gen.* **133**:3–22. [9]

Holmes, E. A., C. Crane, M. J. V. Fennell, and J. M. G. Williams. 2007. Imagery About Suicide in Depression: "Flash-Forwards"? *J. Behav. Ther. Exp. Psychiatry* **38**:423–434. [9]

Holmes, E. A., M. G. Craske, and A. M. Graybiel. 2014. Psychological Treatments: A Call for Mental-Health Science. *Nature* **511**:287–289. [14, 17]

Holmes, E. A., A. Ghaderi, E. Eriksson, et al. 2017. "I Can't Concentrate": A Feasibility Study with Young Refugees in Sweden on Developing Science-Driven Interventions for Intrusive Memories Related to Trauma. *Behav. Cogn. Psychother.* **45**:97–109. [9, 17]

Holmes, E. A., A. Ghaderi, C. J. Harmer, et al. 2018. The Lancet Psychiatry Commission on Psychological Treatments Research in Tomorrow's Science. *Lancet Psychiatry* **5**:237–286. [14, 17]

Holmes, E. A., S. A. Hales, K. Young, and M. Di Simplicio. 2019. Imagery-Based Cognitive Therapy for Bipolar Disorder and Mood Instability. New York: The Guilford Press. [14, 17]

Holmes, E. A., E. L. James, T. Coode-Bate, and C. Deeprose. 2009. Can Playing the Computer Game "Tetris" Reduce the Build-up of Flashbacks for Trauma? A Proposal from Cognitive Science. *PLoS One* **4**:e4153. [9, 17]

Holmes, E. A., E. L. James, E. J. Kilford, and C. Deeprose. 2010. Key Steps in Developing a Cognitive Vaccine against Traumatic Flashbacks: Visuospatial Tetris versus Verbal Pub Quiz. *PLoS One* **5**:e13706. [9, 17]

Holmes, E. A., and A. Mathews. 2010. Mental Imagery in Emotion and Emotional Disorders. *Clin. Psychol. Rev.* **30**:349–362. [8, 9, 14]

Holyoak, K. J., and D. Powell. 2016. Deontological Coherence: A Framework for Commonsense Moral Reasoning. *Psychol. Bull.* **142**:1179–1203. [12]

Homan, P., I. Levy, E. Feltham, et al. 2019. Neural Computations of Threat in the Aftermath of Combat Trauma. *Nat. Neurosci.* **22**:470–476. [9]

Honey, C. J., R. Kotter, M. Breakspear, and O. Sporns. 2007. Network Structure of Cerebral Cortex Shapes Functional Connectivity on Multiple Time Scales. *PNAS* **104**:10240–10245. [10]

Honey, C. J., and O. Sporns. 2008. Dynamical Consequences of Lesions in Cortical Networks. *Hum. Brain Mapp.* **29**:802–809. [10]

Hoogendam, J. M., G. M. Ramakers, and V. Di Lazzaro. 2010. Physiology of Repetitive Transcranial Magnetic Stimulation of the Human Brain. *Brain Stimul.* **3**:95–118. [16]

Höping, W., and R. de Jong-Meyer. 2003. Differentiating Unwanted Intrusive Thoughts from Thought Suppression: What Does the White Bear Suppression Inventory Measure? *Pers. Individ. Dif.* **34**:1049–1055. [9]

Horowitz, M. J. 1969. Psychic Trauma: Return of Images after a Stress Film. *Arch. Gen. Psychiatry* **20**:552–559. [9]

———. 1975. Intrusive and Repetitive Thoughts after Experimental Stress: A Summary. *Arch. Gen. Psychiatry* **32**:1457–1463. [7, 9]

Horowitz, T., and J. Wolfe. 2003. Memory for Rejected Distractors in Visual Search? *Vis. Cogn.* **10**:257–298. [7]

Horsch, A., Y. Vial, C. Favrod, et al. 2017. Reducing Intrusive Traumatic Memories after Emergency Caesarean Section: A Proof-of-Principle Randomized Controlled Study. *Behav. Res. Ther.* **94**:36–47. [6, 14, 17]

Howard, R. A. 1966. Information Value Theory. *IEEE Trans. Systems Sci. Cybern.* **SSC-2**:22–26. [13]

Howland, R. H. 2015. Buspirone: Back to the Future. *J. Psychosoc. Nurs. Ment. Health Serv.* **53**:21–24. [15]

Hu, J., W. Wang, P. Homan, et al. 2018. Reminder Duration Determines Threat Memory Modification in Humans. *Sci. Rep.* **8**:8848. [9]

Hu, Y., B. J. Salmeron, H. Gu, E. A. Stein, and Y. Yang. 2015. Impaired Functional Connectivity Within and between Frontostriatal Circuits and Its Association with Compulsive Drug Use and Trait Impulsivity in Cocaine Addiction. *JAMA Psychiatry* **72**:584. [10]

Huang, Y.-Z., M. J. Edwards, E. Rounis, K. P. Bhatia, and J. C. Rothwell. 2005. Theta Burst Stimulation of the Human Motor Cortex. *Neuron* **45**:201–206. [16]

Huk, A., K. Bonnen, and B. J. He. 2018. Beyond Trial-Based Paradigms: Continuous Behavior, Ongoing Neural Activity, and Natural Stimuli. *J. Neurosci.* **38**:7551–7558. [6]

Huk, A. C., and M. N. Shadlen. 2005. Neural Activity in Macaque Parietal Cortex Reflects Temporal Integration of Visual Motion Signals during Perceptual Decision Making. *J. Neurosci.* **25**:10420–10436. [13]

Hulbert, J. C., and M. C. Anderson. 2018. What Doesn't Kill You Makes You Stronger: Psychological Trauma and Its Relationship to Enhanced Memory Control. *J. Exp. Psychol. Gen.* **147**:1931–1949. [9]

Hulbert, J. C., R. N. Henson, and M. C. Anderson. 2016. Inducing Amnesia through Systemic Suppression. *Nat. Commun.* **7**:11003. [9]

Hur, J., W. Heller, J. L. Kern, and H. Berenbaum. 2017. A Bi-Factor Approach to Modeling the Structure of Worry and Rumination. *J. Exp. Psychopathol.* **8**:252–264. [9]

Hurlemann, R., B. Hawellek, A. Matusch, et al. 2005. Noradrenergic Modulation of Emotion-Induced Forgetting and Remembering. *J. Neurosci.* **25**:6343–6349. [9]

Huys, Q. J. M., N. Eshel, E. O'Nions, et al. 2012. Bonsai Trees in Your Head: How the Pavlovian System Sculpts Goal-Directed Choices by Pruning Decision Trees. *PLoS Comput. Biol.* **8**:e1002410. [9]

Huys, Q. J. M., N. Lally, P. Faulkner, et al. 2015. Interplay of Approximate Planning Strategies. *PNAS* **112**:3098–3103. [9]

Huys, Q. J. M., T. V. Maia, and M. J. Frank. 2016. Computational Psychiatry as a Bridge from Neuroscience to Clinical Applications. *Nat. Neurosci.* **19**:404–413. [9]

Huys, Q. J. M., and D. Renz. 2017. A Formal Valuation Framework for Emotions and Their Control. *Biol. Psychiatry* **82**:413–420. [9]

Hyman, I. E., N. K. Burland, H. M. Duskin, et al. 2013. Going Gaga: Investigating, Creating, and Manipulating the Song Stuck in My Head. *Appl. Cogn. Psychol.* **27**:204–215. [9]

Hyman, I. E., K. I. Cutshaw, C. M. Hall, et al. 2015. Involuntary to Intrusive: Using Involuntary Musical Imagery to Explore Individual Differences and the Nature of Intrusive Thoughts. *Psychomusicol.* **25**:14–27. [9]

Inda, M. C., E. V. Muravieva, and C. M. Alberini. 2011. Memory Retrieval and the Passage of Time: From Reconsolidation and Strengthening to Extinction. *J. Neurosci.* **31**:1635–1643. [7]

IPCC. 2018. Special Report: Global Warming of 1.5°C. https://www.ipcc.ch/sr15/. (accessed Feb. 6, 2020). [9]

Isoda, M., and O. Hikosaka. 2008. Role for Subthalamic Nucleus Neurons in Switching from Automatic to Controlled Eye Movement. *J. Neurosci.* **28**:7209–7218. [11]

Isserles, M., A. Y. Shalev, Y. Roth, et al. 2013. Effectiveness of Deep Transcranial Magnetic Stimulation Combined with a Brief Exposure Procedure in Post-Traumatic Stress Disorder: A Pilot Study. *Brain Stimul.* **6**:377–383. [16]

Ito, M. 2008. Control of Mental Activities by Internal Models in the Cerebellum. *Nat. Rev. Neurosci.* **9**:304–313. [8]

Itti, L., and P. Baldi. 2009. Bayesian Surprise Attracts Human Attention. *Vision Res.* **49**:1295–1306. [10, 13]

Ivins, A., M. Di Simplicio, H. Close, G. M. Goodwin, and E. A. Holmes. 2014. Mental Imagery in Bipolar Affective Disorder versus Unipolar Depression: Investigating Cognitions at Times of 'Positive' Mood. *J. Affect. Disord.* **166**:234–242. [17]

Iwabuchi, S. J., F. Raschke, D. P. Auer, et al. 2017. Targeted Transcranial Theta-Burst Stimulation Alters Fronto-Insular Network and Prefrontal GABA. *NeuroImage* **146**:395–403. [16]

Iyadurai, L., S. E. Blackwell, R. Meiser-Stedman, et al. 2018. Preventing Intrusive Memories after Trauma via a Brief Intervention Involving Tetris Computer Game Play in the Emergency Department: A Proof-of-Concept Randomized Controlled Trial. *Mol. Psychiatry* **23**:674–682. [9, 14, 17]

Iyadurai, L., R. M. Visser, A. Lau-Zhu, et al. 2019. Intrusive Memories of Trauma: A Target for Research Bridging Cognitive Science and Its Clinical Application. *Clin. Psychol. Rev.* **69**:67–82. [14, 15]

Iyer, M. B., N. Schleper, and E. M. Wassermann. 2003. Priming Stimulation Enhances the Depressant Effect of Low-Frequency Repetitive Transcranial Magnetic Stimulation. *J. Neurosci.* **23**:10867–10872. [16]

Jahanshahi, M., I. Obeso, J. C. Rothwell, and J. A. Obeso. 2015. A Fronto-Striato-Subthalamic-Pallidal Network for Goal-Directed and Habitual Inhibition. *Nat. Rev. Neurosci.* **16**:719–732. [9]

James, E. L., M. B. Bonsall, L. Hoppitt, et al. 2015. Computer Game Play Reduces Intrusive Memories of Experimental Trauma via Reconsolidation-Update Mechanisms. *Psychol. Sci.* **26**:1201–1215. [2, 9, 14, 17]

James, E. L., A. Lau-Zhu, I. A. Clark, et al. 2016a. The Trauma Film Paradgm as an Experimental Psychopathology Model of Psychological Trauma: Intrusive Memories and Beyond. *Clin. Psychol. Rev.* **47**:106–142. [9, 14, 17]

James, E. L., A. Lau-Zhu, H. Tickle, A. Horsch, and E. A. Holmes. 2016b. Playing the Computer Game Tetris Prior to Viewing Traumatic Film Material and Subsequent Intrusive Memories: Examining Proactive Interference. *J. Behav. Ther. Exp. Psychiatry* **53**:25–33. [6, 7]

Janes, A. C., M. Datko, A. Roy, et al. 2019. Quitting Starts in the Brain: A Randomized Controlled Trial of App-Based Mindfulness Shows Decreases in Neural Responses to Smoking Cues That Predict Reductions in Smoking. *Neuropsychopharmacology* **44**:1631–1638. [17]

Jannati, A., G. Block, L. M. Oberman, A. Rotenberg, and A. Pascual-Leone. 2017. Interindividual Variability in Response to Continuous Theta-Burst Stimulation in Healthy Adults. *Clin. Neurophysiol.* **128**:2268–2278. [16]

Jaspers, K. 1946. Allgemeine Psychopathologie: Ein Leitfaden für Studierende, Ärzte und Psychologen. Heidelberg: Springer. [8]

Jennings, J. H., C. K. Kim, J. H. Marshel, et al. 2019. Interacting Neural Ensembles in Orbitofrontal Cortex for Social and Feeding Behaviour. *Nature* **565**:645–649. [4]

Jensen, J., A. J. Smith, M. Willeit, et al. 2007. Separate Brain Regions Code for Salience vs. Valence during Reward Prediction in Humans. *Hum. Brain Mapp.* **28**:294–302. [10]

Jeon, H. A., A. Anwander, and A. D. Friederici. 2014. Functional Network Mirrored in the Prefrontal Cortex, Caudate Nucleus, and Thalamus: High-Resolution Functional Imaging and Structural Connectivity. *J. Neurosci.* **34**:9202–9212. [11]

Ji, J. L., S. Burnett Heyes, C. MacLeod, and E. A. Holmes. 2016. Emotional Mental Imagery as Simulation of Reality: Fear and Beyond, a Tribute to Peter Lang. *Behav. Therapy* **47**:702–719. [9]

Ji, J. L., M. Spronk, K. Kulkarni, et al. 2019. Mapping the Human Brain's Cortical-Subcortical Functional Network Organization. *NeuroImage* **185**:35–57. [10, 11]

Joormann, J., and I. H. Gotlib. 2008. Updating the Contents of Working Memory in Depression: Interference from Irrelevant Negative Material. *J. Abnorm. Psychol.* **117**:182–192. [9]

Joormann, J., P. T. Hertel, J. LeMoult, and I. H. Gotlib. 2009. Training Forgetting of Negative Material in Depression. *J. Abnorm. Psychol.* **118**:34–43. [9]

Josselyn, S. A., and P. W. Frankland. 2018. Memory Allocation: Mechanisms and Function. *Annu. Rev. Neurosci.* **41**:389–413. [5]

Josselyn, S. A., C. Shi, W. A. Carlezon, Jr., et al. 2001. Long-Term Memory Is Facilitated by Camp Response Element-Binding Protein Overexpression in the Amygdala. *J. Neurosci.* **21**:2404–2412. [5]

Julien, D., K. P. O'Connor, and F. Aardema. 2007. Intrusive Thoughts, Obsessions, and Appraisals in Obsessive-Compulsive Disorder: A Critical Review. *Clin. Psychol. Rev.* **27**:366–383. [8, 13]

Jun, J. J., N. A. Steinmetz, J. H. Siegle, et al. 2017. Fully Integrated Silicon Probes for High-Density Recording of Neural Activity. *Nature* **551**:232–236. [4]

Jung, W. H., D. H. Kang, J. Y. Han, et al. 2011. Aberrant Ventral Striatal Responses during Incentive Processing in Unmedicated Patients with Obsessive-Compulsive Disorder. *Acta Psychiat. Scand.* **123**:376–386. [10]

Kahneman, D. 2011. Thinking, Fast and Slow. New York: Farrar, Straus and Giroux. [12]

Kahnt, T., L. J. Chang, L. J. Chang, et al. 2012. Connectivity-Based Parcellation of the Human Orbitofrontal Cortex. *J. Neurosci.* **32**:6240–6250. [3]

Kahnt, T., S. Q. Park, J.-D. Haynes, and P. N. Tobler. 2014. Disentangling Neural Representations of Value and Salience in the Human Brain. *PNAS* **111**:5000–5005. [10]

Kahnt, T., and P. N. Tobler. 2017. Reward, Value, and Salience. In: Decision Neuroscience: An Integrative Perspective, ed. J. C. Dreher and L. Tremblay, pp. 109–120. London: Elsevier. [10, 13]

Kaiser, R. H., J. R. Andrews-Hanna, T. D. Wager, and D. A. Pizzagalli. 2015. Large-Scale Network Dysfunction in Major Depressive Disorder. *JAMA Psychiatry* **72**:603. [10]

Kalivas, B. C., and P. W. Kalivas. 2016. Corticostriatal Circuitry in Regulating Diseases Characterized by Intrusive Thinking. *Dialogues Clin. Neurosci.* **18**:65–76. [5, 6, 10]

Kamitani, Y., and F. Tong. 2005. Decoding the Visual and Subjective Contents of the Human Brain. *Nat. Neurosci.* **8**:679–685. [17]

Kane, M. J., L. H. Brown, J. C. McVay, et al. 2007. For Whom the Mind Wanders, and When: an Experience-Sampling Study of Working Memory and Executive Control in Daily Life. *Psychol. Sci.* **18**:614–621. [10, 13]

Kane, R. 1996. The Significance of Free Will. New York: Oxford Univ. Press. [12]

Kapur, S. 2003. Psychosis as a State of Aberrant Salience: A Framework Linking Biology, Phenomenology, and Pharmacology in Schizophrenia. *Am. J. Psychiatry* **160**:13–23. [8]

Kawai, H., R. Lazar, and R. Metherate. 2007. Nicotinic Control of Axon Excitability Regulates Thalamocortical Transmission. *Nat. Neurosci.* **10**:1168–1175. [5]

Keelan, R. E., E. J. Mahoney, M. Sherer, et al. 2019. Neuropsychological Characteristics of the Confusional State Following Traumatic Brain Injury. *J. Int. Neuropsychol. Soc.* **25**:302–313. [13]

Keller, C. J., S. Bickel, C. J. Honey, et al. 2013. Neurophysiological Investigation of Spontaneous Correlated and Anticorrelated Fluctuations of the BOLD Signal. *J. Neurosci.* **33**:6333–6342. [10]

Kelley, A. E., C. A. Schiltz, and C. F. Landry. 2005. Neural Systems Recruited by Drug- and Food-Related Cues: Studies of Gene Activation in Corticolimbic Regions. *Physiol. Behav.* **86**:11–14. [5]

Kemeny, J. G., and J. L. Snell. 1960. Finite Markov Chains. Princeton: Van Nostrand Company, Inc. [14]

Kennerley, S. W., T. E. J. Behrens, and J. D. Wallis. 2011. Double Dissociation of Value Computations in Orbitofrontal and Anterior Cingulate Neurons. *Nat. Neurosci.* **14**:1581–1589. [10, 13]

Keramati, M., A. Dezfouli, and P. Piray. 2011. Speed/Accuracy Trade-Off between the Habitual and the Goal-Directed Processes. *PLoS Comput. Biol.* **7**:e1002055. [9]

Kessler, H., E. A. Holmes, S. E. Blackwell, et al. 2018. Reducing Intrusive Memories of Trauma Using a Visuospatial Interference Intervention with Inpatients with Posttraumatic Stress Disorder (PTSD). *J. Consult. Clin. Psychol.* **86**:1076–1090. [14, 17]

Kessler, H., A.-C. Schmidt, E. L. James, et al. 2020. Visuospatial Computer Game Play after Memory Reminder Delivered Three Days after a Traumatic Film Reduces the Number of Intrusive Memories of the Experimental Trauma. *J. Behav. Ther. Exp. Psychiatry* **67**:101454. [9, 14, 17]

Keynan, J. N., A. Cohen, G. Jackont, et al. 2019. Electrical Fingerprint of the Amygdala Guides Neurofeedback Training for Stress Resilience. *Nat. Hum. Behav.* **3**:63–73. [17]

Khalsa, S. S., R. Adolphs, O. G. Cameron, et al. 2018. Interoception and Mental Health: A Roadmap. *Biol. Psychiatry Cogn. Neurosci. Neuroimaging* **3**:501–513. [10, 13]

Khambhati, A. N., J. D. Medaglia, E. A. Karuza, S. L. Thompson-Schill, and D. S. Bassett. 2018. Subgraphs of Functional Brain Networks Identify Dynamical Constraints of Cognitive Control. *PLoS Comput. Biol.* **14**:e1006234. [11]

Killingsworth, M. A., and D. T. Gilbert. 2010. A Wandering Mind Is an Unhappy Mind. *Science* **330**:932–932. [9, 10]

Kim, K., and D. J. Yi. 2013. Out of Mind, out of Sight: Perceptual Consequences of Memory Suppression. *Psychol. Sci.* **24**:569–574. [9]

Kindt, M., M. Soeter, and B. Vervliet. 2009. Beyond Extinction: Erasing Human Fear Responses and Preventing the Return of Fear. *Nat. Neurosci.* **12**:256–258. [14]

Kira, S., T. Yang, and M. N. Shadlen. 2015. A Neural Implementation of Wald's Sequential Probability Ratio Test. *Neuron* **85**:861–873. [13]

Kirouac, G. J. 2015. Placing the Paraventricular Nucleus of the Thalamus within the Brain Circuits That Control Behavior. *Neurosci. Biobehav. Rev.* **56**:315–329. [5]

Kiverstein, J., E. Rietveld, H. A. Slagter, and D. Denys. 2019. Obsessive Compulsive Disorder: A Pathology of Self-Confidence? *Trends Cogn. Sci.* **23**:369–372. [13]

Klapoetke, N. C., Y. Murata, S. S. Kim, et al. 2014. Independent Optical Excitation of Distinct Neural Populations. *Nat. Methods* **11**:338–346. [4]

Kleckner, I. R., J. Zhang, A. Touroutoglou, et al. 2017. Evidence for a Large-Scale Brain System Supporting Allostasis and Interoception in Humans. *Nat. Hum. Behav.* **1**:0069. [10]

Klein, T. A., M. Ullsperger, and C. Danielmeier. 2013. Error Awareness and the Insula: Links to Neurological and Psychiatric Diseases. *Front. Hum. Neurosci.* **7**:14. [13]

Klinger, E. 1978. Modes of Normal Conscious Flow. In: The Stream of Consciousness: Scientific Investigation into the Flow of Experience, ed. K. S. Pope and J. L. Singer. New York: Plenum. [7]

———. 1999. Thought Flow: Properties and Mechanisms Underlying Shifts in Content. In: At Play in the Fields of Consciousness: Essays in Honor of Jerome Singer, ed. J. Singer and P. Salovey, pp. 29–50. Mahwah, NJ: Erlbaum. [7]

Klinger, E., and W. M. Cox. 2011. Motivation and the Goal Theory of Current Concerns, ed. W. M. Cox and E. Klinger, pp. 3–47. Chichester: Wiley. [9]

Koechlin, E., and A. Hyafil. 2007. Anterior Prefrontal Function and the Limits of Human Decision-Making. *Science* **318**:594–598. [11]

Koechlin, E., C. Ody, and F. Kouneiher. 2003. The Architecture of Cognitive Control in the Human Prefrontal Cortex. *Science* **302**:1181–1185. [11]

Koenigs, M., A. K. Barbey, B. R. Postle, and J. Grafman. 2009. Superior Parietal Cortex Is Critical for the Manipulation of Information in Working Memory. *J. Neurosci.* **29**:14980–14986. [9]

Koizumi, A., K. Amano, A. Cortese, et al. 2016. Fear Reduction without Fear through Reinforcement of Neural Activity That Bypasses Conscious Exposure. *Nat. Hum. Behav.* **1**:6. [17]

Kolling, N., T. E. J. Behrens, M. K. Wittmann, and M. F. S. Rushworth. 2016. Multiple Signals in Anterior Cingulate Cortex. *Curr. Opin. Neurobiol.* **37**:36–43. [10]

Koob, G. F., and N. D. Volkow. 2009. Neurocircuitry of Addiction. *Neuropsychopharmacology* **35**:217–238. [10]

Koolschijn, R. S., U. E. Emir, A. C. Pantelides, et al. 2019. The Hippocampus and Neocortical Inhibitory Engrams Protect against Memory Interference. *Neuron* **101**:528–541. [6]

Korb, F. M., J. Jiang, J. A. King, and T. Egner. 2017. Hierarchically Organized Medial Frontal Cortex-Basal Ganglia Loops Selectively Control Task- and Response-Selection. *J. Neurosci.* **37**:7893–7905. [11]

Korsgaard, C. M. 2009. Self-Constitution: Agency, Identity, and Integrity. New York: Oxford Univ. Press. [13]

Kosslyn, S. M., W. L. Thompson, and G. Ganis. 2006. The Case for Mental Imagery. New York: Oxford Univ. Press. [9]

Kouchaki, M., and F. Gino. 2016. Memories of Unethical Actions Become Obfuscated over Time. *PNAS* **113**:6166–6171. [9]

Kounios, J., and M. Beeman. 2014. The Cognitive Neuroscience of Insight. *Ann. Rev. Psychol.* **65**:71–93. [13]

Kozel, F. A., K. Van Trees, V. Larson, et al. 2019. One Hertz versus Ten Hertz Repetitive TMS Treatment of PTSD: A Randomized Clinical Trial. *Psychiatry Res.* **273**:153–162. [17]

Krans, J., A. D. Brown, and M. L. Moulds. 2018. Can an Experimental Self-Efficacy Induction through Autobiographical Recall Modulate Analogue Posttraumatic Intrusions? *J. Behav. Ther. Exp. Psychiatry* **58**:1–11. [6]

Kraus, S. W., V. Voon, and M. N. Potenza. 2016. Should Compulsive Sexual Behavior Be Considered an Addiction? *Addiction* **111**:2097–2106. [10]

Kress, L., and T. Aue. 2017. The Link between Optimism Bias and Attention Bias: A Neurocognitive Perspective. *Neurosci. Biobehav. Rev.* **80**:688–702. [14]

Kucinski, A., C. Lustig, and M. Sarter. 2018. Addiction Vulnerability Trait Impacts Complex Movement Control: Evidence from Sign-Trackers. *Behav. Brain Res.* **350**:139–148. [5]

Kuhn, B. N., M. S. Klumpner, I. R. Covelo, P. Campus, and S. B. Flagel. 2018. Transient Inactivation of the Paraventricular Nucleus of the Thalamus Enhances Cue-Induced Reinstatement in Goal-Trackers, but Not Sign-Trackers. *Psychopharmacology* **235**:999–1014. [5]

Kühn, S., and J. Gallinat. 2011. Common Biology of Craving across Legal and Illegal Drugs - a Quantitative Meta-Analysis of Cue-Reactivity Brain Response. *Eur. J. Neurosci.* **33**:1318–1326. [10]

Kühn, S., F. Schmiedek, A. Brose, et al. 2013. The Neural Representation of Intrusive Thoughts. *Soc. Cogn. Affect. Neurosci.* **8**:688–693. [6, 8, 15]

Külz, K. K., S. Landmann, B. Cludius, et al. 2014. Mindfulness-Based Cognitive Therapy in Obsessive-Compulsive Disorder: Protocol of a Randomized Controlled Trial. *BMC Psychiatry* **14**:314. [9]

Kumar, S., S. Soren, and S. Chaudhury. 2009. Hallucinations: Etiology and Clinical Implications. *Ind. Psychiatry J.* **18**:119–126. [5]

Kumar, V. J., E. van Oort, K. Scheffler, C. F. Beckmann, and W. Grodd. 2017. Functional Anatomy of the Human Thalamus at Rest. *NeuroImage* **147**:678–691. [10]

Kunishio, K., and S. N. Haber. 1994. Primate Cingulostriatal Projection: Limbic Striatal versus Sensorimotor Striatal Input. *J. Comp. Neurol.* **350**:337–356. [10]

Küpper, C. S., R. G. Benoit, T. Dalgleish, and M. C. Anderson. 2014. Direct Suppression as a Mechanism for Controlling Unpleasant Memories in Daily Life. *J. Exp. Psychol. Gen.* **143**:1443–1449. [9]

Kurth, F., K. Zilles, P. T. Fox, A. R. Laird, and S. B. Eickhoff. 2010. A Link between the Systems: Functional Differentiation and Integration within the Human Insula Revealed by Meta-Analysis. *Brain Struct. Funct.* **214**:519–534. [10]

Lally, N., Q. J. M. Huys, N. Eshel, et al. 2017. The Neural Basis of Aversive Pavlovian Guidance during Planning. *J. Neurosci.* **37**:10215–10229. [9]

Lamichhane, B., B. M. Adhikari, and M. Dhamala. 2016. Salience Network Activity in Perceptual Decisions. *Brain Connect.* **6**:558–571. [10]

Lammel, S., B. K. Lim, C. Ran, et al. 2012. Input-Specific Control of Reward and Aversion in the Ventral Tegmental Area. *Nature* **491**:212–217. [2]

Lang, P. J. 1977. Imagery in Therapy: an Information Processing Analysis of Fear. *Behav. Therapy* **8**:862–886. [9]

Lang, T. J., M. L. Moulds, and E. A. Holmes. 2009. Reducing Depressive Intrusions via a Computerized Cognitive Bias Modification of Appraisals Task: Developing a Cognitive Vaccine. *Behav. Res. Ther.* **47**:139–145. [14]

Langlois, F., M. H. Freeston, and R. Ladouceur. 2000a. Differences and Similarities between Obsessive Intrusive Thoughts and Worry in a Non-Clinical Population: Study 1. *Behav. Res. Ther.* **38**:157–173. [17]

———. 2000b. Differences and Similarities between Obsessive Intrusive Thoughts and Worry in a Non-Clinical Population: Study 2. *Behav. Res. Ther.* **38**:175–189. [17]

Langner, R., and S. B. Eickhoff. 2013. Sustaining Attention to Simple Tasks: A Meta-Analytic Review of the Neural Mechanisms of Vigilant Attention. *Psychol. Bull.* **139**:870–900. [13]

Laplace, P. S. 1951. A Philosophical Essay on Probabilities, trans. & ed., F. W. Truscott and F. L. Emory, series ed. New York: Dover Publications. [12]

Lappin, J. S., and C. W. Eriksen. 1966. Use of a Delayed Signal to Stop a Visual Reaction-Time Response. *J. Exp. Psychol.* **72**:805–811. [9]

Larson, R., and M. Csikszentmihalyi. 1983. The Experience Sampling Method. In: New Directions for Methodology of Social and Behavioral Sciences, ed. H. T. Reis, pp. 21–34. San Francisco: Jossey-Bass. [9]

Latour, B. 2018. Down to Earth: Politics in the New Climatic Regime. Cambridge: Polity Press. [9]

Lau, H. C. 2009. Volition and the Function of Consciousness. In: Downward Causation and the Neurobiology of Free Will: Understanding Complex Systems, ed. N. Murphy et al., pp. 153–172. Heidelberg: Springer. [12]

———. 2019. Is Consciousness a Battle between Your Beliefs and Perceptions? *Aeon*: April 3. [12]

Lau, H. C., and R. E. Passingham. 2006. Relative Blindsight in Normal Observers and the Neural Correlate of Visual Consciousness. *PNAS* **103**:18763–18768. [12]

———. 2007. Unconscious Activation of the Cognitive Control System in the Human Prefrontal Cortex. *J. Neurosci.* **27**:5805–5811. [12]

Lau, H. C., R. D. Rogers, P. Haggard, and R. E. Passingham. 2004. Attention to Intention. *Science* **303**:1208–1210. [12]

Lau, H. C., R. D. Rogers, and R. E. Passingham. 2006. On Measuring the Perceived Onsets of Spontaneous Actions. *J. Neurosci.* **26(27)**:7265–7271. [12]

———. 2007. Manipulating the experienced onset of intention after action execution. *J. Cogn. Neurosci.* **19(1)**:81–90. [12]

Lau, H. C., and D. Rosenthal. 2011. Empirical Support for Higher-Order Theories of Conscious Awareness. *Trends Cogn. Sci.* **15**:365–373. [12]

Laufer, O., D. Israeli, and R. Paz. 2016. Behavioral and Neural Mechanisms of Overgeneralization in Anxiety. *Curr. Biol.* **26**:713–722. [9]

Laurent, V., and B. W. Balleine. 2015. Factual and Counterfactual Action-Outcome Mappings Control Choice between Goal-Directed Actions in Rats. *Curr. Biol.* **25**:1074–1079. [3]

Laurent, V., J. Bertran-Gonzalez, B. C. Chieng, and B. W. Balleine. 2014. δ-Opioid and Dopaminergic Processes in Accumbens Shell Modulate the Cholinergic Control of Predictive Learning and Choice. *J. Neurosci.* **34**:1358–1369. [3]

Laurent, V., B. Leung, N. Maidment, and B. W. Balleine. 2012. μ- and δ-Opioid-Related Processes in the Accumbens Core and Shell Differentially Mediate the Influence of Reward-Guided and Stimulus-Guided Decisions on Choice. *J. Neurosci.* **32**:1875–1883. [3]

Laurent, V., A. K. Morse, and B. W. Balleine. 2015a. The Role of Opioid Processes in Reward and Decision-Making. *Br. J. Pharmacol.* **172**:449–459. [3, 11]

Laurent, V., F. L. Wong, and B. W. Balleine. 2015b. δ-Opioid Receptors in the Accumbens Shell Mediate the Influence of Both Excitatory and Inhibitory Predictions on Choice. *Br. J. Pharmacol.* **172**:562–570. [3]

Lau-Zhu, A., R. N. Henson, and E. A. Holmes. 2019. Intrusive Memories and Voluntary Memory of a Trauma Film: Differential Effects of a Cognitive Interference Task after Encoding. *J. Exp. Psychol. Gen.* **148**:2154–2180. [9, 14, 17]

Lavín, A., and A. A. Grace. 1994. Modulation of Dorsal Thalamic Cell Activity by the Ventral Pallidum: Its Role in the Regulation of Thalamocortical Activity by the Basal Ganglia. *Synapse* **18**:104–127. [3]

Leckman, J. F., M. H. Bloch, M. E. Smith, D. Larabi, and M. Hampson. 2010. Neurobiological Substrates of Tourette's Disorder. *J. Child Adolesc Psychopharmacol* **20**:237–247. [3]

LeDoux, J. E. 2003. The Emotional Brain, Fear, and the Amygdala. *Cell. Mol. Neurobiol.* **23**:727–738. [9]

LeDoux, J. E., and D. Pine. 2016. Using Neuroscience to Help Understand Fear and Anxiety: A Two-System Framework. *Am. J. Psychiatry* **173**:1083–1093. [12]

LeDoux, J. E., and C. Sorrentino. 2019. The Deep History of Ourselves : The Four-Billion-Year Story of How We Got Conscious Brains. New York: Viking. [12]

Lee, H. J., S. H. Lee, H. S. Kim, S. M. Kwon, and M. J. Telch. 2005. A Comparison of Autogenous/Reactive Obsessions and Worry in a Nonclinical Population: A Test of the Continuum Hypothesis. *Behav. Res. Ther.* **43**:999–1010. [17]

Lee, J. L. C. 2008. Memory Reconsolidation Mediates the Strengthening of Memories by Additional Learning. *Nat. Neurosci.* **11**:1264–1266. [2]

———. 2009. Reconsolidation: Maintaining Memory Relevance. *Trends Neurosci.* **32**:413–420. [9]

Lee, J. L. C., A. L. Milton, and B. J. Everitt. 2006. Reconsolidation and Extinction of Conditioned Fear: Inhibition and Potentiation. *J. Neurosci.* **6**:10051–10056. [9, 14]

Lee, J. L. C., K. Nader, and D. Schiller. 2017. An Update on Memory Reconsolidation Updating. *Trends Cogn. Sci.* **21**:531–545. [9, 14]

Lee, W. H., S. H. Lisanby, A. F. Laine, and A. V. Peterchev. 2016. Comparison of Electric Field Strength and Spatial Distribution of Electroconvulsive Therapy and Magnetic Seizure Therapy in a Realistic Human Head Model. *Eur. Psychiatry* **36**:55–64. [16]

Legrand, N., O. Etard, A. Vandevelde, et al. 2019. Does the Heart Forget? Modulation of Cardiac Activity Induced by Inhibitory Control over Emotional Memories. *bioRxiv*376954. [9]

Lerner, T. N., L. Ye, and K. Deisseroth. 2016. Communication in Neural Circuits: Tools, Opportunities, and Challenges. *Cell* **164**:1136–1150. [4]

Leung, B. K., and B. W. Balleine. 2013. The Ventral Striato-Pallidal Pathway Mediates the Effect of Predictive Learning on Choice between Goal-Directed Actions. *J. Neurosci.* **33**:13848–13860. [3]

———. 2015. Ventral Pallidal Projections to Mediodorsal Thalamus and Ventral Tegmental Area Play Distinct Roles in Outcome-Specific Pavlovian-Instrumental Transfer. *J. Neurosci.* **35**:4953–4964. [3]

Levin Stilton, R., W. Heller, A. S. Engels, et al. 2011. Depression and Anxious Apprehension Distinguish Frontocingulate Cortical Activity During Top-Down Attentional Control. *J. Abnorm. Psychol.* **120**:272–285. [9]

Levine, A. Z., and D. M. Warman. 2016. Appraisals of and Recommendations for Managing Intrusive Thoughts: An Empirical Investigation. *Psychiatry Res.* **245**:207–216. [17]

Levine, J. 1983. Materialism and Qualia, the Explanatory Gap. *Pac. Philos. Quart.* **64**:354–361. [12]

Levy, B. J., and M. C. Anderson. 2002. Inhibitory Processes and the Control of Memory Retrieval. *Trends Cogn. Sci.* **6**:299–305. [9]

———. 2008. Individual Differences in the Suppression of Unwanted Memories: The Executive Deficit Hypothesis. *Acta Psychol. (Amst.)* **127**:623–635. [9]

———. 2012. Purging of Memories from Conscious Awareness Tracked in the Human Brain. *J. Neurosci.* **32**:16785–16794. [6, 9]

Levy, N. 2013. Addiction Is Not a Brain Disease (and It Matters). *Front. Psychiatry* **4**:10.3389/fpsyt.2013.00024. [12]

———. 2014. Consciousness and Moral Responsibility. Oxford: Oxford Univ. Press. [12]

———. 2018. Obsessive-Compulsive Disorder as a Disorder of Attention. *Mind Lang.* **33**:3–16. [10]

Lewis-Peacock, J. A., Y. Kessler, and K. Oberauer. 2018. The Removal of Information from Working Memory. *Ann. N.Y. Acad. Sci* **1424**:33–44. [9]

Li, C. S., P. Yan, R. Sinha, and T. W. Lee. 2008. Subcortical Processes of Motor Response Inhibition during a Stop Signal Task. *NeuroImage* **41**:1352–1363. [11]

Li, N. P., M. van Vugt, and S. M. Colarelli. 2018a. The Evolutionary Mismatch Hypothesis: Implications for Psychological Science. *Curr. Dir. Psychol. Sci.* **27**:38–44. [7]

Li, V., E. Michael, J. Balaguer, S. Herce Castañón, and C. Summerfield. 2018b. Gain Control Explains the Effect of Distraction in Human Perceptual, Cognitive, and Economic Decision Making. *PNAS* **115**:E8825–E8834. [10]

Liberzon, I., and B. Martis. 2006. Neuroimaging Studies of Emotional Responses in PTSD. *Ann. N.Y. Acad. Sci* **1071**:87–109. [9]

Libet, B., C. A. Gleason, E. W. Wright, and D. K. Pearl. 1983. Time of Conscious Intention to Act in Relation to Onset of Cerebral-Activity (Readiness-Potential): The Unconscious Initiation of a Freely Voluntary Act. *Brain* **106**:623–642. [12]

Lichtenstein, S., P. Slovic, B. Fischhoff, M. Layman, and B. Combs. 1978. Judged Frequency of Lethal Events. *J. Exp. Psychol. Hum. Learn. Memory* **4**:551–578. [9]

Lieder, F., T. L. Griffiths, Q. J. Huys, and N. D. Goodman. 2018a. The Anchoring Bias Reflects Rational Use of Cognitive Resources. *Psychon. Bull. Rev.* **25**:322–349. [9]

———. 2018b. Empirical Evidence for Resource-Rational Anchoring and Adjustment. *Psychon. Bull. Rev.* **25**:775–784. [9]

Limanowski, J., and K. Friston. 2018. Seeing the Dark: Grounding Phenomenal Transparency and Opacity in Precision Estimation for Active Inference. *Front. Psychol.* **9**:643. [13]

Lin, P., X. Wang, B. Zhang, et al. 2017. Functional Dysconnectivity of the Limbic Loop of Frontostriatal Circuits in First-Episode, Treatment-Naive Schizophrenia. *Hum. Brain Mapp.* **39**:747–757. [10]

Linsker, R. 1990. Perceptual Neural Organization: Some Approaches Based on Network Models and Information Theory. *Annu. Rev. Neurosci.* **13**:257–281. [13]

Litt, A., H. Plassmann, B. Shiv, and A. Rangel. 2010. Dissociating Valuation and Saliency Signals during Decision-Making. *Cereb. Cortex* **21**:95–102. [10]

Liu, K. Y. 2009. Suicide Rates in the World: 1950-2004. *Suicide Life-Threat. Behav.* **39**:204–213. [12]

Logan, G. D., and W. B. Cowan. 1984. On the Ability to Inhibit Thought and Action: A Theory of an Act of Control. *Psychol. Rev.* **91**:295–327. [11]

Logan, G. D., and R. D. Gordon. 2001. Executive Control of Visual Attention in Dual-Task Situations. *Psychol. Rev.* **108**:393–434. [11]

Logue, M. W., S. J. H. van Rooij, E. L. Dennis, et al. 2018. Smaller Hippocampal Volume in Posttraumatic Stress Disorder: A Multisite Enigma-Pgc Study: Subcortical Volumetry Results from Posttraumatic Stress Disorder Consortia. *Biol. Psychiatry* **83**:244–253. [9]

Lonergan, M. H., L. A. Olivera-Figueroa, R. K. Pitman, and A. Brunet. 2013. Propranolol's Effects on the Consolidation and Reconsolidation of Long-Term Emotional Memory in Healthy Participants: A Meta-Analysis. *J. Psychiatry Neurosci.* **38**:222–231. [14]

Lovic, V., B. T. Saunders, L. M. Yager, and T. E. Robinson. 2011. Rats Prone to Attribute Incentive Salience to Reward Cues Are Also Prone to Impulsive Action. *Behav. Brain Res.* **223**:255–261. [5]

Lowe, M. R., D. Arigo, M. L. Butryn, et al. 2016. Hedonic Hunger Prospectively Predicts Onset and Maintenance of Loss of Control Eating among College Women. *Health Psychol.* **35**:238–244. [13]

Lu, L., J. W. Grimm, B. T. Hope, and Y. Shaham. 2004. Incubation of Cocaine Craving after Withdrawal: A Review of Preclinical Data. *Neuropharmacology* **47 Suppl** 1:214–226. [5]

Luhrmann, T. M., R. Padmavati, H. Tharoor, and A. Osei. 2015. Differences in Voice-Hearing Experiences of People with Psychosis in the U.S.A., India and Ghana: Interview-Based Study. *Br. J. Psychiatry* **206**:41–44. [17]

Luo, Y. X., Y. X. Xue, J. F. Liu, et al. 2015. A Novel UCS Memory Retrieval-Extinction Procedure to Inhibit Relapse to Drug Seeking. *Nat. Commun.* **6**:7675. [9]

Ma, S., B. Hangya, C. S. Leonard, W. Wisden, and A. L. Gundlach. 2018. Dual-Transmitter Systems Regulating Arousal, Attention, Learning and Memory. *Neurosci. Biobehav. Rev.* **85**:21–33. [5]

Macleod, C. M., M. D. Dodd, E. D. Sheard, D. E. Wilson, and U. Bibi. 2003. In Opposition to Inhibition. In: The Psychology of Learning and Motivation, ed. B. H. Ross, pp. 163–214, vol. 43. New York: Academic Press. [11]

Magee, J. C., K. P. Harden, and B. A. Teachman. 2012. Psychopathology and Thought Suppression: A Quantitative Review. *Clin. Psychol. Rev.* **32**:189–201. [9]

Maia, T. V., and M. Cano-Colino. 2015. The Role of Serotonin in Orbitofrontal Function and Obsessive-Compulsive Disorder. *Clin. Psychol. Sci.* **3**:460–482. [13]

Maia, T. V., and A. Cleeremans. 2005. Consciousness: Converging Insights from Connectionist Modeling and Neuroscience. *Trends Cogn. Sci.* **9**:397–404. [13]

Maia, T. V., and V. A. Conceição. 2017. The Roles of Phasic and Tonic Dopamine in Tic Learning and Expression. *Biol. Psychiatry* **82**:401–412. [13]

———. 2018. Dopaminergic Disturbances in Tourette Syndrome: An Integrative Account. *Biol. Psychiatry* **84**:332–344. [13]

Maia, T. V., R. E. Cooney, and B. S. Peterson. 2008. The Neural Bases of Obsessive–Compulsive Disorder in Children and Adults. *Dev. Psychopathol.* **20**:1251–1283. [13]

Maia, T. V., and J. L. McClelland. 2012. A Neurocomputational Approach to Obsessive-Compulsive Disorder. *Trends Cogn. Sci.* **16**:14–15. [13]

Maillet, D., and D. L. Schacter. 2016. From Mind Wandering to Involuntary Retrieval: Age-Related Differences in Spontaneous Cognitive Processes. *Neuropsychologia* **80**:142–156. [6]

Majid, D. S. A., W. Cai, J. Corey-Bloom, and A. R. Aron. 2013. Proactive Selective Response Suppression Is Implemented via the Basal Ganglia. *J. Neurosci.* 33:13259–13269. [9]

Majid, D. S. A., W. Cai, J. S. George, F. Verbruggen, and A. R. Aron. 2012. Transcranial Magnetic Stimulation Reveals Dissociable Mechanisms for Global versus Selective Corticomotor Suppression Underlying the Stopping of Action. *Cereb. Cortex* 22:363–371. [9]

Malenka, R. C., and M. F. Bear. 2004. LTP and LTD: An Embarrassment of Riches. *Neuron* 44:5–21. [16]

Marchant, N. J., R. Rabei, K. Kaganovsky, et al. 2014. A Critical Role of Lateral Hypothalamus in Context-Induced Relapse to Alcohol Seeking after Punishment-Imposed Abstinence. *J. Neurosci.* 34:7447–7457. [5]

Mardikian, P. N., S. D. Larowe, S. Hedden, P. W. Kalivas, and R. J. Malcolm. 2007. An Open-Label Trial of N-Acetylcysteine for the Treatment of Cocaine Dependence: A Pilot Study. *Prog. Neuro-Psychopharmacol. Biol. Psychiatry* 31:389–394. [15]

Marks, E. H., A. R. Franklin, and L. A. Zoellner. 2018. Can't Get It out of My Mind: A Systematic Review of Predictors of Intrusive Memories of Distressing Events. *Psychol. Bull.* 144:584–640. [6]

Marneros, A. 1988. Schizophrenic First-Rank Symptoms in Organic Mental Disorders. *Br. J. Psychiatry* 152:625–628. [8]

Marquand, A. F., K. V. Haak, and C. F. Beckmann. 2017. Functional Corticostriatal Connection Topographies Predict Goal Directed Behaviour in Humans. *Nat. Hum. Behav.* 1:0146. [10]

Marr, D. 1982. Vision: A Computational Investigation into the Human Representation and Processing of Visual Information New York: Henry Holt and Co., Inc. [11]

Marsh, R., T. V. Maia, and B. S. Peterson. 2009. Functional Disturbances Within Frontostriatal Circuits across Multiple Childhood Psychopathologies. *Am. J. Psychiatry* 166:664–674. [13]

Mason, A. E., K. Jhaveri, M. Cohn, and J. A. Brewer. 2018. Testing a Mobile Mindful Eating Intervention Targeting Craving-Related Eating: Feasibility and Proof of Concept. *J. Behav. Med.* 41:160–173. [17]

Mathys, C., J. Daunizeau, K. J. Friston, and K. E. Stephan. 2011. A Bayesian Foundation for Individual Learning under Uncertainty. *Front. Hum. Neurosci.* 5:39. [10]

Matsumoto, M., and O. Hikosaka. 2009. Two Types of Dopamine Neuron Distinctly Convey Positive and Negative Motivational Signals. *Nature* 459:837–841. [2]

Mattar, M. G., and N. D. Daw. 2018. Prioritized Memory Access Explains Planning and Hippocampal Replay. *Nat. Neurosci.* 21:1609–1617. [9]

May, J., D. J. Kavanagh, and J. Andrade. 2015. The Elaborated Intrusion Theory of Desire: A 10-Year Retrospective and Implications for Addiction Treatments. *Addict. Behav.* 44:29–34. [13]

May, J. R., and H. J. Johnson. 1973. Physiological Activity to Internally Elicited Arousal and Inhibitory Thoughts. *J. Abnorm. Psychol.* 82:239–245. [6]

Mayberg, H. S. 2009. Targeted Electrode-Based Modulation of Neural Circuits for Depression. *J. Clin. Invest.* 119:717–725. [9]

Mayberg, H. S., S. K. Brannan, R. K. Mahurin, et al. 1997. Cingulate Function in Depression: A Potential Predictor of Treatment Response. *Neuroreport* 8:1057–1061. [5]

McClure, E. A., C. D. Gipson, R. J. Malcolm, P. W. Kalivas, and K. M. Gray. 2014. Potential Role of N-Acetylcysteine in the Management of Substance Use Disorders. *CNS Drugs* 28:95–106. [15]

McClure, S. M., N. D. Daw, and P. R. Montague. 2003. A Computational Substrate for Incentive Salience. *Trends Neurosci.* **26**:423–428. [13]

McCurdy, L. Y., B. Maniscalco, J. Metcalfe, et al. 2013. Anatomical Coupling between Distinct Metacognitive Systems for Memory and Visual Perception. *J. Neurosci.* **33**:1897–1906. [12]

McEwen, B. S., and P. J. Gianaros. 2011. Stress- and Allostasis-Induced Brain Plasticity. *Annu. Rev. Med.* **62**:431–445. [10]

McFarlane, A. C. 1988. The Phenomenology of Posttraumatic Stress Disorders Following a Natural Disaster. *J. Nerv. Ment. Dis.* **176**:22–29. [7]

McGaugh, J. L. 1966. Time-Dependent Processes in Memory Storage. *Science* **153**:1351–1358. [14]

———. 2000. Memory: A Century of Consolidation. *Science* **287**:248–251. [9, 14]

McGuire, P., P. Robson, W. J. Cubala, et al. 2018. Cannabidiol (CBD) as an Adjunctive Therapy in Schizophrenia: A Multicenter Randomized Controlled Trial. *Am. J. Psychiatry* **175**:225–231. [15]

McNab, F., and T. Klingberg. 2008. Prefrontal Cortex and Basal Ganglia Control Access to Working Memory. *Nat. Neurosci.* **11**:103–107. [11]

McRae-Clark, A. L., N. L. Baker, M. M. Maria, and K. T. Brady. 2013. Effect of Oxytocin on Craving and Stress Response in Marijuana-Dependent Individuals: A Pilot Study. *Psychopharmacology* **228**:623–631. [15]

McTeague, L. M., M. S. Goodkind, and A. Etkin. 2016. Transdiagnostic Impairment of Cognitive Control in Mental Illness. *J. Psychiatr. Res.* **83**:37–46. [13]

McTeague, L. M., J. Huemer, D. M. Carreon, et al. 2017. Identification of Common Neural Circuit Disruptions in Cognitive Control across Psychiatric Disorders. *Am. J. Psychiatry* **174**:676–685. [10]

McVay, J. C., M. J. Kane, and T. R. Kwapil. 2009. Tracking the Train of Thought from the Laboratory into Everyday Life: An Experience-Sampling Study of Mind Wandering across Controlled and Ecological Contexts. *Psychon. Bull. Rev.* **16**:857–863. [9]

Medford, N., and H. D. Critchley. 2010. Conjoint Activity of Anterior Insular and Anterior Cingulate Cortex: Awareness and Response. *Brain Struct. Funct.* **214**:535–549. [13]

Meeten, F., G. C. L. Davey, E. Makovac, et al. 2016. Goal Directed Worry Rules Are Associated with Distinct Patterns of Amygdala Functional Connectivity and Vagal Modulation during Perseverative Cognition. *Front. Hum. Neurosci.* **10**:553. [13]

Meister, M. 2016. Physical Limits to Magnetogenetics. *eLife* **5**:e17210. [4]

Menon, V. 2011. Large-Scale Brain Networks and Psychopathology: A Unifying Triple Network Model. *Trends Cogn. Sci.* **15**:483–506. [10]

Menon, V., and L. Q. Uddin. 2010. Saliency, Switching, Attention and Control: A Network Model of Insula Function. *Brain Struct. Funct.* **214**:655–667. [10, 13]

Merlo, E., A. L. Milton, Z. Y. Goozée, D. E. H. Theobald, and B. J. Everitt. 2014. Reconsolidation and Extinction Are Dissociable and Mutually Exclusive Processes: Behavioral and Molecular Evidence. *J. Neurosci.* **34**:2422–2231. [2, 14]

Metzger, C. D. 2010. High Field fMRI Reveals Thalamocortical Integration of Segregated Cognitive and Emotional Processing in Mediodorsal and Intralaminar Thalamic Nuclei. *Front. Neuroanat.* **4**:138. [10]

Metzger, C. D., U. Eckert, J. Steiner, et al. 2010. High Field fMRI Reveals Thalamocortical Integration of Segregated Cognitive and Emotional Processing in Mediodorsal and Intralaminar Thalamic Nuclei. *Front. Neuroanat.* **4**:138. [10]

Meyer, T., T. Smeets, T. Giesbrecht, et al. 2015. The Role of Frontal EEG Asymmetry in Post-Traumatic Stress Disorder. *Biol. Psychol.* **108**:62–77. [6]

Meyer, T. J., M. L. Miller, R. L. Metzger, and T. D. Borkovec. 1990. Development and Validation of the Penn State Worry Questionnaire. *Behav. Res. Ther.* **28**:487–495. [6, 15]

Mickelsen, L. E., M. Bolisetty, B. R. Chimileski, et al. 2019. Single-Cell Transcriptomic Analysis of the Lateral Hypothalamic Area Reveals Molecularly Distinct Populations of Inhibitory and Excitatory Neurons. *Nat. Neurosci.* **22**:642–656. [5]

Mikami, K., Y. Kiyokawa, Y. Takeuchi, and Y. Mori. 2016. Social Buffering Enhances Extinction of Conditioned Fear Responses in Male Rats. *Physiol. Behav.* **163**:123–128. [14]

Milad, M. R., R. K. Pitman, C. B. Ellis, et al. 2009. Neurobiological Basis of Failure to Recall Extinction Memory in Posttraumatic Stress Disorder. *Biol. Psychiatry* **66**:1075–1082. [9]

Milad, M. R., and G. J. Quirk. 2012. Fear Extinction as a Model for Translational Neuroscience: Ten Years of Progress. *Ann. Rev. Psychol.* **63**:129–151. [9]

Miller, E. K., and J. D. Cohen. 2001. An Integrative Theory of Prefrontal Cortex Function. *Annu. Rev. Neurosci.* **24**:167–202. [11]

Miller, M. A., A. Bershad, A. C. King, R. Lee, and H. D. Wit. 2016. Intranasal Oxytocin Dampens Cue-Elicited Cigarette Craving in Daily Smokers. *Behav. Pharmacol.* **27**:697–703. [15]

Milton, A. L., and B. J. Everitt. 2012. The Persistence of Maladaptive Memory: Addiction, Drug Memories and Anti-Relapse Treatments. *Neurosci. Biobehav. Rev.* **36**:1119–1139. [14]

Milton, A. L., and E. A. Holmes. 2018. Of Mice and Mental Health: Facilitating Dialogue and Seeing Further. *Phil. Trans. R. Soc. B* **373**:20170022. [5]

Milton, A. L., J. L. C. Lee, V. J. Butler, R. J. Gardner, and B. J. Everitt. 2008a. Intra-Amygdala and Systemic Antagonism of NMDA Receptors Prevents the Reconsolidation of Drug-Associated Memory and Impairs Subsequently Both Novel and Previously Acquired Drug-Seeking Behaviors. *J. Neurosci.* **28**:8230–8237. [2]

Milton, A. L., J. L. C. Lee, and B. J. Everitt. 2008b. Reconsolidation of Appetitive Memories for Both Natural and Drug Reinforcement Is Dependent on β-Adrenergic Receptors. *Learn. Mem.* **15**:88–92. [2]

Minarini, A., S. Ferrari, M. Galletti, et al. 2017. N-Acetylcysteine in the Treatment of Psychiatric Disorders: Current Status and Future Prospects. *Expert Opin. Drug Metab. Toxicol* **13**:279–292. [15]

Mink, J. W. 1996. The Basal Ganglia: Focused Selection and Inhibition of Competing Motor Programs. *Prog. Neurobiol.* **50**:381–425. [11]

Mirenowicz, J., and W. Schultz. 1996. Preferential Activation of Midbrain Dopamine Neurons by Appetitive Rather Than Aversive Stimuli. *Nature* **379**:449–451. [2]

Mirza, M. B., R. A. Adams, C. D. Mathys, and K. J. Friston. 2016. Scene Construction, Visual Foraging, and Active Inference. *Front. Comput. Neurosci.* **10**:56. [13]

Mithoefer, M. C., A. A. Feduccia, L. Jerome, et al. 2019. MDMA-Assisted Psychotherapy for Treatment of PTSD: Study Design and Rationale for Phase 3 Trials Based on Pooled Analysis of Six Phase 2 Randomized Controlled Trials. *Psychopharmacology* **236**:2735–2745. [17]

Miyata, J. 2019. Toward Integrated Understanding of Salience in Psychosis. *Neurobiol. Dis.* **131**:104414. [10, 13]

Mizrahi, R., M. Kiang, D. C. Mamo, et al. 2006. The Selective Effect of Antipsychotics on the Different Dimensions of the Experience of Psychosis in Schizophrenia Spectrum Disorders. *Schizophr. Res.* **88**:111–118. [15]

Mobbs, D., H. C. Lau, O. D. Jones, and C. D. Frith. 2007. Law, Responsibility, and the Brain. *PLoS Biol.* **5**:e103. [12]

Modell, J. G., J. M. Mountz, G. C. Curtis, and J. F. Greden. 1989. Neurophysiologic Dysfunction in Basal Ganglia/Limbic Striatal and Thalamocortical Circuits as a Pathogenetic Mechanism of Obsessive-Compulsive Disorder. *J. Neuropsychiatr. Clin. Neurosci.* **1**:27–36. [3]

Moeck, E. K., I. E. Hyman, and M. K. T. Takarangi. 2018. Understanding the Overlap between Positive and Negative Involuntary Cognitions Using Instrumental Earworms. *Psychomusicol.* **28**:164–177. [9]

Moeller, S. J., A. B. Konova, M. A. Parvaz, et al. 2014. Functional, Structural, and Emotional Correlates of Impaired Insight in Cocaine Addiction. *JAMA Psychiatry* **71**:61. [13]

Moeller, S. J., and M. P. Paulus. 2018. Toward Biomarkers of the Addicted Human Brain: Using Neuroimaging to Predict Relapse and Sustained Abstinence in Substance Use Disorder. *Prog. Neuro-Psychopharmacol. Biol. Psychiatry* **80**:143–154. [10]

Monfils, M. H., K. K. Cowansage, E. Klann, and J. E. LeDoux. 2009. Extinction-Reconsolidation Boundaries: Key to Persistent Attenuation of Fear Memories. *Science* **324**:951–955. [2, 7, 9, 14]

Monfils, M. H., and E. A. Holmes. 2018. Memory Boundaries: Opening a Window Inspired by Reconsolidation to Treat Anxiety, Trauma-Related, and Addiction Disorders. *Lancet Psychiatry* **5**:1032–1042. [7, 14]

Montague, P. R., P. Dayan, and T. J. Sejnowski. 1996. A Framework for Mesencephalic Dopamine Systems Based on Predictive Hebbian Learning. *J. Neurosci.* **16**:1936–1947. [2]

Moore, J. W., and P. C. Fletcher. 2012. Sense of Agency in Health and Disease: A Review of Cue Integration Approaches. *Conscious. Cogn.* **21**:59–68. [13]

Morand-Beaulieu, S., S. Grot, J. Lavoie, et al. 2017. The Puzzling Question of Inhibitory Control in Tourette Syndrome: A Meta-Analysis. *Neurosci. Biobehav. Rev.* **80**:240–262. [13]

Morecraft, R. J., K. S. Stilwell-Morecraft, P. B. Cipolloni, et al. 2012. Cytoarchitecture and Cortical Connections of the Anterior Cingulate and Adjacent Somatomotor Fields in the Rhesus Monkey. *Brain Res. Bull.* **87**:457–497. [5]

Morecraft, R. J., and J. Tanji. 2009. Cingulofrontal Interactions and the Cingulate Motor Areas. In: Cingulate Neurobiology and Disease, ed. B. A. Vogt, pp. 113–144. Oxford: Oxford Univ. Press. [5]

Morein-Zamir, S., N. A. Fineberg, T. W. Robbins, and B. J. Sahakian. 2010. Inhibition of Thoughts and Actions in Obsessive-Compulsive Disorder: Extending the Endophenotype? *Psychol. Med.* **40**:263–272. [9]

Morein-Zamir, S., S. Shahper, N. A. Fineberg, et al. 2018. Free Operant Observing in Humans: A Translational Approach to Compulsive Certainty Seeking. *Q. J. Exp. Psychol.* **71**:2052–2069. [5]

Morris, L. S., P. Kundu, N. Dowell, et al. 2016. Fronto-Striatal Organization: Defining Functional and Microstructural Substrates of Behavioural Flexibility. *Cortex* **74**:118–133. [16]

Morris, R. W., C. Cyrzon, M. J. Green, M. E. Le Pelley, and B. W. Balleine. 2018. Impairments in Action–Outcome Learning in Schizophrenia. *Transl. Psychiatry* **8**:54. [3]

Morris, R. W., S. Quail, K. R. Griffiths, M. J. Green, and B. W. Balleine. 2015. Corticostriatal Control of Goal-Directed Action Is Impaired in Schizophrenia. *Biol. Psychiatry* **77**:187–195. [3]

Morrow, J. D., S. Maren, and T. E. Robinson. 2011. Individual Variation in the Propensity to Attribute Incentive Salience to an Appetitive Cue Predicts the Propensity to Attribute Motivational Salience to an Aversive Cue. *Behav. Brain Res.* **220**:238–243. [5]

Moscarello, J. M., and C. A. Hartley. 2017. Agency and the Calibration of Motivated Behavior. *Trends Cogn. Sci.* **21**:725–735. [12]

Moulding, R., M. E. Coles, J. S. Abramowitz, et al. 2014. Part 2: They Scare Because We Care—the Relationship between Obsessive Intrusive Thoughts and Appraisals and Control Strategies across 15 Cities. *J. Obsessive Compuls. Relat. Disord.* **3**:280–291. [13]

Mukamel, E. A., A. Nimmerjahn, and M. J. Schnitzer. 2009. Automated Analysis of Cellular Signals from Large-Scale Calcium Imaging Data. *Neuron* **63**:747–760. [4]

Müller, G. E., and A. Pilzecker. 1900. Experimentelle Beitrage Zur Lehre Vom Gedachtniss, vol. 1. Leipzig: J. A. Barth. [9, 14]

Musslick, S., A. Saxe, K. Özcimder, et al. 2017. Multitasking Capability versus Learning Efficiency in Neural Network Architectures. https://pdfs.semanticscholar.org/0bfa/9 3978687f06bbeaddd3d7c258634da36040b.pdf. (accessed Nov. 20, 2019). [11]

Mutz, J., D. R. Edgcumbe, A. R. Brunoni, and C. H. Y. Fu. 2018. Efficacy and Acceptability of Non-Invasive Brain Stimulation for the Treatment of Adult Unipolar and Bipolar Depression: A Systematic Review and Meta-Analysis of Randomised Sham-Controlled Trials. *Neurosci. Biobehav. Rev.* **92**:291–303. [17]

Nader, K. 2003. Memory Traces Unbound. *Trends Neurosci.* **26**:65–72. [9, 14]

Nader, K., and O. Hardt. 2009. A Single Standard for Memory: The Case for Reconsolidation. *Nat. Rev. Neurosci.* **10**:224–234. [9, 14]

Nader, K., G. E. Schafe, and J. E. Le Doux. 2000. Fear Memories Require Protein Synthesis in the Amygdala for Reconsolidation after Retrieval. *Nature* **406**:722–726. [2, 14]

Nagai, Y. 2015. Modulation of Autonomic Activity in Neurological Conditions: Epilepsy and Tourette Syndrome. *Front. Neurosci.* **9**:278. [13]

Nagai, Y., A. Cavanna, and H. D. Critchley. 2009. Influence of Sympathetic Autonomic Arousal on Tics: Implications for a Therapeutic Behavioral Intervention for Tourette Syndrome. *J. Psychosom. Res.* **67**:599–605. [13]

Nagel, T. 1974. What Is It Like to Be a Bat. *Philos. Rev.* **83**:435–450. [12]

Nahmias, E. 2014. Is Free Will an Illusion? Confronting Challenges from the Modern Mind Sciences. In: Moral Psychology, vol. 4: Freedom and Responsibility, ed. W. Sinnott-Armstrong. Cambridge, MA: MIT Press. [12]

———. 2016. Free Will as a Psychological Accomplishment. In: The Oxford Handbook of Freedom, ed. D. Schmidtz and C. E. Pavel, pp. 492–507. Oxford: Oxford Univ. Press. [12]

Najmi, S., B. C. Riemann, and D. M. Wegner. 2009. Managing Unwanted Intrusive Thoughts in Obsessive-Compulsive Disorder: Relative Effectiveness of Suppression, Focused Distraction, and Acceptance. *Behav. Res. Ther.* **47**:494–503. [6]

Nakao, T., A. Nakagawa, T. Yoshiura, et al. 2005. Brain Activation of Patients with Obsessive-Compulsive Disorder during Neuropsychological and Symptom Provocation Tasks before and after Symptom Improvement: A Functional Magnetic Resonance Imaging Study. *Biol. Psychiatry* **57**:901–910. [8]

Namboodiri, V. M. K., J. M. Otis, K. van Heeswijk, et al. 2019. Single-Cell Activity Tracking Reveals That Orbitofrontal Neurons Acquire and Maintain a Long-Term Memory to Guide Behavioral Adaptation. *Nat. Neurosci.* **22**:1110–1121. [5]

Namkung, H., S.-H. Kim, and A. Sawa. 2017. The Insula: An Underestimated Brain Area in Clinical Neuroscience, Psychiatry, and Neurology. *Trends Neurosci.* **40**:200–207. [10]

Naqvi, N. H., and A. Bechara. 2010. The Insula and Drug Addiction: an Interoceptive View of Pleasure, Urges, and Decision-Making. *Brain Struct. Funct.* **214**:435–450. [10, 13]

Naqvi, N. H., N. Gaznick, D. Tranel, and A. Bechara. 2014. The Insula: A Critical Neural Substrate for Craving and Drug Seeking under Conflict and Risk. *Ann. N.Y. Acad. Sci* **1316**:53–70. [13]

National Institute for Health Care Excellence. 2018. Post-Traumatic Stress Disorder: NICE Guideline [NG116]. NICE. https://www.nice.org.uk/guidance/NG116. (accessed Feb. 10, 2020). [14]

Nee, D. E., J. W. Brown, M. K. Askren, et al. 2013. A Meta-Analysis of Executive Components of Working Memory. *Cereb. Cortex* **23**:264–282. [9]

Nee, D. E., and M. D'Esposito. 2016. The Hierarchical Organization of the Lateral Prefrontal Cortex. *eLife* **5**: e12112. [11]

———. 2017. Causal Evidence for Lateral Prefrontal Cortex Dynamics Supporting Cognitive Control. *eLife* **6**: e28040. [11]

Nesse, R. M. 2005. An Evolutionary Framework for Understanding Grief. In: Spousal Bereavement in Late Life, ed. D. Carr et al., pp. 195–226. New York: Springer. [7]

Newby, J. M., and M. L. Moulds. 2011. Characteristics of Intrusive Memories in a Community Sample of Depressed, Recovered Depressed and Never-Depressed Individuals. *Behav. Res. Ther.* **49**:234–243. [6]

———. 2012. A Comparison of the Content, Themes, and Features of Intrusive Memories and Rumination in Major Depressive Disorder. *Br. J. Clin. Psychol.* **51**:197–205. [14]

Newell, A. 1990. Unified Theories of Cognition. Cambridge, MA: Harvard Univ. Press. [11]

Niendam, T. A., A. R. Laird, K. L. Ray, et al. 2012. Meta-Analytic Evidence for a Superordinate Cognitive Control Network Subserving Diverse Executive Functions. *Cogn. Affect. Behav. Neurosci.* **12**:241–268. [13]

Nietzsche, F. W. 1955. Beyond Good and Evil. South Bend: Gateway Editions. [12]

Nieuwenhuys, R. 2012. The Insular Cortex. *Prog. Brain. Res.* **195**:123–163. [13]

Nikolov, S., D. A. Rahnev, and H. C. Lau. 2010. Probabilistic Model of Onset Detection Explains Paradoxes in Human Time Perception. *Front. Psychol.* **1**:37. [12]

Nilsson, J. P., M. Söderström, A. U. Karlsson, et al. 2005. Less Effective Executive Functioning after One Night's Sleep Deprivation. *J. Sleep Res.* **14**:1–6. [9]

Nixon, R. D. V., T. Nehmy, and M. Seymour. 2007. The Effect of Cognitive Load and Hyperarousal on Negative Intrusive Memories. *Behav. Res. Ther.* **45**:2652–2663. [6]

Nolen-Hoeksema, S., and J. Morrow. 1991. A Prospective Study of Depression and Posttraumatic Stress Symptoms after a Natural Disaster: The 1989 Loma Prieta Earthquake. *J. Pers. Soc. Psychol.* **61**:115e121. [7]

Nolen-Hoeksema, S., J. Morrow, and B. L. Fredrickson. 1993. Response Styles and the Duration of Episodes of Depressed Mood. *J. Abnorm. Psychol.* **102**:20 28. [9]

Nolen-Hoeksema, S., B. E. Wisco, and S. Lyubomirsky. 2008. Rethinking Rumination. *Perspect. Psychol. Sci.* **3**:400–424. [8, 9]

Nombela, C., T. Rittman, T. W. Robbins, and J. B. Rowe. 2014. Multiple Modes of Impulsivity in Parkinson's Disease. *PLoS One* **9**:85747. [9]

Nomura, E. M., C. Gratton, R. M. Visser, et al. 2010. Double Dissociation of Two Cognitive Control Networks in Patients with Focal Brain Lesions. *PNAS* **107**:12017–12022. [11]

Noreen, S., R. N. Bierman, and M. D. MacLeod. 2014. Forgiving You Is Hard, but Forgetting Seems Easy: Can Forgiveness Facilitate Forgetting? *Psychol. Sci.* **25**:1295–1302. [9]

Noreen, S., and M. D. MacLeod. 2013. It's All in the Detail: Intentional Forgetting of Autobiographical Memories Using the Autobiographical Think/No-Think Task. *J. Exp. Psychol. Learn. Mem. Cogn.* **39**:375–393. [9]

Norman, K., S. Polyn, G. Detre, and J. Haxby. 2006. Beyond Mind-Reading: Multi-Voxel Pattern Analysis of fMRI Data. *Trends Cogn. Sci.* **10**:424–430. [17]

Norman, L. J., C. Carlisi, S. Lukito, et al. 2016. Structural and Functional Brain Abnormalities in Attention-Deficit/Hyperactivity Disorder and Obsessive-Compulsive Disorder. *JAMA Psychiatry* **73**:815. [13]

Norrholm, S. D., and T. Jovanovic. 2018. Fear Processing, Psychophysiology, and PTSD. *Harv. Rev. Psychiatry* **26**:129–141. [14]

Nour, M. M., T. Dahoun, P. Schwartenbeck, et al. 2018. Dopaminergic Basis for Signaling Belief Updates, but Not Surprise, and the Link to Paranoia. *PNAS* **115**:E10167–E10176. [10]

Oberauer, K. 2013. The Focus of Attention in Working Memory—from Metaphors to Mechanisms. *Front. Hum. Neurosci.* **7**:673. [13]

Obsessive Compulsive Cognitions Working Group. 2001. Development and Initial Validation of the Obsessive Beliefs Questionnaire and the Interpretation of Intrusions Inventory. *Behav. Res. Ther.* **39**:987–1006. [17]

———. 2003. Psychometric Validation of the Obsessive Beliefs Questionnaire and the Interpretation of Intrusions Inventory: Part I. *Behav. Res. Ther.* **41**:863–878. [17]

Odaka, H., S. Arai, T. Inoue, and T. Kitaguchi. 2014. Genetically-Encoded Yellow Fluorescent Camp Indicator with an Expanded Dynamic Range for Dual-Color Imaging. *PLoS One* **9**:e100252. [4]

Oehrn, C. R., J. Fell, C. Baumann, et al. 2018. Direct Electrophysiological Evidence for Prefrontal Control of Hippocampal Processing during Voluntary Forgetting. *Curr. Biol.* **28**:3016–3022. [6, 9]

Offidani, E., J. Guidi, E. Tomba, and G. A. Fava. 2013. Efficacy and Tolerability of Benzodiazepines versus Antidepressants in Anxiety Disorders: A Systematic Review and Meta-Analysis. *Psychother. Psychosom.* **82**:355–362. [15]

Öhman, A., and S. Mineka. 2001. Fears, Phobias, and Preparedness: Toward an Evolved Module of Fear and Fear Learning. *Psychol. Rev.* **108**:483. [7]

Olney, J. J., S. M. Warlow, E. E. Naffziger, and K. C. Berridge. 2018. Current Perspectives on Incentive Salience and Applications to Clinical Disorders. *Curr. Opin. Behav. Sci.* **22**:59–69. [13]

Opie, G. M., and J. Cirillo. 2017. Commentary: Preconditioning tDCS Facilitates Subsequent tDCS Effect on Skill Acquisition in Older Adults. *Front. Aging Neurosci.* **9**:84. [16]

Opie, G. M., E. Vosnakis, M. C. Ridding, U. Ziemann, and J. G. Semmler. 2017. Priming Theta Burst Stimulation Enhances Motor Cortex Plasticity in Young but Not Old Adults. *Brain Stimul.* **10**:298–304. [16]

Optican, L., and B. J. Richmond. 1987. Temporal Encoding of Two-Dimensional Patterns by Single Units in Primate Inferior Cortex, II: Information Theoretic Analysis. *J. Neurophysiol.* **57**:132–146. [13]

Orederu, T., and D. Schiller. 2018. Fast and Slow Extinction Pathways in Defensive Survival Circuits. *Curr. Opin. Behav. Sci.* **24**:96–103. [9]

O'Reilly, R. C. 2006. Biologically Based Computational Models of High-Level Cognition. Science 314:91–94. [13]

O'Reilly, R. C., and M. J. Frank. 2006. Making Working Memory Work: A Computational Model of Learning in the Prefrontal Cortex and Basal Ganglia. Neural Comput. 18:283–328. [11]

Ortiz, V., M. Giachero, P. J. Espejo, V. A. Molina, and I. D. Martijena. 2015. The Effect of Midazolam and Propranolol on Fear Memory Reconsolidation in Ethanol-Withdrawn Rats: Influence of D-Cycloserine. *Int. J. Neuropsychopharm.* **18**:pyu082. [14]

Osman, S., M. Cooper, A. Hackmann, and D. Veale. 2004. Spontaneously Occurring Images and Early Memories in People with Body Dysmorphic Disorder. *Memory* **12**:428–436. [14]

Ostlund, S. B., and B. W. Balleine. 2007a. The Contribution of Orbitofrontal Cortex to Action Selection. *Ann. N.Y. Acad. Sci* **1121**:174–192. [2]

———. 2007b. Selective Reinstatement of Instrumental Performance Depends on the Discriminative Stimulus Properties of the Mediating Outcome. *Learn. Behav.* **35**:43–52. [3]

———. 2008. On Habits and Addiction: An Associative Analysis of Compulsive Drug Seeking. *Drug Discov. Today Dis. Models* **5**:235–245. [3]

Ostlund, S. B., and N. T. Maidment. 2012. Dopamine Receptor Blockade Attenuates the General Incentive Motivational Effects of Noncontingently Delivered Rewards and Reward-Paired Cues without Affecting Their Ability to Bias Action Selection. *Neuropsychopharmacology* **37**:508–519. [2]

Ostlund, S. B., N. T. Maidment, and B. W. Balleine. 2010. Alcohol-Paired Contextual Cues Produce an Immediate and Selective Loss of Goal-Directed Action in Rats. *Front. Integr. Neurosci.* **4**:19. [3]

Ottaviani, C., J. F. Thayer, B. Verkuil, H. D. Critchley, and J. F. Brosschot. 2017. Editorial: Can't Get You out of My Head: Brain-Body Interactions in Perseverative Cognition. *Front. Hum. Neurosci.* **11**:634. [13]

Ottaviani, C., D. R. Watson, F. Meeten, et al. 2016. Neurobiological Substrates of Cognitive Rigidity and Autonomic Inflexibility in Generalized Anxiety Disorder. *Biol. Psychol.* **119**:31–41. [13]

Oudeyer, P. Y., and F. Kaplan. 2007. What Is Intrinsic Motivation? A Typology of Computational Approaches. *Front. Neurorobotics* **1**:6. [13]

Packer, A. M., L. E. Russell, H. W. Dalgleish, and M. Häusser. 2015. Simultaneous All-Optical Manipulation and Recording of Neural Circuit Activity with Cellular Resolution *in Vivo. Nat. Methods* **12**:140–146. [4]

Padoa-Schioppa, C., and J. A. Assad. 2006. Neurons in the Orbitofrontal Cortex Encode Economic Value. *Nature* **441**:223–226. [2]

Pal, P., D. Theisen, M. Datko, et al. 2019. Monte-Carlo Simulation to Reduce Sensor Dimension of EEG Neurofeedback Device. In: APS Meeting Abstracts, ID H23.004. [17]

Palmer, C. J., A. K. Seth, and J. Hohwy. 2015. The Felt Presence of Other Minds: Predictive Processing, Counterfactual Predictions, and Mentalising in Autism. *Conscious. Cogn.* **36**:376–389. [13]

Paolone, G., C. C. Angelakos, P. J. Meyer, T. E. Robinson, and M. Sarter. 2013. Cholinergic Control over Attention in Rats Prone to Attribute Incentive Salience to Reward Cues. *J. Neurosci.* **33**:8321–8335. [5]

Parkes, L., B. Fulcher, M. Yücel, and A. Fornito. 2016. Transcriptional Signatures of Connectomic Subregions of the Human Striatum. *Genes Brain Behav.* **16**:647–663. [10]

Parkinson, L., and S. Rachman. 1981. Part II: The Nature of Intrusive Thoughts. *Adv. Behav. Res. Ther.* **3**:101–110. [13]

Parnaudeau, S., S. S. Bolkan, and C. Kellendonk. 2018. The Mediodorsal Thalamus: An Essential Partner of the Prefrontal Cortex for Cognition. *Biol. Psychiatry* **83**:648–656. [10]

Parr, T., and K. J. Friston. 2017. Working Memory, Attention, and Salience in Active Inference. *Sci. Rep.* **7**:14678. [10, 13]

———. 2019. Attention or Salience? *Curr. Opin. Psychol.* **29**:1–5. [10, 13]

Parsons, M. P., S. Li, and G. J. Kirouac. 2007. Functional and Anatomical Connection between the Paraventricular Nucleus of the Thalamus and Dopamine Fibers of the Nucleus Accumbens. *J. Comp. Neurol.* **500**:1050–1063. [5]

Pascual-Leone, A., J. Valls-Sole, E. M. Wassermann, and M. Hallett. 1994. Responses to Rapid-Rate Transcranial Magnetic Stimulation of the Human Motor Cortex. *Brain* **117 (Pt 4)**:847–858. [16]

Pascual-Vera, B., B. Akin, A. Belloch, et al. 2019. The Cross-Cultural and Transdiagnostic Nature of Unwanted Mental Intrusions. *Int. J. Clin. Health Psychol.* **19**:85–96. [17]

Patriarchi, T., J. R. Cho, K. Merten, et al. 2018. Ultrafast Neuronal Imaging of Dopamine Dynamics with Designed Genetically Encoded Sensors. *Science* **360**:eaat4422. [4]

Paulus, M. P., and M. B. Stein. 2006. An Insular View of Anxiety. *Biol. Psychiatry* **60**:383–387. [13]

———. 2010. Interoception in Anxiety and Depression. *Brain Struct. Funct.* **214**:451–463. [13]

Paulus, M. P., and J. L. Stewart. 2014. Interoception and Drug Addiction. *Neuropharmacology* **76**:342–350. [10]

Pavlov, I. P. 1927. Conditioned Reflexes: An Investigation of the Physiological Activity of the Cerebral Cortex. London: Oxford Univ. Press. [2, 9]

Paz-Alonso, P. M., S. A. Bunge, M. C. Anderson, and S. Ghetti. 2013. Strength of Coupling within a Mnemonic Control Network Differentiates Those Who Can and Cannot Suppress Memory Retrieval. *J. Neurosci.* **33**:5017–5026. [9]

Pearl, J., and D. Mackenzie. 2018. The Book of Why : The New Science of Cause and Effect. London: Allen Lane. [12]

Pearson, J., T. Naselaris, E. A. Holmes, and S. M. Kosslyn. 2015. Mental Imagery: Functional Mechanisms and Clinical Applications. *Trends Cogn. Sci.* **19**:590–602. [9, 14]

Pepper, J., M. Hariz, and L. Zrinzo. 2015. Deep Brain Stimulation versus Anterior Capsulotomy for Obsessive-Compulsive Disorder: A Review of the Literature. *J. Neurosurg.* **122**:1028–1037. [3]

Pereboom, D. 2001. Living without Free Will. Cambridge Studies in Philosophy. Cambridge: Cambridge Univ. Press. [12]

Pergola, G., L. Danet, A.-L. Pitel, et al. 2018. The Regulatory Role of the Human Mediodorsal Thalamus. *Trends Cogn. Sci.* **22**:1011–1025. [10]

Perry, E. K., and R. H. Perry. 1995. Acetylcholine and Hallucinations: Disease-Related Compared to Drug-Induced Alterations in Human Consciousness. *Brain. Cogn.* **28**:240–258. [5]

Persaud, N., M. Davidson, B. Maniscalco, et al. 2011. Awareness-Related Activity in Prefrontal and Parietal Cortices in Blindsight Reflects More Than Superior Visual Performance. *NeuroImage* **58**:605–611. [12]

Pes, R., S. C. Godar, A. T. Fox, et al. 2017. Pramipexole Enhances Disadvantageous Decision-Making: Lack of Relation to Changes in Phasic Dopamine Release. *Neuropharmacology* **114**:77–87. [5]

Peters, A., B. S. McEwen, and K. Friston. 2017. Uncertainty and Stress: Why It Causes Diseases and How It Is Mastered by the Brain. *Prog. Neurobiol.* **156**:164–188. [13]

Peters, S. K., K. Dunlop, and J. Downar. 2016. Cortico-Striatal-Thalamic Loop Circuits of the Salience Network: A Central Pathway in Psychiatric Disease and Treatment. *Front. Syst. Neurosci.* **10**:104. [10]

Pettorruso, M., P. A. Spagnolo, L. Leggio, et al. 2018. Repetitive Transcranial Magnetic Stimulation of the Left Dorsolateral Prefrontal Cortex May Improve Symptoms of Anhedonia in Individuals with Cocaine Use Disorder: A Pilot Study. *Brain Stimul.* **11**:1195–1197. [5]

Pezzulo, G., P. Iodice, L. Barca, et al. 2018. Increased Heart Rate after Exercise Facilitates the Processing of Fearful but Not Disgusted Faces. *Sci. Rep.* **8**:398. [13]

Phillips, P. E. M., G. D. Stuber, M. L. A. V. Heien, R. M. Wightman, and R. M. Carelli. 2003. Subsecond Dopamine Release Promotes Cocaine Seeking. *Nature* **422**:614–618. [5]

Picciotto, M. R., M. J. Higley, and Y. S. Mineur. 2012. Acetylcholine as a Neuromodulator: Cholinergic Signaling Shapes Nervous System Function and Behavior. *Neuron* **76**:116–129. [5]

Pievsky, M. A., and R. E. McGrath. 2018. The Neurocognitive Profile of Attention-Deficit/Hyperactivity Disorder: A Review of Meta-Analyses. *Arch. Clin. Neuropsychol.* **33**:143–157. [13]

Pischedda, D., K. Gorgen, J. D. Haynes, and C. Reverberi. 2017. Neural Representations of Hierarchical Rule Sets: The Human Control System Represents Rules Irrespective of the Hierarchical Level to Which They Belong. *J. Neurosci.* **37**:12281–12296. [11]

Pitman, R. K. 1987. Pierre Janet on Obsessive-Compulsive Disorder (1903). *Arch. Gen. Psychiatry* **44**:226. [13]

Pitman, R. K., A. M. Rasmusson, K. C. Koenen, et al. 2012. Biological Studies of Posttraumatic Stress Disorder. *Nat. Rev. Neurosci.* **13**:769–787. [9]

Pittig, A., M. Treanor, R. T. LeBeau, and M. G. Craske. 2018. The Role of Associative Fear and Avoidance Learning in Anxiety Disorders: Gaps and Directions for Future Research. *Neurosci. Biobehav. Rev.* **88**:117–140. [14]

Pitts, E. G., E. T. Barfield, E. P. Woon, and S. L. Gourley. 2020. Action-Outcome Expectancies Require Orbitofrontal Neurotrophin Systems in Naive and Cocaine-Exposed Mice. *Neurotherapeutics* **17**:165–177. [5]

Pitts, E. G., D. C. Li, and S. L. Gourley. 2018. Bidirectional Coordination of Actions and Habits by TrkB in Mice. *Sci. Rep.* **8**:4495. [5]

Pizzagalli, D. A. 2011. Frontocingulate Dysfunction in Depression: Toward Biomarkers of Treatment Response. *Neuropsychopharmacology* **36**:183–206. [5]

Pockett, S., W. P. Banks, and S. Gallagher. 2006. Does Consciousness Cause Behavior? Cambridge, MA: MIT Press. [12]

Poldrack, R. A. 2011. Inferring Mental States from Neuroimaging Data: From Reverse Inference to Large Scale Decoding. *Neuron* **72**:692–697. [6]

Pollan, M. 2018. How to Change Your Mind: What the New Science of Psychedelics Teaches Us About Consciousness, Dying, Addiction, Depression, and Transcendence. New York: Penguin Books. [17]

Popa, I., C. Donos, A. Barborica, et al. 2016. Intrusive Thoughts Elicited by Direct Electrical Stimulation during Stereo-Electroencephalography. *Front. Neurol.* **7(114)**:1–6. [6, 15]

Porcheret, K., E. A. Holmes, G. M. Goodwin, R. G. Foster, and K. Wulff. 2015. Psychological Effect of an Analogue Traumatic Event Reduced by Sleep Deprivation. *Sleep* **38**:1017–1025. [14]

Porcheret, K., L. Iyadurai, M. B. Bonsall, et al. 2020. Sleep and Intrusive Memories Immediately after a Traumatic Event in Emergency Department Patients. *Sleep* doi: 10.1093/sleep/zsaa1033. [14]

Porcheret, K., D. van Heugten, G. M. Goodwin, et al. 2019. Investigation of the Impact of Total Sleep Deprivation at Home on the Number of Intrusive Memories to an Analogue Trauma. *Transl. Psychiatry* **9**:104. [14]

Potenza, M. N. 2015. Perspective: Behavioural Addictions Matter. *Nature* **522**:S62–S62. [10]

Power, J. D., A. L. Cohen, S. M. Nelson, et al. 2011. Functional Network Organization of the Human Brain. *Neuron* **72**:665–678. [10, 11, 13]

Powers, A. R., C. Mathys, and P. R. Corlett. 2017. Pavlovian Conditioning–Induced Hallucinations Result from Overweighting of Perceptual Priors. *Science* **357**:596–600. [13, 17]

Purdon, C., and D. A. Clark. 1993. Obsessive Intrusive Thoughts in Nonclinical Subjects: Part I: Content and Relation with Depressive, Anxious and Obsessional Symptoms. *Behav. Res. Ther.* **31**:713–720. [6, 9, 17]

———. 1994. Obsessive Intrusive Thoughts in Nonclinical Subjects: Part II: Cognitive Appraisal, Emotional Response and Thought Control Strategies. *Behav. Res. Ther.* **32**:403–410. [6, 17]

Putnam, H. 1967. Psychological Predicates. In: Art, Mind, and Religion, ed. W. H. Capitan and D. D. Merrill, pp. 37–48. Pittsburgh: Univ. of Pittsburgh Press. [11]

Quadt, L., H. D. Critchley, and S. N. Garfinkel. 2018. The Neurobiology of Interoception in Health and Disease. *Ann. N.Y. Acad. Sci* **1428**:112–128. [13]

Rachman, S. 1981. Part 1: Unwanted Intrusive Cognitions. *Adv. Behav. Res. Ther.* **3**:89–99. [7, 15]

———. 2014. Global Intrusive Thoughts: A Commentary. *J. Obsessive Compuls. Relat. Disord.* **3**:300–302. [17]

Rachman, S., and P. de Silva. 1978. Abnormal and Normal Obsessions. *Behav. Res. Ther.* **16**:233–248. [17]

Rachman, S. J., and R. J. Hodgson. 1980. Obsessions and Compulsions. Hillsdale, NJ: Prentice-Hall. [13]

Radomsky, A. S., G. M. Alcolado, J. S. Abramowitz, et al. 2014. Part 1—You Can Run but You Can't Hide: Intrusive Thoughts on Six Continents. *J. Obsessive Compuls. Relat. Disord.* **3**:269–279. [17]

Rae, C. L., H. D. Critchley, and A. K. Seth. 2019a. A Bayesian Account of the Sensory-Motor Interactions Underlying Symptoms of Tourette Syndrome. *Front. Psychiatry* **10**:29. [13]

Rae, C. L., D. E. O. Larsson, J. A. Eccles, J. Ward, and H. D. Critchley. 2018. Subjective Embodiment during the Rubber Hand Illusion Predicts Severity of Premonitory Sensations and Tics in Tourette Syndrome. *Conscious. Cogn.* **65**:368–377. [13]

Rae, C. L., D. E. O. Larsson, S. N. Garfinkel, and H. D. Critchley. 2019b. Dimensions of Interoception Predict Premonitory Urges and Tic Severity in Tourette Syndrome. *Psychiatry Res.* **271**:469–475. [13]

Rahnev, D. A., E. Huang, and H. Lau. 2012. Subliminal Stimuli in the near Absence of Attention Influence Top-Down Cognitive Control. *Atten. Percept. Psychophys.* **74**:521–532. [12]

Raichle, M. E., A. M. MacLeod, A. Z. Snyder, et al. 2001. A Default Mode of Brain Function. *PNAS* **98**:676–682. [10, 13]

Rainer, G., W. F. Asaad, and E. K. Miller. 1998. Memory Fields of Neurons in the Primate Prefrontal Cortex. *PNAS* **95**:15008–15013. [11]

Ramsay, D. S., and S. C. Woods. 2014. Clarifying the Roles of Homeostasis and Allostasis in Physiological Regulation. *Psychol. Rev.* **121**:225–247. [13]

Ranti, C., C. H. Chatham, and D. Badre. 2015. Parallel Temporal Dynamics in Hierarchical Cognitive Control. *Cognition* **142**:205–229. [11]

Rao, R. P. N., and D. H. Ballard. 1999. Predictive Coding in the Visual Cortex: A Functional Interpretation of Some Extra-Classical Receptive-Field Effects. *Nat. Neurosci.* **2**:79–87. [10, 13]

Rao-Ruiz, P., D. C. Rotaru, R. J. van der Loo, et al. 2011. Retrieval-Specific Endocytosis of GluA2-AMPARs Underlies Adaptive Reconsolidation of Contextual Fear. *Nat. Neurosci.* **14**:1302. [5]

Rapinesi, C., G. Kotzalidis, S. Ferracuti, et al. 2019. Brain Stimulation in Obsessive-Compulsive Disorder (OCD): A Systematic Review. *Current neuropharmacology* **17**:787–807. [17]

Rassin, E. 2003. The White Bear Suppression Inventory (WBSI) Focuses on Failing Suppression Attempts. *Eur. J. Personality* **17**:285–298. [9]

RCIF. 2007. The International Intrusive Thoughts Interview Schedule [Version 6]. Barcelona: Research Consortium on Intrusive Fear. [6]

Rebetez, M. M., L. Rochat, C. Barsics, and M. V. Linden. 2018. Procrastination as a Self-Regulation Failure: The Role of Impulsivity and Intrusive Thoughts. *Psychologic. Rep.* **121**:26–41. [15]

Reddan, M. C., T. D. Wager, and D. Schiller. 2018. Attenuating Neural Threat Expression with Imagination. *Neuron* **100**:994–1005. [12]

Redish, A. D., and J. A. Gordon, eds. 2016. Computational Psychiatry: New Perspectives on Mental Illness, Strüngmann Forum Reports, J. R. Lupp, series ed. Cambridge, MA: MIT Press. [9]

Reichle, E. D., A. E. Reineberg, and J. W. Schooler. 2010. Eye Movements During Mindless Reading. *Psychol. Sci.* **21**:1300–1310. [6]

Remijnse, P. L., M. M. Nielen, A. J. van Balkom, et al. 2006. Reduced Orbitofrontal-Striatal Activity on a Reversal Learning Task in Obsessive-Compulsive Disorder. *Arch. Gen. Psychiatry* **63**:1225–1236. [3]

Rescorla, R. A. 1994. Transfer of Instrumental Control Mediated by a Devalued Outcome. *Anim. Learn. Behav.* **22**:27–33. [3]

Rescorla, R. A., and P. C. Holland. 1982. Behavioral Studies of Associative Learning in Animals. *Ann. Rev. Psychol.* **33**:265–308. [9]

Reynolds, M., and A. Wells. 1999. The Thought Control Questionnaire – Psychometric Properties in a Clinical Sample, and Relationships with PTSD and Depression. *Psychol. Med.* **29**:1089–1099. [6]

Ribeiro, C. M., G. Sanacora, R. Hoffman, and R. Ostroff. 2016. The Use of Ketamine for the Treatment of Depression in the Context of Psychotic Symptoms. *Biol. Psychiatry* **79**:e65–e66. [15]

Ribot, T. A. 1882. Diseases of Memory: an Essay in the Positive Psychology. New York: D. Appleton and Co. [9]

Rigotti, M., O. Barak, M. R. Warden, et al. 2013. The Importance of Mixed Selectivity in Complex Cognitive Tasks. *Nature* **497**:585–590. [11]

Rizio, A. A., and N. A. Dennis. 2013. The Neural Correlates of Cognitive Control: Successful Remembering and Intentional Forgetting. *J. Cogn. Neurosci.* **25**:297–312. [9]

Robbins, T. W. 2005. Chemistry of the Mind: Neurochemical Modulation of Prefrontal Cortical Function. *J. Comp. Neurol.* **493**:140–146. [13]

Robbins, T. W., C. M. Gillan, D. G. Smith, S. de Wit, and K. D. Ersche. 2012. Neurocognitive Endophenotypes of Impulsivity and Compulsivity: Towards Dimensional Psychiatry. *Trends Cogn. Sci.* **16**:81–91. [8, 9]

Robbins, T. W., M. M. Vaghi, and P. Banca. 2019. Obsessive-Compulsive Disorder: Puzzles and Prospects. *Neuron* **102**:27–47. [3, 8, 13]

Roberts, A. C., M. A. De Salvia, L. S. Wilkinson, et al. 1994. 6-Hydroxydopamine Lesions of the Prefrontal Cortex in Monkeys Enhance Performance on an Analog of the Wisconsin Card Sort Test: Possible Interactions with Subcortical Dopamine. *J. Neurosci.* **14**:2531–2544. [13]

Roberts-Wolfe, D., A.-C. Bobadilla, J. A. Heinsbroek, D. Neuhofer, and P. W. Kalivas. 2018. Drug Refraining and Seeking Potentiate Synapses on Distinct Populations of Accumbens Medium Spiny Neurons. *J. Neurosci.* **38**:7100–7107. [10]

Robinson, T. E., and K. C. Berridge. 1993. The Neural Basis of Drug Craving: an Incentive-Sensitization Theory of Addiction. *Brain Res. Brain Res. Rev.* **18**:247–291. [8]

Rodriguez-Romaguera, J., B. D. Greenberg, S. A. Rasmussen, and G. J. Quirk. 2016. An Avoidance-Based Rodent Model of Exposure with Response Prevention Therapy for Obsessive-Compulsive Disorder. *Biol. Psychiatry* **80**:534–540. [5]

Rogers, C. R. 1959. A Theory of Therapy, Personality, and Interpersonal Relationships: As Developed in the Client-Centered Framework. In: Psychology: A Study of a Science. Study 1, Volume 3: Formulations of the Person and the Social Context, ed. S. Koch, pp. 184–256. New York: McGraw-Hill. [13]

Roh, D., J.-G. Chang, S. W. Yoo, J. Shin, and C.-H. Kim. 2017. Modulation of Error Monitoring in Obsessive-Compulsive Disorder by Individually Tailored Symptom Provocation. *Psychol. Med.* **47**:2071–2080. [6]

Roiser, J. P., O. D. Howes, C. A. Chaddock, E. M. Joyce, and P. McGuire. 2012. Neural and Behavioral Correlates of Aberrant Salience in Individuals at Risk for Psychosis. *Schizophr. Bull.* **39**:1328–1336. [10]

Roitman, M. F., R. A. Wheeler, R. M. Wightman, and R. M. Carelli. 2008. Real-Time Chemical Responses in the Nucleus Accumbens Differentiate Rewarding and Aversive Stimuli. *Nat. Neurosci.* **11**:1376–1377. [2]

Rolls, E. T. 2012. Glutamate, Obsessive–Compulsive Disorder, Schizophrenia, and the Stability of Cortical Attractor Neuronal Networks. *Pharmacol. Biochem. Behav.* **100**:736–751. [13]

Rolls, E. T., M. Loh, and G. Deco. 2008. An Attractor Hypothesis of Obsessive-Compulsive Disorder. *Eur. J. Neurosci.* **28**:782–793. [13]

Romero-Sanchiz, P., R. Nogueira-Arjona, A. Godoy-Ávila, A. Gavino-Lázaro, and M. H. Freeston. 2017. Assessing Transdiagnostic Intrusive Thoughts: Factor Structure, Reliability and Validity of the Cognitive Intrusions Questionnaire-Transdiagnostic Version in a Spanish Sample. *Pers. Individ. Dif.* **114**:181–186. [17]

Root, D. H., R. I. Melendez, L. Zaborszky, and T. C. Napier. 2015. The Ventral Pallidum: Subregion-Specific Functional Anatomy and Roles in Motivated Behaviors. *Prog. Neurobiol.* **130**:29–70. [3]

Rosenbaum, D., M. Thomas, P. Hilsendegen, et al. 2018. Stress-Related Dysfunction of the Right Inferior Frontal Cortex in High Ruminators: An fNIRS Study. *NeuroImage Clin.* **18**:510–517. [6]

Rosenthal, D. M. 2004. Varieties of Higher-Order Theory. In: Higher-Order Theories of Consciousness: An Anthology, ed. R. J. Gennaro. Amsterdam: John Benjamins. [12]

———. 2005. Consciousness and Mind. Oxford: Oxford Univ. Press. [12]

———. 2012. Higher-Order Awareness, Misrepresentation and Function. *Philos. Trans. R. Soc. Lond. B. Biol. Sci.* **367**:1424–1438. [12]

Ross, S., A. Bossis, J. Guss, et al. 2016. Rapid and Sustained Symptom Reduction Following Psilocybin Treatment for Anxiety and Depression in Patients with Life-Threatening Cancer: A Randomized Controlled Trial. *J. Psychopharmacol.* **30**:1165–1180. [15]

Roth, B. L. 2016. DREADDs for Neuroscientists. *Neuron* **89**:683–694. [4]

Rounis, E., B. Maniscalco, J. C. Rothwell, R. E. Passingham, and H. Lau. 2010. Theta-Burst Transcranial Magnetic Stimulation to the Prefrontal Cortex Impairs Metacognitive Visual Awareness. *Cogn. Neurosci.* **1**:165–175. [12]

Rubin, D. C., D. Berntsen, C. M. Ogle, S. A. Deffler, and J. C. Beckham. 2016a. Scientific Evidence versus Outdated Beliefs: A Response to Brewin (2016). *J. Abnorm. Psychol.* **125**:1018–1021. [9]

Rubin, D. C., S. A. Deffle, C. M. Ogle, et al. 2016b. Participant, Rater, and Computer Measures of Coherence in Posttraumatic Stress Disorder. *J. Abnorm. Psychol.* **125**:11–25. [9]

Rugg, M. D., and T. Curran. 2007. Event-Related Potentials and Recognition Memory. *Trends Cogn. Sci.* **11**:251–257. [6]

Russek, E. M., I. Momennejad, M. M. Botvinick, S. J. Gershman, and N. D. Daw. 2017. Predictive Representations Can Link Model-Based Reinforcement Learning to Model-Free Mechanisms. *PLoS Comput. Biol.* **13**:e1005768–e1005768. [9]

Russell, S., and E. Wefald. 1991. Do the Right Thing: Studies in Limited Rationality. Cambridge, MA: MIT Press. [9]

Ryan, R. M., and E. L. Deci. 1985. Intrinsic Motivation and Self-Determination in Human Behavior. New York: Plenum. [13]

Saalmann, Y. B. 2014. Intralaminar and Medial Thalamic Influence on Cortical Synchrony, Information Transmission and Cognition. *Front. Syst. Neurosci.* **8**:83. [10]

Salkovskis, P. M. 1988. Intrusive Thoughts and Obsessional Disorders. In: Current Issues in Clinical Psychology, ed. D. Glasgow and N. Eisenberg, vol. 4. London: Gower. [7]

Salkovskis, P. M., and J. Harrison. 1984. Abnormal and Normal Obsessions: A Replication. *Behav. Res. Ther.* **22**:549–552. [13]

Salters-Pedneault, K., V. Vine, M. A. Mills, C. Park, and B. T. Litz. 2009. The Experience of Intrusions Scale: A Preliminary Examination. *Anxiety Stress Coping* **22**: 27–37. [6, 17]

Salvato, G., F. Richter, L. Sedeno, G. Bottini, and E. Paulesu. 2020. Building the Bodily Self-Awareness: Evidence for the Convergence between Interoceptive and Exteroceptive Information in a Multilevel Kernel Density Analysis Study. *Hum. Brain Mapp.* **41**:401–418. [13]

Sara, S. J. 2000. Retrieval and Reconsolidation: Toward a Neurobiology of Remembering. *Learn. Mem.* **7**:73–84. [9, 14]

Saricicek, A., I. Esterlis, K. H. Maloney, et al. 2012. Persistent β2*-Nicotinic Acetylcholinergic Receptor Dysfunction in Major Depressive Disorder. *Am. J. Psychiatry* **169**:851–859. [5]

Sartor, G. C., and G. Aston-Jones. 2014. Post-Retrieval Extinction Attenuates Cocaine Memories. *Neuropsychopharmacology* **39**:1059–1065. [9]

Sartorius, N., A. Jablensky, and R. Shapiro. 1977. Two-Year Follow-up of the Patients Included in the WHO International Pilot Study of Schizophrenia. *Psychol. Med.* **7**:529–541. [8]

Sartory, G., J. Cwik, H. Knuppertz, et al. 2013. In Search of the Trauma Memory: A Meta-Analysis of Functional Neuroimaging Studies of Symptom Provocation in Posttraumatic Stress Disorder (PTSD). *PLoS One* **8**:e58150. [6]

Saunders, A., E. Z. Macosko, A. Wysoker, et al. 2018. Molecular Diversity and Specializations among the Cells of the Adult Mouse Brain. *Cell* **174**:1015–1030. e1016. [5]

Saunders, B. T., and T. E. Robinson. 2010. A Cocaine Cue Acts as an Incentive Stimulus in Some but Not Others: Implications for Addiction. *Biol. Psychiatry* **67**:730–736. [5]

Saxena, S., A. L. Brody, J. M. Schwartz, and L. R. Baxter. 1998. Neuroimaging and Frontal-Subcortical Circuitry in Obsessive-Compulsive Disorder. *Br. J. Psychiatry (Suppl.)* **35**:26–37. [13]

Schaul, T., J. Quan, I. Antonoglou, and D. Silver. 2016. Prioritized Experience Replay. *ArXiv* 1511.05952v05954–01511.05952v05954. [9]

Schiller, D., and M. R. Delgado. 2010. Overlapping Neural Systems Mediating Extinction, Reversal and Regulation of Fear. *Trends Cogn. Sci.* **14**:268–276. [9]

Schiller, D., J. W. Kanen, J. E. Ledoux, M.-H. Monfils, and E. A. Phelps. 2013. Extinction during Reconsolidation of Threat Memory Diminishes Prefrontal Cortex Involvement. *PNAS* **110**:20040–20045. [14]

Schiller, D., M.-H. Monfils, C. M. Raio, et al. 2010. Preventing the Return of Fear in Humans Using Reconsolidation Update Mechanisms. *Nature* **463**:49–53. [2, 7, 9, 14]

Schmidhuber, J. 2010. Formal Theory of Creativity, Fun, and Intrinsic Motivation (1990-2010). *IEEE Trans. Autonom. Mental Dev.* **2**:230–247. [13]

Schmidt, A., M. Antoniades, P. Allen, et al. 2016. Longitudinal Alterations in Motivational Salience Processing in Ultra-High-Risk Subjects for Psychosis. *Psychol. Med.* **47**:243–254. [10]

Schmidt, R., D. K. Leventhal, N. Mallet, F. Chen, and J. D. Berke. 2013. Canceling Actions Involves a Race between Basal Ganglia Pathways. *Nat. Neurosci.* **16**:1118–1124. [11]

Schmidt, R. E., P. Gay, D. Courvoisier, et al. 2009. Anatomy of the White Bear Suppression Inventory (WBSI): A Review of Previous Findings and a New Approach. *J. Person. Assess.* **91**:323–330. [6]

Schmitz, T. W., M. M. Correia, C. S. Ferreira, A. P. Prescot, and M. C. Anderson. 2017. Hippocampal GABA Enables Inhibitory Control over Unwanted Thoughts. *Nat. Commun.* **8**:1311. [6, 9]

Schneider, K. 1959. Clinical Psychopathology. New York: Grune and Stratton. [8]

Schoofs, N., and A. Heinz. 2013. Pathological Gambling: Impulse Control Disorder, Addiction or Compulsion? *Nervenarzt* **84**:629–634. [8]

Schooler, J. W. 2002. Re-Representing Consciousness: Dissociations between Experience and Meta-Consciousness. *Trends Cogn. Sci.* **6**:339–344. [9]

Schooler, J. W., J. Smallwood, K. Christoff, et al. 2011. Meta-Awareness, Perceptual Decoupling and the Wandering Mind. *Trends Cogn. Sci.* **15**:319–326. [9]

Schou Andreassen, C., J. Billieux, M. D. Griffiths, et al. 2016. The Relationship between Addictive Use of Social Media and Video Games and Symptoms of Psychiatric Disorders: A Large-Scale Cross-Sectional Study. *Psychol. Addict. Behav.* **30**:252–262. [17]

Schultz, W., P. Dayan, and P. R. Montague. 1997. A Neural Substrate of Prediction and Reward. *Science* **275**:1593–1599. [2]

Schulz, A., J. H. Matthey, C. Vögele, et al. 2016. Cardiac Modulation of Startle Is Altered in Depersonalization-/Derealization Disorder: Evidence for Impaired Brainstem Representation of Baro-Afferent Neural Traffic. *Psychiatry Res.* **240**:4–10. [13]

Schurger, A., J. D. Sitt, and S. Dehaene. 2012. An Accumulator Model for Spontaneous Neural Activity Prior to Self-Initiated Movement. *PNAS* **109**:E2904–2913. [12]

Scofield, M. D., J. A. Heinsbroek, C. D. Gipson, et al. 2016. The Nucleus Accumbens: Mechanisms of Addiction across Drug Classes Reflect the Importance of Glutamate Homeostasis. *Pharmacol. Rev.* **68**:816–871. [5]

Sebold, M., L. Deserno, S. Nebe, et al. 2014. Model-Based and Model-Free Decisions in Alcohol Dependence. *Neuropsychobiology* **70**:122–131. [8]

Sebold, M., S. Nebe, M. Garbusow, et al. 2017. When Habits Are Dangerous: Alcohol Expectancies and Habitual Decision Making Predict Relapse in Alcohol Dependence. *Biol. Psychiatry* **82**:847–856. [8]

Sedikides, C., and J. D. Green. 2009. Memory as a Self-Protective Mechanism. *Soc. Personal. Psychol. Compass* **3**:1055–1068. [9]

Sedikides, C., J. D. Green, J. Saunders, J. J. Skowronski, and B. Zengel. 2016. Mnemic Neglect: Selective Amnesia of One's Faults. *Eur. Rev. Social Psychol.* **27**:1–62. [9]

Seeley, W. W., V. Menon, A. F. Schatzberg, et al. 2007. Dissociable Intrinsic Connectivity Networks for Salience Processing and Executive Control. *J. Neurosci.* **27**:2349–2356. [10, 13]

Seery, M. D., R. J. Leo, S. P. Lupien, C. L. Kondrak, and J. L. Almonte. 2013. An Upside to Adversity?: Moderate Cumulative Lifetime Adversity Is Associated with Resilient Responses in the Face of Controlled Stressors. *Psychol. Sci.* **24**:1181–1189. [9]

Segerstrom, S. C., A. L. Stanton, L. E. Alden, and B. E. Shortridge. 2003. A Multidimensional Structure for Repetitive Thought: What's on Your Mind, and How, and How Much? *J. Pers. Soc. Psychol.* **85**:909–921. [10]

Seli, P., J. S. A. Carriere, and D. Smilek. 2015. Not All Mind Wandering Is Created Equal: Dissociating Deliberate from Spontaneous Mind Wandering. *Psychol. Res.* **79**:750–758. [9]

Sell, A. N. 2011. The Recalibrational Theory and Violent Anger. *Aggr. Violent Behav.* **16**:381–389. [7]

Sergent, C., and S. Dehaene. 2004. Neural Processes Underlying Conscious Perception: Experimental Findings and a Global Neuronal Workspace Framework. *J. Physiol.* **98**:374–384. [13]

Seth, A. K. 2013. Interoceptive Inference, Emotion, and the Embodied Self. *Trends Cogn. Sci.* **17**:565–573. [10, 13]

Seth, A. K., K. Suzuki, and H. D. Critchley. 2012. An Interoceptive Predictive Coding Model of Conscious Presence. *Front. Psychol.* **2**:395. [13]

Sevenster, D., T. Beckers, and M. Kindt. 2014. Prediction Error Demarcates the Transition from Retrieval, to Reconsolidation, to New Learning. *Learn. Mem.* **21**:580–584. [14]

Sha, Z., T. D. Wager, A. Mechelli, and Y. He. 2019. Common Dysfunction of Large-Scale Neurocognitive Networks across Psychiatric Disorders. *Biol. Psychiatry* **85**:379–388. [10]

Shalev, A. Y. 1992. Posttraumatic Stress Disorder among Injured Survivors of a Terrorist Attack: Predictive Value of Early Intrusion and Avoidance Symptoms. *J. Nerv. Ment. Dis.* **180**:505–509. [7]

Shanks, D. R., and A. Dickinson. 1991. Instrumental Judgment and Performance under Variations in Action-Outcome Contingency and Contiguity. *Mem. Cogn.* **19**:353–360. [3]

Sharpe, M. J., C. Y. Chang, M. A. Liu, et al. 2017. Dopamine Transients Are Sufficient and Necessary for Acquisition of Model-Based Associations. *Nat. Neurosci.* **20**:735–742. [2]

Shehzad, Z., C. Kelly, P. T. Reiss, et al. 2014. A Multivariate Distance-Based Analytic Framework for Connectome-Wide Association Studies. *NeuroImage* **93**:74–94. [10]

Shenhav, A., M. M. Botvinick, and J. D. Cohen. 2013. The Expected Value of Control: an Integrative Theory of Anterior Cingulate Cortex Function. *Neuron* **79**:217–240. [11]

Shenhav, A., J. D. Cohen, and M. M. Botvinick. 2016. Dorsal Anterior Cingulate Cortex and the Value of Control. *Nat. Neurosci.* **19**:1286–1291. [10]

Shepherd, J. 2017. The Folk Psychological Roots of Free Will. In: Experimental Metaphysics, ed. D. Rose. London: Bloomsbury Academic. [12]

Shiba, Y., L. Oikonomidis, S. J. Sawiak, et al. 2017. Converging Prefronto-Insula-Amygdala Pathways in Negative Emotion Regulation in Marmoset Monkeys. *Biol. Psychiatry* **82**:895–903. [13]

Shibata, K., G. Lisi, A. Cortese, et al. 2018. Toward a Comprehensive Understanding of the Neural Mechanisms of Decoded Neurofeedback. *NeuroImage* **188**:539–556. [17]

Shields, G. S., M. A. Sazma, and A. P. Yonelinas. 2016. The Effects of Acute Stress on Core Executive Functions: A Meta-Analysis and Comparison with Cortisol. *Neurosci. Biobehav. Rev.* **68**:651–668. [9]

Shin, G., A. M. Gomez, R. Al-Hasani, et al. 2017. Flexible near-Field Wireless Optoelectronics as Subdermal Implants for Broad Applications in Optogenetics. *Neuron* **93**:509–521.e503. [4]

Shin, N. Y., T. Y. Lee, E. Kim, and J. S. Kwon. 2014. Cognitive Functioning in Obsessive-Compulsive Disorder: A Meta-Analysis. *Psychol. Med.* **44**:1121–1130. [13]

Shipp, S. 2016. Neural Elements for Predictive Coding. *Front. Psychol.* **7**:1792. [13]

Silvanto, J., Z. Cattaneo, L. Battelli, and A. Pascual-Leone. 2008a. Baseline Cortical Excitability Determines Whether TMS Disrupts or Facilitates Behavior. *J. Neurophysiol.* **99**:2725–2730. [16]

Silvanto, J., N. Muggleton, A. Cowey, and V. Walsh. 2007. Neural Adaptation Reveals State-Dependent Effects of Transcranial Magnetic Stimulation. *Eur. J. Neurosci.* **25**:1874–1881. [16]

Silvanto, J., N. Muggleton, and V. Walsh. 2008b. State-Dependency in Brain Stimulation Studies of Perception and Cognition. *Trends Cogn. Sci.* **12**:447–454. [16]

Simons, M., C. Bernaards, and J. Slinger. 2012. Active Gaming in Dutch Adolescents: A Descriptive Study. *Int. J. Behav. Nutr. Phys. Act.* **9**:118. [17]

Simpson, E. H., C. Kellendonk, and E. Kandel. 2010. A Possible Role for the Striatum in the Pathogenesis of the Cognitive Symptoms of Schizophrenia. *Neuron* **65**:585–596. [3]

Sinnott-Armstrong, W., ed. 2008. The Neuroscience of Morality: Emotion, Brain Disorders, and Development, Moral Psychology, vol. 3. Cambridge, MA: MIT Press. [12]

Sinopoli, V. M., C. L. Burton, S. Kronenberg, and P. D. Arnold. 2017. A Review of the Role of Serotonin System Genes in Obsessive-Compulsive Disorder. *Neurosci. Biobehav. Rev.* **80**:372–381. [13]

Sitaram, R., T. Ros, L. Stoeckel, et al. 2016. Closed-Loop Brain Training: The Science of Neurofeedback. *Nat. Rev. Neurosci.* **18**:86–100. [17]

Siuda, E. R., J. G. McCall, R. Al-Hasani, et al. 2015. Optodynamic Simulation of β-Adrenergic Receptor Signalling. *Nat. Commun.* **6**:8480. [4]

Skårderud, F. 2007. Eating One's Words, Part I: 'Concretised Metaphors' and Reflective Function in Anorexia Nervosa—an Interview Study. *Eur. Eat. Disord. Rev.* **15**:163–174. [13]

Skewes, J. C., E. M. Jegindo, and L. Gebauer. 2014. Perceptual Inference and Autistic Traits. *Autism* **19**:301–307. [13]

Skinner, B. F. 1971. Beyond Freedom and Dignity. Middlesex: Penguin. [12]

Smallwood, J., and J. Schooler. 2006. The Restless Mind. *Psychol. Bull.* **132**:946–958. [9]

———. 2015. The Science of Mind Wandering: Empirically Navigating the Stream of Consciousness. *Ann. Rev. Psychol.* **66**:487–518. [9]

Smith, A. M. 2005. Responsibility for Attitudes: Activity and Passivity in Mental Life. *Ethics* **115**:236–271. [12]

Smith, S. M., P. T. Fox, K. L. Miller, et al. 2009. Correspondence of the Brain's Functional Architecture during Activation and Rest. *PNAS* **106**:13040–13045. [10]

Soeter, M., and M. Kindt. 2015. An Abrupt Transformation of Phobic Behavior after a Post-Retrieval Amnesic Agent. *Biol. Psychiatry* **78**:880–886. [14]

Solinas, M., C. Chauvet, N. Thiriet, R. El Rawas, and M. Jaber. 2008. Reversal of Cocaine Addiction by Environmental Enrichment. *PNAS* **105**:17145–17150. [12]

Sommer, I. E., C. W. Slotema, Z. J. Daskalakis, et al. 2012. The Treatment of Hallucinations in Schizophrenia Spectrum Disorders. *Schizophr. Bull.* **38**:704–714. [15]

Soon, C. S., M. Brass, H. J. Heinze, and J. D. Haynes. 2008. Unconscious Determinants of Free Decisions in the Human Brain. *Nat. Neurosci.* **11**:543–545. [12]

Southwick, S. M., M. Davis, B. Horner, et al. 2002. Relationship of Enhanced Norepinephrine Activity During Memory Consolidation to Enhanced Long-Term Memory in Humans. *Am. J. Psychiatry* **159**:1420–1422. [9]

Spangler, S. M., and M. R. Bruchas. 2017. Optogenetic Approaches for Dissecting Neuromodulation and GPCR Signaling in Neural Circuits. *Curr. Opin. Pharm.* **32**:56–70. [4]

Speckens, A. E. M., A. Ehlers, A. Hackmann, F. A. Ruths, and D. M. Clark. 2007. Intrusive Memories and Rumination in Patients with Post-Traumatic Stress Disorder: A Phenomenological Comparison. *Memory* **15**:249–257. [6]

Spencer, S., C. Garcia-Keller, D. Roberts-Wolfe, et al. 2017. Cocaine Use Reverses Striatal Plasticity Produced During Cocaine Seeking. *Biol. Psychiatry* **81**:616–624. [10]

Spinella, M. 2003. Evolutionary Mismatch, Neural Reward Circuits, and Pathological Gambling. *Int. J. Neurosci.* **113**:503–512. [7]

Sporns, O. 2011. Networks of the Brain. Cambridge, MA: MIT Press. [5]

Sridharan, D., D. J. Levitin, and V. Menon. 2008. A Critical Role for the Right Fronto-Insular Cortex in Switching between Central-Executive and Default-Mode Networks. *PNAS* **105**:12569–12574. [10]

Srinivasan, M. V., S. B. Laughlin, and A. Dubs. 1982. Predictive Coding: A Fresh View of Inhibition in the Retina. *Proc. R. Soc. Lond. B* **216**:427–459. [13]

Stagg, C. J., M. Wylezinska, P. M. Matthews, et al. 2009. Neurochemical Effects of Theta Burst Stimulation as Assessed by Magnetic Resonance Spectroscopy. *J. Neurophysiol.* **101**:2872–2877. [16]

Stalnaker, T. A., B. Berg, N. Aujla, and G. Schoenbaum. 2016. Cholinergic Interneurons Use Orbitofrontal Input to Track Beliefs About Current State. *J. Neurosci.* **36**:6242–6257. [13]

Stalnaker, T. A., N. K. Cooch, M. A. McDannald, et al. 2014. Orbitofrontal Neurons Infer the Value and Identity of Predicted Outcomes. *Nat. Commun.* **5**:3926–3926. [2]

Stanley, M. L., and F. De Brigard. 2019. Moral Memories and the Belief in the Good Self. *Curr. Dir. Psychol. Sci.* **28**:387–391. [9]

Stanley, S. A., J. Sauer, R. S. Kane, J. S. Dordick, and J. M. Friedman. 2015. Remote Regulation of Glucose Homeostasis in Mice Using Genetically Encoded Nanoparticles. *Nat. Med.* **21**:92–98. [4]

Steinfurth, E. C. K., J. W. Kanen, C. M. Raio, et al. 2014. Young and Old Pavlovian Fear Memories Can Be Modified with Extinction Training during Reconsolidation in Humans. *Learn. Mem.* **21**:338–341. [14]

Stephan, K. E., Z. M. Manjaly, C. D. Mathys, et al. 2016. Allostatic Self-Efficacy: A Metacognitive Theory of Dyshomeostasis-Induced Fatigue and Depression. *Front. Hum. Neurosci.* **10**:550. [13]

Stephens, G., and G. Graham. 2000. When Self-Consciousness Breaks: Alien Voices and Inserted Thoughts. Cambridge, MA: MIT Press. [8]

Sterling, P. 2012. Allostasis: A Model of Predictive Regulation. *Physiol. Behav.* **106**:5–15. [13]

Sterzer, P., R. A. Adams, P. Fletcher, et al. 2018. The Predictive Coding Account of Psychosis. *Biol. Psychiatry* **84**:634–643. [10]

Sterzer, P., A. L. Mishara, M. Voss, and A. Heinz. 2016. Thought Insertion as a Self-Disturbance: An Integration of Predictive Coding and Phenomenological Approaches. *Front. Hum. Neurosci.* **10**:502. [8]

Stokes, M. G., M. Kusunoki, N. Sigala, et al. 2013. Dynamic Coding for Cognitive Control in Prefrontal Cortex. *Neuron* **78**:364–375. [11]

Stone, C. B., A. J. Barnier, J. Sutton, and W. Hirst. 2013. Forgetting Our Personal Past: Socially Shared Retrieval-Induced Forgetting of Autobiographical Memories. *J. Exp. Psychol. Gen.* **142**:1084–1099. [9]

Stone, C. B., A. Coman, A. D. Brown, J. Koppel, and W. Hirst. 2012. Toward a Science of Silence: The Consequences of Leaving a Memory Unsaid. *Perspect. Psychol. Sci.* **7**:39–53. [9]

Storm, B. C., and T. A. Jobe. 2012. Retrieval-Induced Forgetting Predicts Failure to Recall Negative Autobiographical Memories. *Psychol. Sci.* **23**:1356–1363. [9]

Storm, B. C., and B. J. Levy. 2012. A Progress Report on the Inhibitory Account of Retrieval-Induced Forgetting. *Mem. Cogn.* **40**:827–843. [9]

Strafella, A. P., T. Paus, J. Barrett, and A. Dagher. 2001. Repetitive Transcranial Magnetic Stimulation of the Human Prefrontal Cortex Induces Dopamine Release in the Caudate Nucleus. *J. Neurosci.* **21**:RC157. [16]

Strange, B. A., and R. J. Dolan. 2004. B-Adrenergic Modulation of Emotional Memory-Evoked Human Amygdala and Hippocampal Responses. *PNAS* **101**:11454–11458. [9]

Strawson, G. 1994. The Impossibility of Moral Responsibility. *Philos. Stud.* **75**:5–24. [12]

Streb, M., A. Mecklinger, M. C. Anderson, L.-H. Johanna, and T. Michael. 2016. Memory Control Ability Modulates Intrusive Memories after Analogue Trauma. *J. Affect. Disord.* **192**:134–142. [6, 9]

Stuss, D. T., and D. F. Benson. 1987. The Frontal Lobes and Control of Cognition and Memory. In: The Frontal Lobes Revisited, ed. E. Perecman, pp. 141–158. New York: The IRBN Press. [11]

Sun, F., J. Zeng, M. Jing, et al. 2018. A Genetically Encoded Fluorescent Sensor Enables Rapid and Specific Detection of Dopamine in Flies, Fish, and Mice. *Cell* **174**:481–496. [4]

Sun, Y., F. Gomez, and J. Schmidhuber. 2011. Planning to Be Surprised: Optimal Bayesian Exploration in Dynamic Environments. In: Intl. Conf. on Artificial General Intelligence 2011. Lecture Notes in Computer Science, vol. 6830, ed. J. Schmidhuber et al. Heidelberg: Springer. [13]

Sutherland, M. T., A. J. Carroll, B. J. Salmeron, T. J. Ross, and E. A. Stein. 2013. Insula's Functional Connectivity with Ventromedial Prefrontal Cortex Mediates the Impact of Trait Alexithymia on State Tobacco Craving. *Psychopharmacology* **228**:143–155. [10]

Sutherland, M. T., M. J. McHugh, V. Pariyadath, and E. A. Stein. 2012. Resting State Functional Connectivity in Addiction: Lessons Learned and a Road Ahead. *NeuroImage* **62**:2281–2295. [10]

Suto, N., A. Laque, G. L. De Ness, et al. 2016. Distinct Memory Engrams in the Infralimbic Cortex of Rats Control Opposing Environmental Actions on a Learned Behavior. *eLife* **5**:e21920. [5]

Sutton, R. S., and A. G. Barto. 1998. Reinforcement Learning: An Introduction. Adaptive Computation and Machine Learning. Cambridge, MA: MIT Press. [2]

Suzuki, A., S. A. Josselyn, P. W. Frankland, et al. 2004. Memory Reconsolidation and Extinction Have Distinct Temporal and Biochemical Signatures. *J. Neurosci.* **24**:4787–4795. [9]

Swanson, A. M., A. G. Allen, L. P. Shapiro, and S. L. Gourley. 2015. Gaba$_A$α1-Mediated Plasticity in the Orbitofrontal Cortex Regulates Context-Dependent Action Selection. *Neuropsychopharmacology* **40**:1027–1036. [5]

Swedo, S. E., P. Pietrini, H. L. Leonard, et al. 1992. Cerebral Glucose Metabolism in Childhood-Onset Obsessive-Compulsive Disorder: Revisualization during Pharmacotherapy. *Arch. Gen. Psychiatry* **49**:690–694. [8]

Szechtman, H., and E. Woody. 2004. Obsessive-Compulsive Disorder as a Disturbance of Security Motivation. *Psychol. Rev.* **111**:111–127. [13]

Takarangi, M. K. T., D. Nayda, D. Strange, and R. D. V. Nixon. 2017. Do Meta-Cognitive Beliefs Affect Meta-Awareness of Intrusive Thoughts About Trauma? *J. Behav. Ther. Exp. Psychiatry* **54**:292–300. [17]

Takarangi, M. K. T., D. Strange, and D. S. Lindsay. 2014. Self-Report May Underestimate Trauma Intrusions. *Conscious. Cogn.* **27**:297–305. [9]

Tamber-Rosenau, B. J., M. Esterman, Y. C. Chiu, and S. Yantis. 2011. Cortical Mechanisms of Cognitive Control for Shifting Attention in Vision and Working Memory. *J. Cogn. Neurosci.* **23**:2905–2919. [9]

Tang, W., S. Jbabdi, Z. Zhu, et al. 2019. A Connectional Hub in the Rostral Anterior Cingulate Cortex Links Areas of Emotion and Cognitive Control. *eLife* **8**:e43761. [5]

Tasan, R., D. Verma, J. Wood, et al. 2016. The Role of Neuropeptide Y in Fear Conditioning and Extinction. *Neuropeptides* **55**:111–126. [15]

Taschereau-Dumouchel, V., A. Cortese, T. Chiba, et al. 2018a. Towards an Unconscious Neural Reinforcement Intervention for Common Fears. *PNAS* **115**:3470–3475. [12, 17]

Taschereau-Dumouchel, V., K. Y. Liu, and H. C. Lau. 2018b. Unconscious Psychological Treatments for Physiological Survival Circuits. *Curr. Opin. Behav. Sci.* **24**:62–68. [12]

Taubenfeld, A., M. C. Anderson, and D. A. Levy. 2019. The Impact of Retrieval Suppression on Conceptual Implicit Memory. *Memory* **27**:686–697. [9]

Taylor, J. R., and M. M. Torregrossa. 2015. Pharmacological Disruption of Maladaptive Memory. *Handb. Exp. Pharmacol.* **228**:381–415. [15]

Taylor, S. E., and J. D. Brown. 1988. Illusion and Well-Being: A Social Psychological Perspective on Mental-Health. *Psychol. Bull.* **103**:193–210. [9]

Telch, M. J., J. York, C. L. Lancaster, and M. H. Monfils. 2017. Use of a Brief Fear Memory Reactivation Procedure for Enhancing Exposure Therapy. *Clin. Psychol. Sci.* **5**:367–378. [7, 9, 14]

Terraneo, A., L. Leggio, M. Saladini, et al. 2016. Transcranial Magnetic Stimulation of Dorsolateral Prefrontal Cortex Reduces Cocaine Use: A Pilot Study. *Eur. Neuropsychopharmacol.* **26**:37–44. [5]

Tervo, D. G., J. B. Tenenbaum, and S. J. Gershman. 2016. Toward the Neural Implementation of Structure Learning. *Curr. Opin. Neurobiol.* **37**:99–105. [13]

Théberge, F. R., A. L. Milton, D. Belin, J. L. C. Lee, and B. J. Everitt. 2010. The Basolateral Amygdala and Nucleus Accumbens Core Mediate Dissociable Aspects of Drug Memory Reconsolidation. *Learn. Mem.* **17**:444–453. [2]

Thickbroom, G. W. 2007. Transcranial Magnetic Stimulation and Synaptic Plasticity: Experimental Framework and Human Models. *Exp. Brain Res.* **180**:583–593. [16]

Thorpe, S. J., E. T. Rolls, and S. Maddison. 1983. The Orbitofrontal Cortex: Neuronal Activity in the Behaving Monkey. *Exp. Brain Res.* **49**:93–115. [2]

Thorsteinsson, E. B., and J. E. James. 1999. A Meta-Analysis of the Effects of Experimental Manipulations of Social Support during Laboratory Stress. *Psychol. Health* **14**:869–886. [14]

Thrailkill, E. A., and M. E. Bouton. 2015. Contextual Control of Instrumental Actions and Habits. *J. Exp. Psychol. Anim. Learn. Cogn.* **41**:69–80. [3]

Tiffany, S. T., and B. L. Carter. 1998. Is Craving the Source of Compulsive Drug Use? *J. Psychopharmacol.* **12**:23–30. [2]

Tomie, A. 1996. Locating Reward Cue at Response Manipulandum (CAM) Induces Symptoms of Drug Abuse. *Neurosci. Biobehav. Rev.* **20**:505–535. [5]

Tooby, J., and L. Cosmides. 2008. The Evolutionary Psychology of the Emotions and Their Relationship to Internal Regulatory Variables. In: Handbook of Emotions, 3rd Ed., ed. M. Lewis et al., pp. 114–137. New York: Guilford. [7]

Tozzi, A., A. Tscherter, V. Belcastro, et al. 2007. Interaction of A2A Adenosine and D2 Dopamine Receptors Modulates Corticostriatal Glutamatergic Transmission. *Neuropharmacology* **53**:783–789. [3]

Treanor, M., L. A. Brown, J. Rissman, and M. G. Craske. 2017. Can Memories of Traumatic Experiences or Addiction Be Erased or Modified? A Critical Review of Research on the Disruption of Memory Reconsolidation and Its Applications. *Perspect. Psychol. Sci.* **12**:290–305. [14]

Treynor, W., R. Gonzalez, and S. Nolen-Hoeksema. 2003. Rumination Reconsidered: A Psychometric Analysis. *Cogn. Ther. Res.* **27**:247–259. [6, 9]

Tschentscher, N., D. Mitchell, and J. Duncan. 2017. Fluid Intelligence Predicts Novel Rule Implementation in a Distributed Frontoparietal Control Network. *J. Neurosci.* **37**:4841–4847. [11]

Tse, P. 2013. The Neural Basis of Free Will Criterial Causation. Cambridge, MA: MIT Press. [12]

Tsunoda, N., M. Hashimoto, T. Ishikawa, et al. 2018. Clinical Features of Auditory Hallucinations in Patients with Dementia with Lewy Bodies: A Soundtrack of Visual Hallucinations. *J. Clin. Psychiatry* **79**: 17m11623. [5]

Twenge, J. M., T. E. Joiner, M. L. Rogers, and G. N. Martin. 2017. Increases in Depressive Symptoms, Suicide-Related Outcomes, and Suicide Rates Among U.S. Adolescents after 2010 and Links to Increased New Media Screen Time. *Clin. Psychol. Sci.* **6**:3–17. [17]

Tyagi, H., A. M. Apergis-Schoute, H. Akram, et al. 2019. A Randomized Trial Directly Comparing Ventral Capsule and Anteromedial Subthalamic Nucleus Stimulation in Obsessive-Compulsive Disorder: Clinical and Imaging Evidence for Dissociable Effects. *Biol. Psychiatry* **85**:726–734. [17]

Uddin, L. Q. 2014. Salience Processing and Insular Cortical Function and Dysfunction. *Nat. Rev. Neurosci.* **16**:55–61. [10, 13]

Ungless, M. A. 2004. Dopamine: The Salient Issue. *Trends Neurosci.* **27**:702–706. [2]

Vaccaro, A. G., and S. M. Fleming. 2018. Thinking About Thinking: A Coordinate-Based Meta-Analysis of Neuroimaging Studies of Metacognitive Judgements. *Brain Neurosci. Adv.* **2**:2398212818810591. [12]

van den Heuvel, M. P., and O. Sporns. 2013. Network Hubs in the Human Brain. *Trends Cogn. Sci.* **17**:683–696. [5, 11]

van den Heuvel, O. A., D. J. Veltman, H. J. Groenewegen, et al. 2005. Disorder-Specific Neuroanatomical Correlates of Attentional Bias in Obsessive-Compulsive Disorder, Panic Disorder, and Hypochondriasis. *Arch. Gen. Psychiatry* **62**:922–933. [8]

Vander Weele, C. M., C. A. Siciliano, G. A. Matthews, et al. 2018. Dopamine Enhances Signal-to-Noise Ratio in Cortical-Brainstem Encoding of Aversive Stimuli. *Nature* **563**:397–401. [2]

Van der Werf, Y. D., M. P. Witter, and H. J. Groenewegen. 2002. The Intralaminar and Midline Nuclei of the Thalamus: Anatomical and Functional Evidence for Participation in Processes of Arousal and Awareness. *Brain Res. Rev.* **39**:107–140. [10]

van Gaal, S., and V. A. Lamme. 2012. Unconscious High-Level Information Processing: Implication for Neurobiological Theories of Consciousness. *Neuroscientist* **18**:287–301. [12]

van Lutterveld, R., S. D. Houlihan, P. Pal, et al. 2017. Source-Space EEG Neurofeedback Links Subjective Experience with Brain Activity during Effortless Awareness Meditation. *NeuroImage* **151**:117–127. [17]

van Schie, K., and M. C. Anderson. 2017. Successfully Controlling Intrusive Memories Is Harder When Control Must Be Sustained. *Memory* **25**:1201–1216. [9, 15]

Vanvossen, A. C., M. A. M. Portes, R. Scoz-Silva, et al. 2017. Newly Acquired and Reactivated Contextual Fear Memories Are More Intense and Prone to Generalize after Activation of Prelimbic Cortex NMDA Receptors. *Neurobiol. Learn. Mem.* **137**:154–162. [2]

Venniro, M., D. Caprioli, and Y. Shaham. 2016. Animal Models of Drug Relapse and Craving: From Drug Priming-Induced Reinstatement to Incubation of Craving after Voluntary Abstinence. *Prog. Brain. Res.* **224**:25–52. [5]

Verbruggen, F., and G. D. Logan. 2008. Response Inhibition in the Stop-Signal Paradigm. *Trends Cogn. Sci.* **12**:418–424. [9]

Verbruggen, F., R. McLaren, M. Pereg, and N. Meiran. 2018. Structure and Implementation of Novel Task Rules: A Cross-Sectional Developmental Study. *Psychol. Sci.* **29**:1113–1125. [11]

Vernet, M., S. Bashir, W. K. Yoo, et al. 2014. Reproducibility of the Effects of Theta Burst Stimulation on Motor Cortical Plasticity in Healthy Participants. *Clin. Neurophysiol.* **125**:320–326. [16]

Verstynen, T. D., D. Badre, K. Jarbo, and W. Schneider. 2012. Microstructural Organizational Patterns in the Human Corticostriatal System. *J. Neurophysiol.* **107**:2984–2995. [11]

Vidal-Pineiro, D., P. Martin-Trias, C. Falcon, et al. 2015. Neurochemical Modulation in Posteromedial Default-Mode Network Cortex Induced by Transcranial Magnetic Stimulation. *Brain Stimul.* **8**:937–944. [16]

Vidaurre, D., S. M. Smith, and M. W. Woolrich. 2017. Brain Network Dynamics Are Hierarchically Organized in Time. *PNAS* **114**:12827–12832. [17]

Visser, R. M., A. Lau-Zhu, R. N. Henson, and E. A. Holmes. 2018. Multiple Memory Systems, Multiple Time Points: How Science Can Inform Treatment to Control the Expression of Unwanted Emotional Memories. *Phil. Trans. R. Soc. B* **373**:20170209. [6, 8, 9, 14, 17]

Vogel, E. K., A. W. McCollough, and M. G. Machizawa. 2005. Neural Measures Reveal Individual Differences in Controlling Access to Working Memory. *Nature* **438**:500–503. [9]

Volkow, N. D., G.-J. Wang, J. S. Fowler, et al. 2010. Addiction: Decreased Reward Sensitivity and Increased Expectation Sensitivity Conspire to Overwhelm the Brain's Control Circuit. *Bioessays* **32**:748–755. [10]

Voon, V., K. Derbyshire, C. Ruck, et al. 2015. Disorders of Compulsivity: A Common Bias Towards Learning Habits. *Mol. Psychiatry* **20**:345–352. [8, 9]

Vosgerau, G., and M. Synofzik. 2010. A Cognitive Theory of Thoughts. *Am. Philos. Quart.* **47**:205–222. [8]

Vosgerau, G., and M. Voss. 2014. Authorship and Control over Thoughts. *Mind Lang.* **29**:534–565. [8]

Waldhauser, G. T., M. J. Dahl, M. Ruf-Leuschner, et al. 2018. The Neural Dynamics of Deficient Memory Control in Heavily Traumatized Refugees. *Sci. Rep.* **8**:13132. [6, 9]

Waldum, E. R., and L. Sahakyan. 2012. Putting Congeniality Effects into Context: Investigating the Role of Context in Attitude Memory Using Multiple Paradigms. *J. Mem. Lang.* **66**:717–730. [9]

Walker, S. C., T. W. Robbins, and A. C. Roberts. 2009. Differential Contributions of Dopamine and Serotonin to Orbitofrontal Cortex Function in the Marmoset. *Cereb. Cortex* **19**:889–898. [13]

Walker, W. R., J. J. Skowronski, and C. P. Thompson. 2003. Life Is Pleasant--and Memory Helps to Keep It That Way! *Rev. Gen. Psychol.* **7**:203–210. [9]

Wallace-Wells, D. 2019. The Uninhabitable Earth: Life after Warming. New York: Tim Duggan Books. [9]

Wang, X. J. 2001. Synaptic Reverberation Underlying Mnemonic Persistent Activity. *Trends Neurosci.* **24**:455–463. [13]

Wang, X. J., J. Tegnér, C. Constantinidis, and P. S. Goldman-Rakic. 2004. Division of Labor among Distinct Subtypes of Inhibitory Neurons in a Cortical Microcircuit of Working Memory. *PNAS* **101**:1368–1373. [13]

Warburton, D. M., K. Wesnes, J. Edwards, and D. Larrad. 1985. Scopolamine and the Sensory Conditioning of Hallucinations. *Neuropsychobiology* **14**:198–202. [5]

Watanabe, T., Y. Sasaki, K. Shibata, and M. Kawato. 2017. Advances in fMRI Real-Time Neurofeedback. *Trends Cogn. Sci.* **21**:997–1010. [12]

Watkins, E. R. 2008. Constructive and Unconstructive Repetitive Thought. *Psychol. Bull.* **134**:163–206. [10]

Watson, H. J., Z. Yilmaz, L. M. Thornton, et al. 2019. Genome-Wide Association Study Identifies Eight Risk Loci and Implicates Metabo-Psychiatric Origins for Anorexia Nervosa. *Nat. Genet.* **51**:1207–1214. [13]

Watson, P., R. W. Wiers, B. Hommel, and S. de Wit. 2018. Motivational Sensitivity of Outcome-Response Priming: Experimental Research and Theoretical Models. *Psychon. Bull. Rev.* **25**:2069–2082. [9]

Weber, M. 1930. The Protestant Ethic and the Spirit of Capitalism. New York: Scribner. [12]

Wegner, D. M. 1997. When the Antidote Is the Poison: Ironic Mental Control Processes. *Psychol. Sci.* **8**:148–150. [9]

———. 2002. The Illusion of Conscious Will. Cambridge, MA: MIT Press. [12]

Wegner, D. M., D. J. Schneider, S. R. Carter, and T. L. White. 1987. Paradoxical Effects of Thought Suppression. *J. Pers. Soc. Psychol.* **53**:5–13. [6, 9]

Wegner, D. M., and T. Wheatley. 1999. Apparent Mental Causation: Sources of the Experience of Will. *Am. Psychol.* **54**:480–492. [12]

Wegner, D. M., and S. Zanakos. 1994. Chronic Thought Suppression. *J. Pers. Soc. Psychol.* **62**:615–640. [6, 9]

Wei, X. F., and W. M. Grill. 2005. Current Density Distributions, Field Distributions and Impedance Analysis of Segmented Deep Brain Stimulation Electrodes. *J. Neural. Eng.* **2**:139–147. [16]

Weiskrantz, L. 1997. Consciousness Lost and Found: A Neuropsychological Exploration. Oxford: Oxford Univ. Press. [12]

Weiss, D. S. 1997. The Impact of Event Scale – Revised. In: Assessing Psychological Trauma and PTSD: A Handbook for Practitioners, ed. J. P. Wilson and T. M. Keane pp. 168–189. New York: The Guilford Press. [6]

Wells, A., and M. I. Davies. 1994. The Thought Control Questionnaire: A Measure of Individual Differences in the Control of Unwanted Thoughts. *Behav. Res. Ther.* **32**:871–878. [6]

Wenzlaff, R. M., and D. M. Wegner. 2000. Thought Suppression. *Ann. Rev. Psychol.* **51**:59–91. [6, 9]

Wernicke, K. 1885/1994. Some New Studies on Aphasia. In: Reader in the History of Aphasia, ed. P. Eling, pp. 90–98. Amsterdam: John Benjamins. [5]

Wessel, J. R., and A. R. Aron. 2017. On the Globality of Motor Suppression: Unexpected Events and Their Influence on Behavior and Cognition. *Neuron* **93**:259–280. [9, 11]

Wessel, J. R., A. Ghahremani, K. Udupa, et al. 2016. Stop-Related Subthalamic Beta Activity Indexes Global Motor Suppression in Parkinson's Disease. *Movement Disord.* **31**:1846–1853. [9]

Wheeler, M. A., C. J. Smith, M. Ottolini, et al. 2016. Genetically Targeted Magnetic Control of the Nervous System. *Nat. Neurosci.* **19**:756–761. [4]

Whitaker, L. R., B. L. Warren, M. Venniro, et al. 2017. Bidirectional Modulation of Intrinsic Excitability in Rat Prelimbic Cortex Neuronal Ensembles and Non-Ensembles after Operant Learning. *J. Neurosci.* **37**:8845–8856. [5]

Whitton, A. E., M. T. Treadway, and D. A. Pizzagalli. 2015. Reward Processing Dysfunction in Major Depression, Bipolar Disorder and Schizophrenia. *Curr. Opin. Psychiatry* **28**:7–12. [10]

Whyte, A. J., H. W. Kietzman, A. M. Swanson, et al. 2019. Reward-Related Expectations Trigger Dendritic Spine Plasticity in the Mouse Ventrolateral Orbitofrontal Cortex. *J. Neurosci.* **39**:4595–4605. [5]

Wickens, J. 1993. A Theory of the Striatum. Oxford: Pergamon Press. [11]

Widge, A. S., K. K. Ellard, A. C. Paulk, et al. 2017. Treating Refractory Mental Illness with Closed-Loop Brain Stimulation: Progress Towards a Patient-Specific Transdiagnostic Approach. *Exp. Neurol.* **287**:461–472. [16]

Wiegert, J. S., M. Mahn, M. Prigge, Y. Printz, and O. Yizhar. 2017. Silencing Neurons: Tools, Applications, and Experimental Constraints. *Neuron* **95**:504–529. [4]

Wiener, N. 1948. Cybernetics: Or Control and Communication in the Animal and the Machine. New York: John Wiley & Sons, Inc. [11]

Wierzba, M., M. Riegel, M. Wypych, et al. 2018. Cognitive Control over Memory-Individual Differences in Memory Performance for Emotional and Neutral Material. *Sci. Rep.* **8**:3808 [9]

Wijkstra, J., J. Lijmer, H. Burger, J. Geddes, and W. A. Nolen. 2013. Pharmacological Treatment for Psychotic Depression. *Cochrane Database Syst. Rev.* **11**:CD004044. [15]

Wilkinson, S., G. Dodgson, and K. Meares. 2017. Predictive Processing and the Varieties of Psychological Trauma. *Front. Psychol.* **8**:1840. [13]

Williams, A. D., and M. L. Moulds. 2007. Cognitive Avoidance of Intrusive Memories: Recall Vantage Perspectives Associations with Depression. *Behav. Res. Ther.* **45**:1145–1153. [9]

Williams, G. C. 1966. Natural Selection, the Costs of Reproduction, and a Refinement of Lack's Principle. *Am. Natural.* **100**:687–690. [7]

Williams, J. M. G., A. Mathews, and C. MacLeod. 1996. The Emotional Stroop Task and Psychopathology. *Psychol. Bull.* **120** 3–24. [9]

Williams, W. A., and M. N. Potenza. 2008. The Neurobiology of Impulse Control Disorders. *Braz. J. Psychiatry* **30 (Suppl 1)**:S24–S30. [15]

Willuhn, I., L. M. Burgeno, B. J. Everitt, and P. E. M. Phillips. 2012. Hierarchical Recruitment of Phasic Dopamine Signaling in the Striatum during the Progression of Cocaine Use. *PNAS* **109**:20703–20708. [5]

Willuhn, I., L. M. Burgeno, P. A. Groblewski, and P. E. M. Phillips. 2014. Excessive Cocaine Use Results from Decreased Phasic Dopamine Signaling in the Striatum. *Nat. Neurosci.* **17**:704. [5]

Wilmer, H. H., L. E. Sherman, and J. M. Chein. 2017. Smartphones and Cognition: A Review of Research Exploring the Links between Mobile Technology Habits and Cognitive Functioning. *Front. Psychol.* **8**:605. [17]

Wilson, R. C., Y. K. Takahashi, G. Schoenbaum, and Y. Niv. 2014. Orbitofrontal Cortex as a Cognitive Map of Task Space. *Neuron* **81**:267–279. [13]

Wimmer, G. E., and D. Shohamy. 2012. Preference by Association: How Memory Mechanisms in the Hippocampus Bias Decisions. *Science* **338**:270–273. [2]

Winn, J., and C. M. Bishop. 2005. Variational Message Passing. *J. Machine Learn. Res* 6:661–694. [13]

Wolman, D. 2019. The Split Brain: A Tale of Two Halves. *Nature* **483**:260–263. [12]

Womelsdorf, T., J. M. Schoffelen, R. Oostenveld, et al. 2007. Modulation of Neuronal Interactions through Neuronal Synchronization. *Science* **316**:1609–1612. [13]

Woodward, M. J., J. Eddinger, A. V. Henschel, et al. 2015. Social Support, Posttraumatic Cognitions, and PTSD: The Influence of Family, Friends, and a Close Other in an Interpersonal and Non-Interpersonal Trauma Group. *J. Anxiety Disord.* **35**:60–67. [14]

Woolgar, A., J. Jackson, and J. Duncan. 2016. Coding of Visual, Auditory, Rule, and Response Information in the Brain: 10 Years of Multivoxel Pattern Analysis. *J. Cogn. Neurosci.* **28**:1433–1454. [11]

Woolgar, A., A. Parr, R. Cusack, et al. 2010. Fluid Intelligence Loss Linked to Restricted Regions of Damage within Frontal and Parietal Cortex. *PNAS* **107**:14899–14902. [11]

Woolgar, A., R. Thompson, D. Bor, and J. Duncan. 2011. Multi-Voxel Coding of Stimuli, Rules, and Responses in Human Frontoparietal Cortex. *NeuroImage* **56**:744–752. [11]

Wyland, C. L., W. M. Kelley, C. N. Macrae, H. L. Gordon, and T. F. Heatherton. 2003. Neural Correlates of Thought Suppression. *Neuropsychologia* **41**:1863–1867. [6]

Xia, L., S. K. Nygard, G. G. Sobczak, N. J. Hourguettes, and M. R. Bruchas. 2017. Dorsal-CA1 Hippocampal Neuronal Ensembles Encode Nicotine-Reward Contextual Associations. *Cell Rep.* **19**:2143–2156. [4]

Xu, Y., V. Ramanathan, and D. G. Victor. 2018. Global Warming Will Happen Faster Than We Think. *Nature* **564**:30–32. [9]

Xue, Y. X., Y. X. Luo, P. Wu, et al. 2012. A Memory Retrieval-Extinction Procedure to Prevent Drug Craving and Relapse. *Science* **336**:241–245. [9]

Yahata, N., J. Morimoto, R. Hashimoto, et al. 2016. A Small Number of Abnormal Brain Connections Predicts Adult Autism Spectrum Disorder. *Nat. Commun.* **7**:11254. [17]

Yao, S. N., J. Cottraux, and R. Martin. 1999. A Controlled Study of Irrational Interpretations of Intrusive Thoughts in Obsessive-Compulsive Disorder. *L'Encephale* **25**:461–469. [13]

Yeo, B. T., F. M. Krienen, J. Sepulcre, et al. 2011. The Organization of the Human Cerebral Cortex Estimated by Intrinsic Functional Connectivity. *J. Neurophysiol.* **106**:1125–1165. [11]

Yeterian, E. H., and D. N. Pandya. 1993. Striatal Connections of the Parietal Association Cortices in Rhesus Monkeys. *J. Comp. Neurol.* **332**:175–197. [5]

Yin, H. H., B. J. Knowlton, and B. W. Balleine. 2004. Lesions of Dorsolateral Striatum Preserve Outcome Expectancy but Disrupt Habit Formation in Instrumental Learning. *Eur. J. Neurosci.* **19**:181–189. [3]

———. 2005a. Blockade of NMDA Receptors in the Dorsomedial Striatum Prevents Action-Outcome Learning in Instrumental Conditioning. *Eur. J. Neurosci.* **22**:505–512. [3]

———. 2006. Inactivation of Dorsolateral Striatum Enhances Sensitivity to Changes in the Action-Outcome Contingency in Instrumental Conditioning. *Behav. Brain Res.* **166**:189–196. [3]

Yin, H. H., S. B. Ostlund, B. J. Knowlton, and B. W. Balleine. 2005b. The Role of the Dorsomedial Striatum in Instrumental Conditioning. *Eur. J. Neurosci.* **22**:513–523. [3]

Yizhar, O., L. E. Fenno, T. J. Davidson, M. Mogri, and K. Deisseroth. 2011. Optogenetics in Neural Systems. *Neuron* **71**:9–34. [4]

Young, A. M. J. 2004. Increased Extracellular Dopamine in Nucleus Accumbens in Response to Unconditioned and Conditioned Aversive Stimuli: Studies Using 1 Min Microdialysis in Rats. *J. Neurosci. Methods* **138**:57–63. [2]

Zald, D. H., M. McHugo, K. L. Ray, et al. 2014. Meta-Analytic Connectivity Modeling Reveals Differential Functional Connectivity of the Medial and Lateral Orbitofrontal Cortex. *Cereb. Cortex* **24**:232–248. [3]

Zedelius, C. M., and J. W. Schooler. 2017. What Are People's Lay Theories About Mind Wandering and How Do Those Beliefs Affect Them?, pp. 71–93. Heidelberg: Springer. [9]

Zeigarnik, B. 1938. On Finished and Unfinished Tasks, ed. W. D. Ellis, pp. 300–314. London: Kegan Paul, Trench, Trubner & Company. [9]

Zetsche, U., T. Ehring, and A. Ehlers. 2009. The Effects of Rumination on Mood and Intrusive Memories after Exposure to Traumatic Material: An Experimental Study. *J. Behav. Ther. Exp. Psychiatry* **40**:499–514. [6]

Zhang, M., D. S. S. Fung, and H. Smith. 2019. Variations in the Visual Probe Paradigms for Attention Bias Modification for Substance Use Disorders. *Int. J. Environ. Res. Pub. Health* **16**:3389. [17]

Zhang, R., X. Geng, and T. M. C. Lee. 2017. Large-Scale Functional Neural Network Correlates of Response Inhibition: an fMRI Meta-Analysis. *Brain Struct. Funct.* **222**:3973–3990. [13]

Zhang, Z., L. E. Russell, A. M. Packer, O. M. Gauld, and M. Häusser. 2018. Closed-Loop All-Optical Interrogation of Neural Circuits *in Vivo*. *Nat. Methods* **15**:1037–1040. [4]

Zhou, T., G. Hong, T.-M. Fu, et al. 2017. Syringe-Injectable Mesh Electronics Integrate Seamlessly with Minimal Chronic Immune Response in the Brain. *PNAS* **114**:5894–5899. [4]

Zhu, H., I. Paschalidis, and M. Hasselmo. 2018. Neural Circuits for Learning Context-Dependent Associations of Stimuli. *Neural Netw.* **107**:48–60. [13]

Zhu, Y., Q. Fan, H. Zhang, et al. 2016. Altered Intrinsic Insular Activity Predicts Symptom Severity in Unmedicated Obsessive-Compulsive Disorder Patients: A Resting State Functional Magnetic Resonance Imaging Study. *BMC Psychiatry* **16**:104. [13]

Ziemann, U., and J. C. Rothwell. 2000. I-waves in Motor Cortex. *J. Clin. Neurophysiol.* **17**:397–405. [16]

Zimmermann, K. S., C. C. Li, D. G. Rainnie, K. J. Ressler, and S. L. Gourley. 2018. Memory Retention Involves the Ventrolateral Orbitofrontal Cortex: Comparison with the Basolateral Amygdala. *Neuropsychopharmacology* **43**:373–383. [5]

Zimmermann, K. S., J. A. Yamin, D. G. Rainnie, K. J. Ressler, and S. L. Gourley. 2017. Connections of the Mouse Orbitofrontal Cortex and Regulation of Goal-Directed Action Selection by Brain-Derived Neurotrophic Factor. *Biol. Psychiatry* **81**:366–377. [5]

Zink, C. F., G. Pagnoni, J. Chappelow, M. Martin-Skurski, and G. S. Berns. 2006. Human Striatal Activation Reflects Degree of Stimulus Saliency. *NeuroImage* **29**:977–983. [10]

Subject Index

Strüngmann Forum Report Series*

Intrusive Thinking: From Molecules to Free Will
Edited by Peter W. Kalivas and Martin P. Paulus
ISBN: 9780262542371

Deliberate Ignorance: Choosing Not to Know
edited by Ralph Hertwig and Christoph Engel
ISBN 9780262045599

Youth Mental Health: A Paradigm for Prevention and Early Intervention
Edited by Peter J. Uhlhaas and Stephen J. Wood
ISBN: 9780262043977

The Neocortex
Edited by Wolf Singer, Terrence J. Sejnowski and Pasko Rakic
ISBN: 9780262043243

Interactive Task Learning: Humans, Robots, and Agents Acquiring New Tasks through Natural Interactions
Edited by Kevin A. Gluck and John E. Laird
ISBN: 9780262038829

Agrobiodiversity: Integrating Knowledge for a Sustainable Future
Edited by Karl S. Zimmerer and Stef de Haan
ISBN: 9780262038683

Rethinking Environmentalism: Linking Justice, Sustainability, and Diversity
Edited by Sharachchandra Lele, Eduardo S. Brondizio, John Byrne,
Georgina M. Mace and Joan Martinez-Alier
ISBN: 9780262038966

Emergent Brain Dynamics: Prebirth to Adolescence
Edited by April A. Benasich and Urs Ribary
ISBN: 9780262038638

The Cultural Nature of Attachment: Contextualizing Relationships and Development
Edited by Heidi Keller and Kim A. Bard
ISBN (Hardcover): 9780262036900   ISBN (ebook): 9780262342865
Winner of the Ursula Gielen Global Psychology Book Award

Investors and Exploiters in Ecology and Economics: Principles and Applications
edited by Luc-Alain Giraldeau, Philipp Heeb and Michael Kosfeld
ISBN (Hardcover): 9780262036122   ISBN (eBook): 9780262339797

Computational Psychiatry: New Perspectives on Mental Illness
edited by A. David Redish and Joshua A. Gordon
ISBN: 9780262035422

---

*   Available at https://mitpress.mit.edu/books/series/strungmann-forum-reports

Complexity and Evolution: Toward a New Synthesis for Economics
edited by David S. Wilson and Alan Kirman
ISBN: 9780262035385

The Pragmatic Turn: Toward Action-Oriented Views in Cognitive Science
edited by Andreas K. Engel, Karl J. Friston and Danica Kragic
ISBN: 9780262034326

Translational Neuroscience: Toward New Therapies
edited by Karoly Nikolich and Steven E. Hyman
ISBN: 9780262029865

Trace Metals and Infectious Diseases
edited by Jerome O. Nriagu and Eric P. Skaar
ISBN 9780262029193

Pathways to Peace: The Transformative Power of Children and Families
edited by James F. Leckman, Catherine Panter-Brick and Rima Salah,
ISBN 9780262027984

Rethinking Global Land Use in an Urban Era
edited by Karen C. Seto and Anette Reenberg
ISBN 9780262026901

Schizophrenia: Evolution and Synthesis
edited by Steven M. Silverstein, Bita Moghaddam and Til Wykes,
ISBN 9780262019620

Cultural Evolution: Society, Technology, Language, and Religion
edited by Peter J. Richerson and Morten H. Christiansen,
ISBN 9780262019750

Language, Music, and the Brain: A Mysterious Relationship
edited by Michael A. Arbib
ISBN 9780262019620

Evolution and the Mechanisms of Decision Making
edited by Peter Hammerstein and Jeffrey R. Stevens
ISBN 9780262018081

Cognitive Search: Evolution, Algorithms, and the Brain
edited by Peter M. Todd, Thomas T. Hills and Trevor W. Robbins,
ISBN 9780262018098

Animal Thinking: Contemporary Issues in Comparative Cognition
edited by Randolf Menzel and Julia Fischer
ISBN 9780262016636

Disease Eradication in the 21st Century: Implications for Global Health
edited by Stephen L. Cochi and Walter R. Dowdle
ISBN 9780262016735

Better Doctors, Better Patients, Better Decisions: Envisioning Health Care 2020
edited by Gerd Gigerenzer and J. A. Muir Gray
ISBN 9780262016032

Dynamic Coordination in the Brain: From Neurons to Mind
edited by Christoph von der Malsburg, William A. Phillips and Wolf Singer,
ISBN 9780262014717

Linkages of Sustainability
edited by Thomas E. Graedel and Ester van der Voet
ISBN 9780262013581

Biological Foundations and Origin of Syntax
edited by Derek Bickerton and Eörs Szathmáry
ISBN 9780262013567

*Clouds in the Perturbed Climate System: Their Relationship to Energy Balance,
Atmospheric Dynamics, and Precipitation*
edited by Jost Heintzenberg and Robert J. Charlson
ISBN 9780262012874
Winner of the Atmospheric Science Librarians International Choice Award

*Better Than Conscious? Decision Making, the Human Mind, and Implications
For Institutions*
edited by Christoph Engel and Wolf Singer
ISBN 978-0-262-19580-5